Nelson Yuan-sheng Kiang

CEREBRAL CORTEX

Volume 8A

Comparative Structure and Evolution of Cerebral Cortex, Part I

CEREBRAL CORTEX

Edited by Edward G. Jones and Alan Peters

Volume 1:	Cellular Components of the Cerebral Cortex
Volume 2:	Functional Properties of Cortical Cells
Volume 3:	Visual Cortex
Volume 4:	Association and Auditory Cortices
Volume 5:	Sensory-Motor Areas and Aspects of Cortical Connectivity
Volume 6:	Further Aspects of Cortical Function, Including Hippocampus
Volume 7:	Development and Maturation of Cerebral Cortex
Volume 8A:	Comparative Structure and Evolution of Cerebral Cortex, Part I
Volume 8B:	Comparative Structure and Evolution of Cerebral Cortex, Part II
Volume 9:	Normal and Altered States of Function

CEREBRAL CORTEX

Volume 8A
Comparative Structure and Evolution of Cerebral Cortex, Part I

Edited by

EDWARD G. JONES
California College of Medicine
University of California, Irvine
Irvine, California

and

ALAN PETERS
Boston University College of Medicine
Boston, Massachusetts

Plenum Press • New York and London

Library of Congress Cataloging in Publication Data

(Revised for vol. 8a—8b)
Cerebral cortex.

Vol. 2, 5- edited by Edward G. Jones and Alan Peters.
Includes bibliographies and indexes.
Contents: v. 1. Cellular components of the cerebral cortex—v. 2. Functional properties of cortial cells—[etc.]—v. 8. Comparative structure and evolution of cerebral cortex, Parts I and II.
1. Cerebral cortex. I. Peters, Alan, 1929- . II. Jones, Edward G., 1939- . [DNLM: 1. Cerebral Cortex—anatomy and histology. 2. Cerebral Cortex—physiology.
QP383.C45 1984 612'.825 84-1982

ISBN 0-306-43477-6

A Division of Plenum Publishing Corporation
233 Spring Street, New York, N.Y. 10013

Printed in the United States of America

Contributors

John Allman Division of Biology, California Institute of Technology, Pasadena, California 91125

Harry J. Jerison Department of Psychiatry and Biobehavioral Sciences, University of California, School of Medicine, Los Angeles, California 90024

Edward G. Jones Department of Anatomy and Neurobiology, University of California, Irvine, Irvine, California 92717

Daniel Margoliash Department of Organismal Biology and Anatomy, and Committees on Neurobiology and Biopsychology, University of Chicago, Chicago, Illinois 60637

J. Meek Department of Anatomy and Embryology, University of Nijmegen, 6500 HB Nijmegen, The Netherlands

Timothy J. Neary Department of Anatomy, Creighton University School of Medicine, Omaha, Nebraska 68178

R. Nieuwenhuys Department of Anatomy and Embryology, University of Nijmegen, 6500 HB Nijmegen, The Netherlands

Wilhelmus J. A. J. Smeets Department of Anatomy, Free University, 1007 MC Amsterdam, The Netherlands

Philip S. Ulinski Department of Organismal Biology and Anatomy, and Committee on Neurobiology, University of Chicago, Chicago, Illinois 60637

Preface

The cerebral cortex, especially that part customarily designated "neocortex," is one of the hallmarks of mammalian evolution and reaches its greatest size, relatively speaking, and its widest structural diversity in the human brain. The evolution of this structure, as remarkable for the huge numbers of neurons that it contains as for the range of behaviors that it controls, has been of abiding interest to many generations of neuroscientists. Yet few theories of cortical evolution have been proposed and none has stood the test of time. In particular, no theory has been successful in bridging the evolutionary gap that appears to exist between the pallium of nonmammalian vertebrates and the neocortex of mammals. Undoubtedly this stems in large part from the rapid divergence of nonmammalian and mammalian forms and the lack of contemporary species whose telencephalic wall can be seen as having transitional characteristics. The monotreme cortex, for example, is unquestionably mammalian in organization and that of no living reptile comes close to resembling it. Yet anatomists such as Ramón y Cajal, on examining the finer details of cortical structure, were struck by the similarities in neuronal form, particularly of the pyramidal cells, and their predisposition to laminar alignment shared by representatives of all vertebrate classes. In later years, other anatomists, in studying the connectivity of the pallium in nonmammals, have been able to identify putative homologues of the mammalian hippocampal formation and possibly of other cortical territories as well. These observations lend credence to the belief in a fundamental commonality from which has emerged in the course of evolution the radical differences in structure of the telencephalic wall exhibited by fish, amphibians, birds, and mammals.

The present pair of volumes cannot be regarded as a complete study in the evolution of the cerebral cortex. However, they do set out in some detail the range of telencephalic and especially cortical structure and connectivity exhibited by the five major classes of vertebrates, and contain many insights into the special characteristics of each. Volume 8A deals largely with nonmammalian vertebrates. It starts with three chapters by Wilhelmus J. A. J. Smeets and by R. Nieuwenhuys and J. Meek on the brains of fish, in which telencephalic evolution seems to have emphasized the development of a large central nucleus at the expense of a pallium. It then includes chapters by Timothy J. Neary and Philip S. Ulinski on the telencephalic wall of amphibia and reptiles, in which a rudimentary and, in some regions, relatively advanced cortical structure is seen. This part concludes with an essay by Philip S. Ulinski and Daniel Margoliash on birds, in which the functions of a cortex per se seem to have been largely subsumed by the enormous expansion of that unique structure, the hyperstriatum. The latter part of Volume 8A serves as an introduction to the detailed studies of mammals presented in Volume 8B. It includes our analysis of the fossil record of neocortical evolution by Harry J. Jerison, and essays by John Allman and Edward G. Jones on the fundamental elements of cortical structure and cortical development upon which the evolutionary process must exert its effects. Volume 8B deals solely with the cerebral cortex of mammals and includes chapters on comparative structure of particular regions, such as the olfactory cortex (by Lewis B. Haberly) and the hippocampal formation (by Christine Gall), and considerations by Peter J. Morgane, Ilya I. Glezer, and Myron S. Jacobs, and Mark Rowe, of cortical structure in orders as diverse as marsupials and dolphins. The evolutionary development of some particular functional areas of the neocortex is pursued by Mary Carlson and John Irwin Johnson in their chapters. A study by Wally Welker on the determinants of sulci and gyri in living forms nicely complements that by Harry J. Jerison in Volume 8A.

The editors are again grateful for the efforts and forbearance displayed by the contributors to this wide-ranging pair of volumes, and feel confident that the high quality of their chapters will be as widely recognized.

Edward G. Jones
Alan Peters

Irvine and Boston

Contents

Volume 8A

I. Nonmammalian Vertebrates

Chapter 3

The Telencephalon of Sarcopterygian Fishes

R. Nieuwenhuys and J. Meek

Chapter 4

The Pallium of Anuran Amphibians

Timothy J. Neary

II. Introduction to Mammalian Cortical Evolution

Chapter 7

Evolution of Neocortex

John Allman

Chapter 8

Fossil Evidence on the Evolution of the Neocortex

Harry J. Jerison

Chapter 9

Modulatory Events in the Development and Evolution of Primate Neocortex

Edward G. Jones

Volume 8B

III. Mammals

I

Nonmammalian Vertebrates

1

The Telencephalon of Cartilaginous Fishes

WILHELMUS J. A. J. SMEETS

1. Introduction

The fishes are the largest group of vertebrates. Two subdivisions can be recognized: the Osteichthyes or bony fishes including about 25,000 species, and the Chondrichthyes or cartilaginous fishes comprising only about 800 species. The latter is separated from the former by several features but the most distinctive is their cartilaginous endoskeleton.

Early in their history the cartilaginous fishes separated into two major divisions, the Elasmobranchii and the Holocephali. The living elasmobranchs (sharks, skates, and rays) comprise four major superorders: Squalomorphii, Galeomorphii, Squatinomorphii, and Batoidea (Compagno, 1977). The sharks included in the Squalomorphii include about 25% of all shark species. The group contains the primitive and relatively uncommon six- and seven-gilled forms (*Chlamydoselachus, Hexanchus, Heptranchias*) as well as several better-known species (*Centrophorus, Deania, Centroscymnus, Squalus*). The Galeomorphii, constituting 75% of the sharks, differ in many respects from the other groups and include most of the well-known sharks (e.g., *Carcharodon, Carcharinus, Negaprion, Sphyrna, Ginglymostoma, Mustelus, Scyliorhinus*). The Squatinomorphii contain only a single genus, *Squatina*. This genus has some similarities with the Batoidea but resembles the Squalomorphii more closely in cranial and brain structure. The Batoidea includes the rajiformes (skates), rhinobatiformes (guitarfishes), pristiformes (sawfishes), torpediniformes (electric rays), and myliobatiformes (e.g., sting-, eagle-, and devil-rays).

WILHELMUS J. A. J. SMEETS • Department of Anatomy, Free University, 1007 MC Amsterdam, The Netherlands.

The other subclass, the Holocephali, abundant in the Carboniferous, is represented today by only six living genera. These fish differ from elasmobranchs in several ways. The name Holocephali refers to the distinctive autostylic suspension of the jaws of these fishes, in contrast to the hyostylic and amphistylic jaw suspension of elasmobranchs. There are also important differences in skeletal organization and in the structure of the brain.

Until recently, cartilaginous fishes were considered to be generalized primitive vertebrates and so were important to morphologists for typological comparisons with other vertebrates. As our understanding of these fishes has progressed, it appears, however, that they are less simple in organization than had been supposed and that they possess, e.g., relatively large brains, elegant sensory systems, and sophisticated reproductive mechanisms.

A remarkable finding of the last decade is the relatively high brain weight/body weight ratio in cartilaginous fishes. Investigations of Bauchot *et al.* (1976), Ebbesson and Northcutt (1976), and Northcutt (1977, 1978, 1981) have revealed that the brain/body ratio in some elasmobranchs is comparable to that of birds and mammals and exceeds that of most other nonmammalian species. Traditionally, large brains have been thought to reflect complex behavior, but in the elasmobranchs it is difficult to make any statements about behavior, because so little is known. It is worthwhile to mention that squalomorph sharks have low brain/body ratios, whereas galeomorph sharks have brain/body ratios that are two to six times larger. A similar variation can be recognized within the superorder Batoidea, for the rajiformes have low brain/body ratios, whereas the more advanced myliobatiformes show the highest brain/body ratios known for elasmobranchs. Furthermore, telencephalic data on cartilaginous fishes (Northcutt, 1978) show that primitive sharks and skates possess telencephalic volumes comparable to those of bony fishes and amphibians, whereas advanced sharks and rays reveal a 5- to 13-fold increase in telencephalic volumes.

A second, remarkable finding was that, contrary to what had been assumed, only a restricted part of the telencephalon in cartilaginous fishes receives olfactory information. It was thought that the chondrichthian telencephalon was dominated by the olfactory system and that its efferents projected to epithalamic as well as hypothalamic centers integrating olfactory and gustatory information. It was believed that the tectum of the midbrain constituted the highest relay center where somatic and visual inputs were integrated into a few stereotyped behavioral responses. However, there is presently good evidence that only a restricted area of the chondrichthian telencephalon receives olfactory information and that visual, auditory, as well as mechanoreceptive and electroreceptive lateral line information reach the telencephalon (for review see Smeets *et al.,* 1983). Therefore, the telencephalon of cartilaginous fishes appears to play a more important role in integrating various sensory modalities than was previously supposed.

The aim of the present chapter is to provide a review on the structure and function of the telencephalon of cartilaginous fishes. The core of this chapter is constituted by the telencephala of four species representing the major groups within the chondrichthians (Fig. 1): the squalomorph shark *Squalus acanthias,* the galeomorph shark *Scyliorhinus canicula,* the batoid *Raja clavata,* and the holocephalian *Hydrolagus collei.* I am well aware that these genera selected do not represent the total range of cartilaginous fishes, but they do exemplify the main features of the major groups. A comprehensive study of the entire brain of the same species has been published recently (Smeets *et al.,* 1983), to which the reader is referred for more details.

2. Gross Morphology

The chondrichthian brain is large and well developed (Figs. 2, 3). Except for its rostral pole, in most sharks the brain is tightly enclosed by the neurocranium. The same holds true for the brain of holocephalians, but in batoids the brain occupies only a part of the neurocranial cavity. If we compare the gross morphology of the brains of various chondrichthians, it appears that two brain divisions differ considerably (Northcutt, 1978; Sato *et al.*, 1983): cerebellum and telencephalon. The brain stem is similar in all chondrichthians studied, except for the extraordinarily developed electric lobe in *Torpedo* and the fusion of the dorsal alar plate in *Hydrolagus*. Of the cerebellum it is particularly the corpus

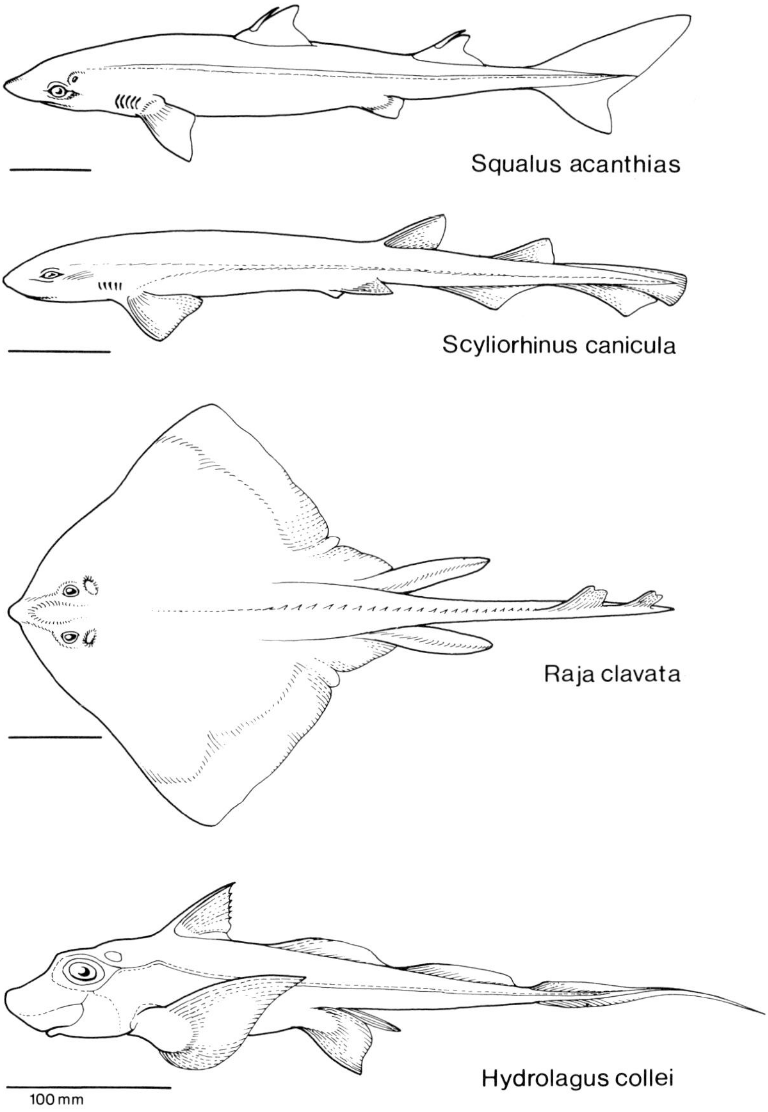

Figure 1. General body form of the four species studied. (From Smeets *et al.*, 1983, by permission of Springer-Verlag.)

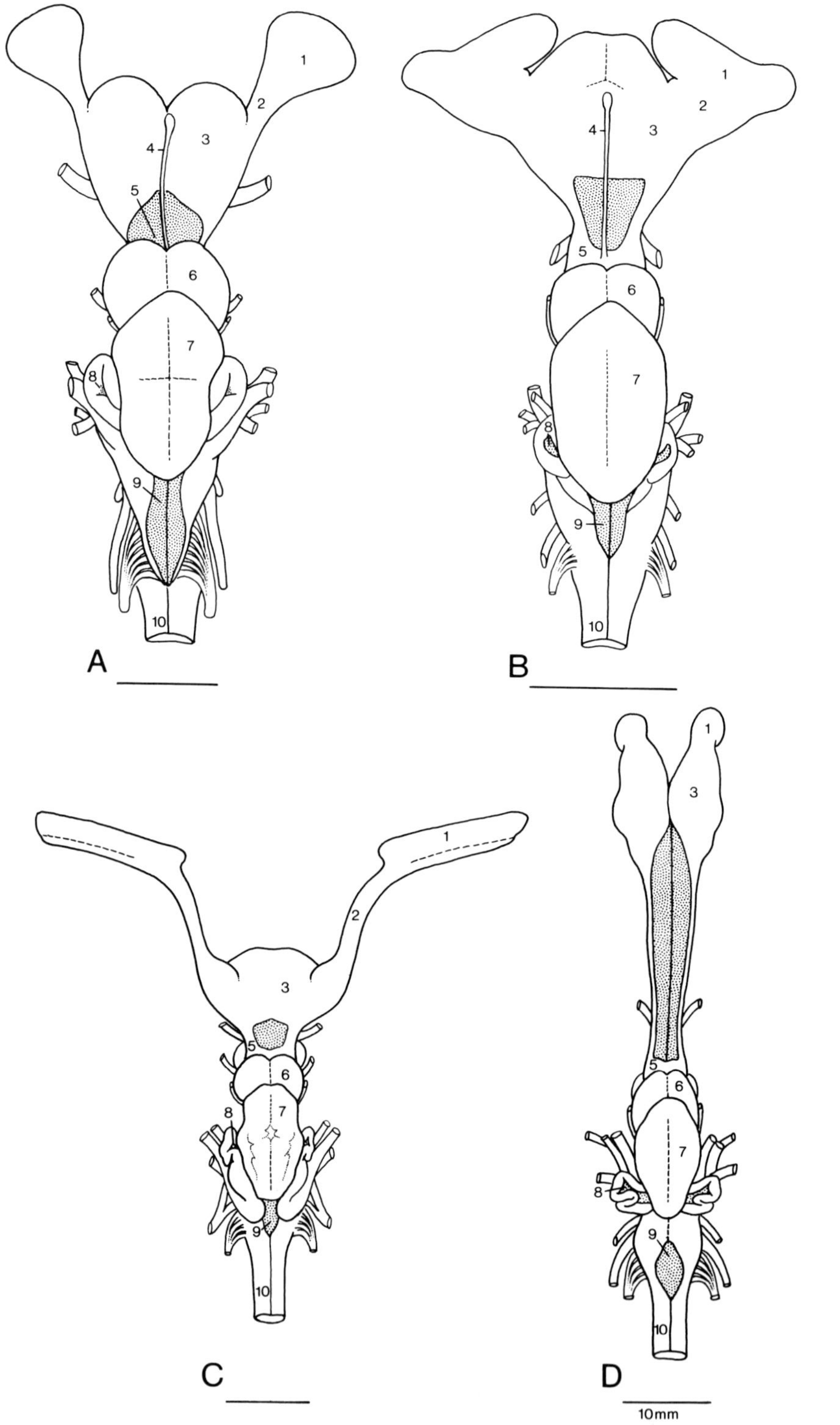

Figure 2. (A–D) Dorsal views of the brain of the four species studied. (A) *Squalus acanthias,* (B) *Scyliorhinus canicula,* (C) *Raja clavata,* (D) *Hydrolagus collei.* 1, bulbus olfactorius; 2, pedunculus bulbi olfactorii; 3, lobus hemisphericus telencephali; 4, epiphysis; 5, diencephalon; 6, tectum mesencephali; 7, corpus cerebelli; 8, recessus lateralis ventriculi quarti; 9, fossa rhomboidea; 10, medulla spinalis. (From Smeets *et al.,* 1983, by permission of Springer-Verlag.)

cerebelli that varies considerably in external form and size between species. In general, it can be stated that holocephalians and squalomorph sharks have a nonconvoluted corpus cerebelli, whereas galeomorph sharks have a convoluted cerebellar body with hypertrophy that results in asymmetry. Within the group of batoids, the more primitive rajiformes and torpediniformes have a slightly convoluted cerebellum (Figs. 2, 3), whereas the more advanced batoids like myliobatiformes possess a complexly convoluted, asymmetrical cerebellum (Fig. 3).

There is a considerable variation in size and morphology between the telencephalon of the chondrichthian fishes studied. With respect to size, the telencephalon of *Squalus* constitutes 24% of total brain weight, in *Scyliorhinus, Raja,* and *Hydrolagus* approximately 30%, whereas in *Carcharinus* and *Sphyrna* more than 50% of the total brain weight is achieved by the telencephalon (Northcutt, 1978). The biological meaning of these increased values for cerebellar and telencephalic subdivisions is unknown, but there seems to be no linear relationship between these two structures.

The telencephala of the various chondrichthian species also differ remarkably in respect to olfactory bulb insertion, length of olfactory peduncle, degree

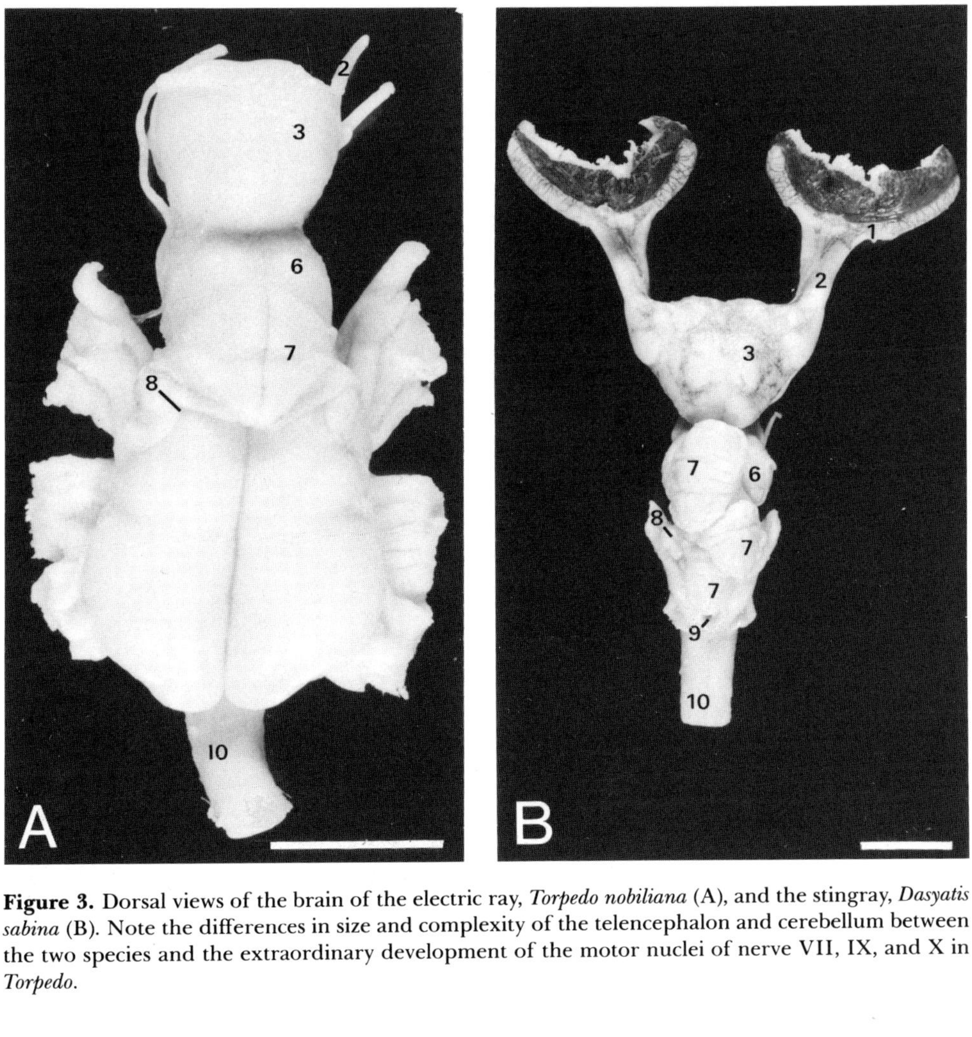

Figure 3. Dorsal views of the brain of the electric ray, *Torpedo nobiliana* (A), and the stingray, *Dasyatis sabina* (B). Note the differences in size and complexity of the telencephalon and cerebellum between the two species and the extraordinary development of the motor nuclei of nerve VII, IX, and X in *Torpedo*.

of fusion between the hemispheric lobes, and degree of reduction of the telencephalic ventricle.

The insertion of the olfactory peduncle varies from a rostral position in *Hydrolagus* to a lateral one in *Scyliorhinus* (Fig. 2). The length of the olfactory peduncle is long in batoids and some sharks such as *Squalus* and *Ginglymostoma,* but short in *Scyliorhinus.* In the holocephalian *Hydrolagus,* the bulb passes imperceptibly into the telencephalic hemisphere.

A characteristic feature of the chondrichthian forebrain is the fusion between the medial walls of both hemispheres. Pallial as well as subpallial zones are involved in this fusion. Johnston (1911) supposed that the connection of both hemispheres is the result of a secondary fusion of the meningeal surfaces of the medial hemispheric walls. However, Bäckström (1924) concluded from his embryological material that the fusion of the hemispheres results largely from a migration of nervous elements in the initially membranous structures that interconnect the two telencephalic halves. The degree of fusion varies from species to species. In primitive sharks such as *Chlamydoselachus* and *Hexanchus* it is only very restricted, but in sharks such as *Mustelus, Scyliorhinus,* and *Ginglymostoma,* and batoids the medial hemisphere walls are completely united. *Squalus* occupied an intermediate position in this respect, whereas in the Holocephali a fusion is completely lacking (Figs. 16, 17, 19).

The early development of the telencephalon of batoids closely resembles that of the sharks. However, later in development, the two groups diverge with regard to the extent of the ventricular cavities. Whereas in most sharks the lateral ventricles remain wide, in batoids they become gradually reduced to two short, horn-shaped rudiments (Fig. 18). Solid hemispheres, due to a massive development of pallial cell masses, are also found in advanced galeomorph sharks species such as the nurse shark *Ginglymostoma.*

3. Telencephalon: Cytoarchitecture and Fiber Connections

Studies on the structure of the telencephalon of cartilaginous fishes, based on normal material, have been made by Johnston (1911), Ariëns Kappers and Carpenter (1911), Holmgren (1922), Bäckström (1924), Gerlach (1947), Faucette (1969a,b), Kuhlenbeck (1973, 1977), Williams (1973), Northcutt (1978, 1981), Ebbesson (1980), Schnitzlein (1982), and Smeets *et al.* (1983). The telencephalon has been subdivided by all these students of the chondrichthian forebrain into three parts. These are, from rostral to caudal: (1) the bulbus olfactorius, (2) the lobus hemisphericus telencephali, and (3) the telencephalon impar.

In all cartilaginous fishes studied, the *olfactory bulbs* are well-developed structures, but they vary considerably in shape (Fig. 2). In sharks, such as *Notorynchus, Squalus,* and *Ginglymostoma,* the bulbi are ovoid, but in most batoids they have a flattened saucer-shaped appearance. Other species like *Scyliorhinus* and *Mustelus* lie between these extremes. In holocephalians, the olfactory bulbs are also ovoid, but connect directly with the telencephalic hemispheres. Despite this variation in shape, four layers can always be distinguished histologically in the olfactory bulbs of cartilaginous fishes (Nieuwenhuys, 1967): (1) a stratum nervosum, consisting of intertwined primary olfactory fibers; (2) a stratum glomerulosum, containing several layers of well-defined glomeruli; (3) a stratum mitrale, built up of large polymorph elements, and (4) a wide stratum granulare. It has been shown by Golgi studies (Catois, 1901; Sterzi, 1909) and electron microscopic studies (Andres, 1970, 1975) that the dendrites of the mitral cells generally enter into the formation of several glomeruli. The axons of the mitral cells traverse the granular layer obliquely and assemble in the periventricular zone of the bulb forming the secondary olfactory tracts.

The *lobus hemisphericus* has been subdivided by all students of the chondrichthian forebrain into a pallium (roof) and a subpallium (floor). However, there is no general agreement with respect to the boundaries of these two principal parts. According to Johnston (1911) and Holmgren (1922), medial and lateral cell-free zonae limitantes indicate the pallio-subpallial boundaries, whereas Kuhlenbeck (1929) and Gerlach (1947) considered ventricular sulci together with zonae limitantes as important landmarks. Northcutt (1978, 1981), on the other hand, left both sulci and zonae limitantes out of consideration and paid more attention to fiber connections. The study by Smeets *et al.* (1983) has revealed that in both *Squalus* and *Scyliorhinus,* more or less distinct cell-free zones exist, but only in the caudal part of the telencephalic hemispheres (Figs. 4, 5, 7, 8). Both a zona limitans lateralis in the lateral wall and a zona limitans medialis in the medial wall could be recognized. The positions of these zones are marked on the ventricular surface by two distinct sulci, i.e., sulcus limitans lateralis and medialis, whereas in *Scyliorhinus* the position of the zona limitans lateralis is also marked by an external groove. In *Hydrolagus* the pallio-subpallial boundary is marked throughout the extent of the hemispheric lobe by ventricular sulci, but distinct zonae limitantes are absent, though at some levels a trace of zona limitans can be recognized (Fig. 14). In *Raja,* as in other batoids and sharks that have reduced ventricles, the subdivision of the telencephalon into a pallium and a subpallium is rather difficult, since distinct zonae limitantes and sulci limitantes do not exist except for the most caudal portion (Fig. 10). Therefore, it may be concluded that cell-free zones and sulci limitantes have only limited value in establishing homologies between the telencephalic structures of chondrichthian fishes.

The third telencephalic subdivision, the *telencephalon impar,* is that portion of the forebrain that is not involved in the lateral evagination of the hemispheric lobes. Following Johnston (1911), its caudal boundary is defined as a plane passing from the velum transversum to the caudal pole of the optic chiasm. This definition implies that the preoptic area is considered a part of the telencephalon impar.

3.1. Cell Masses

The cells in the chondrichthian forebrain are in general small (< 20 μm) (Smeets *et al.*, 1983) and are in most places spread through the entire width of the neuronal wall. Although the telencephalic gray matter consists mainly of a continuum of diffusely arranged cells, about 20 cell masses could be delineated within this brain part of each of the four species studied.

Apart from neuronal elements, the chondrichthian brain comprises at least two types of glial elements—tanycytes and astrocytes (Horstmann, 1954; Klatzo, 1967). The tanycytes, a term introduced by Horstmann, are characterized by long processes that span the entire width of the neuronal wall. The cell bodies of

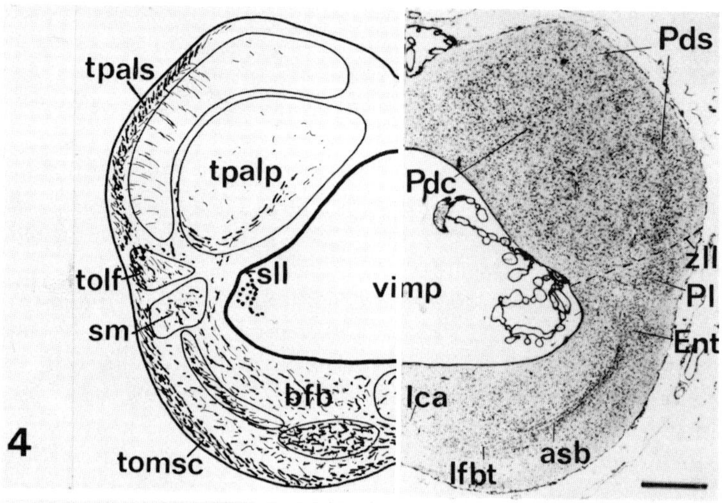
tpals
tpalp
tolf
sll
vimp
sm
bfb
lca
4
tomsc
lfbt
asb
Pds
Pdc
zll
Pl
Ent

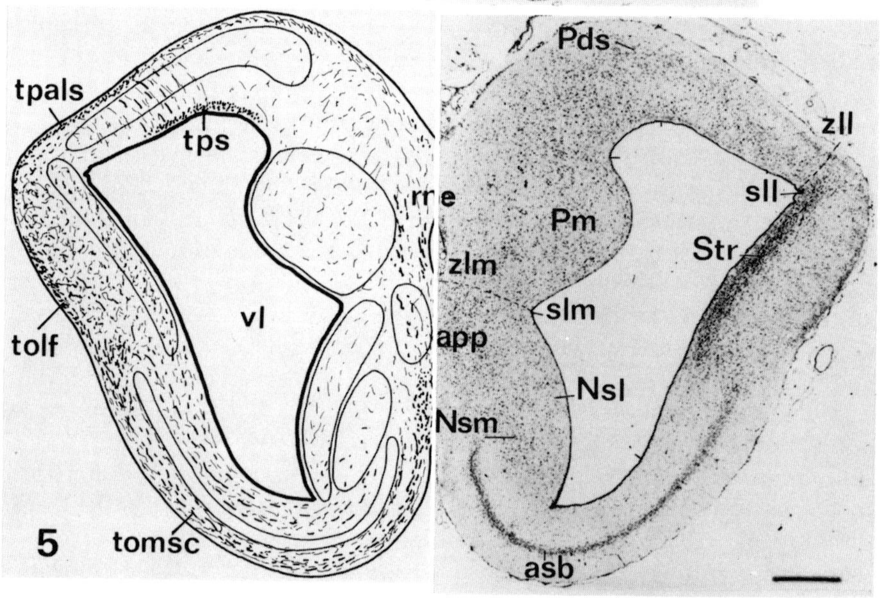
tpals
tps
rme
zlm
vl
tolf
app
Nsm
5
tomsc
asb
Pds
zll
sll
Pm
Str
slm
Nsl

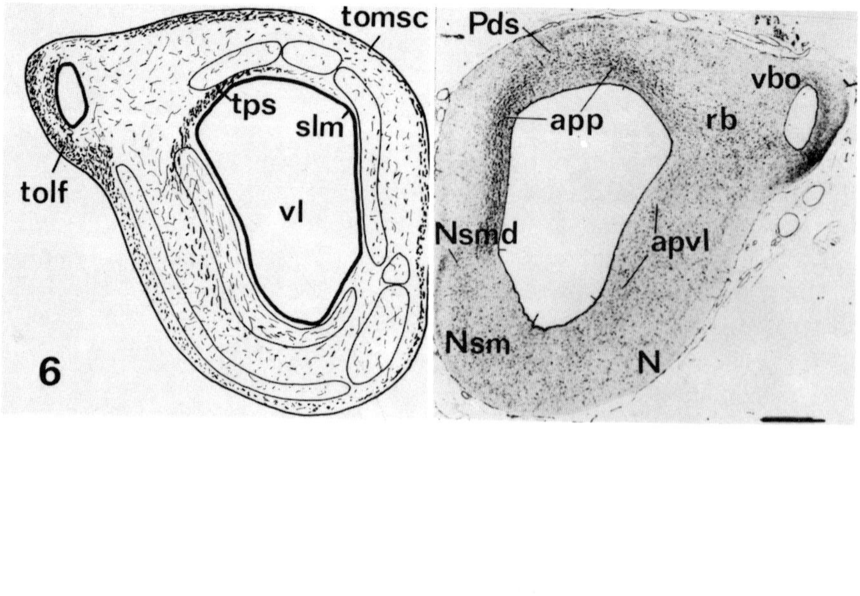
tomsc
tps
slm
tolf
vl
6
Pds
app
rb
vbo
Nsmd
apvl
Nsm
N

the tanycytes lie mostly in the ependymal layer and their processes show an overall radial pattern. Astrocytes do not possess such long processes, but have numerous, short ramifications. In contrast to tanycytes, the majority of astrocytes do not lie in the ependymal layer, but in the proximity of neurons, blood vessels, and meningeal surface. Both tanycytes and astrocytes lie adjacent to blood vessels or form characteristic end feet that entwine around the capillaries. In addition to the two glia types mentioned, Klatzo (1967) identified oligodendrocytes and microgliacytes. The former have an oval or round cell body and slender branching processes. The microgliacytes have an irregular shape, and short, thorny processes. Whereas the oligodendrocytes appear to be very abundant in the various parts of the chondrichthian brain, the microgliacytes lie mainly in the caudal part of the brain. A striking feature observed in certain telencephalic areas of the lemon shark (*Negaprion* (Klatzo, 1967) and the thornback ray *Raja* (Smeets *et al.*, 1983) is satellitosis, i.e., a clustering of glial cells including astrocytes, oligodendrocytes, and microgliacytes around nerve cells (Fig. 12, SP3).

3.1.1. Cell Masses in the Pallium

At the beginning of this century, it was thought that the entire roof of the chondrichthian telencephalon represented a primordium hippocampi (Johnston, 1911). However, embryological studies by Holmgren (1922) and Bäckström (1924) revealed that during early ontogenesis, three pallial formations can be recognized, viz., the medial, or hippocampal pallium, the dorsal, or general pallium, and the lateral, or piriform pallium. During further development, these subdivisions become less distinct, but can still be recognized in the adult brain. However, much disagreement exists about the extent and boundaries of these pallial subdivisions. For clarity I will first describe the pallial subdivisions as recognized in the sharks and, subsequently, these data will be compared to those obtained for *Raja* and *Hydrolagus*.

The caudal part of the telencephalic roof in the sharks *Squalus* and *Scyliorhinus* is almost completely occupied by the *dorsal pallium* (Figs. 4, 7, 16, 17). This cell mass can be subdivided into inner and outer laminae separated from each other by a cell-poor zone. the innermost layer is termed the pallium dorsale, pars centralis, whereas the outermost layer has been indicated as the pallium dorsale, pars superficialis. In squalomorph sharks, like *Notorynchus* and *Squalus*, the central portion of the dorsal pallium is relatively poorly developed, when compared to the greatly enlarged nucleus in galeomorph sharks, such as *Ginglymostoma*. In this respect, *Scyliorhinus* cannot be considered the most advanced representative of its group. In *Ginglymostoma* the central part of the dorsal pallium extends rostrocaudally over almost half of the total length of the telencephalon and is obviously a heterogeneous structure (Luiten, 1981). Whereas the

←

Figures 4–6. Transverse sections through the caudal (4), intermediate (5), and rostral (6) portions of the telencephalon of the shark *Squalus acanthias*. The left half of each figure is a drawing of myelinated (thick interrupted lines) and unmyelinated (thin interrupted lines) fibers as seen in Klüver-Barrera and Bodian stained material, whereas the right half is a photomicrograph taken from the adjacent Nissl stained section. The levels of the sections are indicated in the topographical reconstruction (Fig. 16). Scales = 1 mm.

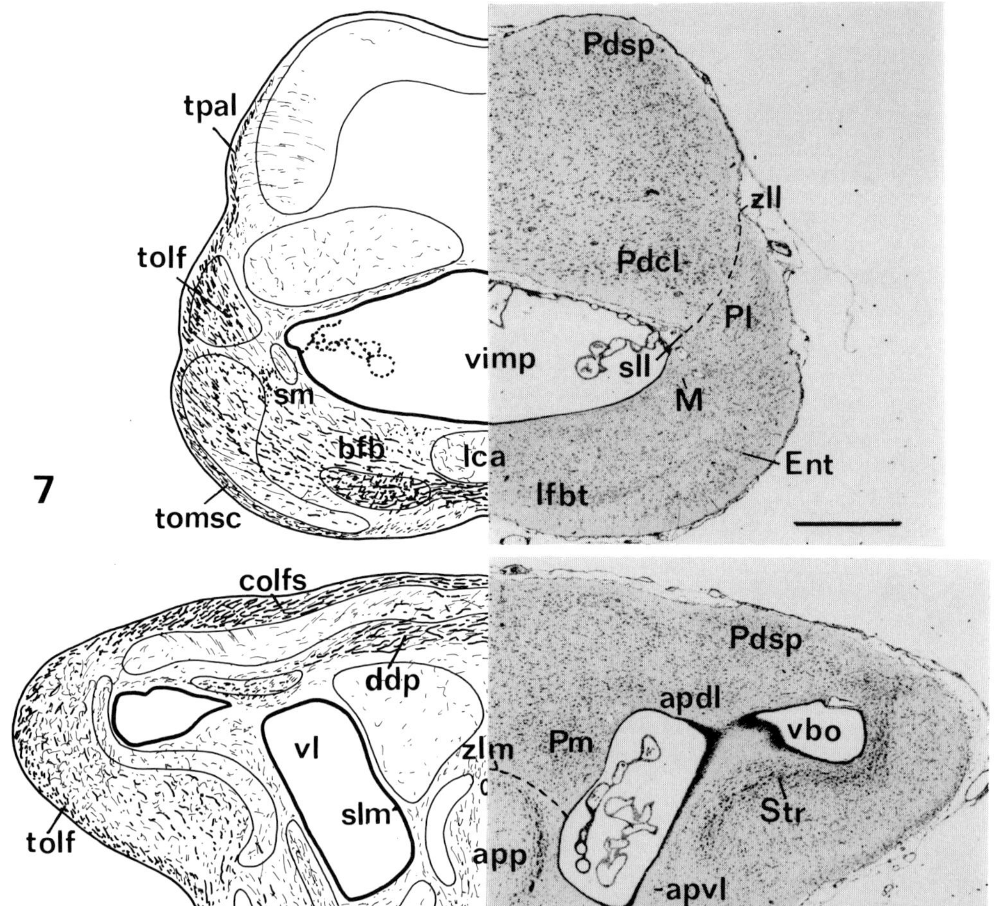

Figures 7–9. Transverse sections through the caudal (7), intermediate (8), and rostral (9) portions of the telencephalon of the shark *Scyliorhinus canicula.* The levels of the transverse sections are indicated in the topographical reconstruction (Fig. 17). For further explanation see Figs. 4–6.

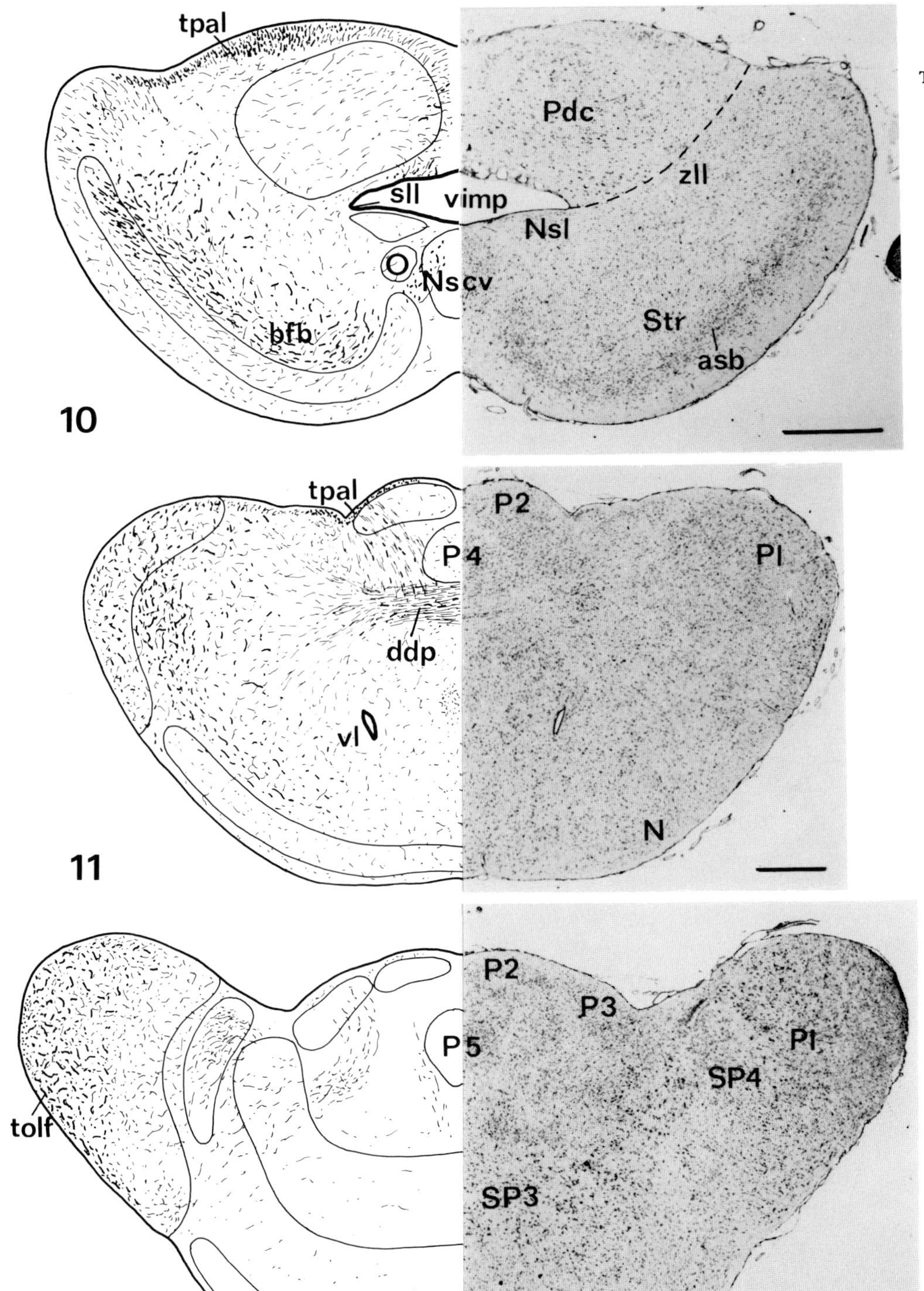

Figures 10–12. Transverse sections through the caudal (10), intermediate (11), and rostral (12) portions of the telencephalon of the ray *Raja clavata*. The levels of the transverse sections are indicated in the topographical reconstruction (Fig. 18). For further explanation see Figs. 4–6.

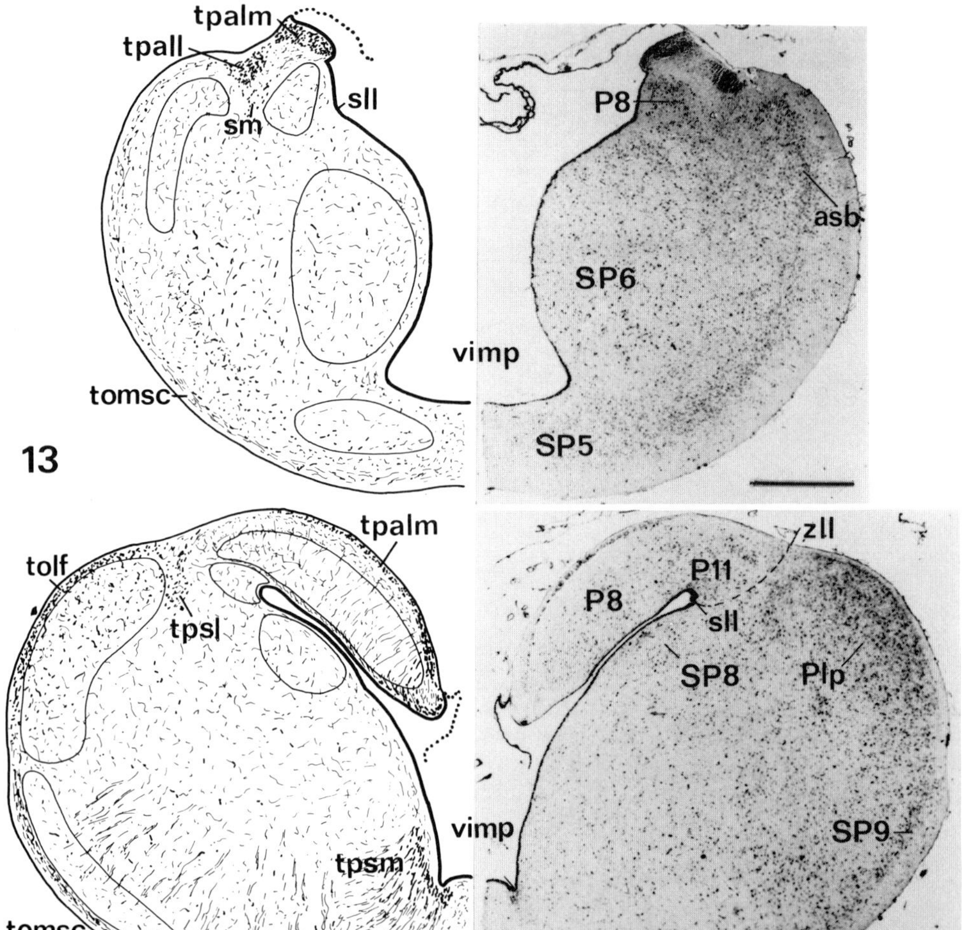

Figures 13–15. Transverse sections through the caudal (13), intermediate (14), and rostral (15) portions of the telencephalon of the ratfish *Hydrolagus collei.* The levels of the transverse sections are indicated in the topographical reconstruction (Fig. 19). For further explanation see Figs. 4–6.

central subdivision of the dorsal pallium occupies a restricted area in the caudal part of the telencephalic roof in *Squalus* and *Scyliorhinus*, the superficial portion of the dorsal pallium extends rostrally beyond the level of the external neuropore (Figs. 16, 17). A remarkable feature of the caudal part of the pallium dorsale, pars superficialis, is the presence of fibers that traverse the lateral portion of this cell mass to join the pallial tract (see Section 3.2.3).

Besides the superficial and central subdivisions, the dorsal pallium of both sharks studied contains two other cell masses: the area periventricularis pallialis and the area periventricularis dorsolateralis. The periventricular pallial area lies in the rostral part of the pallium and consists of clusters of granular cells that show a laminar arrangement (Figs. 5, 6, 8, 9, 16, 17). This cell mass covers the ventricle on the medial and dorsal side. The medial portion of the cell mass that extends farther caudally than the dorsal portion was interpreted by Johnston (1911) and Northcutt (1981) as a septal nucleus. The second remaining cell mass in the pallium dorsale, i.e., the area periventricularis dorsolateralis, is particularly well developed in *Squalus* and has a dorsolateral, periventricular position in

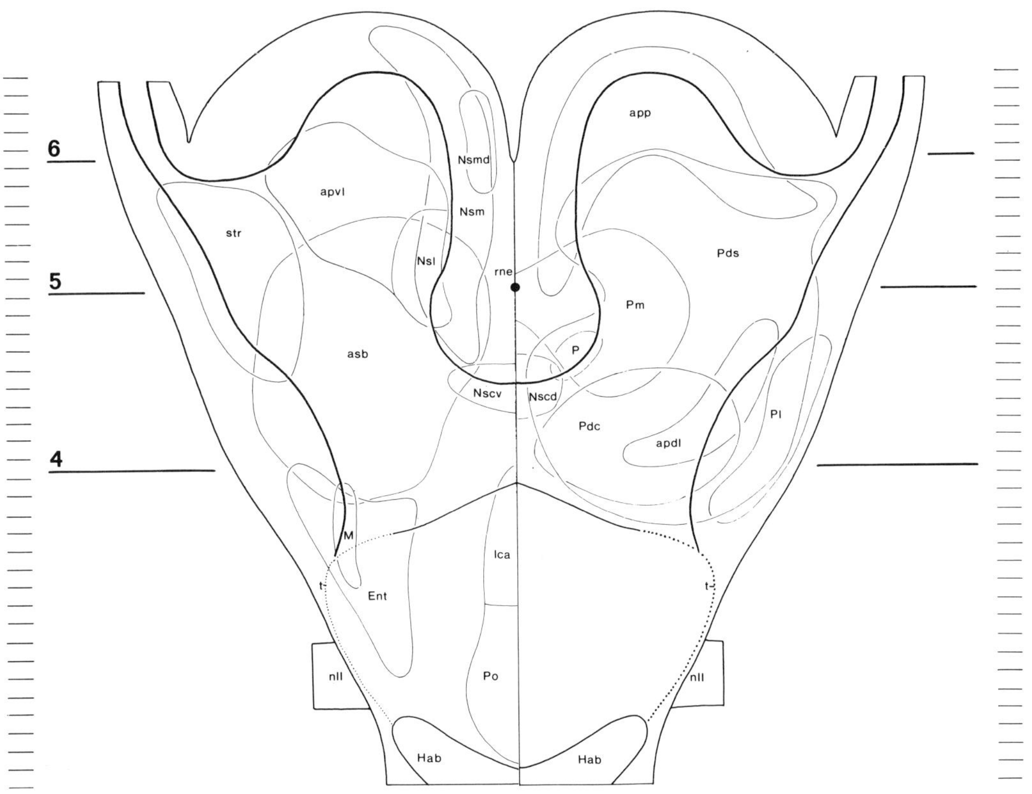

Figure 16. Topographical reconstruction of the cell masses of the telencephalon of *Squalus acanthias* as projected upon a horizontal plane. For clarity, the left half of the figure presents the cell masses of the subpallium and the telencephalon impar, whereas in the right half only the pallial cell masses are drawn. Heavy continuous lines represent the outlines of the ventricle, medium continuous curves represent the contours of the external telencephalic features and the dotted lines indicate the taeniae. The remaining thin lines represent the cell masses. Numbers at the side of the drawing refer to the numbers of the selected transverse sections. Distance between sections, indicated by small horizontal bars, is 300 μm. (Slightly modified from Smeets *et al.*, 1983, by permission of Springer-Verlag.)

the caudal half of the pallium causing a slight protuberance into the ventricle (Figs. 16, 17).

The *pallium mediale* or *hippocampale* in both sharks studied consists of rather small, ellipsoid cells and fuses with its counterpart across the midline (Figs. 5, 8, 16, 17). Compared to the dorsal pallium, it is relatively small and causes a protrusion of the dorsomedial wall into the telencephalic ventricle at midpallial level.

The third pallial subdivision, i.e., the *pallium laterale,* is triangular and lies ventral to the zona limitans lateralis in both sharks studied (Figs. 4, 7). Despite its subpallial position, I consider this cell mass as the lateral pallium in agreement with Northcutt (1977). The principal reason for doing so is that this cell mass receives the main olfactory input (Smeets, 1983), just like the lateral pallium of tetrapods. In older studies, this cell mass was indicated as nucleus olfactorius lateralis, whereas the lateral part of the dorsal pallium was considered as pallium laterale (Johnston, 1911; Holmgren, 1922).

The pallial structures as described for the two shark species studied can only partly be compared to pallial structures found in batoids and holocephalians. A pallium dorsale, pars centralis can be recognized in *Raja* (Fig. 10), whereas the pallial areas P1, P2, and the caudal part of P3 (Figs. 11, 18) may be considered as homologous to the caudal part of the pallium dorsale, pars superficialis of sharks. The latter assumption is based on their position and on the characteristic

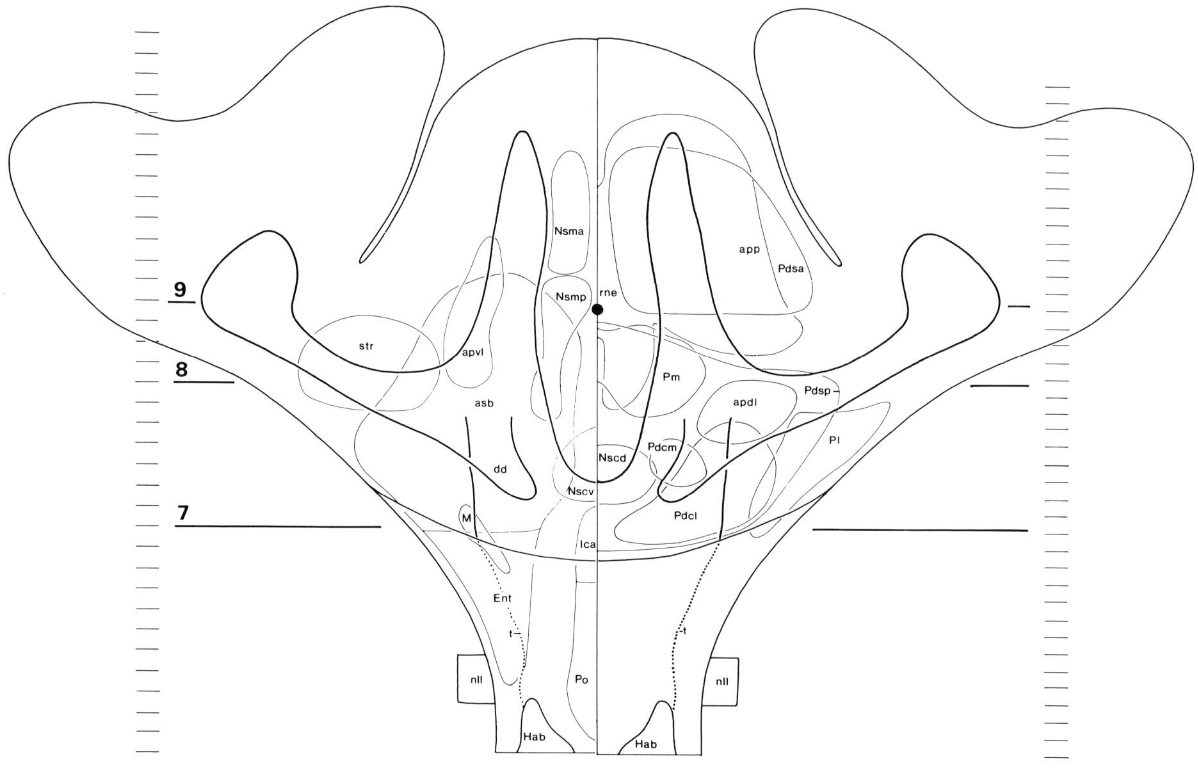

Figure 17. Topographical reconstruction of the cell masses of the telencephalon of *Scyliorhinus canicula* as projected upon a horizontal plane. For further explanation see Fig. 16.

course of fibers that traverse these areas before joining the pallial tract. Because of the same reasons, the areas labeled P8 and P9m in *Hydrolagus* (Figs. 13–15, 19) may be considered as a pallium dorsale, pars superficialis. Pallial area P7 and the rostral part of P3 in *Raja* (Fig. 18), as well as pallial area P10 in *Hydrolagus* (Fig. 19) may, on account of their position and lack of traversing fibers, be compared with the rostral part of the superficial part of the dorsal pallium of sharks.

A presumed homologue of the medial pallium in sharks is observed in the holocephalian telencephalon at the level where the pallium meets the subpallium (Figs. 15, 19). In *Raja clavata,* Smeets *et al.* (1983) recognized small clusters of neurons and glial elements that form a flattened plate above the reduced ventricles. They called this cell mass, which could be hardly delineated, the inner zone of the dorsal pallium. However, in other species (*Raja eglanteria*), Smeets and Boord (1985) could delineate a medial pallium in the caudal part of the roof lying immediately dorsal to the impaired telencephalic ventricle and extending rostrally above the reduced ventricles.

The lateral pallium of *Raja* is more extensive than in both sharks studied (Figs. 11, 12, 18). It is worth mentioning that it consists of numerous large neurons that are surrounded by satellite glial cells. In *Hydrolagus* the lateral pallium is constituted by a large cell mass that lies in the lateral hemispheric wall and extends caudally into the telencephalon impar (Figs. 14, 15, 19). On the basis of the size of its constituent neurons, it can be subdivided into rostral and caudal portions.

Apart from the pallial structures that may probably be homologized with

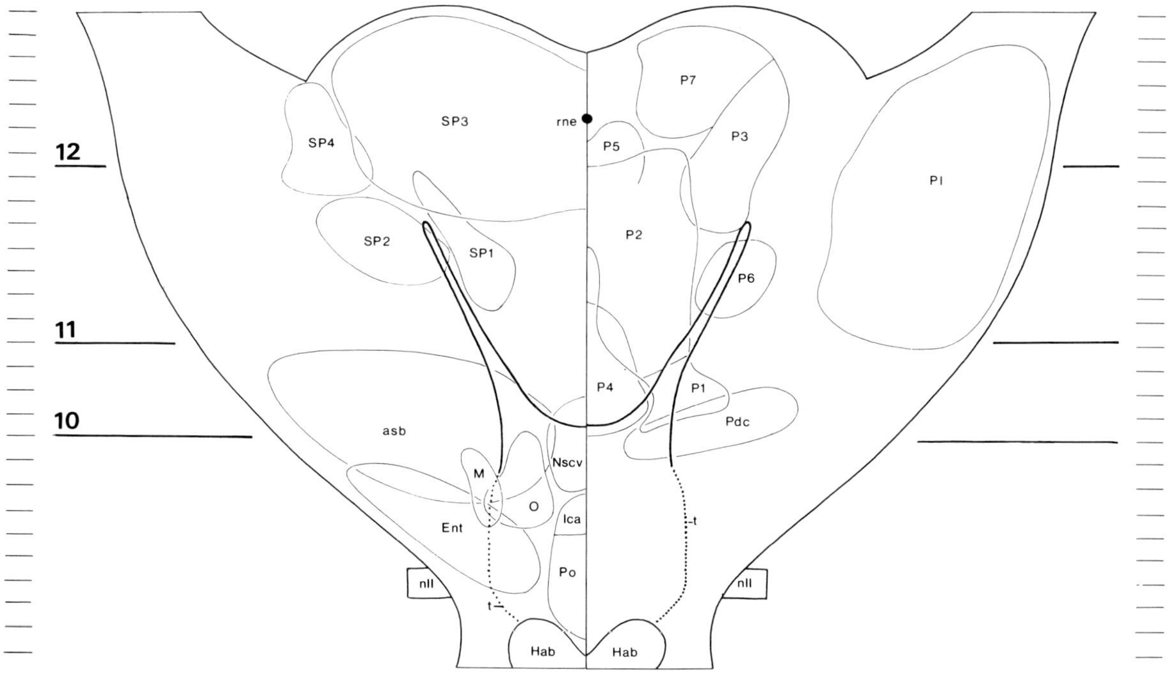

Figure 18. Topographical reconstruction of the cell masses of the telencephalon of *Raja clavata* as projected upon a horizontal plane. For further explanation see Fig. 16.

those found in sharks, the telencephalic roof of batoids and *Hydrolagus* contains several additional cell masses that, because of lack of any information about their connections, have been labeled with arabic numerals (Figs. 18, 19).

3.1.2. Cell Masses in the Subpallium and Telencephalon Impar

Because some subpallial cell masses extend beyond the caudal boundary of the telencephalic hemispheres, viz., the foramen of Monro, I shall include in this section the cell masses of the telencephalon impar as well as those of the subpallial division. As in the previous section, I will first deal with the structures in both sharks and subsequently compare them with the corresponding structures in the subpallium and telencephalon impar of *Raja* and *Hydrolagus*.

The most caudal telencephalic cell mass is the nucleus preopticus that lies

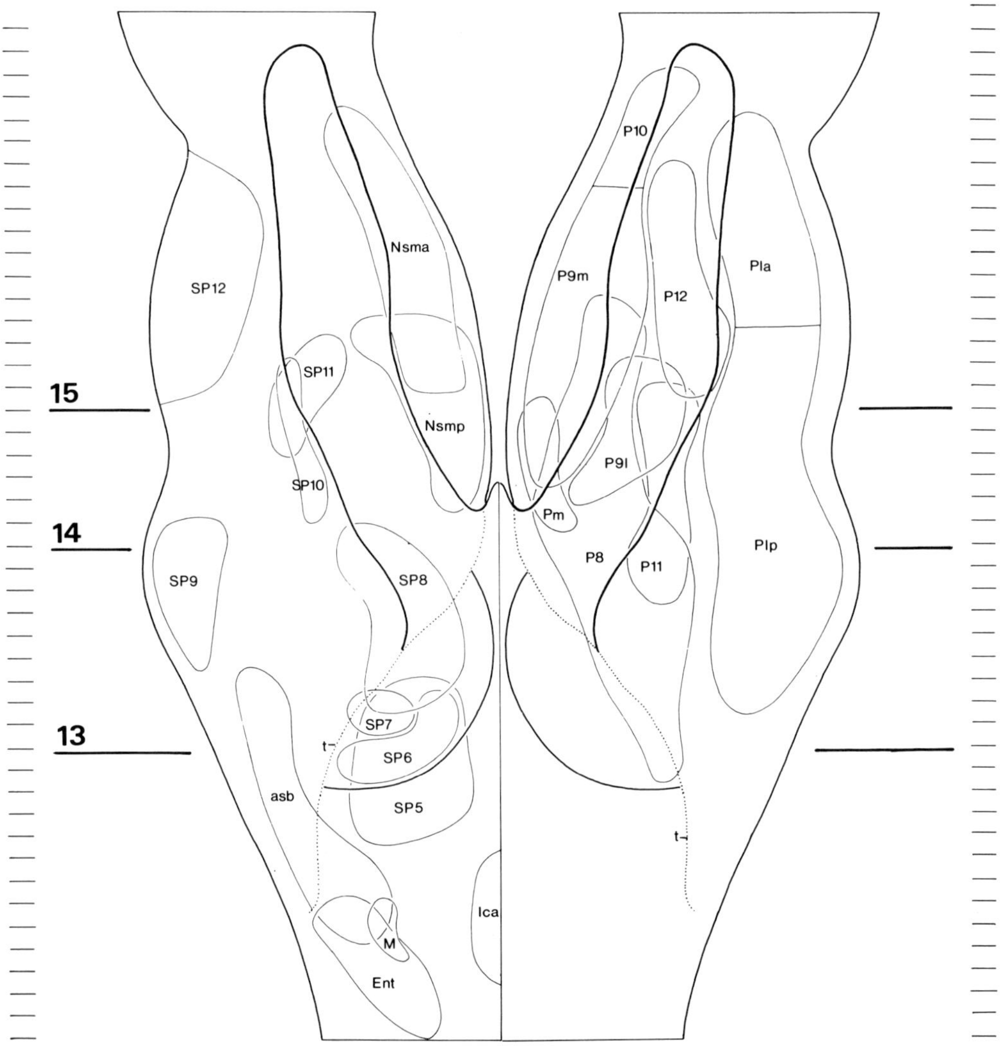

Figure 19. Topographical reconstruction of the cell masses of the telencephalon of *Hydrolagus collei* as projected upon a horizontal plane. For further explanation see Fig. 16.

dorsal to the optic chiasm. Rostrally, this nucleus merges into the nucleus interstitialis of the anterior commissure (Figs. 4, 7, 16, 17). The latter cell mass is identified as a cluster of rather densely packed cells lying dorsal to the anterior commissure. Throughout its extent, the basal forebrain bundle contains numerous cells. In the rostral part of the telencephalon impar, these cells increase in number and this accumulation of cells has been indicated as nucleus interstitialis fasciculi basalis telencephali (Figs. 4, 7, 16, 17). Other nuclei in the telencephalon impar are the nucleus entopeduncularis and nucleus M. The term *nucleus entopeduncularis* is used in the strictly descriptive sense of "lying within the peduncle" and does not imply, for the moment, any homology with the nucleus of the same name in tetrapods. Dorsal to the rostral pole of the entopeduncular nucleus, a small cell mass can be recognized that has been labeled nucleus M. This nucleus has a periventricular position adjacent to the stria medullaris and taenia chorioidea.

The telencephalon impar of the sharks can be easily compared with the corresponding region in *Raja* and *Hydrolagus*. In all species studied, this part of the forebrain contains a preoptic nucleus, an interstitial nucleus of the basal forebrain bundle, an interstitial nucleus of the anterior commissure, an entopeduncular nucleus, and a nucleus M (Figs. 18, 19). One striking difference is that in the holocephalians the preoptic region is extremely elongated due to the far rostral position of the telencephalic lobes.

The subpallial region of sharks can be subdivided into a lateral division containing the area superficialis basalis, the striatum, and the area periventricularis ventrolateralis, and a medial division comprising the septal nuclei. The area superficialis basalis or tuberculum olfactorium (Holmgren, 1922; Bäckström, 1924) merges caudally with the entopeduncular nucleus. However, it can be easily distinguished from the latter nucleus, since its cells are more densely packed and show a laminar organization. The basal superficial area is U-shaped and occupies the entire ventral outer zone of the subpallium (Figs. 4, 5, 8, 9, 16, 17). At rostral telencephalic levels, the area superficialis basalis is replaced by another superficial cell layer. Several authors (e.g., William, 1973; Northcutt, 1978) considered this layer, labeled as nucleus N (Figs. 6, 9), as a rostral continuation of the area superficialis basalis. However, the cells in nucleus N are somewhat larger and more diffusely arranged than those of the area superficialis basalis and, moreover, the two cell masses differ with respect to their fiber connections. In both sharks studied, a conspicuous cell mass is found in the ventrolateral wall of the hemisphere, where it causes a slight protrusion (Figs 5, 8, 16, 17). This cell mass, labeled as striatum, lies rostral to the foramen of Monro and consists of numerous clusters of small granular cells that show a laminar organization. Likewise in the ventrolateral wall, but medial to the striatum, a cell mass could be delineated that has been indicated descriptively as area periventricularis ventrolateralis (Figs. 6, 8, 9, 16, 17). Based on its position, this cell mass might be compared to the nucleus accumbens of tetrapods.

An area superficialis basalis can also be recognized in the subpallium of *Raja* and *Hydrolagus* (Figs. 10, 13, 18, 19). However, a striatum could not be identified with certainty in these two species. In *Raja,* Northcutt (1978) labeled an area as striatum that corresponds to the U-shaped area subpallialis 3 (SP3, Figs. 12, 18) of Smeets *et al.* (1983). Recent experimental studies (Smeets and Boord, 1985; Meredith and Smeets, 1987) provide some evidence that the striatum of batoids is an ill-defiend area lying internal to the area superficialis basalis (Fig. 10). In holocephalian fishes, the large intraventricular protrusion of the lateral telencephalic wall (Figs. 13, 14) has been considered as striatal swellings (Ariëns Kappers and Carpenter, 1911; Holmgren, 1922; Faucette, 1969a,b). However, in *Hydrolagus* this prominent bulge contains at least three distinct cell masses, i.e., area subpallialis 6, 7, and 8 (Fig. 19), and it is still uncertain whether any of these three cell masses can be compared with the shark's striatum.

The septal region comprises the medial as well as the ventromedial portion of the subpallium in both sharks studied. It is worth mentioning that both in *Squalus* and *Scyliorhinus*, a part of the septal region arches around the foramen Monroi and occupies a position in the dorsomedial wall of the cerebral hemisphere (Figs. 16, 17). The septal region is not a homogeneous structure in sharks, but comprises four (*Scyliorhinus*) or five (*Squalus*) subdivisions (Figs. 5, 6, 8, 9, 16, 17). The presumed septal area in *Raja* contains numerous, diffusely arranged neurons. Apart from a nucleus septi caudoventralis (Figs. 10, 18), no septal nuclei could be clearly delineated. In *Hydrolagus*, a medial septal nucleus can be recognized extending throughout the rostral half of the forebrain (Figs. 15, 19). It can be subdivided into an anterior and a posterior part, of which the former contains somewhat larger cells. No other distinct septal subdivisions could be recognized in *Hydrolagus*.

In addition to the structures mentioned above, the subpallial areas of *Raja* and *Hydrolagus* contain a number of cell masses that cannot be easily homologized with any structure in the telencephalon of sharks. These cell masses are labeled SP1–12 and the reader is referred to the study by Smeets *et al.* (1983) for further details.

3.2. Fiber Systems

The majority of the fibers in the forebrain of cartilaginous fishes are unmyelinated. Apart from the main fiber streams, the chondrichthian telencephalon contains a huge number of unmyelineated fibers that intermingle and form an extensive, diffuse meshwork. In normal, Bodian stained material, it is, therefore, impossible to trace precisely a given fiber system from the cells of origin to the sites of termination. For this reason, I will deal in the present section only with the main fiber systems and with experimental verified connections. For a review on the fiber connections as seen in normal material, the reader is referred to Smeets *et al.* (1983). In the following account, I will describe first the secondary olfactory tracts (Section 3.2.1), then the intrinsic telencephalic connections (Section 3.2.2), and finally the telencephalic connections with more caudally situated brain regions (Section 3.2.3).

3.2.1. Secondary Olfactory Tracts

In the classical studies on the chondrichthian forebrain (e.g., Johnston, 1911; Bäckström, 1924; Ariëns Kappers *et al.*, 1936), and also in some more recent studies (Williams, 1973; Kuhlenbeck, 1977), almost the entire telencephalon has been thought to be dominated by the olfactory system, and, hence, considered to be a rhinencephalon. The most detailed study on the fiber anatomy of the elasmobranch telencephalon is that of Bäckström (1924), who recognized no fewer than seven secondary olfactory tracts distributing their fibers to ipsilateral as well as to contralateral telencephalic areas. The first evidence that only a restricted area of the telencephalon receives secondary olfactory fibers was obtained from an experimental study by Ebbesson and Heimer (1970). The latter study revealed that, after unilateral transection of the olfactory peduncle, degenerated fibers could be traced only to the ipsilateral hemisphere and that the secondary olfactory projections appear to be restricted to the lateral pallium and to a region superficial to the lateral part of the area superficialis basalis. Subsequently, electrophysiological studies (Bruckmoser, 1973; Bruckmoser and Dieringer, 1973; Vesselkin and Kovačević, 1973) confirmed the restricted distribution of olfactory input to the forebrain in *Torpedo* and *Scyliorhinus*. A degeneration study of secondary olfactory projections in *Scyliorhinus* and *Raja* by Smeets (1983) yielded similar results, but also demonstrated that the secondary olfactory projections in *Scyliorhinus* are more extensive than in other chondrichthians studied experimentally. In addition to a strong ipsilateral projection, a weak projection could be demonstrated to the contralateral pallium and olfactory bulb. However, even in *Scyliorhinus* the extent of the regions that receive olfactory input are more limited than had been thought previously (see Bäckström, 1924). In *Raja,* the secondary olfactory projections are restricted to the ipsilateral lateral pallium. The lack of evidence for a projection to the lateral part of the area superficialis basalis, as reported in all sharks studied experimentally, is especially notable. Although no experimental data are available, it does seem unlikely that the olfactory projections in *Hydrolagus,* like those in *Raja,* reach the area superficialis basalis.

3.2.2. Intrinsic Telencephalic Connections

The telencephalic hemispheres contain numerous fibers that can be classified as neither direct afferent nor efferent fibers. The majority of these fibers are unmyelinated and form a dense meshwork within the telencephalon. Despite this chaotic arrangement, some fibers collect as rather distinct bundles that can be traced for some distance within the forebrain. However, the exact sites of origin and termination cannot be determined in our material. In classical studies (e.g., Johnston, 1911; Bäckström, 1924), these fiber bundles had already been recognized and labeled. Although I do not agree fully with the terminology of Bäckström (1924), I will follow his nomenclature to avoid a further accumulation of novel names.

According to Bäckström, the tractus olfactorius medialis septi cruciatus consists of secondary olfactory fibers that separate from the medial olfactory tract and curve along the rostral pole of the hemisphere until they reach the midplane. From there, the fibers arch caudoventrally through the septum and cross, ventral to the anterior commissure, to the opposite side. Next, the bundle turns dorsolaterally and courses again rostrally issuing its fibers to striatum, area superficialis basalis, and lateral pallium. In both sharks studied, a fiber system is found that in its course resembles the tractus olfactorius medialis septi cruciatus of Bäckström (Figs. 4–9). In *Squalus* this tract is well developed and contains numerous myelinated fibers, but in *Scyliorhinus* it consists almost exclusively of unmyelinated fibers. A similar pathway could be recognized in *Hydrolagus* (Figs. 13–15), but not in *Raja.* However, the olfactory nature of this tract may be seriously questioned, since experimental studies on secondary olfactory projections in cartilaginous fibers have, hitherto, not provided any evidence of the existence of such a pathway.

A tractus pallialis septi can be recognized in *Squalus, Scyliorhinus,* and *Hydrolagus,* but not in *Raja.* It consists of numerous myelinated fibers that form a longitudinal bundle in the telencephalic roof. In *Squalus* its fibers intermingle caudally with those of the pallial tract, whereas rostrally its fibers fan out in the ventrolateral hemispheric wall (Figs. 5, 6). In *Scyliorhinus* a comparable fiber system could be recognized only at rostral telencephalic levels (Fig. 9) and in *Hydrolagus* only at caudal levels where it consists of a medial and a lateral component (Fig. 14).

A strongly developed decussatio dorsalis pallii can be recognized in *Scyliorhinus* as well as in *Raja.* Although the various components of the decussatio dorsalis pallii are hard to trace to their sites of origin or termination, it may be well assumed that this complex consists of both decussating and true commissural fibers that interconnect dorsal pallium, lateral pallium, striatum, and area superficialis basalis. A small, pallial commissure may exist in *Squalus* (Johnston, 1911), but I was unable to confirm its existence. Since the pallial regions of both hemispheres of *Hydrolagus* do not fuse, pallial commissures are absent in holocephalians.

3.2.3. Telencephalic Connections with Caudal Brain Regions

In all cartilaginous fishes studied, three distinct fiber tracts connecting the telencephalon with more caudally situated brain regions can be recognized. These are: (1) the fasciculus basalis telencephali, (2) the tractus pallii, and (3) the stria medullaris.

Unlike amniotes, the fasciculus basalis telencephali or basal forebrain bundle in cartilaginous fishes does not show distinct medial and lateral components. The bundle is well developed in the four species studied and consists of both myelinated and unmyelinated fibers. In normal material the basal forebrain bundle can be traced from the level of the caudal pole of the area superficialis basalis (Figs. 4, 7, 9) rostrally, to the midbrain caudally. The majority of its fibers decussate in the anterior commissure. The basal forebrain bundle contains ascending as well as descending fibers. Experimental studies (Smeets, 1983; Smeets and Boord, 1985) have revealed that fibers originating from striatum and area superficialis basalis project to the hypothalamic inferior lobe and more caudal brain regions reaching as far as the rhombencephalic reticular formation. These studies have confirmed that the majority of the fibers in the basal forebrain bundle cross to the opposite side, via the anterior commissure, before terminating in diencephalic and mesencephalic regions. The ascending component of the basal forebrain bundle consists, according to Johnston (1911), of fibers that connect thalamic with pallial regions. These ascending fibers, collectively labeled as the tractus thalamocorticalis by Johnston, course through the lateral part of the basal forebrain bundle. At present, there is experimental evidence that fibers reaching the telencephalon by way of the basal forebrain bundle originate from the thalamus, the hypothalamus, and the tegmentum of the midbrain (Schroeder and Ebbesson, 1974; Smeets and Boord, 1985; Meredith and Smeets, 1987).

The tractus pallii, which is the main pathway between the pallium and the hypothalamus, is strongly developed in the four species studied and consists of both myelinated and unmyelinated fibers. In *Scyliorhinus* and *Raja,* only a single pallial tract can be identified (Figs. 7, 10, 11), whereas in *Squalus* two components can be recognized at telencephalic levels, which, on the basis of their position, are termed pars superficialis and pars periventricularis (Figs. 4, 5). In *Hydrolagus* this tract also consists of two separate bundles that have been labeled pars medialis and pars lateralis on account of their position (Figs. 13–15). As mentioned in Section 3.1.1, the pallial tract is constituted by fibers that traverse the superficial part of the dorsal pallium. It can be traced caudally, keeping a position lateral to the basal forebrain bundle, and, on reaching the level of the optic chiasm, the fibers of the pallial tract run superficial to, or intermingle with, lateral fibers of the optic tract. Finally, the pallial tract fibers decussate partly caudoventral to the optic chiasm and mostly terminate in the rostral two-thirds of the hypothalamic inferior lobe (Smeets and Boord, 1985).

The stria medullaris, which is the most important connection of the habenular ganglion, is considerably smaller than the basal forebrain bundle and the pallial tract. Although the exact site of origin of its fibers cannot be determined in normal material, it may be assumed that the stria medullaris collects its fibers from the caudal parts of both pallium and subpallium (Figs. 4, 7, 13). Various components of this tract have been described in normal material (for review see Smeets *et al.,* 1983), but none of these components have been confirmed experimentally.

4. Functional Correlations and Concluding Remarks

In the previous sections, a brief review of the structural organization of the telencephalon of cartilaginous fishes has been presented. In this final section, I will discuss the possible role that the chondrichthian forebrain may play in gathering and integrating information received by the various sensory systems and in influencing motor behavior.

Although it is well known that the olfactory receptors of cartilaginous fishes are highly efficient detectors of waterborne chemicals (for review see Kleerekoper, 1978), it is now well established that only a restricted area in the chondrichthian telencephalon receives olfactory input (Fig. 20A) and that even tertiary projections seem to be limited in character (Ebbesson, 1972). So the term *rhinencephalon* so many times used to typify the chondrichthian telencephalon has to be abandoned. The finding of a restricted olfactory input raises the important question: what could be the function of those large telencephalic regions that apparently do not receive olfactory information? Evidence is increasing that the telencephalon of cartilaginous fishes, in addition to olfactory input, receives auditory, lateral line, and visual information.

Although olfaction plays a very important role in detection of food sources and reproductive and social behavior, evidence is now available that hearing also is involved in locating food sources and possibly even other objects such as competitors and predators (for review see Myrberg, 1978). Auditory information perceived by the auditory receptors in the maculae sacculi and neglecta of the inner ear (Corwin, 1981a,b) is relayed, together with vestibular input, to all brain-stem octaval nuclei (Koester, 1983; Montgomery and Roberts, 1979). Although we know that auditory information reaches the telencephalon (Bullock and Corwin, 1979; Corwin and Northcutt, 1982), we know almost nothing about the higher-order projections of the octaval nuclei.

The lateral line system comprises two sensory modalities: mechanoreceptive and electroreceptive (for review see Roberts, 1978). Evoked potential recordings (Platt *et al.,* 1974; Bodznick and Northcutt, 1984) have provided evidence that both lateral line sensory modalities reach the chondrichthian forebrain. Experimental anatomical (McCready and Boord, 1976; Boord and Northcutt, 1982) and electrophysiological studies (Bodznick and Northcutt, 1980; Schweitzer, 1983, 1986) have demonstrated that there is a distinct separation of mechanoreceptive from electroreceptive inputs of lateral line receptors not only in their primary central targets, i.e., the intermediate and dorsal nuclei of the octavolateral area, but also in their higher-order projections to midbrain and diencephalon centers (Fig. 20A). Nothing is known about the way along which the mechanoreceptive lateral line information reaches the telencephalon. On the contrary, a fairly complete picture exists of the higher-order projections of the electrosensory lateral line component, which is known to play a role in detecting living food and in determining the direction and polarity of electric fields in the ocean (Kalmijn, 1978, 1982). After having been relayed in the posterior lateral thalamic nucleus, electrosensory lateral line input reaches the medial pallium by

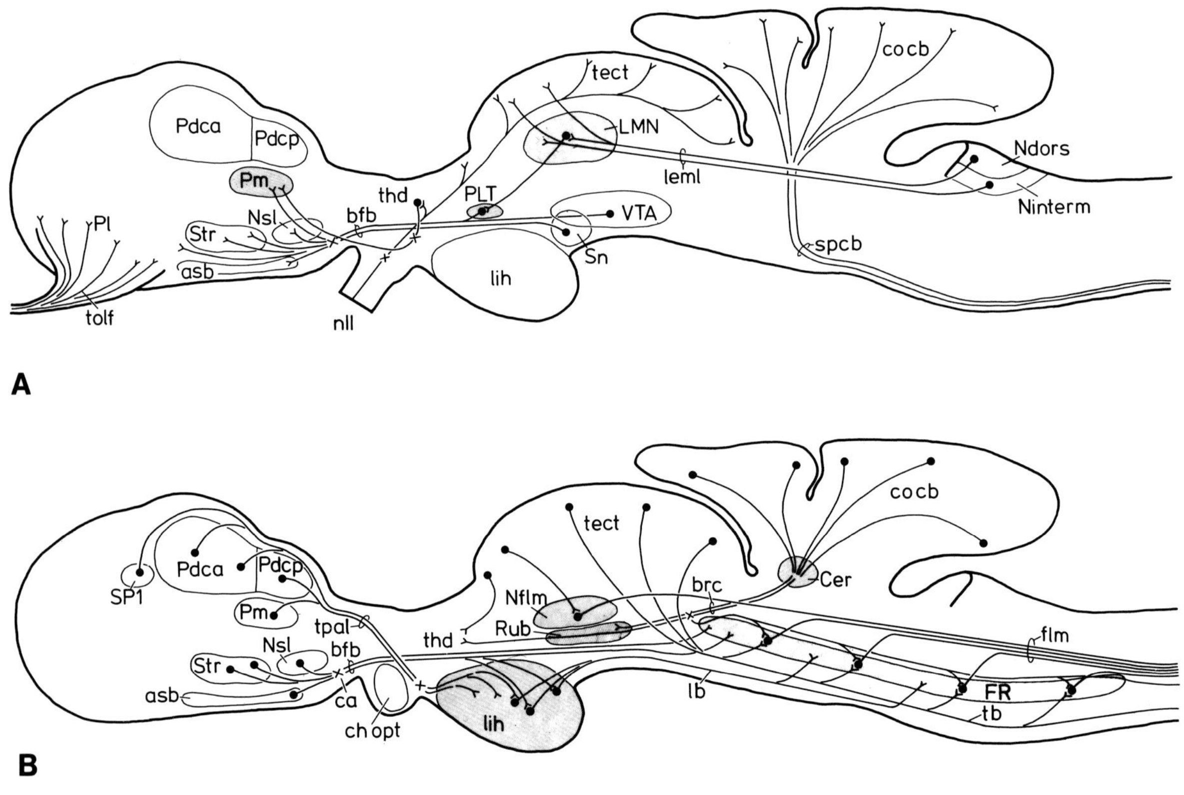

Figure 20. Schematic diagrams showing the main afferent (A) and efferent (B) fiber systems of the chondrichthian forebrain and midbrain as revealed by experimental studies.

way of the basal forebrain bundle (Bodznick and Northcutt, 1984; Smeets and Northcutt, 1987; Fig. 20A).

Experimental anatomical evidence that the chondrichthian telencephalon could have a visual function has been provided by Ebbesson and Schroeder (1971), Schroeder and Ebbesson (1974), and Luiten (1981). It has been shown that, at least in some chondrichthians, two visual pathways exist: (1) a retino-thalamotelencephalic pathway (Fig. 20A) and (2) a retino-tecto-thalamotelencephalic pathway. Electrophysiological evidence in support of the telencephalic visual function in cartilaginous fishes has been presented by Cohen *et al.* (1973), Vesselkin and Kovačević (1973), and Bodznick and Northcutt (1984). The importance of telencephalic visual function has been dramatically underscored by the behavioral experiments of Graeber and his colleagues (Graeber *et al.*, 1983, 1978; Graeber, 1978, 1980). They found that free-swimming nurse sharks (*Ginglymostoma*) with complete tectal ablations could still learn to perform simple visual discriminations such as the distinction between black and white, or horizontal and vertical stripes, but partial or complete lesioning of the visual area in the telencephalon led to an inability to perform these discriminations. In *Ginglymostoma* the visual telencephalic area has been labeled as the central nucleus and has been considered as the deep portion of the pallium dorsale. However, a recent study on the electrosensory and visual projections to the telencephalon of the little skate *Raja erinacea* (Bodznick and Northcutt, 1984) has provided evidence that these projections terminate in the medial pallium.

In addition to the olfactory, auditory, lateral line, and visual information, the chondrichthian telencephalon may receive inputs from the tectum (Northcutt and Boord, 1981; Smeets, 1981), cerebellum (Ebbesson and Campbell, 1973), and spinal cord (Ebbesson and Hodde, 1981) that relay in the dorsal thalamus.

Because of its multimodal input, the chondrichthian telencephalon should be considered as a higher-order coordinating center. However, almost nothing is known about the way in which the telencephalon integrates the various kinds of information, but it is obvious that the central nucleus (or medial pallium) is the place where integration takes place, since the various sensory modalities converge to this telencephalic structure (Bodznick and Northcutt, 1984).

Only a few data are available on the way in which the chondrichthian telencephalon may influence motor behavior. A direct connection of the telencephalon to the spinal cord, like in mammals, has not been demonstrated in cartilaginous fishes. Efferent telencephalic pathways in the nurse shark *Ginglymostoma* can be traced as far as the obex issuing fibers to the reticular formation, but in *Raja* it seems that the descending pathways do not pass beyond the isthmic level. However, the pallial areas that receive the various sensory modalities may project by way of the pallial tract and the basel forebrain bundle to the hypothalamic inferior lobe (Fig. 20B). The latter structure gives rise to a strong lobobulbar tract, the fibers of which terminate in the lateral reticular formation of the brain stem. We do not know how far the telencephalon may influence motor behavior, but it is obvious from the schematic diagram (Fig. 20B) that the midbrain tectum as well as the cerebellum, both of which also receive multimodal sensory inputs (Boord and Northcutt, 1982; Hayle, 1973a,b; Ebbesson and Hodde, 1981; Platt *et al.*, 1974; Smeets, 1982; Tong and Bullock, 1982), may have a more direct influence on motor behavior by way of their strong efferent connections with the brain-stem medial reticular formation (Ebbesson, 1972; Ebbesson and Campbell, 1973; Smeets, 1981), which in turn projects heavily onto the spinal motoneurons (Smeets and Timerick, 1981; Roberts and Williamson, 1983).

In conclusion, it may be stated that the telencephalon of cartilaginous fishes constitutes the highest relay center for multimodal sensory information. The way in which the forebrain integrates this information is still unknown, but, as far as we know, all ascending thalamic projections converge to the medial pallium. The chondrichthian forebrain has no direct projections to the spinal cord, but may influence motor behavior via its connections with the hypothalamic inferior lobe, the midbrain reticular formation, and, at least in the nurse shark, the rhombencephalic reticular formation. Although the chondrichthian forebrain shares most of these features with that of other anamniotes (see other chapters in this volume), two specific features have to be mentioned. First, in contrast to other anamniotes, no direct olfactory projections to diencephalic areas nor contralateral projections via the habenular commissure have been demonstrated in cartilaginous fishes. Second, elasmobranchs have two distinct mesencephalic dopaminergic cell groups that project heavily to basal forebrain regions (Meredith and Smeets, 1987). Such cell groups are apparently absent in the midbrain of bony fishes and anurans, whereas in urodeles only one, relatively small group of catecholaminergic cells can be recognized (Parent *et al.*, 1984). From the foregoing, it may be clear that the telencephalon of cartilaginous fishes shows some features that may be of particular interest to comparative anatomists and evolutionary biologists.

Note Added in Proof

Since the final submission of this chapter (December 1986), several new data about forebrain organization in the cartilaginous fishes have been collected. The most important are Bleckmann *et al.* (1987, 1989) and Northcutt, Reiner, and Karten (1988).

5. Abbreviations

apdl	Area periventricularis dorsolateralis
app	Area periventricularis pallialis
apvl	Area periventricularis ventrolateralis
asb	Area superficialis basalis
bfb	Basal forebrain bundle
brc	Brachium conjunctivum
ca	Commissura anterior
Cer	Nucleus cerebelli
ch opt	Chiasma opticum
co cb	Corpus cerebelli
colfs	Commissura olfactoria superior
cvt	Commissura ventralis telencephali
dd	Diverticulum dorsale telencephali
ddp	Decussatio dorsalis pallii
Ent	Nucleus entopeduncularis
flm	Fasciculus longitudinalis medialis
FR	Formatio reticularis
gc	Granular cells of the olfactory bulb
glom	Glomeruli olfactorii
Hab	Ganglion habenulae
Ica	Nucleus interstitialis commissurae anterioris
Ifbt	Nucleus interstitialis fasciculi basalis telencephali
lb	Tractus lobobulbaris

leml	Lemniscus lateralis
lih	Lobus inferior hypothalami
LMN	Lateral mesencephalic nucleus
M	Nucleus M
N	Nucleus N
Ndors	Nucleus dorsalis areae octavolateralis
Nflm	Nucleus fasciculi longitudinalis medialis
Ninterm	Nucleus intermedius areae octavolateralis
Nscd	Nucleus septi caudodorsalis
Nscv	Nucleus septi caudoventralis
Nsl	Nucleus septi lateralis
Nsm	Nucleus septi medialis
Nsma	Nucleus septi medialis, pars anterior
Nsmd	Nucleus septi medialis, pars dorsalis
Nsmp	Nucleus septi medialis, pars posterior
nII	Nervus opticus
O	Nucleus O
P	Nucleus P
P1–12	Area pallialis 1–12
P9l	Area pallialis 9, pars lateralis
P9m	Area pallialis 9, pars medialis
Pdc	Pallium dorsale, pars centralis
Pdca	Pallium dorsale, pars centralis anterior
Pdcl	Pallium dorsale, pars centralis lateralis
Pdcm	Pallium dorsale, pars centralis medialis
Pdcp	Pallium dorsale, pars centralis posterior
Pds	Pallium dorsale, pars superficialis
Pdsa	Pallium dorsale, pars superficialis anterior
Pdsp	Pallium dorsale, pars superficialis posterior
Pl	Pallium laterale
Pla	Pallium laterale, pars anterior
Plp	Pallium laterale, pars posterior
PLT	Posterior lateral thalamic nucleus
Pm	Pallium mediale
Po	Nucleus preopticus
rb	Area retrobulbaris
rne	Recessus neuroporius externus
Rub	Nucleus ruber
sll	Sulcus limitans lateralis
slm	Sulcus limitans medialis
sm	Stria medullaris
Sn	Substantia nigra
SP1–12	Area subpallialis 1–12
spcb	Tractus spinocerebellaris
Str	Striatum
t	Taenia
tect	Tectum mesencephali
thd	Thalamus dorsalis
tolf	Tractus olfactorius
tomsc	Tractus olfactorius medialis septi cruciatus
tpal	Tractus pallii
tpall	Tractus pallii, pars lateralis
tpalm	Tractus pallii, pars medialis
tpalp	Tractus pallii, pars periventricularis
tpals	Tractus pallii, pars superficialis
tps	Tractus pallialis septi
tpsl	Tractus pallialis septi, pars lateralis
tpsm	Tractus pallialis septi, pars medialis
vbo	Ventriculus bulbi olfactorii
vimp	Ventriculus impar telencephali
vl	Ventriculus lateralis telencephali
VTA	Ventral tegmental area
zll	Zona limitans lateralis
zlm	Zona limitans medialis

ACKNOWLEDGMENTS. I thank Dr. G. E. Meredith and Dr. B. L. Roberts for critically reading the manuscript, Mr. D. de Jong for preparing the photomicrographs, Mrs. N. Nauta-Kaat and Mrs. R. Stoevelaar-de Zeeuw for typing the manuscript, and Springer-Verlag for permission to reproduce several illustrations. Financial support was provided by grants from the Napels-Roscoff Committee of the Royal Dutch Academy of Sciences and the Netherlands Organization for the Advancement of Pure Research (Z.W.O.).

6. References

Andres, K. H., 1970, Anatomy and ultrastructure of the olfactory bulb in fish, amphibia, reptiles and mammals, *Ciba Found. Symp. Taste and Smell in Vertebrates,* pp. 177–196.

Andres, K. H., 1975, Neue morphologische Grundlagen zur Physiologie des Riechens und Schmeckens, *Arch. Oto-Rhino-Laryngol.* **210:**1–41.

Ariëns Kappers, C. U., and Carpenter, F. W., 1911, Das Gehirn von *Chimaera monstrosa, Folia Neurobiol.* **5:**127–160.

Ariëns Kappers, C. U., Huber, G. C., and Crosby, E. C., 1936, *The Comparative Anatomy of the Nervous System of Vertebrates, Including Man,* Volume 1, Macmillan Co., New York, p. 864.

Bäckström, K., 1924, Contributions to the forebrain morphology in selachians, *Acta Zool.* **5:**123–240.

Bauchot, R., Platel, R., and Ridet, J.-M., 1976, Brain–body weight relationships in Selachii, *Copeia* 1976:305–309.

Bleckmann, H., Bullock, T. H., and Jorgensen, J. M., 1987, The lateral line mechanoreceptive mesencephalic, diencephalic, and telencephalic regions in the thornback ray, *Platyrhinoidis triseriata* (Elasmobranchii), *J. Comp. Physiol. A* **161:**67–84.

Bleckmann, H., Weiss, O., and Bullock, T. H., 1989, Physiology of lateral line mechanoreceptive regions in the elasmobranch brain, *J. Comp. Physiol. A* **164:**459–474.

Bodznick, D., and Northcutt, R. G., 1980, Segregation of electro- and mechanoreceptive inputs to the elasmobranch medulla, *Brain Res.* **195:**313–321.

Bodznick, D., and Northcutt, R. G., 1984, An electrosensory area in the telencephalon of the little skate, *Raja erinacea, Brain Res.* **298:**117–124.

Boord, R. L., and Northcutt, R. G., 1982, Ascending lateral line pathways to the midbrain of the clearnose skate, *Raja eglanteria, J. Comp. Neurol.* **207:**274–282.

Bruckmoser, P., 1973, Beziehungen zwischen Struktur und Funktion in der Evolution des Telencephalon, *Verh. Dtsch. Zool. Ges.* **66:**219–229.

Bruckmoser, P., and Dieringer, N., 1973, Evoked potentials in the primary and secondary olfactory projection areas of the forebrain in Elasmobranchia, *J. Comp. Physiol.* **87:**65–74.

Bullock, T. H., and Corwin, J. T., 1979, Acoustic evoked activity in the brain of sharks, *J. Comp. Physiol.* **129:**223–234.

Catois, E. M., 1901, Recherches sur l'histologie et l'anatomie microscopique de l'encephale chez les poissons, *Bull. Sci. Fr. Belg.* **36:**81–83.

Cohen, D. H., Duff, T. A., and Ebbesson, S. O. E., 1973, Electrophysiological identification of a visual area in shark telencephalon, *Science* **182:**492–494.

Compagno, L. J. V., 1977, Phyletic relationships of living sharks and rays, *Am. Zool.* **17:**303–322.

Corwin, J. T., 1981a, Peripheral auditory physiology in the lemon shark; evidence of parallel otolithic and non-otolithic sound detection, *J. Comp. Physiol. A* **142:**379–390.

Corwin, J. T., 1981b, Audition in elasmobranchs, in: *Hearing and Sound Communication in Fishes* (W. N. Tavolga, A. N. Popper, and R. R. Fay, eds.), Springer, Berlin, pp. 81–105.

Corwin, J. T., and Northcutt, R. G., 1982, Auditory centers in the elasmobranch brainstem: Deoxyglucose autoradiography and evoked potential recording, *Brain Res.* **236:**261–273.

Ebbesson, S. O. E., 1972, New insights into the organization of the shark brain, *Comp. Biochem. Physiol. A* **42:**121–129.

Ebbesson, S. O. E., 1980, On the organization of telencephalon in elasmobranchs, in: *Comparative Neurology of the Telencephalon* (S. O. E. Ebbesson, ed.), Plenum Press, New York, pp. 1–16.

Ebbesson, S. O. E., and Campbell, C. B. G., 1973, On the organization of cerebellar efferent pathways in the nurse shark (*Ginglymostoma cirratum*), *J. Comp. Neurol.* **152:**233–254.

Ebbesson, S. O. E., and Heimer, L., 1970, Projections of the olfactory tract fibers in the nurse shark (*Ginglymostoma cirratum*), *Brain Res.* **17:**47–55.

Ebbesson, S. O. E., and Hodde, K. C., 1981, Ascending spinal systems in the nurse shark, *Ginglymostoma cirratum, Cell Tissue Res.* **216:**313–331.

Ebbesson, S. O. E., and Northcutt, R. G., 1976, Neurology of anamniotic vertebrates, in: *Evolution of Brain and Behavior in Vertebrates* (R. B. Masteron, M. E. Bitterman, and C. B. G. Campbell, eds.), Erlbaum, Hillsdale, N.J., pp. 115–146.

Ebbesson, S. O. E., and Schroeder, D., 1971, Connections of the nurse shark's telencephalon, *Science* **173:**254–256.

Faucette, J. R., 1969a, The olfactory bulb and medial hemisphere wall of the rat-fish, *Chimaera, J. Comp. Neurol.* **137:**377–406.

Faucette, J. R., 1969b, The accessory olfactory bulbs and the lateral telencephalic wall of the rat-fish, *Chimaera, J. Comp. Neurol.* **137:**407–432.

Gerlach, J., 1947, Beiträge zur vergleichenden Morphologie des Selachierhirnes, *Anat. Anz.* **96:**79–165.

Graeber, R. C., 1978, Behavioral studies correlated with central nervous system integration of vision in sharks, in: *Sensory Biology of Sharks, Skates and Rays* (E. S. Hodgson and R. F. Mathewson, eds.), U.S. Government Printing Office, Washington, D.C., pp. 195–225.

Graeber, R. C., 1980, Telencephalic function in elasmobranchs. A behavioral perspective, in: *Com-*

parative Neurology of the Telencephalon (S. O. E. Ebbesson, ed.), Plenum Press, New York, pp. 17–39.

Graeber, R. C., Ebbesson, S. O. E., and Jane, J. A., 1973, Visual discrimination in sharks without optic tectum, *Science* **180:**413–415.

Graeber, R. C., Schroeder, D. M., Jane, J. A., and Ebbesson, S. O. E., 1978, Visual discrimination following partial telencephalic ablations in nurse sharks (*Ginglymostoma cirratum*), *J. Comp. Neurol.* **180:**325–344.

Hayle, T. H., 1973a, A comparative study of spinal projections to the brain (except cerebellum) in three classes of poikilothermic vertebrates, *J. Comp. Neurol.* **149:**463–476.

Hayle, T. H., 1973b, A comparative study of spinocerebellar systems in three classes of poikilothermic vertebrates, *J. Comp. Neurol.* **149:**477–495.

Holmgren, N., 1922, Points of view concerning forebrain morphology in lower vertebrates, *J. Comp. Neurol.* **34:**391–459.

Horstmann, E., 1954, Die Faserglia des Selachiergehirns, *Z. Zellforsch.* **30:**588–617.

Johnston, J. B., 1911, The telencephalon of selachians, *J. Comp. Neurol.* **21:**1–113.

Kalmijn, A. J., 1978, Electric and magnetic sensory world of sharks, skates and rays, in: *Sensory Biology of Sharks, Skates, and Rays* (E. S. Hodgson and R. F. Mathewson, eds.), U.S. Government Printing Office, Washington, D.C., pp. 507–528.

Kalmijn, A. J., 1982, Electric and magnetic field detection in elsamobranch fishes, *Science* **218:**916–917.

Klatzo, I., 1967, Cellular morphology of the lemon shark brain, in: *Sharks, Skates and Rays* (P. W. Gilbert, R. F. Mathewson, and D. P. Rall, eds.), Johns Hopkins Press, Baltimore, pp. 341–359.

Kleerekoper, H., 1978, Chemoreception and its interaction with flow and light perception in the locomotion and orientation of some elasmobranchs, in: *Sensory Biology of Sharks, Skates, and Rays* (E. S. Hodgson and R. F. Mathewson, eds.), U.S. Government Printing Office, Washington, D.C., pp. 269–329.

Koester, D. M., 1983, Central projections of the octavolateralis nerves of the clearnose skate, *Raja eglanteria, J. Comp. Neurol.* **221:**199–215.

Kuhlenbeck, H., 1929, Die Grundbestandteile des Endhirns im Lichte der Bauplanlehre, *Anat. Anz.* **67:**1–51.

Kuhlenbeck, H., 1973, Overall morphologic pattern, in: *The Central Nervous System of Vertebrates,* Volume 3, Part 2, Karger, Basel, p. 950.

Kuhlenbeck, H., 1977, Derivatives of the prosencephalon: Diencephalon and telencephalon, in: *The Central Nervous System of Vertebrates,* Volume 5, Part 1, Karger, Basel, p. 888.

Luiten, P. G. M., 1981, Two visual pathways to the telencephalon in the nurse shark (*Ginglymostoma cirratum*). II. Ascending thalamo-telencephalic connections, *J. Comp. Neurol.* **196:**539–548.

McCready, P. J., and Boord, R. L., 1976, The topography of the superficial roots and ganglia of the anterior lateral line nerve of the smooth dogfish, *Mustelus canis, J. Morphol.* **150:**527–538.

Meredith, G. E., and Smeets, W. J. A. J., 1987, Immunocytochemical analysis of the forebrain and midbrain dopaminergic systems of *Raja radiata:* Evidence for a substantia nigra and ventral tegmental area in cartilaginous fish, *J. Comp. Neurol.* **265:**530–548.

Montgomery, J. C., and Roberts, B. L., 1979, Organization of vestibular afferents to the vestibular nuclei of the dogfish. *Brain Behav. Evol.* **16:**81–98.

Myrberg, A. A., Jr., 1978, Underwater sound—Its effects on the behavior of sharks, in: *Sensory Biology of Sharks, Skates, and Rays* (E. S. Hodgson and R. F. Mathewson, eds.), U.S. Government Printing Office, Washington, D.C., pp. 391–417.

Nieuwenhuys, R., 1967, Comparative anatomy of olfactory centres and tracts, *Prog. Brain Res.* **23:**1–64.

Northcutt, R. G., 1977, Elasmobranch central nervous system organization and its possible evolutionary significance, *Am. Zool.* **17:**411–429.

Northcutt, R. G., 1978, Brain organization in the cartilaginous fishes, in: *Sensory Biology of Sharks, Skates, and Rays* (E. S. Hodgson and R. F. Mathewson, eds.), U.S. Government Printing Office, Washington, D.C., pp. 117–193.

Northcutt, R. G., 1981, Evolution of the telencephalon in nonmammals, *Annu. Rev. Neurosci.* **4:**301–350.

Northcutt, R. G., and Boord, R. L., 1981, Efferent projections of the optic tectum in the clearnose skate, *Raja eglanteria, Anat. Rec.* **199:**185A.

Northcutt, R. G., Reiner, A., and Karten, H. J., 1988, Immunohistochemical study of the telencephalon of the spiny dogfish, *Squalus acanthias, J. Comp. Neurol.* **277:**250–267.

Parent, A., Poitras, D., and Dubé, L., 1984, Comparative anatomy of central monoaminergic systems, in: *Handbook of Chemical Neuroanatomy,* Vol. 2: *Classical Transmitters in the CNS,* Part I (A. Björklund and T. Hökfelt, eds.), Elsevier, Amsterdam, pp. 409–439.

Platt, C. J., Bullock, T. H., Czéh, G., Kovačević, N., Konjević, D., and Gojković, M., 1974, Comparison of electroreceptor, mechanoreceptor and optic evoked potentials in the brain of some rays and sharks, *J. Comp. Physiol.* **95:**323–355.

Roberts, B. L., 1978, Mechanoreceptors and the behaviour of elasmobranch fishes with special reference to the acoustico-lateralis system, in: *Sensory Biology of Sharks, Skates, and Rays* (E. S. Hodgson and R. F. Mathewson, eds.), U.S. Government Printing Office, Washington, D.C., pp. 331–390.

Roberts, B. L., and Williamson, R. M., 1983, Motor pattern formation in the dogfish spinal cord, in: *Neural Origin of Rhythmic Movements* (A. Roberts and B. Roberts, eds.), Cambridge University Press, London, pp. 331–350.

Sato, Y., Takatsuji, K., and Masai, H., 1983, Brain organization of sharks, with special reference to archaic species, *J. Hirnforsch.* **24:**289–295.

Schnitzlein, H. N., 1982, Telencephalon of fishes, in: *Comparative Correlative Neuroanatomy of the Vertebrate Telencephalon* (E. C. Crosby and H. N. Schnitzlein, eds.), Macmillan Co., New York, pp. 3–160.

Schroeder, D. M., and Ebbesson, S. O. E., 1974, Nonolfactory telencephalic afferents in the nurse shark (*Ginglymostoma cirratum*), *Brain Behav. Evol.* **9:**121–155.

Schweitzer, J., 1983, The physiological and anatomical localization of two electroreceptive diencephalic nuclei in the thornback ray, *Platyrhinoidis triseriata, J. Comp. Physiol. A* **153:**331–341.

Schweitzer, J., 1986, Functional organization of the electroreceptive midbrain in an elasmobranch (*Platyrhinoidis triseriata*), *J. Comp. Physiol.* **158:**43–58.

Smeets, W. J. A. J., 1981, Efferent tectal pathways in two chondrichthyans, the shark *Scyliorhinus canicula* and the ray *Raja clavata, J. Comp. Neurol.* **195:**13–23.

Smeets, W. J. A. J., 1982, The afferent connections of the tectum mesencephali in two chondrichthyans, the shark *Scyliorhinus canicula* and the ray *Raja clavata, J. Comp. Neurol.* **205:**139–152.

Smeets, W. J. A. J., 1983, The secondary olfactory connections in two chondrichthyans, the shark *Scyliorhinus canicula* and the ray *Raja clavata, J. Comp. Neurol.* **218:**334–344.

Smeets, W. J. A. J., and Boord, R. L., 1985, Connections of the lobus inferior hypothalami of the clearnose skate *Raja eglanteria* (Chondrichthyes), *J. Comp. Neurol.* **234:**380–392.

Smeets, W. J. A. J., and Northcutt, R. G., 1987, At least one thalamotelencephalic pathway in cartilaginous fishes projects to the medial pallium, *Neurosci. Lett.* **78:**277–282.

Smeets, W. J. A. J., and Timerick, S. J. B., 1981, Cells of origin of pathways descending to the spinal cord in two chondrichthyans, the shark *Scyliorhinus canicula* and the ray *Raja clavata, J. Comp. Neurol.* **202:**473–491.

Smeets, W. J. A. J., Nieuwenhuys, R., and Roberts, B. L., 1983, *The Central Nervous System of Cartilaginous Fishes: Structure and Functional Correlations,* Springer, Berlin, p. 266.

Sterzi, G., 1909, *Il sistema nervoso centrale dei vertebrate,* Volume 2, Pesci I. Selaci, Draghi, Padova.

Tong, S.-L., and Bullock, T. H., 1982, The sensory functions of the cerebellum of the thornback ray, *Platyrhinoidis triseriata, J. Comp. Physiol. A* **148:**399–410.

Vesselkin, N. P., and Kovačević, N., 1973, Nonolfactory telencephalic afferent projections in elasmobranch fishes. *Zh. Evol. Biokhim. Fiziol.* **9:**585–592.

Williams, H. T., 1973, The telencephalon of the newborn dogfish shark, *Squalus acanthias, J. Hirnforsch.* **14:** 261–285.

Wourms, J. P., 1977, Reproduction and development in chondrichthyan fishes, *Am. Zool.* **17:**379–410.

2

The Telencephalon of Actinopterygian Fishes

R. NIEUWENHUYS and J. MEEK

1. Introduction

The class of bony fishes or Osteichthyes is usually subdivided into two subclasses, the Actinopterygii or ray-finned fishes and the Sarcopterygii or lobe-finned fishes. The subclass last mentioned, the telencephalon of which will be treated in the next chapter, can further be subdivided into the Dipnoi or lungfishes and the Crossopterygii or tassel-finned fishes. The Actinopterygii constitute by far the most abundant group of recent vertebrates. They are usually subdivided into three superorders, the Chondrostei, Holostei, and Teleostei. Interestingly, these three superorders may be considered as representing three subsequent stages or gradations of actinopterygian evolution. The Chondrostei are considered as descendants of the Palaeoniscoidei, a primitive group of paleozoic fishes. During the early and middle Mesozoic times, the Chondrostei were supplanted by the Holostei, which in turn were largely replaced in the last phase of the Mesozoic era and the Cenozoic era by the expanding and now-abundant Teleostei.

One group of Osteichthyes, viz., the Polypteriformes, present us with taxonomic difficulties. This is a small group of African freshwater fish, having only some ten extant species. Formerly, it was believed that the polypteriform fishes were crossopterygians, but at present they are either assigned to the Actinopterygii (Romer, 1962; Moy-Thomas and Miles, 1971) or considered as an isolated group (Stensiö, 1921; Daget, 1958; Jarvik, 1968a,b; Nelson, 1969). Because the telencephalon of the polypteriform fishes has several typical features in common with that of the chondrosteans, holosteans, and teleosts, its structure will be dealt with in the present chapter.

R. NIEUWENHUYS and J. MEEK • Department of Anatomy and Embryology, University of Nijmegen, 6500 HB Nijmegen, The Netherlands.

The present chapter is mainly based on examination of the following species: the reedfish *Erpetoichthys calabaricus,* a representative of the polypteriform fishes; the shovelnose sturgeon *Scaphirhynchus platorynchus,* a typical chondrostean; the bowfin *Amia calva,* a species belonging to the Holostei; and three teleosts: the trout *Salmo gairdneri,* the goldfish *Carassius carassius,* and the osteoglossomorph *Xenomystis nigri.* The trout and the salmonids in general may be considered as rather primitive, "generalized" teleosts. The goldfish exemplifies the "typical" teleost, whereas *Xenomystis* is selected because of the high degree of differentiation of its telencephalon. Prior to discussion of the macrostructure and microstructure of the telencephalon of these fishes, it is felt necessary to comment briefly on the morphogenesis of the forebrain.

2. Morphogenesis of the Telencephalon: Eversion versus Evagination

It may be recalled that the forebrain in an early embryonic stage is a simple tubelike or vesicular structure, consisting of a pair of thickened side plates that are connected dorsally, rostrally, and ventrally by a thin, epithelial membrane (Fig. 1). The dorsal part of this membrane represents the telencephalic portion of the neural roof plate or lamina supraneuroporica, whereas the ventral connection of the two side plates is called the lamina terminalis. During further development there occurs in most vertebrates a bulging out or evagination of the side walls. This gives rise to two hemispheres, the central ventricular spaces of which are surrounded by nervous tissue (Figs. 1–3).

Although evagination is the principal morphogenetic event in the forebrain of most vertebrates, the hemispheres which are the products of this process may show widely different shapes. The hemispheres of urodele amphibians remain relatively thin-walled and are evaginated in a rostral, lateral, as well as caudal direction. The rostral evagination is well-marked, but the caudally directed evagination is only slight (Fig. 3). The lateral evagination brings the dorsal (pallial) portions of the telencephalic side plates in a position as to cap the ventral (basal) portion (Figs. 1, 2). This part of the evagination process has been described by Holmgren (1922) and others as the bending inward or inversion of the pallium.

The thin-walled telencephalon of lungfishes, which will be discussed in the next chapter, has much in common with that of amphibians. In this group the evagination takes place particularly in a rostral and ventral direction; a caudal evagination is practically absent, however.

In amniotes the evagination process is complicated by the formation of one or two ventrolaterally situated thickenings which are going to protrude into the ventricular cavity (Fig. 1). The development of these intraventricular ridges, from which among other structures the basal ganglia or corpus striatum develop, may lead to a considerable reduction of the size of the ventricular cavity. In mammals the evagination is further complicated by an inrolling of the most medial (hippocampal) portion of the dorsal telencephalic wall.

A primary olfactory center, i.e., a region that receives a direct input from the olfactory organ, forms an integral part of the telencephalon. In most vertebrates this center is bilaterally represented by clearly evaginated, hollow structures. It is, however, important to note that the process leading to the formation of these structures is a separate evagination that occurs later in development than the evagination of the hemispheres.

Whereas evagination or bulging out is the chief morphogenetic event in the forebrain of almost all groups of vertebrates, the actinopterygian hemispheres undergo a developmental process of quite a different nature. In this group the ventral parts of the telencephalic walls maintain essentially the position that they occupy in the early tube-shaped forebrain anlage, but the dorsal portions show marked changes in form and position. These parts separate gradually and finally recurve laterally, a process known as eversion (Gage, 1893; Studnicka, 1896; Nieuwenhuys, 1962a, 1963). As a result of this eversion the ventricular surface of the side plates increases considerably in size and the initially narrow telencephalic roof plate is transformed into a wide membrane that envelops the dorsal parts of the hemispheres (Fig. 1). In polypteriform fishes the telencephalic walls remain relatively thin (Figs. 7, 8), but in the chondrosteans (Fig. 9), holosteans (Fig. 10), and teleosts (Figs. 11–13) the eversion is accompanied by a marked thickening of the dorsal portions of the side walls. It is important to note that due to the eversion, lateral ventricles in the proper sense are entirely lacking in the actinopterygian hemispheres.

The amount of outward bending of the telencephalic wall differs among the various actinopterygian groups. In chondrosteans (Fig. 9) and in salmonid teleosts (Fig. 11) the eversion is only slight, but in polypteriforms (Fig. 8), the bowfin

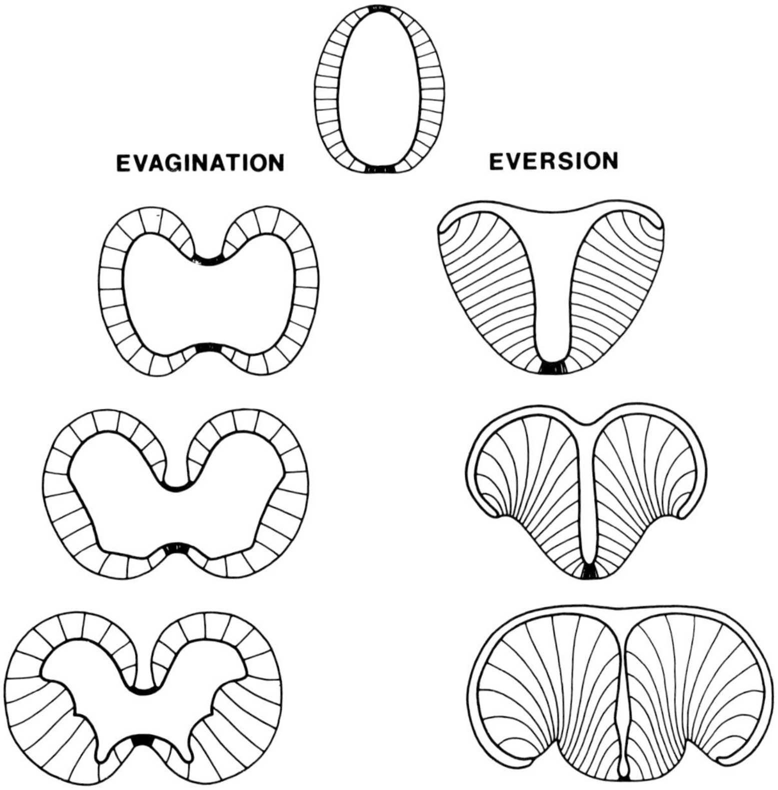

Figure 1. Diagrammatic transverse sections illustrating how evaginated as well as everted telencephala develop from an early embryonic tube-shaped anlage.

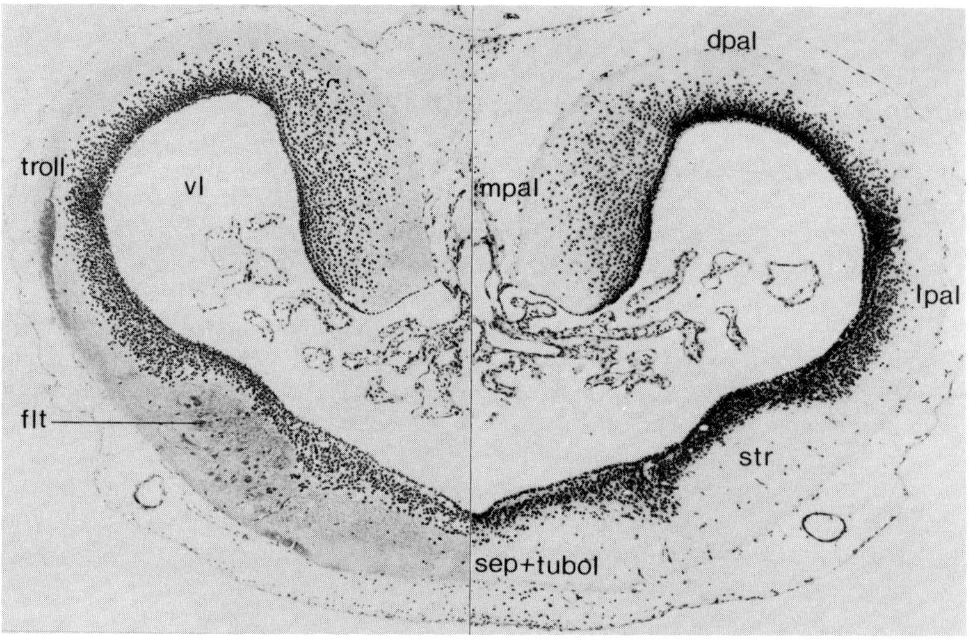

Figure 2. Transverse section through the telencephalon of the tiger salamander, *Ambystoma tigrinum*. The right half is a photomicrograph of a Nissl-stained section, whereas the left half is a photomicrograph of the adjacent Bodian-stained section.

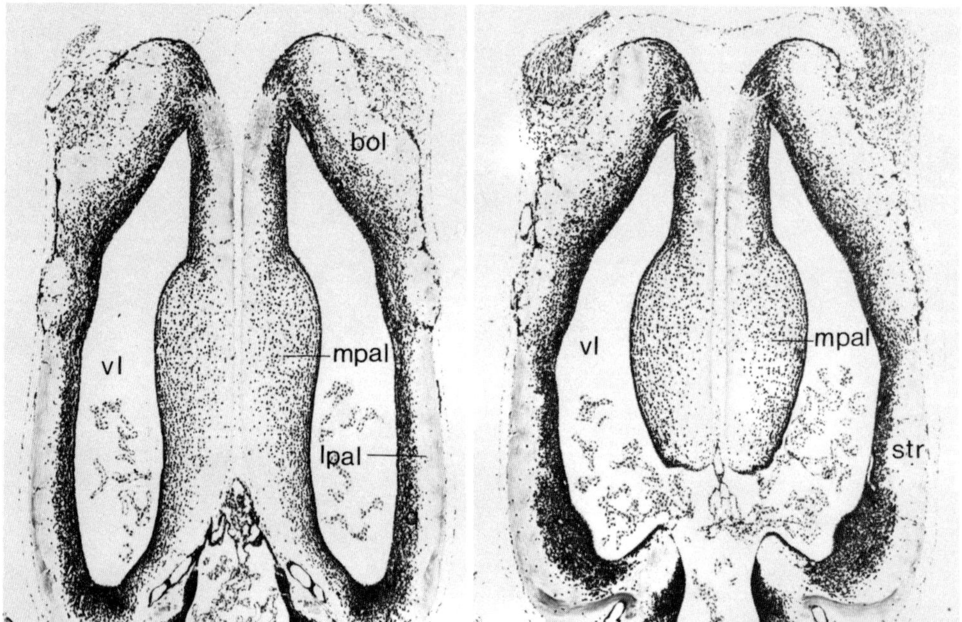

Figure 3. Horizontal sections through the telencephalon of the tiger salamander, *Ambystoma tigrinum*, at a dorsal (A) and at a more ventral (B) level. Bodian stain.

(Fig. 10), and most groups of teleosts (Figs. 12, 13) the eversion is very pronounced.

The morphogenesis of the actinopterygian telencephalon has led to much controversy. A number of authors (Gage, 1893; Studnicka, 1896; Holmgren, 1920, 1922; Kuhlenbeck, 1929; Miller, 1940; Källén, 1950a,b, 1951; Nieuwenhuys, 1962a,b, 1963, 1964, 1967a) considered eversion as the principal developmental process in this telencephalon, but others (Droogleever Fortuyn, 1961; Story, 1964; Crosby *et al.*, 1966; Schnitzlein, 1968; Morgan, 1975) denied the existence and/or significance of an eversion process, whereas still others (Sheldon, 1912; Herrick, 1921, 1922), although not denying that some eversion may occur, considered thickening of the telencephalic side plates combined with a complicated shifting and plastic rearrangement of the cell material to be the most important developmental processes.

According to the observations of one of us (Nieuwenhuys, 1962a, 1964), the eversion does not only appear from the absence of lateral ventricles and the widening of the telencephalic roof plate, but also from the orientation of the blood vessels and the ependymal gliocytes. In the evaginated and inverted cerebral hemispheres of amphibians, the blood vessels pass through the whole external surface and converge toward the central ventricular cavities (Fig. 4A). In the actinopterygian forebrain, however, an entirely different pattern is seen; the blood vessels enter and leave the forebrain only at the laterobasal side and fan out radially in the direction of the large ependymal surface (Fig. 4B). The processes of the ependymal cells show a pattern conforming exactly to that of the blood vessels. In the evaginated amphibian forebrain the long processes of the ependymal cells diverge from the smaller ventricular surface toward the larger meningeal surface (Fig. 5A). In the forebrain of bony fishes, on the contrary, the processes of the ependymal gliocytes converge from the greatly elaborated ventricular surface toward the small, basal, meningeal surface (Fig. 5B). It stands to reason that recognition and appreciation of the eversion process is of paramount importance for a correct interpretation of the parts of the actinopterygian forebrain.

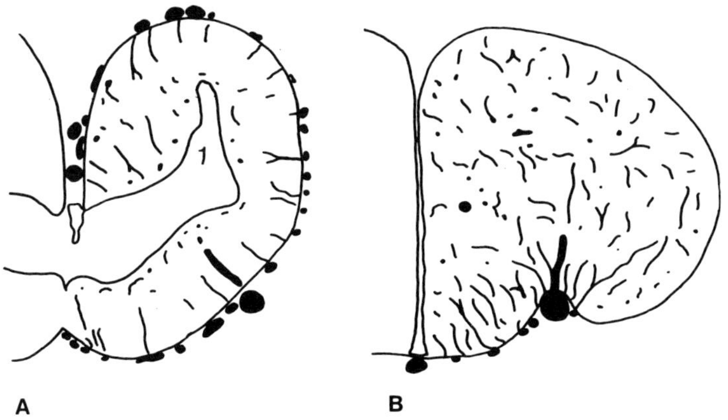

Figure 4. The pattern of the blood vessels, as seen in a transverse hemisection through the telencephalon of the tiger salamander, *Ambystoma tigrinum* (A), and the teleost *Gasterosteus aculeatus* (B). From Nieuwenhuys (1966).

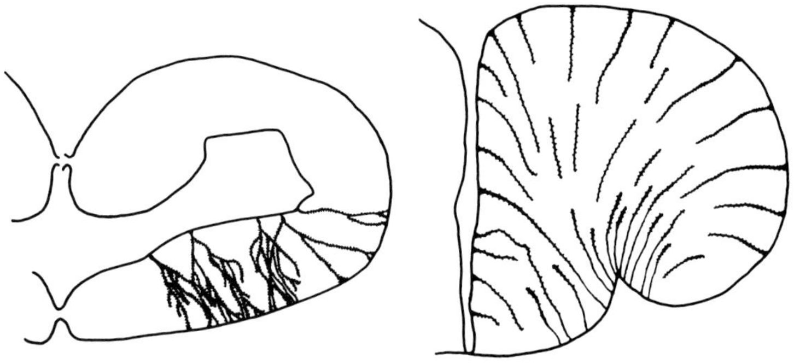

Figure 5. Transverse hemisections through the telencephalon of the mud puppy, *Necturus maculatus* (A), and the teleost *Gasterosteus aculeatus* (B), showing the ependymal gliocytes. Golgi technique. (A, redrawn from Herrick, 1933.) From Nieuwenhuys (1966).

3. Gross Morphology

In its overall appearance the slender, elongated brain of polypteriforms (*Polypterus* and *Erpetoichthys*) somewhat resembles that of the dipnoans (Fig. 6A). The rhombencephalon is well developed. The cerebellum has invaginated into the ventricular cavity. Due to this unusual configuration, only the lateral and morphologically most caudal parts of this structure are externally visible. Its central part protrudes into the fourth ventricle and its rostral part thrusts forward under the roof of the midbrain. The mesencephalon and the diencephalon are medium-sized. The roof of the midbrain is differentiated into bilateral lobes. The telencephalon is relatively large and elongated. Its rostral part consists of two well-developed, hollow olfactory bulbs (Fig. 7B). The ventral parts of the telencephalon proper form the lateral walls of a median telencephalic ventricle (Fig. 7C). However, as already mentioned, the dorsal parts of the telencephalic walls are everted, i.e., folded laterally. Interestingly, the dorsocaudal parts of the telencephalic walls are recurved in such a fashion as to surround the extraencephalic space. The resultant "pseudoventricles" are clearly visible in Fig. 7A and B. Due to the eversion the telencephalic roof plate has become transformed into a wide membranous structure that covers the dorsal parts of the telencephalon and is attached to the lateral margins of the walls. This tela telencephali forms a long median fold, the septum ependymale, which largely divides the telencephalic ventricle into two parts (Fig. 8).

The brain of chondrosteans (*Acipenser, Scaphirhynchus, Polyodon*) shows many points of resemblance with that of the polypteriforms (Fig. 6B). However, in the former the roof of the midbrain is smaller and the hypothalamus is much larger than in the latter. The telencephalon is clearly everted but, because this recurvature is coupled with a thickening of the lateral walls, the chondrostean hemispheres have become solid structures (Fig. 9).

The brain of holosteans (*Amia, Lepisosteus*) is somewhat more compact than that of chondrosteans (Fig. 6C). The central part of the cerebellum, i.e., the corpus cerebelli, does not protrude into the fourth ventricle, but is represented

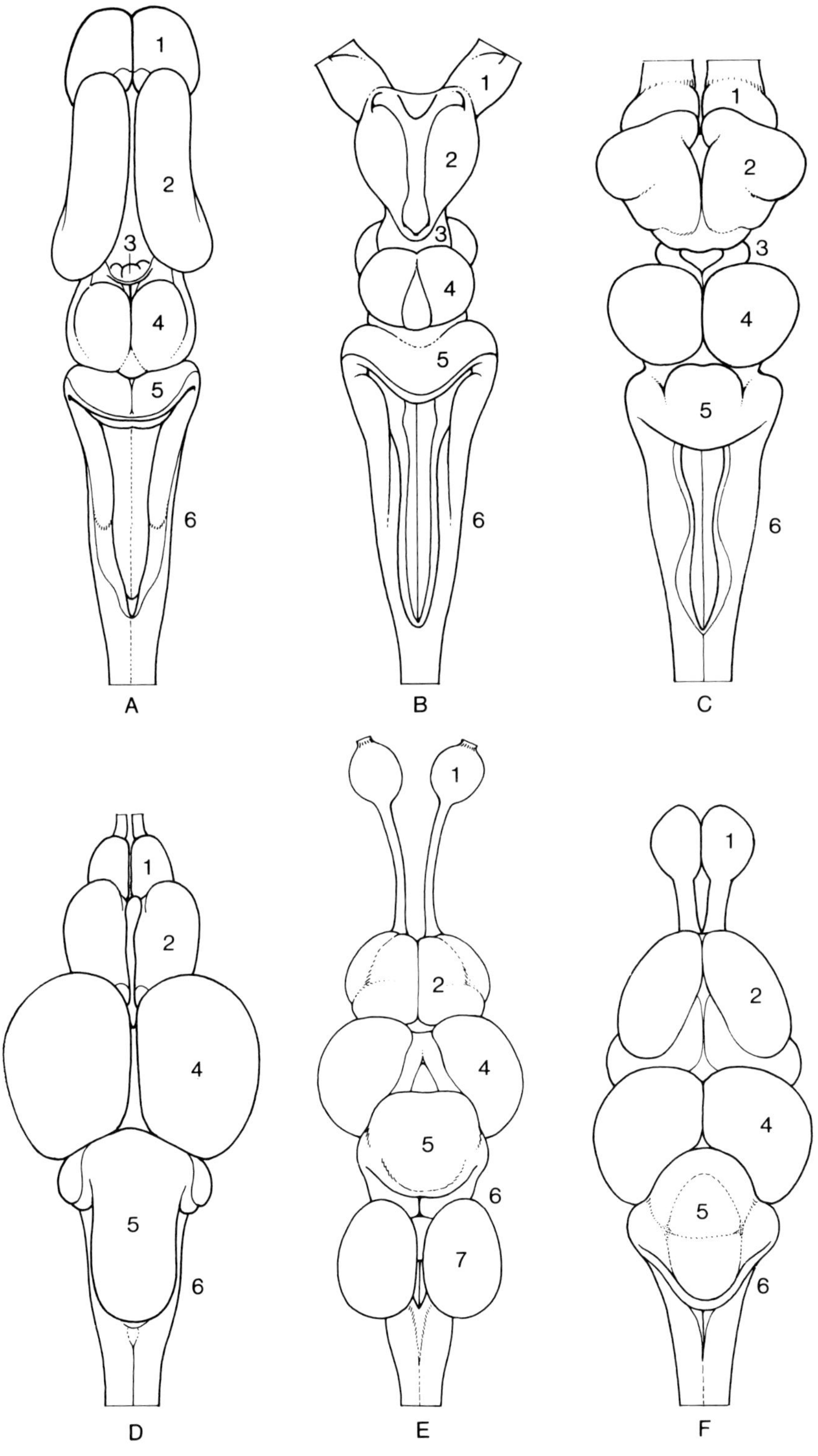

Figure 6. Dorsal view of the brains of: A, the reed fish *Erpetoichthys calabaricus* (Polypteriformes); B, the shovelnose sturgeon *Scaphirhynchus platorynchus* (Chondrostei); C, the bowfin *Amia calva* (Holostei); D, the trout *Salmo gairdneri;* E, the goldfish *Carassius auratus;* F, the osteoglossomorph *Xenomystis nigri* (C, adapted from McCormick, 1978). 1, bulbus olfactorius; 2, lobus hemisphericus telencephali; 3, diencephalon; 4, tectum mesencephali; 5, cerebellum; 6, rhombencephalon; 7, lobus vagi.

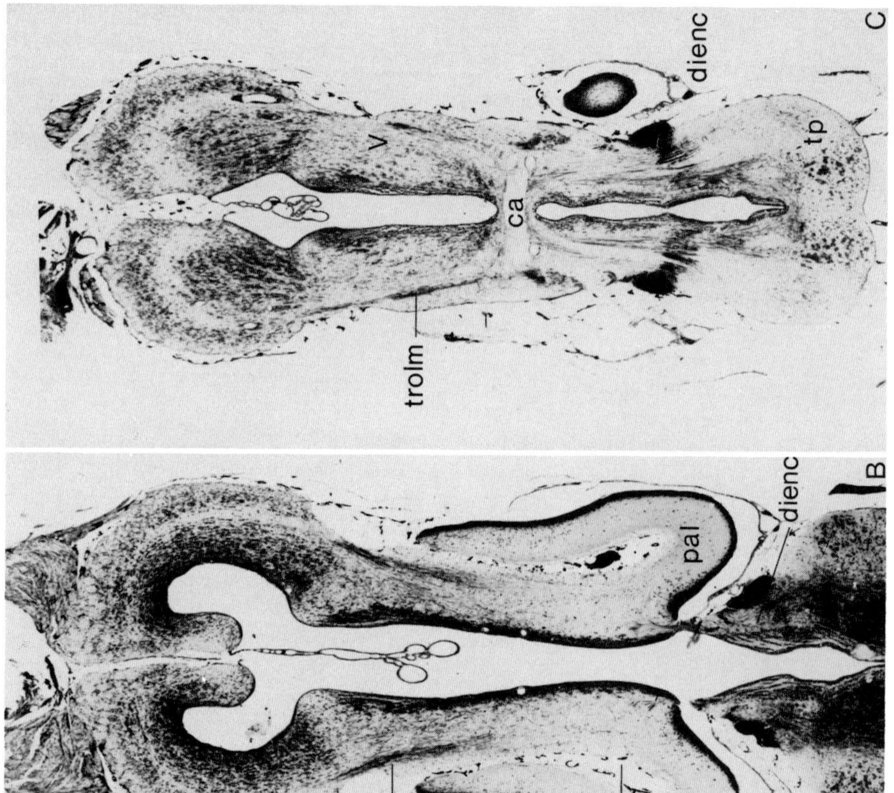

Figure 7. Horizontal sections through the rostral parts of the brain of the reed fish *Erpetoichthys calabaricus* (Polypteriformes), stained according to Klüver-Barrera; A, B, C: sections through the ventral, intermediate, and dorsal parts of the brain, respectively.

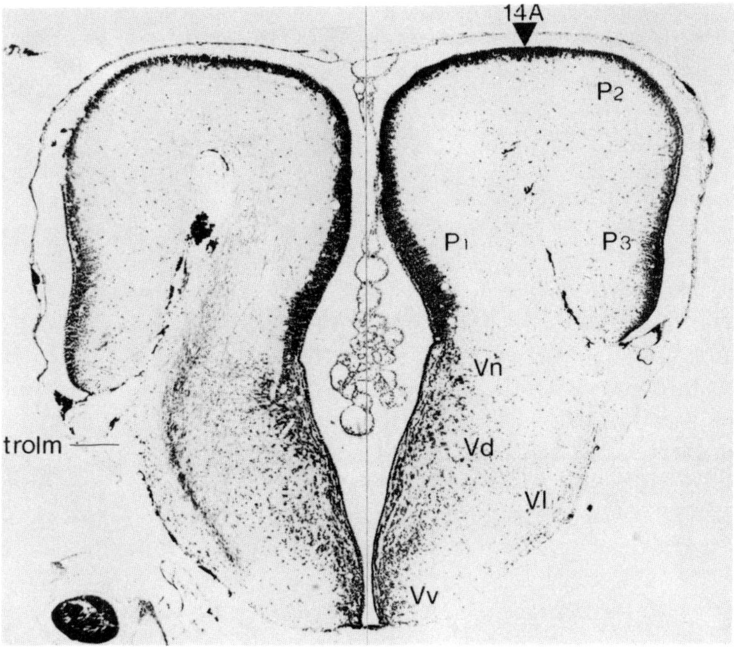

Figure 8. Transverse section through the telencephalon of the reed fish (Polypteriformes). The right half of this and the following four figures represents a photomicrograph of a Nissl-stained section, whereas the left half represents a photomicrograph of the adjacent Klüver-Barrera- (Figs. 8, 10) or Bodian-stained (Figs. 9, 11, 12) section.

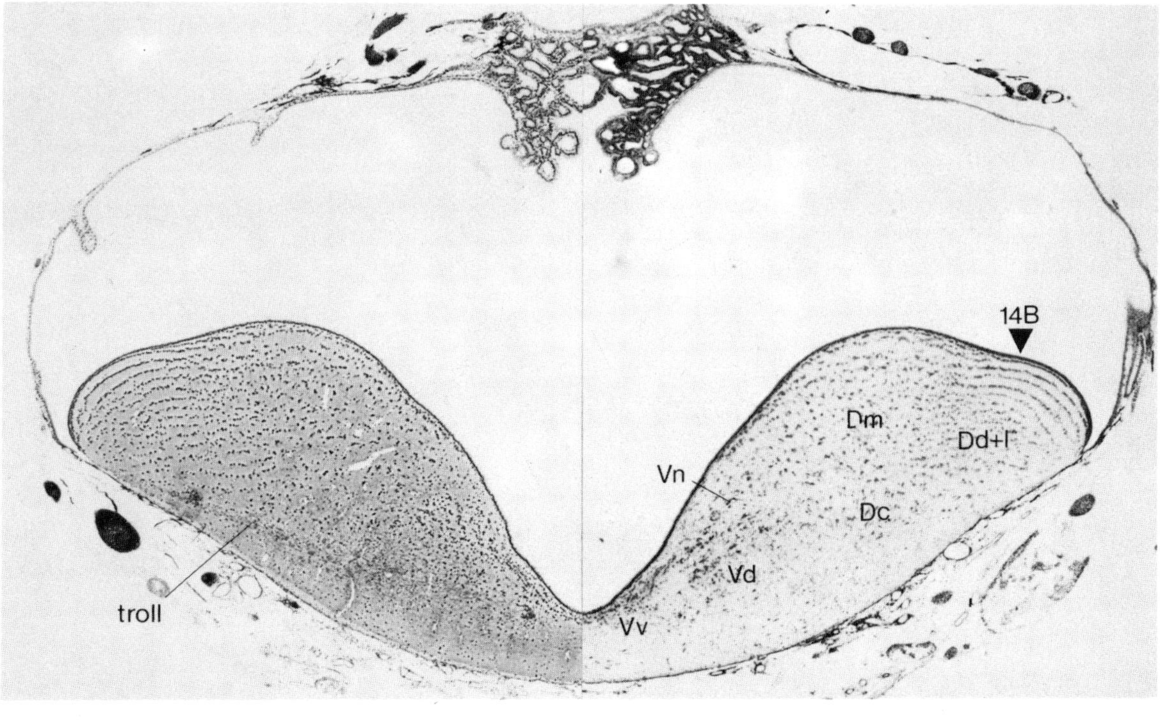

Figure 9. Transverse section through the telencephalon of the shovelnose sturgeon (Chondrostei).

by a dorsally evaginated, dome-shaped structure. The tectum mesencephali is clearly differentiated into bilateral lobes. As in chondrosteans and teleosts, eversion and thickening of the lateral walls characterize the morphogenesis of the holostean forebrain. It is, however, noteworthy that in *Amia* (Fig. 10) the eversion is much more pronounced than in *Lepisosteus* (Fig. 15D,E).

The gross structure of the brain varies greatly among teleosts. Some groups of this large superorder show strong development of a particular sense system with corresponding enlargement of its recipient center in the rhombencephalon. Thus, in mormyrids the lateral line system is strongly developed and its sense organs have become extensively transformed into electroreceptor organs. Consequently, the lateral line nerves are very large and their center of termination in the rhombencephalon is an enormous lobe (Nieuwenhuys and Nicholson, 1969). In cyprinids and silurids the taste buds are particularly numerous and are not confined to the mouth but spread over the head and body. As a result, the sensory components of the VIIth, IXth, and Xth nerves and their rhombencephalic centers of termination are very large (Herrick, 1901; Fig. 6E). The roof of the midbrain varies considerably in size; it is relatively very large in highly visual teleostean groups, like the Salmonidae (Fig. 6D) and the Holocentridae. The size of the olfactory bulbs reflects the development of the olfactory system. In most teleosts these structures are well developed, though smaller than in the polypteriforms, chondrosteans, and holosteans. Whereas in most actinopterygians the olfactory bulbs are closely apposed to the cerebral hemispheres, in some groups of teleosts, e.g., the silurids, cyprinids (Fig. 6E), and osteoglossomorphs (Fig. 6F), the bulbs and the hemispheres are interconnected by olfactory tracts or peduncles. The teleostean cerebral hemispheres show also a considerable variation with regard to both size and degree of eversion. To mention only the extremes: in salmonids these structures are relatively small (Fig. 6D) and only slightly everted (Fig. 11), but in osteoglossomorphs the hemispheres are large and differentiated into several lobes (Fig. 6F), whereas the

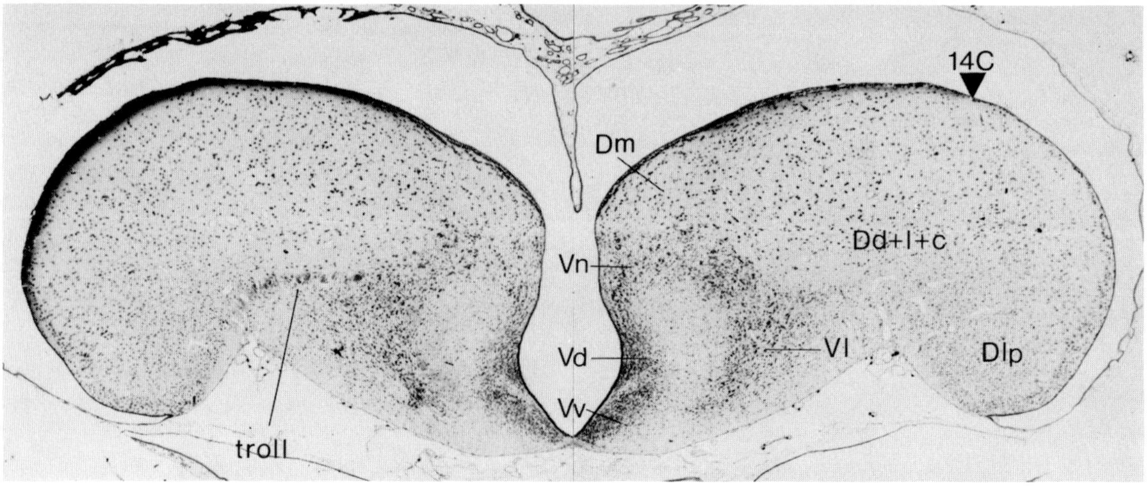

Figure 10. Transverse section through the telencephalon of the bowfin (Holostei).

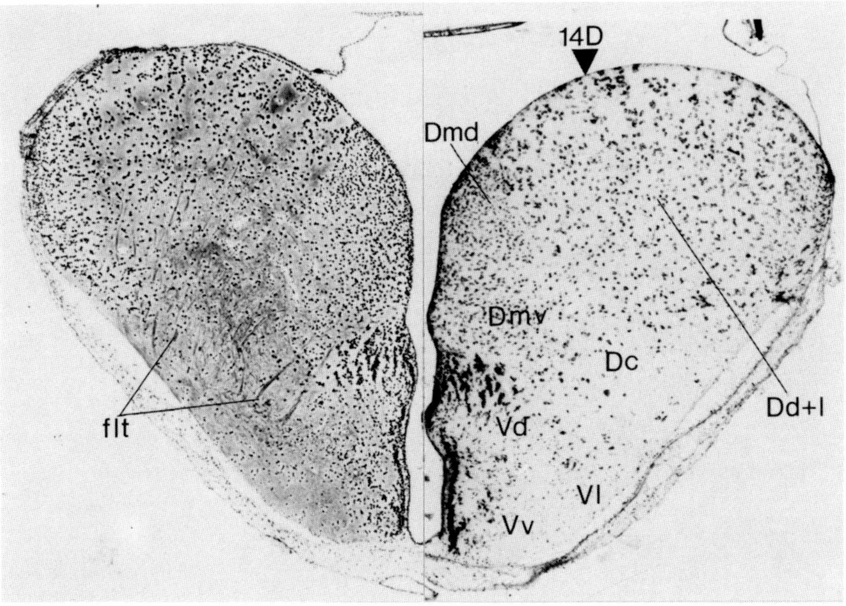

Figure 11. Transverse section through the telencephalon of the trout (Teleostei).

eversion is far advanced (Fig. 13). The cerebellum is by far the most variable part of the brain of teleosts. Without going into details it may be mentioned that in most teleosts the cerebellum arches caudally over the fourth ventricle [e.g., in *Salmo* (Fig. 6D) and *Xenomystis* (Fig. 6F)]. However, in silurids this structure extends rostrally over the tectum and in mormyrids it has grown out to such an extent that it covers all other parts of the brain (Nieuwenhuys, 1967b, 1982).

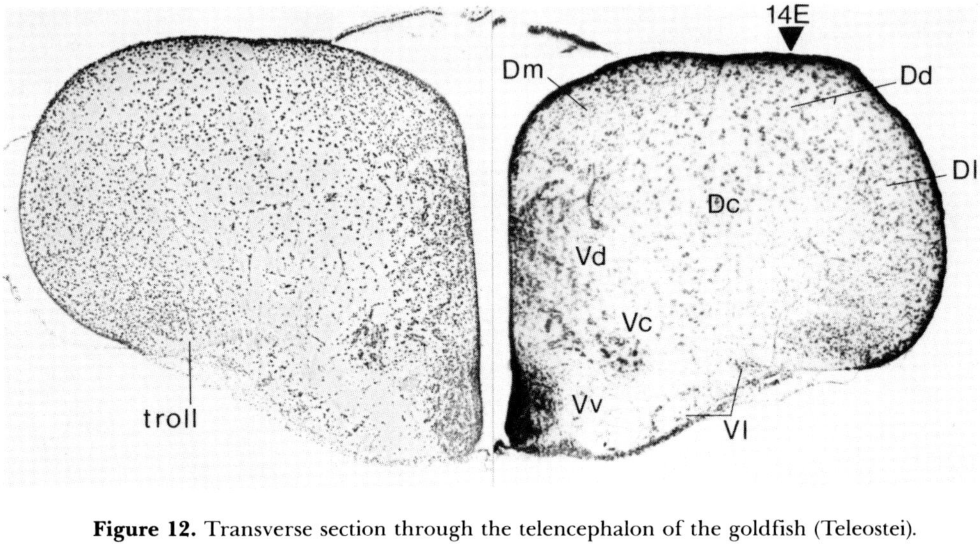

Figure 12. Transverse section through the telencephalon of the goldfish (Teleostei).

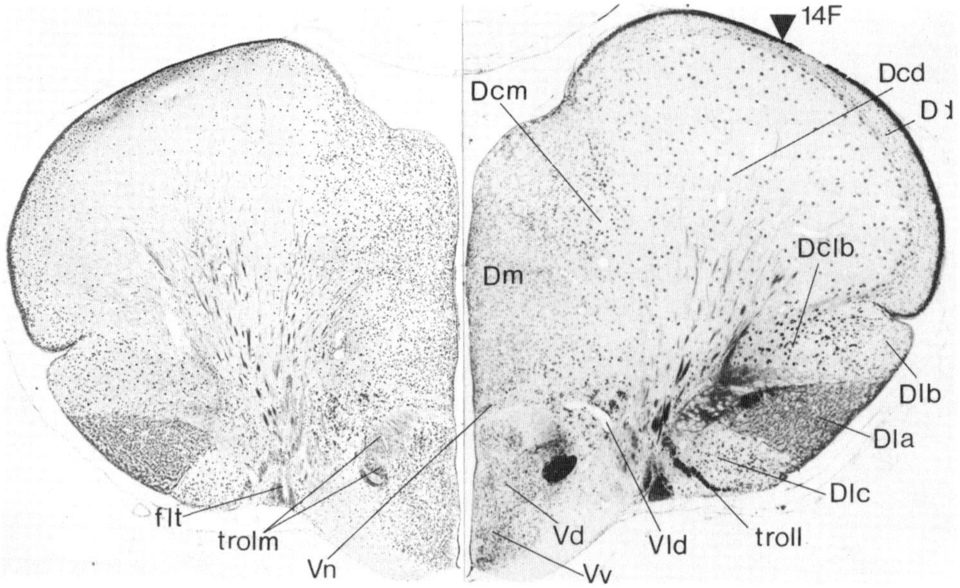

Figure 13. Transverse section through the telencephalon of *Xenomystis nigri* (Teleostei). The right half represents a photomicrograph of a Klüver-Barrera-stained section, whereas the left half represents a photomicrograph of the adjacent Bodian-stained section.

4. Structure of the Telencephalon

4.1. Introductory Notes

The structure of the telencephalon of actinopterygian fishes has been studied by a large number of observers, among whom may be mentioned Johnston (1911), Sheldon (1912), Holmgren (1920), Ariëns Kappers *et al.* (1936), Weston (1937), Meader (1939), Nieuwenhuys (1962a,b, 1963, 1966, 1983), Northcutt and Braford (1980), Bass (1981a,b,c), Ito *et al.* (1982), Murakami *et al.* (1983), Northcutt and Davis (1983), and Levine and Dethier (1985). These studies have yielded numerous different terminologies, many of which reflect the views of the pertinent authors on the question as to how the various parts of the actinopterygian forebrain should be interpreted. The following account is mainly based on the publications of Nieuwenhuys (1962b, 1963), Northcutt and Braford (1980), and Northcutt and Davis (1983), and on a study of serial sections through the brains of the six species mentioned in the introduction and depicted in Fig. 6. The nomenclature used is that of one of us (Nieuwenhuys, 1963), which has been adopted by practically all later students of the actinopterygian forebrain. The question as to how the various cell masses in this forebrain can be homologized with telencephalic grisea in other groups of vertebrates will be addressed in the final section of the present survey.

4.2. The Olfactory Bulb

In the telencephalic region of most vertebrates two different evagination processes occur, viz. (1) hemispheric evagination and (2) evagination giving rise to the formation of the olfactory bulbs. As already discussed, the former process is absent in actinopterygians, but the latter does occur in the polypteriforms, chondrosteans, and holosteans. In these groups, and also in some teleosts (e.g., *Salmo, Anguilla*), the rostroventral portions of the forebrain walls bulge out to form primary olfactory centers with distinct ventricular cavities (Fig. 7B). However, in most teleosts these bulbar ventricles are represented merely by shallow pits located on the caudomedial side of the bulbi. Here the olfactory bulbs develop as solid thickenings of the telencephalic walls.

In most actinopterygians the olfactory bulbs are closely apposed to the cerebral hemispheres (Fig. 6A–D), but in some teleostean groups (e.g., Cypriniformes, Osteoglossiformes; Fig. 6E, F) the primary olfactory centers are connected with the hemispheres through long stalks or peduncles. These peduncles are hollow structures; their thickened base contains compact bundles of secondary olfactory fibers; their roof is, on the other hand, thin and forms a direct continuation of the widened telencephalic roof plate.

The olfactory bulbs of actinopterygian fishes display, as in all vertebrates, a more or less distinct laminar structure (Nieuwenhuys, 1963, 1967a). The layers, which are arranged concentrically around the bulbar ventricle or its vestige, are, from the outside toward the inside, the following: (1) a stratum nervosum, made up of intertwined primary olfactory fibers; (2) a stratum glomerulosum, consisting of several layers of distinct glomeruli; (3) a stratum cellulare externum, which often contains, apart from large mitral cells, also smaller elements; (4) a stratum fibrosum, in which the secondary olfactory fibers assemble; and (5) a stratum cellulare internum or stratum granulare. In many species there is a gradual transition between layers 3 and 4, and 4 and 5, with intermingling of their constituent elements. The stratum cellulare internum is often designated as nucleus olfactorius anterior (e.g., Sheldon, 1912). For a more detailed description of the actinopterygian olfactory bulb, the reader is referred to Nieuwenhuys (1967a).

The large mitral cells are generally considered as the principal source of secondary olfactory fibers, but the retrogade tracer experiments of Bass (1981b) have shown that smaller cells located in the external and internal cell layers also contribute to the olfactory tracts.

It is noteworthy that the primary olfactory fibers which originate from the nasal mucosa do not constitute the only source of bulbar afferents. Experimental neuroanatomical studies have shown that elements located in the mitral cell layer of the contralateral olfactory bulb (Bass, 1981c) and in different areas of the cerebral hemispheres project to the olfactory bulb (Bass, 1981c; von Bartheld *et al.*, 1984; Levine and Dethier, 1985). Bulbopetal cells have also been demonstrated in the diencephalic nucleus preglomerulosus and in the rhombencephalic locus coeruleus (Levine and Dethier, 1985).

Finally, it should be mentioned that large, conspicuous ganglion cells of the small terminal nerve have been found in the olfactory nerve as well as in the caudomedial zone of the olfactory bulb (Sheldon and Brookover, 1909; Bass,

1981a; Demski and Northcutt, 1983). These cells, which are indicated as terminal nerve ganglion or nucleus olfacto-retinalis, send peripheral processes into the olfactory epithelium. Their central processes constitute a fascicle, which passes caudally through the most superficial portion of the area ventralis telencephali. Several authors (Münz and Claas, 1981; Springer, 1983; Demski and Northcutt, 1983; Levine and Dethier, 1985) have reported that a limited number of terminal nerve fibers project to the two retinas (see below).

4.3. The Area Ventralis

Histologically, two principal regions can be distinguished in the actinopterygian telencephalon. These regions, which were described by Johnston (1911) under the names *area olfactoria* and *primordium hippocampi,* occupy in the adult forebrain a ventromedial and dorsolateral position, respectively. The same regions were also recognized by Holmgren (1920, 1922), Källén (1950a,b, 1951), Nieuwenhuys (1959, 1962b, 1963), Northcutt and Braford (1980), and many others. The terms *area ventralis* and *area dorsalis,* which are now commonly used for these two principal regions, were introduced by Källén (1950a,b, 1951).

The area ventralis telencephali, hereafter to be abbreviated to V, shows with regard to structure and degree of differentiation a great constancy throughout the entire group. Caudally this area grades into the preoptic region. Two periventricularly situated cell masses, designated as the ventral and dorsal parts of V, have been delineated in all actinopterygians studied (Figs. 8–13).

Vv—the ventral part of V—is a relatively undifferentiated cell-zone, situated along the basalmost part of the ventricle. At the level of the anterior commissure, this cell zone is largely intersected by decussating fibers. Caudally to this level, Vv continues as the nucleus preopticus.

Vd—the dorsal part of V—is in most though not all species the largest cell zone of the area ventralis. This cell zone borders dorsally on the area dorsalis, but in some teleosts as well as in the more primitive actinopterygian groups, the precommissural portion of Vd is partly separated from the area dorsalis by Vn, another cell mass belonging to V. The dorsal part of V continues above the anterior commissure, forming the supracommissural part of V, Vs; caudally to this level, it extends to the tel–diencephalic border as the postcommissural part of V, Vp. This latter cell mass shows in many forms a lateralward extension, known as the pars intermedia: Vi (Sheldon, 1912; Fig. 15). Whereas the parts of the cell column under discussion, situated behind the anterior commissure, generally do not display a particular cell arrangement, the precommissural portion is commonly characterized by a more or less distinct lamination of its constituent elements. The lateralmost part of Vd is often formed by one or a few somewhat isolated cell clusters. In silurids (Bass, 1981a) and cyprinids (Levine and Dethier, 1985) Vd is laterally flanked by a group of cells that has been designated as the central part of V, Vc (Fig. 12).

As already mentioned, the dorsalmost portion of the area ventralis is in the more primitive actinopterygian groups partly occupied by a separate cell mass, Vn. In polypteriforms this area has the appearance of a compact agglomeration of cells (Fig. 8). In chondrosteans area Vn is ill-defined and contains merely scattered elements (Fig. 9). In holosteans, on the contrary, Vn is strongly devel-

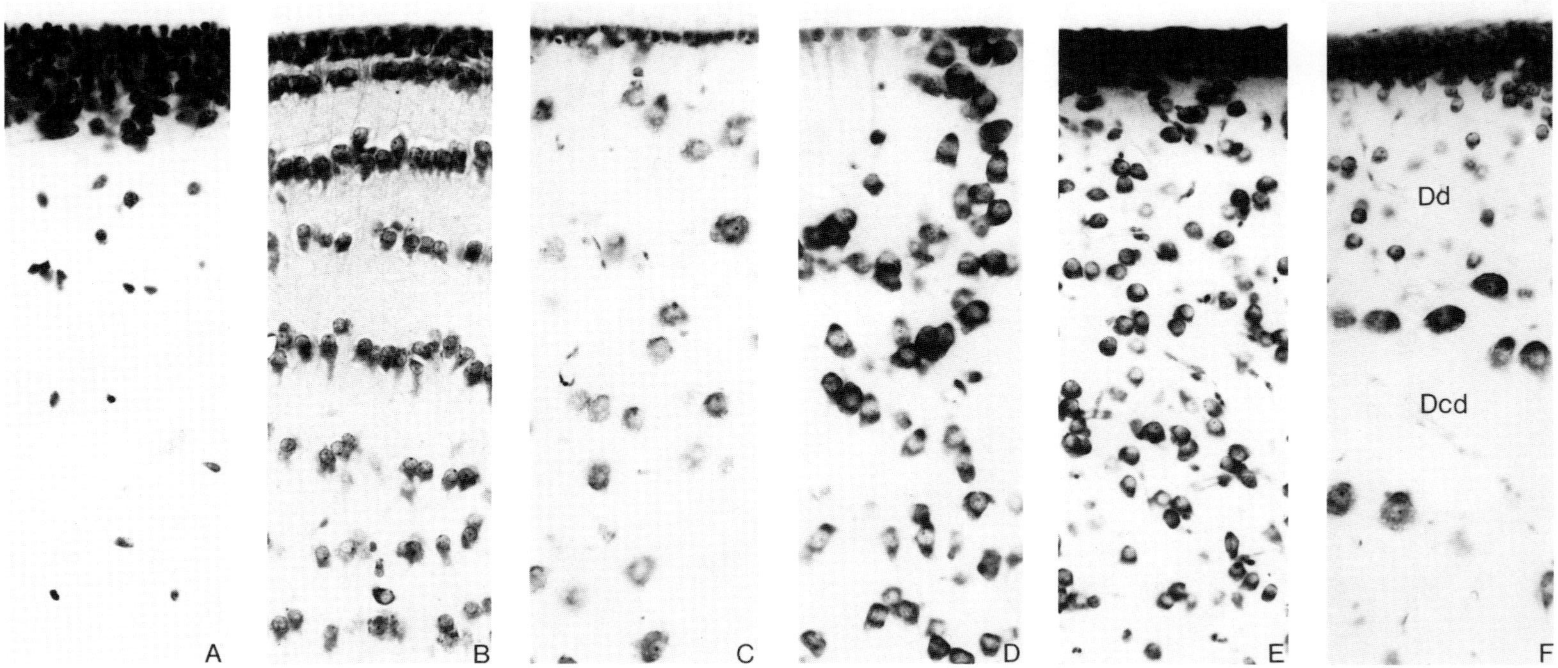

Figure 14. The cytoarchitecture of certain parts of the area dorsalis of actinopterygian fishes as seen in transverse Nissl-stained sections. A, field P1 of the reed fish, *Erpetoichthys calabaricus;* B, field Dd+l of the shovelnose sturgeon, *Scaphirhynchus platorynchus;* C, field Dd+l+c of the bowfin, *Amia calva;* D, field Dd+l of the trout, *Salmo gairdneri;* E, field Dd of the goldfish, *Carassius auratus;* F, fields Dd and Dcd of the osteoglossomorph *Xenomystis nigri.* The positions of these sections are indicated in Figs. 8–13.

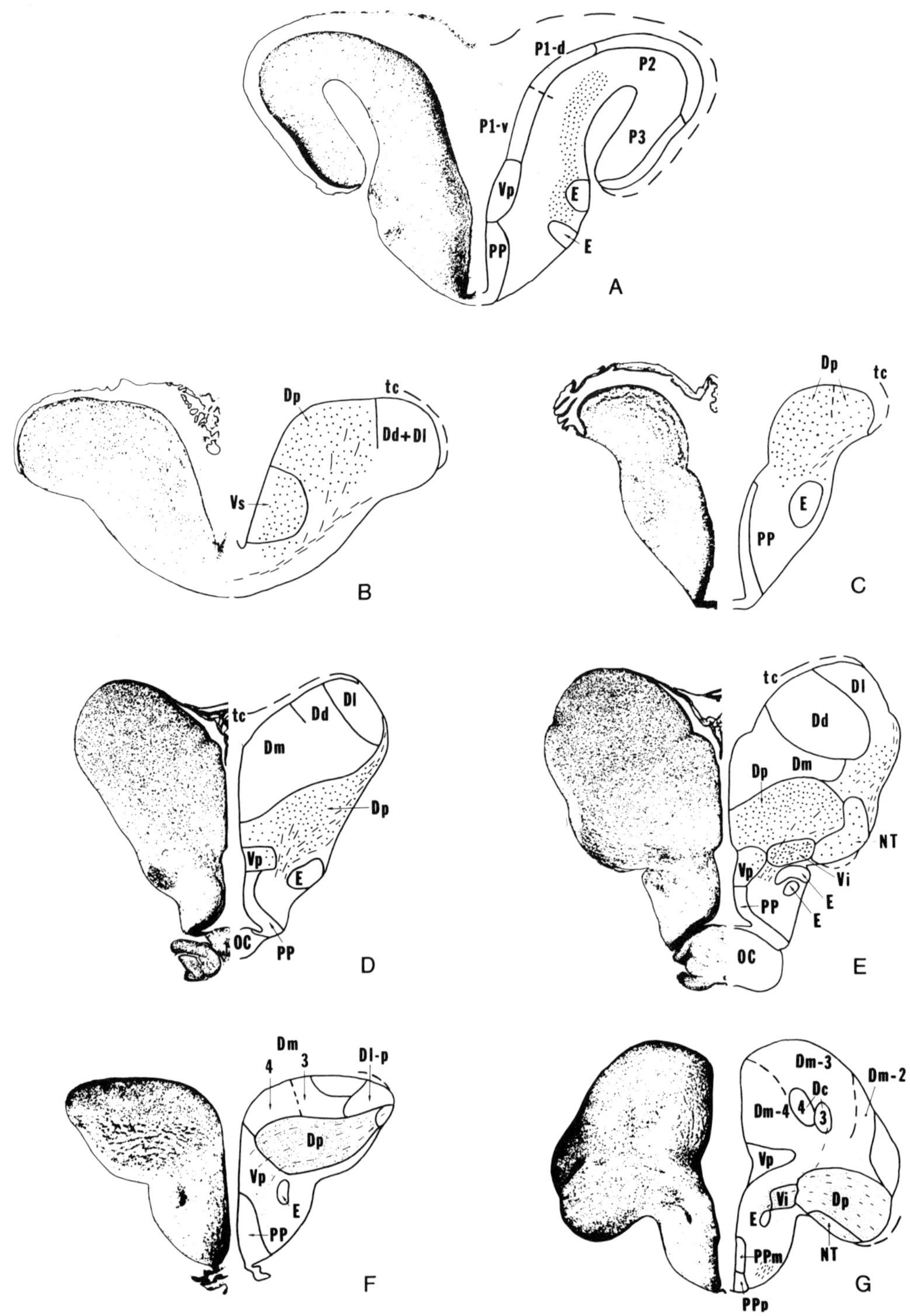

Figure 15. Transverse sections through the caudal parts of the telencephalon of actinopterygians, illustrating the cytoarchitecture, the olfactory terminal fields (dots), and the position of the field Dp, as delineated by Northcutt and Braford (1980). A, *Polypterus;* B, C, *Scaphirhynchus;* D, E, *Lepisosteus;* F, *Salmo;* G, *Lepomis.* Modified from Northcutt and Braford (1980) and Northcutt and Davis (1983).

oped, forming a distinct nucleus, which contains small groups of closely packed cells (Fig. 10). In some teleosts, as for instance *Xenomystis*, Vn can also be delineated (Fig. 13).

The submeningeal region of the area ventralis contains, apart from diffusely spread cells, in many forms a separate nuclear mass, which is named the lateral part of V, Vl. In polypteriforms this cell mass has the appearance of a thin layer, which is situated directly laterally to the large medial olfactory tract (Fig. 8). In chondrosteans Vl could not be discerned by us. Northcutt and Braford (1980) labeled in this group the lateral zone of the areas Vd and Vn of our subdivision as Vl. In holosteans Vl is strongly developed and touches dorsally the most lateral part of Vn (Fig. 10). A superficially situated nucleus has also been found in the area ventralis of numerous teleosts (Nieuwenhuys, 1963), yet this cell mass attains here neither the density nor the size of the holostean nucleus Vl (Figs. 11, 12). In osteoglossomorphs, like *Xenomystis* and *Gnathonemus*, a conspicuous, large-celled nucleus is situated laterally to Vn, occupying a position corresponding to that of the dorsomedial part of the holostean nucleus Vl (Fig. 13).

At or behind the level of the anterior commissure the place of Vl is taken up by a compact, small-celled nucleus, the nucleus entopeduncularis (Fig. 15E). This cell group, which has been found in all actinopterygians studied, is in most species situated in the area between the medial and lateral forebrain bundles, but in a number of teleosts certain components of the bundle last mentioned pass through the nucleus.

4.4. The Area Dorsalis

The eversion of the telencephalic walls occurs chiefly in the region from which the area dorsalis develops. It has already been mentioned that the degree to which the eversion proceeds differs considerably among the various actinopterygian groups; hence, the area dorsalis shows a great diversity in shape and position (Figs. 8–13). Ventromedially, the area dorsalis is in contact with the area ventralis. Cytoarchitectural differences render a demarcation of these two areas from each other possible. Moreover, the boundary plane between the dorsal and ventral areas is in many species marked by a more or less distinct cell-free zona limitans (Figs. 10–12). Due to the eversion the ependymal surface of the area dorsalis is very large; the dorsomedial, dorsal, and a varying part of the lateral surface are all bounded by the laterally expanded ventricular cavity. The meningeal surface of the area dorsalis is on the contrary only small. Not only the shape but also the histological structure of the area dorsalis show profound differences among the various actinopterygian groups, as may appear from the following survey.

The area dorsalis of polypteriform fishes, which, contrary to that of all other actinopterygians, is thin-walled, displays an exceptionally simple histological organization. The majority of the neuronal perikarya is concentrated in a narrow, subependymal layer. Beyond this cell zone only sparsely distributed small cells are found (Figs. 7A,B, 8, 14A). In several previous publications, Nieuwenhuys (1962b, 1963, 1966, 1967a, 1969) expressed the opinion that the layer of pallial gray shows a uniform structure throughout. However, more recently that author (Nieuwenhuys, 1983) affirmed the correctness of the view of Holmgren (1922) and Braford and Northcutt (1974, 1978) that this layer is subdivisible into three zones. Following the nomenclature of Braford and Northcutt (1974), these zones, plus their adjacent neuropil and fiber fields are designated here as P1, P2, and P3 (Fig. 8). The medial (P1) and lateral (P3) zones contain small, darkly staining cells, whereas the cell layer in the intervening zone (P2) is composed of slightly larger, somewhat lighter staining elements.

The structure of the area dorsalis differs in chondrosteans considerably from that of the polypteriforms. The greater part of the area contains fairly large triangular cells that are arranged in a number of layers parallel to the ependymal surface (Fig. 14B). The region showing this arrangement of its neurons occupies the dorsal and lateral parts of the area dorsalis and has been designated accordingly as Dd+1 (Fig. 9). In the rostromedial region of the area dorsalis the neurons are less regularly arranged than in the other parts of the ventricle-bordering zone; this area has been designated as Dm. Northcutt and Braford (1980) reported that Dm in chondrosteans, unlike the situation in holosteans and teleosts, does not extend the full length of the area dorsalis, since they indicate the mediocaudal part of D as Dp because of the termination of secondary olfactory fibers (Fig. 15). For this reason they also include the caudal part of our Dd+1 in Dp, in spite of the continuation of the characteristic lamination pattern of Dd+1 in this region. In our opinion the olfactory target area designated as Dp by Northcutt and Braford is not a cytoarchitectonic unit; rather it is composed of several subfields, some of which are continuous with the more rostrally located fields Dm and Dd+1. The central part of D in chondrosteans is at midtelencephalic levels occupied by a group of large, scattered cells that has been designated as Dc (Fig. 9).

The area dorsalis of holosteans can also be subdivided into some fields (Figs. 10, 14C). The medial part—Dm—is characterized by the presence of some laminae of cells in its superficial part, whereas its deeper part contains rather small, scattered elements. The most ventrolateral zone of the dorsal area, which is situated in the vicinity of the line of attachment of the ependymal telencephalic roof, is occupied by a zone of densely packed, rather small cells. This zone corresponds to the area olfactoria dorsalis pars lateralis of Meader (1939) and to the area dorsalis, pars lateralis posterior—Dlp—which one of us (Nieuwenhuys, 1963) observed in a large number of teleosts. The area in question presumably corresponds to the field Dl, which Northcutt and Braford (1980) delineated in *Lepisosteus*. The remainder of the holostean area dorsalis is occupied by fairly large, scattered cells. This undifferentiated zone presumably includes the homologues of the teleostean areas Dd, Dl, and Dc (Figs. 10, 14C).

In teleosts the area dorsalis shows a remarkable structural diversity. However, in most species three longitudinally arranged periventricular fields, surrounding a central field, can be distinguished (Holmgren, 1920; Ariëns Kappers *et al.*, 1936; Meader, 1939; Nieuwenhuys, 1959, 1962b, 1963). These fields are described under different names in the literature. Here they are designated as Dm, Dd, Dl, and Dc, i.e., the medial, dorsal, lateral, and central parts of D (Figs. 11–13).

Dm occupies in most species a considerable part of the medial wall of the hemispheres and usually also extends over some distance along the dorsal surface. It consists of densely packed small cells and receives a separate medial branch of the lateral forebrain bundle. Differences in cell size and cell density render it possible to subdivide Dm in some teleosts into separate dorsal and ventral subfields Dmd and Dmv [e.g., *Salmo* (Fig. 11)], while further subdivisions into Dm-1–4 have also been proposed (Fig. 15).

Dd varies considerably in size and structure. In most teleosts it is small and confined to the intermediate part of the telencephalon. It generally contains small neurons, but in some species, e.g., the goldfish, larger elements prevail in this field (Figs. 12, 14E). In osteoglossomorphs, like *Gnathonemus* and *Xenomystis*, Dd is strongly developed and shows a high cell density in its most superficial zone (Figs. 13, 14F). In the trout, on the other hand, no separate Dd can be distinguished. In this species we designated the large, uniform field situated laterally to Dm as Dd+1 (Figs. 11, 14D).

Dl occupies in most teleosts the entire lateral portion of the everted area dorsalis. The cells in this area are often arranged in parallel layers, which, however, do not show such a regular appearance as those found in the ventricle-bordering zone of the Chondrostei. The structure of the caudoventral part of Dl differs somewhat from that found in the other parts of this field. In many species this part of D is characterized by the presence of a number of irregular zones of closely packed, small neurons. This arrangement of cells is similar to that of the cell zone Vd, with which the cell mass under discussion shares the property of receiving a strong secondary olfactory projection. In a previous publication, one of us (Nieuwenhuys, 1963) labeled the subfield just discussed as the posterior portion of Dl: Dlp. Northcutt and Braford (1980) and Northcutt and Davis (1983) used the designation Dlp for another cell mass and indicate that they consider the subfield Dlp of our designation as a separate, posteriorly located entity within the dorsal area, which was accordingly labeled Dp. The description and interpretation of these parts of the actinopterygian telencephalon presented by Northcutt and Braford (1980) and Northcutt and Davis (1983) are predominantly based on tracing of secondary olfactory projections in four species, the teleosts *Salmo* and *Lepomis*, the holostean *Lepisosteus*, and the chondrostean *Scaphirhynchus*. In all of these species they labeled a certain olfactorecipient part of the area dorsalis as Dp (Fig. 15). We should like to emphasize that, with regard to their relative position and their histological organization, all of these areas represent dissimilar structures. In *Lepomis* Dp represents Dlp of our description but in *Salmo* Dp forms part of Dc; in *Lepisosteus* Dp is largely a cell-poor transitional zone whereas, as already mentioned, in *Scaphirhynchus* Dp represents parts of the fields Dm, Dc, and Dd+1.

The intermediate portion of Dl shows in osteoglossomorph teleosts a remarkable differentiation into three well-defined subfields. Because the fields Dm and Dd are also well differentiated in this group, the ventricle-bordering zone of the area dorsalis is occupied here by at least five clear-cut cytoarchitectonic fields, as is shown in Fig. 13. The three subfields that comprise Dl are designated by us as Dla, Dlb, Dlc, in conformity with Weston (1937) who first described these cell masses and named them provisionally: nucleus a, b, and c.

The central portion of the teleostean area dorsalis—Dc—also shows considerable structural variety. This region, considered by several authors to represent a primordial corpus striatum (Goldstein, 1905; Sheldon, 1912; Holmgren, 1920), contains scattered large cells that are generally more sparsely distributed than those in the zone bordering the ventricle. The chief phenomena responsible for the structural variety of Dc include: (1) the formation of local concentrations of cells and (2) a segregation into different cell zones. Local concentrations of neurons can be found in many different locations within the central portion of the teleostean telencephalon (Weston, 1937; Northcutt and Braford, 1980; Bass, 1981a). In some forms these cell masses have the appearance of separate nuclei indicated as Dc1, Dc2, etc. (Fig. 15); more commonly, however, the compact portions of Dc associate closely with one of the overlying, ventricle-bordering fields. Segregation of different cell areas within Dc can be clearly observed in osteoglossomorphs. In this group the central cell field can be clearly divided into three subfields, each of which corresponds to one of the overlying fields Dm, Dd, and Dlb. In accordance with the nomenclature introduced by one of us (Nieuwenhuys, 1962a,b, 1963), we have designated these centrally located subfields as Dcm, Dcd, and Dclb (Figs. 13, 14F).

Golgi studies (Sheldon, 1912; Holmgren, 1920; Nieuwenhuys, 1959) have shown that in the periventricular zone of the teleostean area dorsalis, numerous neurons occur that extend their dendrites mainly in layers parallel to the surface. These dendrites issue branches that are directed centrally. The axons of these elements are also directed away from the ventricular surface and join the lateral forebrain bundle. The large neurons in the central part of the area dorsalis extend their long, spiny dendrites in all direction. Experimental neuroanatomical studies have shown that a certain proportion of the large central neurons project to the deeper layers of the tectum (Ito and Kishida, 1978; Bass, 1981c). Sheldon (1912) reported that the perikarya of the central elements are surrounded by a dense network formed by the terminal arborizations of axons that ascend with the lateral forebrain bundle.

Finally, the nucelus taeniae, a cell mass occurring in holosteans as well as teleosts, should be mentioned. This cell mass, which consists of small, densely packed cells, constitutes a narrow band in the caudoventral part of D, ventromedial to Dlp. Its most lateral part lies in the immediate vicinity of the taenia, i.e., the line of attachment of the membranous telencephalic roof.

5. Connections

5.1. Introductory Notes

With respect to the fiber connections of the actinopterygian telencephalon, by far most experimental tracer studies deal with teleosts, whereas only few experimental data are available for Polypteriformes, Chondrostei, and Holostei. Therefore, the present survey will mainly describe our present knowledge for teleosts, followed by some comparative remarks on other actinopterygians. Subjects dealt with are successively: olfactory bulb projections (i.e., secondary olfactory and terminal nerve projections), ascending telencephalic afferents (i.e., diencephalotelencephalic and other projections), telencephalic efferents (both descending and ascending to the olfactory bulb), and intrinsic (nonolfactory) telencephalic connections.

5.2. Olfactory Bulb Projections

Since the olfactory bulbs receive their principal input from the olfactory epithelium, their efferents are generally indicated as secondary olfactory fibers. However, it should be recalled that the olfactory bulbs are not exclusively olfactory relay stations, since they receive additional input from other telencephalic areas (see Section 5.4), from the diencephalic nucleus preglomerulosus, and from the rhombencephalic locus coeruleus (see above). Moreover, the olfactory bulb gives rise to projections of the ganglion cells of the nervus terminalis (also indicated as nucleus olfacto-retinalis).

The most recent and complete experimental study of teleostean olfactory bulb projections deals with the goldfish (Levine and Dethier, 1985). In this fish, secondary olfactory fibers descend via the lateral and medial olfactory tracts to telencephalic as well as diencephalic targets. The lateral olfactory tract terminates in a so-called posterior terminal field, including ventral parts of Dl and associated portions of Dc (together indicated by Levine and Dethier as Dlc, i.e., dorsolateral nuclear complex), as well as to Dlp, which they indicate as Dp. The medial olfactory tract terminates in all subdivisions of the area ventralis, although not throughout the entire extent of these parts. As the medial tract enters the telencephalon, it separates into two branches—the ventromedial and dorsolateral fascicles. The ventromedial fascicle gives rise to a periventral and a subventral terminal field around Vv. The dorsolateral fascicle, which is the more substantial of the two, constitutes a medial terminal field, which encompasses parts of Vd, Vc, Vs, and Vp, and also gives rise to descending olfactory projections that terminate in the preoptic and tuberal diencephalic regions. Moreover, a certain proportion of the fibers of the dorsolateral fascicle of the medial olfactory tract decussate in the anterior commissure and contribute to the contralateral posterior terminal field. Most terminal regions receive a strong ipsilateral and a sparse contralateral projection, except for the medial terminal field and the preoptic region, which are only ipsilaterally innervated by secondary olfactory fibers. It is noteworthy that both fascicles of the medial olfactory tract contain fibers that, after having decussated in the anterior commissure, pass to the contralateral olfactory bulb, thus forming a true interbulbar commissure.

Similar secondary olfactory projections, although slightly different in some detail, have been demonstrated by von Bartheld *et al.* (1984), who also used goldfish; Bass (1981b), using catfish; Davis *et al.* (1981), using a perciform teleost; Ebbesson *et al.* (1981), using the piranha; and Prasada Rao and Finger (1984), using the winter flounder. All these authors mention the presence of secondary olfactory projections to most subregions of V and to the posterolateroventral part of D, including always an area designated as Dp, and often also ventral parts of Dl and Dc, as well as a nucleus taeniae, when distinguished. The field Dp, as delineated by these authors, corresponds to Dlp of the present description. Previous degeneration studies had revealed basically similar, although more restricted projections as well (see Northcutt and Davis, 1983, for review). In summary, it may be stated that teleostean secondary olfactory projections terminate in all subdivisions of V, although not always throughout the whole extent of these subdivisions, and in ventral parts of Dc and Dl, including Dlp. In contrast, Dm, Dd, and large dorsal parts of Dl and Dc do not receive secondary olfactory projections.

In chondrosteans and holosteans the bulbar projections have so far only been studied with degeneration techniques (Northcutt and Braford, 1980; Northcutt and Davis, 1983). As in teleosts these projections terminate in several parts of V as well as in a certain portion of D. According to the authors mentioned the principal target within the area dorsalis of *Lepisosteus* is a terminal field that they designated as Dp. We are of the opinion that this field only partly corresponds to Dlp of *Amia* and of teleosts (cf. Fig. 10 with Fig. 15D,E). It is particularly noteworthy that the area Dp, as delineated by Northcutt and collaborators in *Lepisosteus,* contrary to the similarly labeled teleostean area, directly borders the median ventricular space (cf. Fig. 15D,E with Fig. 15F,G). The secondary olfactory target area in the area dorsalis of the chondrostean *Scaphirhynchus* has also been labeled as Dp by Northcutt and collaborators. In our opinion this terminal field encompasses parts of Dm, Dd+1, and Dc (cf. Fig. 9 with Fig. 15B,C).

For *Polypterus* both a degeneration study (Braford and Northcutt, 1974; see also Northcutt and Braford, 1980) and a tracer study with HRP and cobalt (von Bartheld and Meyer, 1986b) are available. Both studies agree in describing a large medial olfactory tract, occupying a lateral position in the area ventralis and extending dorsally throughout the superficial zone of region P1 (Figs. 8, 15A). During its course this tract does not give rise to projections to the various cell masses within V, but only to the areas lateral to Vv, Vd, Vn, and Vs and medial to Vl. Most fibers of the medial olfactory tract terminate in P1, i.e., the homologue of field Dm in other actinopterygians.

The lateral olfactory tract, not detected in degeneration experiments (Braford and Northcutt, 1974), appears to course through and to project to the ventral portion of region P3, i.e., the zone situated close to the line of attachment of the membranous telencephalic roof (von Bartheld and Meyer, 1986b). The significance of the different secondary olfactory projection patterns within actinopterygians for establishing homologies will be discussed below.

With respect to the terminal nerve, most teleostean studies agree in describing projections to Vs and/or Vv (Davis *et al.,* 1981; Bass, 1981b; Demski and Northcutt, 1983; Levine and Dethier, 1985; von Bartheld and Meyer, 1986b), as well as to the retina (Münz and Claas, 1981; Münz, 1982; Springer, 1983; Demski and Northcutt, 1983; Levine and Dethier, 1985; von Bartheld and Meyer, 1986c). Because of the latter projection, the terminal nerve ganglion cells are indicated by some authors as nucleus olfacto-retinalis. According to several authors, the projection to V arises from ganglion cells with processes into the olfactory epithelium (distal type), whereas the projections to the retina arise from a different population of neurons that have no rostral processes (proximal type; see Levine and Dethier, 1985, and von Bartheld and Meyer, 1986a). Additional projections to a variety of other areas have recently been described by von Bartheld and Meyer (1986b), using cobalt application to the olfactory bulb of goldfish. These include Dc, the preoptic region (see also Springer, 1983), the diencephalic nucleus preglomerulosus anterior as well as the mesencephalic tectum, torus semicircularis, and torus longitudinalis. It should be noted that the ganglion cells of the nervus terminalis do not only receive olfactory input. Thus, in *Navodon* these cells receive additional afferent projections from Vs and Vp (Matsutani *et al.,* 1986), whereas in the goldfish an input from a midbrain nucleus located just rostral to the locus coeruleus has been reported (von Bartheld *et al.,* 1986).

A recent study of von Bartheld and Meyer (1986a) revealed projections of the "distal type" of nervus terminalis ganglion cells in *Polypterus* that resemble those in teleosts very much in terminating in Vv, Vs, as well as the preoptic region, with additional projections to several di- and mesencephalic targets. A "proximal type" of projection to the retina could not be demonstrated in *Polypterus* (von Bartheld and Meyer, 1986a). For chondrosteans and holosteans, no experimental data dealing with the terminal nerve are available.

5.3. Ascending Telencephalic Afferents

Experimental research dealing with the ascending (nonolfactory) afferent connections of the actinopterygian telencephalon has been restricted to teleostean species, except for a short communication dealing with *Polypterus* (Northcutt, 1981a). Since most of the telencephalic connections demonstrated in teleosts turn out to arise in the diencephalon, a short comment on the relevant diencephalic nuclei will be presented first.

5.3.1. Some Diencephalic Nuclei

Our present knowledge about the teleostean diencephalon and the interpretation of the cell masses encountered in this region has recently been reviewed by Braford and Northcutt (1983), and for details the reader is referred to that treatise. For the purpose of the present chapter we have to focus on several nuclei that are included by Braford and Northcutt (1983) in the tuberculum posterius, a region located ventrolaterally to the thalamus, the homology of which is uncertain. The central part of this tuberculum is occupied by a nucleus termed nucleus glomerulosus. In teleosts belonging to the group of acanthopterygians, e.g., the squirrelfish *Holocentrus rufus*, the filefish *Navodon modestus*, and the coastal rockfish *Sebastiscus marmoratus*, this nucleus indeed contains glomeruli (Sakamoto and Ito, 1982; Braford and Northcutt, 1983). However, in other teleosts, like the goldfish, *Carassius* sp., carp, *Cyprinus carpio*, and trout, *Salmo* sp., glomeruli are absent in this region. Nevertheless, many authors indicate a central mass of small granule cells in this region as nucleus glomerulosus. Following Braford and Northcutt (1983), we will indicate the latter nucleus as "glomerulosus" (between quotation marks). The connections of both types of nuclei seem to be different since Sakamoto and Ito (1982) found in *Navodon* for the true nucleus glomerulosus only afferent connections from pretectal nuclei (nucleus intermedialis of Brickner and nucleus corticalis), whereas Morita *et al.* (1980, 1983) described tertiary gustatory connections for the nucleus "glomerulosus" in the crucian carp *Carassius carassius*. This makes the homology of nucleus glomerulosus and "glomerulosus," and consequently also that of surrounding nuclei uncertain.

Rostrally, medially, and laterally to the nucleus glomerulosus or "glomerulosus" a more sparsely celled nucleus preglomerulosus is distinguished with at least a medial, a lateral, and an anterior subdivision. Together with the glomerular or "glomerular" nucleus this cell mass is usually indicated as the glomerulosus complex, which shows many variations in different teleosts. Since different sets of connections have been described for the nucleus preglomerulosus in teleosts with and without a true glomerular nucleus (Ito and Kishida, 1978; Murakami *et al.*, 1986a; Wullimann, 1988), we will distinguish likewise a nucleus preglomerulosus and "preglomerulosus." In the teleosts *Holocentrus* and *Sebastiscus* a nucleus prethalamicus (of Meader) is distinguished (Ito and Vanegas, 1983, 1984; Morita *et al.*, 1983), composed of a medial zone of small cells and a lateral zone of large cells. According to Braford and Northcutt (1983) this nucleus might be homologuous to the nucleus "preglomerulosus" pars lateralis.

In the thalamus of teleosts, which is located dorsomedial to the glomerulosus complex, a ventral and a dorsal part are distinguished (Braford and Northcutt, 1983). The most prominent cell mass in the ventral thalamus is the medioventral thalamic nucleus (Ito *et al.*, 1986). In the dorsal thalamus several cytoarchitectural units may be distinguished, which by most authors, on the basis of their topographic position, are indicated as the anterior, dorsomedial, dorsolateral, dorsal posterior, and central posterior nuclei.

5.3.2. Diencephalotelencephalic Connections (Fig. 16A)

In teleosts several interesting diencephalotelencephalic projections, related to different functional systems, have been recently described. Visual information is relayed to the telencephalon by the nucleus prethalamicus, also termed central optic nucleus, which receives a tectal input (Ebbesson, 1980). Acoustic and lateral line input is relayed to the telencephalon via the nucleus preglomerulosus, which receives afferents from the torus semicircularis. Somatosensory input from the spinal cord and from the sensory trigeminal nuclei may be relayed to the teleos-

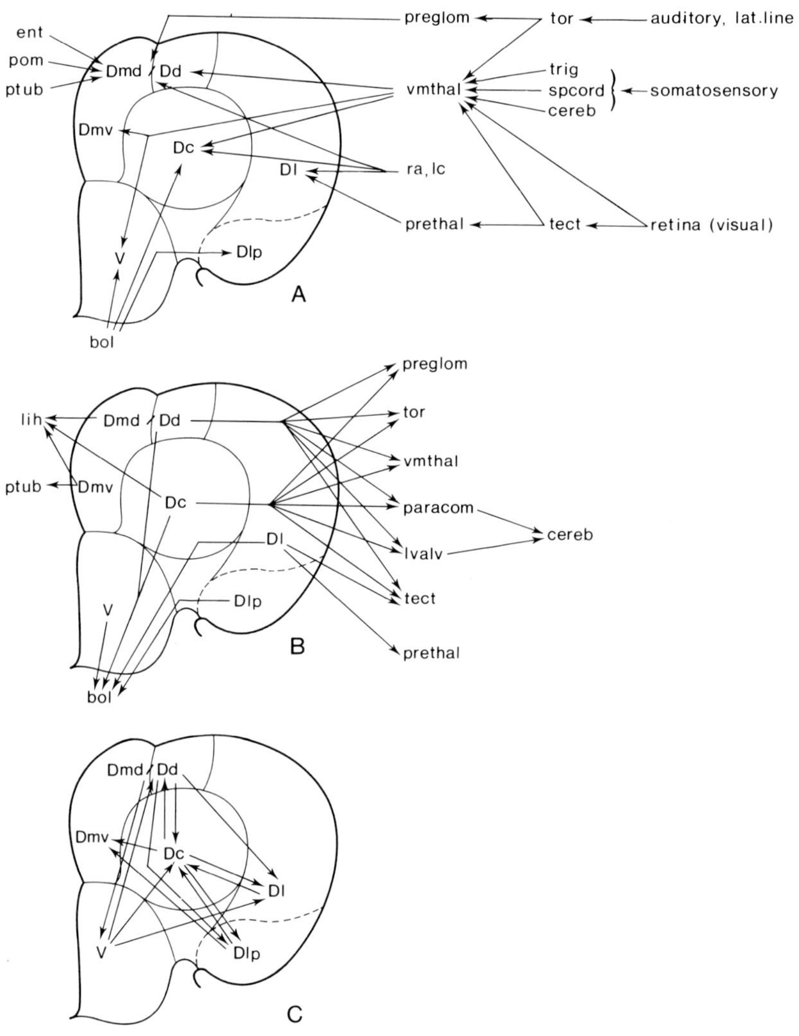

Figure 16. Diagram summarizing the afferent (A), efferent (B), and intrinsic connections (C) of the cell masses in the cerebral hemispheres of acanthopterygian teleosts.

tean telencephalon by the medioventral thalamic nucleus, and the same center has also been reported to be a link in a cerebello-thalamo-telencephalic projection. The literature on the connections just indicated may be summarized as follows.

HRP injections in the telencephalon of *Holocentrus* and *Sebastiscus* cause retrograde labeling of the large cells in the lateral zone of the prethalamic nucleus (Ebbesson, 1980; Ito *et al.*, 1980a, 1982). More detailed investigations revealed that these neurons project in both species mentioned to Dl (Murakami *et al.*, 1983; Ito and Vanegas, 1984). In turn, the parvocellular medial zone of the nucleus prethalamicus receives projections from Dc, Dd, and Dmd (Murakami *et al.*, 1983; Ito and Vanegas, 1983, 1984) and in *Sebastiscus* possibly also some fibers from Dl (Murakami *et al.*, 1983). Since the medial small cells of the nucleus prethalamicus probably project to the large laterally located cells of this nucleus (Ito and Vanegas, 1983), and since Dl projects to Dc in these teleosts, the following closed circuit appears to be present: nucleus prethalamicus, pars magnocellularis–Dl–Dc–nucleus prethalamicus, pars parvocellularis–nucleus prethalamicus, pars magnocellularis. Because the magnocellular part of the prethalamic nucleus also receives a major projection from the tectum (Ebbesson and Vanegas, 1976; Ito and Vanegas, 1983, 1984), the connectivity pattern of this center resembles that of the reptilian nucleus rotundus and of the mammalian lateralis posterior-pulvinar complex (Ito *et al.*, 1980a). Ebbesson (1980) proposed for all of these centers the common name *central optic nucleus*. A teleostean nucleus that might resemble the mammalian lateral geniculate nucleus in relaying direct retinal input to the telencephalon is not present in the teleosts studied to date. However, Northcutt (1981a) described such a nucleus in *Polypterus,* where the nucleus anterior thalami receives retinal input and projects to the telencephalon. In the same species, the dorsal posterior thalamic nucleus has to some extent similar connections as the teleostean nucleus prethalamicus, i.e., a tectal input and a projection to the telencephalon (Northcutt, 1981a). A nucleus prethalamicus has not been described for nonacanthopterygian teleosts like the goldfish, carp, and trout. In these teleosts it is uncertain by which pathways, if any, visual input might reach the telencephalon.

The telencephalic input from the nucleus preglomerulosus has most extensively been studied in *Sebastiscus*. In this teleost, the nucleus preglomerulosus projects to Dd and the associated part of Dm (Dmd), and receives in turn projections from the same telencephalic areas and from Dc, as well as a substantial input from the torus semicircularis and the nucleus of the lateral lemniscus (Ito *et al.*, 1982; Murakami *et al.*, 1983, 1986b). The torus semicircularis is a midbrain center homologous to the inferior colliculus of mammals, which receives acoustic and lateral line input from the rhombencephalic alar plate by way of the lateral lemniscus (Echteler, 1984; Murakami *et al.*, 1986b). Telencephalic projections from the nucleus preglomerulosus have also been described for other acanthopterygian teleosts, including *Eugerres, Navodon,* and several holocentrids (Vanegas and Ebbesson, 1976; Ito *et al.*, 1982). In teleosts without a true nucleus glomerulosus, the connections of the nucleus "preglomerulosus" are only partly similar to those of the acanthopterygian nucleus preglomerulosus. Finger (1980) and Finger and Bullock (1982) have described in the catfish, *Ictalurus,* a mechanoreceptive thalamic zone (MTZ), with comparable connections, i.e., input from the torus semicircularis and a projection to the telencephalon, in particular the medial part of Dc. In cyprinids predominantly the lateral part of the nucleus "preglomerulosus" projects to the telencephalon (ito and Kishida, 1978; Echteler and Saidel, 1981; Murakami *et al.,* 1986a). This projection terminates in the goldfish, as in *Sebastiscus,* in Dm and Dd (Echteler and Saidel, 1981), but in the carp in Dc (Murakami *et al.,* 1986a). In the goldfish, Dm projects back to the nucleus "preglomerulosus" (Echteler and Saidel, 1981). With respect to its afferent connections, the nucleus "preglomerulosus" resembles the nucleus preglomerulosus in receiving input from the torus semicircularis (Murakami *et al.,* 1986a). However, the nucleus "preglomerulosus" does not receive input from the nucleus of the lateral lemniscus (Murakami *et al.,* 1986a) and receives its main input from the hypothalamic nucleus anterior tuberis as well as additional projections from the ventromedial and central posterior thalamic nuclei. Since these nuclei all receive to some extent input from the torus semicircularis, Murakami *et al.* (1986a) suggest that the main function of the nucleus "preglomerulosus" pars lateralis, like that of the nucleus preglomerulosus, is to relay acoustic and lateral line impulses to the telencephalon, in a way reminiscent of the medial geniculate nucleus of mammals (see Murakami *et al.,* 1986b).

A third diencephalic region that projects to the telencephalon of *Sebastiscus* is the nucleus ventromedialis thalami, which projects to Dc, Dm, Dd as well as to part of the area ventralis telencephali (Murakami *et al.,* 1986b; Ito *et al.,* 1986). A telencephalic projection from the nucleus ventromedialis thalami has also been described for the carp (Murakami *et al.,* 1986a), but in this animal the precise region of termination is uncertain. Ito *et al.* (1986) characterized the nucleus ventromedialis thalami as a visual, somatosensory, octaval, and cerebellar relay center to the telencephalon on the basis of demonstrated input from the retina, spinal cord, trigeminal nuclei, torus semicircularis, and cerebellum, while Dc also projects back to this nucleus. The nucleus ventromedialis thalami does not only project to the telencephalon, but also to the nucleus prethalamicus, the dorsomedial thalamic nucleus, the tectum, and the torus semicircularis, thus entertaining a very complicated connectivity pattern.

Although telencephalic afferents originating in the dorsal thalamus have not been found in acanthopterygian teleosts (e.g., Ito *et al.,* 1982, 1986; Murakami *et al.,* 1983, 1986b; Ito and Vanegas, 1984; see Fig. 16), several sources of such projections have been reported for cyprinids (Echteler and Saidel, 1981; Murakami *et al.,* 1986a). Echteler and Saidel (1981) described in the goldfish the presence of a projection from the anterior nucleus of the dorsal thalamus to Dm and from the dorsal posterior nucleus of the same entity to Dl. In contrast, Murakami *et al.* (1986a) reported that the central posterior and the posterior nucleus of the dorsal thalamus project to the carp telencephalon. The central posterior nucleus is, like the ventromedial thalamic nucleus and nucleus "preglomerulosus" lateralis of the carp, involved in octavo-lateral line processing (Echteler, 1984, 1985; Murakami *et al.,* 1986a). Braford and Northcutt (1983) have suggested that the anterior thalamic nucleus might receive a retinal projection and the dorsal posterior thalamic nucleus a tectal projection, on the basis of connections found in holostean actinopterygians. However, experimental tracer studies did not show such connections in the goldfish (see Springer and Gaffney, 1981; Grover and Sharma, 1981) or other teleosts (see Vanegas and Ito, 1983, for review). In our opinion the fiber connections of teleostean dorsal thalamic nuclei require further study.

Finally, it should be mentioned that Murakami *et al.* (1986a) found two small additional telencephalic projections originating in diencephalic nuclei, i.e., a nucleus subrotundus and nucleus "preglomerulosus" pars anterior. There are presently no data about the other connections of these nuclei nor about their homology and functional significance.

With respect to nonteleostean actinopterygians, only one study is available, dealing with *Polypterus* (Northcutt, 1981a). In this fish, thalamo-telencephalic projections arise from the anterior, the dorsal posterior, as well as the central posterior nucleus in the dorsal thalamus, while also a periventricular and a median nucleus of the posterior tuberculum give rise to telencephalic afferents. It is unknown to which telencephalic regions these nuclei project. For chondrosteans and holosteans, no experimental data are available.

5.3.3. Remaining Telencephalic Afferents

In addition to the descending olfactory and the ascending thalamo-telencephalic input described above, the following telencephalic afferents have to be mentioned. Both in *Sebastiscus* (Murakami *et al.*, 1983) and in goldfish (Echteler and Saidel, 1981) the telencephalic entopeduncular nucleus projects to Dd and the dorsal part of Dm. In *Sebastiscus*, the medial preoptic area and the hypothalamic nucleus posterior tuberis project to the same telencephalic area as well (Murakami *et al.*, 1983). The most caudally located sources of telencephalic input in *Sebastiscus* and other teleosts are the nucleus raphes superior and the locus coeruleus (Ito *et al.*, 1982; Murakami *et al.*, 1983), the latter of which has been found to project in goldfish also to the olfactory bulbs (Levine and Dethier, 1985). Telencephalic projections from the hypothalamus (i.e., the torus lateralis) as well as from the nucleus raphes superior and the locus coeruleus have also been reported for *Polypterus* (Northcutt, 1981a).

5.4. Efferent Telencephalic Connections (Fig. 16B)

Similar to afferent telencephalic connections, efferents of the actinopterygian forebrain have almost exclusively been studied experimentally in teleosts. In this group of fishes, descending telencephalic efferents have been described to terminate in the thalamus, hypothalamus, pretectal region, torus semicircularis, and midbrain tectum and tegmentum. In addition, telencephalic efferents from various sources ascend to the olfactory bulbs.

Several reciprocal telencephalo-thalamic projections have already been documented in the previous paragraphs. In summary, these include the following four projections. First, in *Sebastiscus* and *Holocentrus* the nucleus prethalamicus receives input from Dd, Dmd, Dc and in *Sebastiscus* possibly also from Dl (Ito *et al.*, 1980a, 1982; Ito and Vanegas, 1983, 1984; Murakami *et al.*, 1983). Second, projections to the nucleus preglomerulosus or "preglomerulosus" have been described in *Sebastiscus* to arise from Dd, Dmd, and Dc (Murakami *et al.*, 1983, 1986b), and in goldfish from Dm (Echteler and Saidel, 1981). Also in other teleosts, telencephalo-preglomerulosar projections have been found, but without specification of their precise origin (Ito *et al.*, 1982). However, Murakami *et al.* (1986a) did not find such a projection in the carp. Third, a projection form Dc to the nucleus medioventralis thalami has been reported in *Sebastiscus* by Ito *et al.* (1986). Finally, a projection from Dd to the dorsal thalamic region has been described by Echteler and Saidel (1981) in the goldfish.

Telencephalic projections to the lobus inferior hypothalami have been described in the goldfish (Oka and Ueda, 1981; Echteler and Saidel, 1981), *Holocentrus* (Vanegas and Ebbesson, 1976), *Sebastiscus* (Ito *et al.*, 1982; Murakami *et al.*, 1983), and several other teleostean species (Ito *et al.*, 1982). According to Murakami *et al.* (1983), this projection has in *Sebastiscus* its origin in Dm, Dd, as well as Dc, whereas Echteler and Saidel (1981) claim that in the goldfish Dm is its only source. Telencephalo-hypothalamic fibers are not confined to the inferior lobe. A projection to the medial hypothalamus has been described in the goldfish (Oka and Ueda, 1981), and a projection to the nucleus posterior tuberis, originating from Dm, has been reported in *Sebastiscus* (Murakami *et al.*, 1983).

A pretectal target of telencephalic efferents has thus far only been described for acanthopterygian teleosts, including *Sebastiscus, Navodon,* and holocentrids (Ito *et al.*, 1982). In these teleosts, a pretectal nucleus located laterally to the posterior commissure, termed nucleus paracommissuralis, receives projections from Dd, Dmd, as well as Dc (Murakami *et al.*, 1983) and projects in turn to the cerebellum, thus establishing a telencephalo-cerebellar relay (Ito *et al.*, 1982). In contrast, in mormyrid teleosts a telencephalo-cerebellar relay nucleus is located in the midbrain tegmentum (see below).

Telencephalo-tectal projections have been described for a variety of teleosts, including cyprinids (e.g., Ito and Kishida, 1977; Oka and Ueda, 1981; Luiten, 1981; Grover and Sharma, 1981; Echteler and Saidel, 1981; Airhart and Kriebel, 1985), catfish (Bass, 1981c), piranha (Fiebig *et al.*, 1983), *Holocentrus* (Vanegas and Ebbesson, 1976; Ito *et al.*, 1980b), and *Sebastiscus* (Ito *et al.*, 1982; Murakami *et al.*, 1983). All authors mentioned describe an exclusive or at least major projection from Dc, except for Echteler and Saidel (1981), who claim an origin from Dm in the goldfish. In the catfish, Bass (1981c) observed that the population of Dc neurons that project to the tectum is different from those projecting to the olfactory bulb (see below). According to Murakami *et al.* (1983), in *Sebastiscus* additional projections to the tectum arise from Dl, Dd, and Dmd.

A telencephalic projection to the midbrain torus semicircularis, i.e., the homologue of the mammalian inferior colliculus, has been described in *Sebastiscus* (Murakami *et al.*, 1983, 1986b), where this projection originates in Dc as well as Dd and Dmd, and in the carp, where only Dc projects to this midbrain center (Echteler, 1984). In *Sebastiscus,* Dm, Dd, and Dmd project in addition to a medial zone of the midbrain tegmentum (Murakami *et al.*, 1983, 1986b), a projection also described for the goldfish (Echteler and Saidel, 1981) and *Holocentrus* (Vanegas and Ebbesson, 1976). Our own, partly unpublished, observations are that this medial mesencephalic projection is particularly well developed in mormyrids and *Xenomystis,* where a massive projection from Dcd terminates in the nucleus lateralis valvulae, a midbrain nucleus projecting to the cerebellum (see Nieuwenhuys and Verrijdt, 1983; Meek *et al.*, 1986a,b). Telencephalic projections descending beyond the midbrain have not been documented experimentally in teleosts.

Ascending telencephalic efferents, i.e., fibers projecting to the olfactory bulb, arise in the goldfish from four groups of neurons, a ventromedial group located in V, and dorsal, anterolateral, and posterior bulbopetal groups in D (von Bartheld *et al.*, 1984; Levine and Dethier, 1985). The ventromedial group is mainly located in the dorsal part of Vv, but also includes cells in Vs, Vd (Levine and Dethier, 1985), and Vl (von Bartheld *et al.*, 1984) and lies within the confines of the medial terminal field of secondary olfactory fibers (see above). The dorsal bulbopetal group includes cells in the ventrolateral part of Dm and the associated region of Dc. The anterolateral group lies within the confines of the precommissural part of the posterior secondary olfactory terminal field, including ventral parts of Dl and Dc, while the posterior bulbopetal group lies within the confines of the postcommissural part of the posterior olfactory terminal field, i.e., Dlp (i.e., Dp of most authors) and the neighboring nucleus taeniae and Vi (Levine and Dethier, 1985). Thus, only the dorsal bulbopetal group does not maintain reciprocal connections with the olfactory bulb. In the catfish, similar ascending telencephalic efferent cell groups have been described (Bass, 1981c), whereas in some other teleosts only the ventromedial and posterior groups have been recognized (Murakami *et al.,* 1983; Prasada Rao and Finger, 1984), which might be due to technical differences. Whereas Bass (1981c), Murakami *et al.* (1983), and Levine and Dethier (1985) report exclusively ipsilateral bulbopetal connections from the telencephalon, von Bartheld *et al.* (1984) and Prasada Rao and Finger (1984) have described smaller contralateral projections as well.

In *Polypterus,* bulbopetal cells are restricted to the lateral part of the ipsilateral Vd, with some surrounding neurons in Vv, Vs, and Vp, possibly even continuing into the preoptic region (von Bartheld and Meyer, 1986a), thus resembling the medioventral bulbopetal group of teleosts. Bulbopetal neurons in the area dorsalis or pallium of *Polypterus* were not observed by von Bartheld and Meyer (1986b). For Chondrostei and Holostei, no experimental data on ascending projections to the olfactory bulb and their cells of origin are available.

5.5. Intrinsic Telencephalic Connections

Intrinsic telencephalic connections, i.e., connections between different parts of V and D, have only been described on the basis of tracer experiments in the teleost *Sebastiscus* (Murakami *et al.,* 1983), and are summarized in Fig. 16C. For other teleosts or actinopterygians no experimental data are available.

6. The Interpretation of the Various Centers

6.1. Introductory Notes

Although the nomenclature of the various centers in the actinopterygian cerebral hemisphere as presented in the preceding sections, is widely accepted, the interpretation of these centers, i.e., their homology with particular grisea in the telencephalon of other vertebrate groups, has given rise to much controversy among different authors. A full discussion of all of the interpretations presented over the last 100 years is clearly beyond the scope of the present chapter. However, an attempt wil be made to highlight the most important interpretational trends under the following headings: (6.2) positional relations: topology versus topography, (6.3) histochemistry: the location of a striatal primordium, (6.4) fiber connections: secondary olfactory and ascending telencephalic afferents, and (6.5) structure: presence or absence of cortical equivalents.

6.2. Positional Relations: Topology versus Topography

Holmgren (1920, 1922) was of the opinion that in the evaginated and inverted telencephalic hemispheres of chondrichthians, dipnoans, and amphibians, a ventral subpallium and a dorsal pallium can be distinguished and that in all of these groups cell-free zonae limitantes generally mark the location of the pallio-subpallial boundary. Holmgren also pointed out that in the subpallium as well as in the pallium of the groups mentioned, three longitudinally arranged cellular zones or grisea can be delineated, which he termed (1) septum, (2) tuberculum olfactorium, (3) nucleus olfactorius lateralis, i.e., the striatal primordium or striatum of most other authors, (4) piriform pallium, (5) general pallium, and (6) hippocampal pallium (Fig. 17A). Taking into consideration that the actnopterygian cerebral hemispheres are everted rather than inverted, Holmgren interpreted the cell masses in the telencephalon of this group as indicated in Fig. 17B, i.e., Vv = septum, Vd = tuberculum olfactorium, Vn = nucleus olfactorius lateralis or striatum, Dm = piriform pallium, Dd = general pallium, and Dl = hippocampal pallium. Dc belongs according to Holmgren to the general pallium.

The interpretation presented by Holmgren (1920, 1922) has been principally accepted by Källén (1951) and Nieuwenhuys (1963, 1966, 1969). However, these authors were of the opinion that, although the actinopterygian area dorsalis is homologous to the pallium of other vertebrates, the cell masses found within the former structures are the result of a differentiation that has taken place entirely within the actinopterygian line, and hence cannot be compared to separate pallial grisea of other vertebrate groups. Although Nieuwenhuys (1983) later agreed with a tripartitioning of the pallial gray of polypteriform fishes as proposed by Holmgren (1922), Braford and Northcutt (1974), and Northcutt and Braford (1980), the present authors are still reluctant to accept Holmgren's interpretation throughout for other actinopterygians. Reasons for this are not only the absence of clear cytoarchitectonic landmarks indicating a tripartitioning in most actinopterygians, but also the observation that only in one of the three major nonactinopterygian fish groups, viz. the chondrichthians, separate medial, dorsal, and lateral pallial fields can be delineated, but that in the other two,

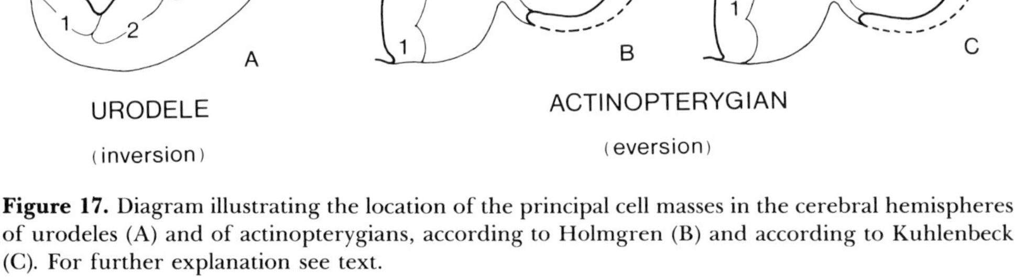

Figure 17. Diagram illustrating the location of the principal cell masses in the cerebral hemispheres of urodeles (A) and of actinopterygians, according to Holmgren (B) and according to Kuhlenbeck (C). For further explanation see text.

i.e., the dipnoans and the crossopterygians, such a tripartitioning of the pallium is not possible (cf. Nieuwenhuys, 1969, and the next chapter in the present volume).

Kuhlenbeck (1929) and Miller (1940) followed in their interpretation of the teleostean telencephalon a procedure similar to that of Holmgren (1922) and advocated a corresponding sequence of grisea along the ventricular surface, but yet these authors arrived at a different conclusion, in that they placed the pallio-subpallial boundary between the cell groups Dm and Dd (Fig. 17C). Accordingly, they believed that Dm in teleosts is homologous to the striatum of other vertebrates and they argued furthermore that Dd is homologous to the piriform pallium and Dl to the dorsal or general and the hippocampal pallium. So far as the interpretation of Dm is concerned, Northcutt and Braford (1980) supported the interpretation of Kuhlenbeck (1929) and Miller (1940), primarily on the basis of histochemical considerations (see below).

In the interpretations discussed so far the relative or topological position of the various grisea has been the criterion used for the establishment of their homology. It is, however, interesting to note that several authors have compared the centers in the actinopterygian telencephalon with those of land vertebrates on a purely topographical basis. Thus, Rabl-Rückhard (1882), who compared the brains of teleosts with those of mammalian embryos, contended that the wide membranous roof of the teleostean forebrain represents the pallium of mammals, whereas the thickened "basal lobes" are homologous to the striatum (Fig. 18). Several other authors (Droogleever Fortuyn, 1961; Crosby *et al.*, 1966; Schnitzlein, 1968; Wright, 1967; Morgan, 1975) believed that all principal parts of the telencephalon of land vertebrates are also represented in the actinopterygian forebrains in which they occupy topographically corresponding positions. Their interpretation includes the following homologies: Vv plus Vd: septum; Dc: striatum; Dm: hippocampal pallium; Dd: dorsal or general pallium; and Dl: piriform pallium. It is difficult to understand how these authors arrived at their interpretation. All of them ignore or deny the significance of the eversion process and one (Droogleever Fortuyn, 1961) states explicitly that his interpretation was based primarily on topographical relations. Remarkably, all adherents of a direct topographical comparison considered the sulcus externus, a longitudinal groove on the basal surface of the actinopterygian forebrain, homologous with the mammalian sulcus endorhinalis. In our opinion this groove comes about by the eversion and is as such typical for actinopterygians (Figs. 1, 4b, 5b, 12, 13).

It is noteworthy that Sheldon (1912) arrived at an interpretation of the teleostean telencephalon that closely corresponds to the one just discussed. However, contrary to the adherents of a direct topographical comparison, Sheldon was fully aware of the occurrence of an eversion during the development of the teleostean telencephalon. As a possible explanation for the fact that, notwithstanding the eversion, the teleostean hippocampal pallium is situated dorsomedially and the piriform pallium ventrolaterally, Sheldon suggested that during ontogenesis a complex rearrangement of the various telencephalic cell groups would occur, leading to a transposition of the hippocampal and piriform primordia. Studies of the ontogenesis of the actinopterygian telencephalon (Källén, 1950a,b, 1951; Nieuwenhuys, 1962a, 1964; Nieuwenhuys *et al.*, 1969) have brought no confirmation of Sheldon's suppositions.

In conclusion, we are of the opinion that on the basis of topological com-

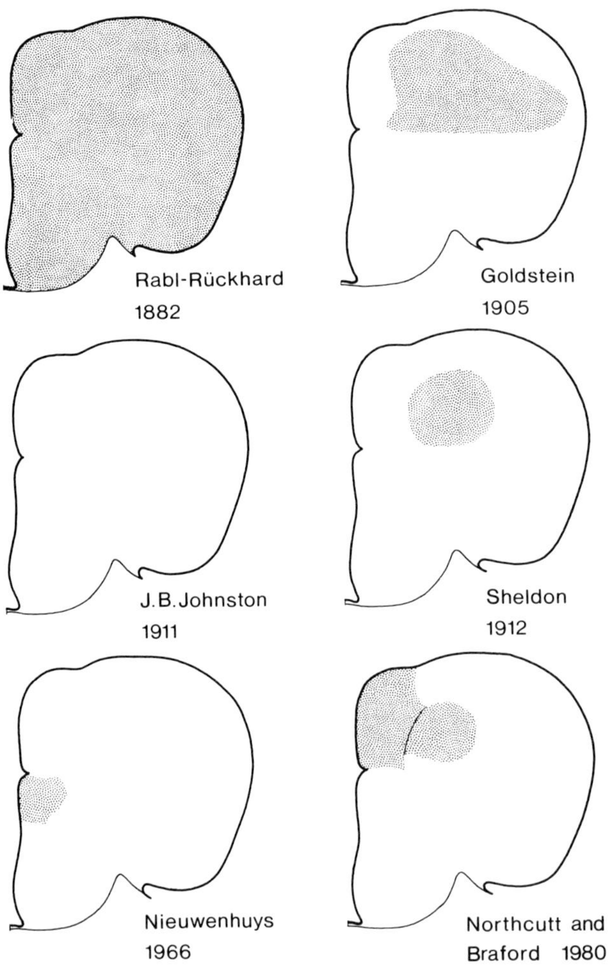

Figure 18. The location of the striatum in the actinopterygian cerebral hemispheres according to different authors.

parison a pallium and subpallium may be distinguished, i.e., area dorsalis and ventralis, respectively, but that a further subdivision of the area dorsalis into pallial fields homologous to those in tetrapods (piriform, general and hippocampal pallium) is not obvious in most actinopterygians on the basis of the relative position of various subfields of D.

6.3. Histochemistry: The Location of a Striatal Primordium

Although it has been argued that the results of histochemical investigations often indicate functional rather than morphological similarity, and hence analogy rather than homology (Gans, 1969), similarities in histochemical properties have during the last 20 years frequently been used to institute homologies or at least to help in the establishment of homologies. In the actinopterygian telencephalon, this type of finding has been employed particularly in attempts to identify the homologue of the striatum of amniotes.

Reference to Fig. 18 shows that the actinopterygian striatum has been located quite differently by different authors. As already mentioned, Rabl-Rückhard (1882) held that the wide tela choroidea of the teleostean telencephalon is the homologue of the pallium of land vertebrates and that the solid "basal lobes" represent the striatum. Goldstein (1905) regarded the superficial zone of the dorsal half of the solid hemispheres as a thickened ependyma and the central zone as the striatum. Johnston (1911) held that the actinopterygian telencephalon consists merely of a ventromedial area olfactoria and a dorsolateral primordium hippocampi. A striatum or striatal primordium is not present according to that author. Sheldon (1912) designated the central portion of the area dorsalis as palaeostriatum (see above), and Nieuwenhuys (1966) considered the most dorsal part of the area ventralis on topological grounds as the homologue of the striatum of other groups of vertebrates.

The mammalian striatum or caudatoputamen complex has two outstanding features that can be visualized with histochemical techniques: (1) it is heavily innervated by dopaminergic fibers that originate from a mesencephalic center, the substantia nigra, and (2) it contains a population of large intrinsic cholinergic neurons that gives rise to an extremely dense meshwork of axonal ramifications. Hence, the presence of high concentrations of dopamine (or catecholamines) and acetylcholinesterase (AChE) have frequently been used as criteria for the identification of the striatum in other vertebrate groups. Thus, Northcutt and Braford (1980; see also Parent *et al.*, 1978) found that in the teleost *Lepomis* the caudal part of Dm displays high AChE concentrations and is heavily innervated by catecholaminergic fibers. They suggested that these data support the homology of the caudal part of Dm with the striatum of land vertebrates (Fig. 18) and argued that the adjacent part of Dc on topological grounds might be homologized with the globus pallidus. We are reluctant to accept these interpretations for the following reasons:

1. In the holostean *Lepisosteus* the caudal part of Dm does not show a high AChE activity (Northcutt and Braford, 1980) and does not receive a significantly denser catecholaminergic innervation than the remaining superficial portions of the area dorsalis (Parent and Northcutt, 1982).
2. Because in *Polypterus* the area Vn does show AChE activity, whereas the areas P1, P2, and P3 do not, Northcutt and Braford suggested that this area is homologous to the caudal part of Dm of other actinopterygians. This interpretation leaves us with the problem of finding another interpretation for Vn in chondrosteans, holosteans, and teleosts.
3. High levels of AChE combined with dense catecholaminergic innervation are good criteria for the identification of the striatum in birds and mammals, but do not lead to clear and unequivocal results at the level of amphibians and reptiles (see Parent, 1986). Hence, it is not to be expected that these histochemical features will be of much help in identifying a striatal homologue in the various groups of fish.

In our opinion the question—which part of the actinopterygian telencephalon does represent the striatum—cannot be answered at present.

6.4. Fiber Connections: Secondary Olfactory and Ascending Telencephalic Afferents

Apart from positional relations, fiber connections are considered by most comparative neuroanatomists as important guides for the institution of homologies (e.g., Ariëns Kappers *et al.*, 1936; Campbell and Hodos, 1970). In the attempts to interpret the various parts of the actinopterygian telencephalon, two sets of fibers, viz. the secondary olfactory tracts and the projections ascending to the telencephalon from more caudally located centers, have played a crucial role. These connections and the considerations attached to them will now be briefly discussed.

In most vertebrate groups with evaginated forebrains a distinct lateral olfactory tract is present, which terminates in the piriform primordium or lobe, a pallial field situated close to the pallio-subpallial boundary. In chondrosteans, holosteans, and teleosts a similarly named tract is found that, after having arched over the external sulcus (when present), terminates mainly in the caudobasolateral portion of the telencephalon (Figs. 10, 12, 13). Course and termination of this tract show a striking resemblance to the situation seen in primitive mammals and it is not surprising that many previous students of the actinopterygian telencephalon have designated the main target of this tract as a piriform primordium. A major problem is, however, that, as we have seen above, this interpretation is incompatible with the fact that the actinopterygian telencephalon is everted rather than evaginated (Figs. 1, 17). Braford and Northcutt (1974), Northcutt and Braford (1980), and Northcutt and Davis (1983) have studied the secondary olfactory connections in representatives of the polypteriforms, the chondrosteans, the holosteans, and the teleosts with axon degeneration techniques (see Fig. 15). They found that in *Polypterus* the secondary olfactory fibers passing to the area dorsalis or pallium terminate in the medial pallial field P1. In the chondrostean *Scaphirhynchus* the terminal field included parts of the areas labeled by us as Dm, Dd+1, and Dc, whereas in the holostean *Lepisosteus* the fibers under discussion appeared to terminate in an area encompassing Dlp and the caudoventral parts of Dm and Dd+1+c of the present description. Northcutt and Braford, however, considered the main olfactory terminal field within the chondrostean and holostean area dorsalis as a single entity that they labeled as Dp (Fig. 15). Comparing their findings with the results of some experimental studies on the olfactory projections of teleosts, they argued that in the series polypteriforms–chondrosteans–holosteans–teleosts the pallial olfactory area, while retaining its integrity, has migrated laterally until it has come to occupy a lateral and caudal position in advanced teleosts. The findings of Northcutt and Braford (1980), if confirmed by experimental studies using more sensitive techniques, pose a comparative neuroanatomical dilemma. Either the target area of the olfacto-pallial projection remains constant throughout the actinopterygian series, but shows considerable positional changes, or the interpretation of the various parts of the actinopterygian area dorsalis as suggested by the present authors is correct, which implies that the target area of the olfacto-pallial projection differs from group to group. It is in our opinion not possible to decide which of these two cases is the most likely one. It is, however, noteworthy that studies of normal (Nieuwenhuys, 1963, 1983; Fig. 7A) as well as experimental material (von Bartheld and Meyer, 1986b) have shown that in polypteriforms the lateral pallial field P3 receives a distinct secondary olfactory projection. We believe that this projection is at least partly homologous to the lateral olfactory tract of other actinopterygians.

An earlier generation of comparative neuroanatomists (e.g., Ariëns Kappers, 1929; Ariëns Kappers *et al.*, 1936; Herrick, 1948) assumed that in anamniotes practically the entire telencephalon is dominated by olfactory projections, and that nonolfactory pathways, originating from the dorsal thalamus, first reach the pallium in amniotes. Consequently, they considered the existence of thalamo-telencephalic connections, also termed lemniscal systems, of paramount importance for the recognition of neocortical primordia.

Recent experimental neuroanatomical studies have revealed that in actinopterygians several ascending fiber systems project to various parts of the area dorsalis and the question arises as to whether these connections have some relevance for the establishment of homologies with parts of the pallium of tetrapods. However, it should be noted that even in tetrapods, sensory projections from dorsal thalamic nuclei are not unfailing guides for neocortical primordia, since in reptiles it is not the neocortical primordium, but quite another pallial derivative, viz. the dorsal ventricular ridge, that receives somesthetic, visual, as well as auditory information from distinct dorsal thalamic centers (e.g., Balaban and Ulinski, 1981; Bruce and Butler, 1984a,b). Morphologically, the mammalian neocortex and the reptilian dorsal ventricular ridge have evolved independently and their similarity in connections is probably the product of parallel evolution. They are homoplaseous rather than homologous (Northcutt, 1981b).

We have seen that in all of the four actinopterygian groups a certain portion of the area dorsalis or pallium does not receive any secondary olfactory projections. Experimental studies, reviewed a previous section (5.3), have shown that this nonolfactory pallial area may receive ascending projections from three different diencephalic sources, the dorsal thalamus, the ventral thalamus, and the posterior tubercular region.

As regards the dorsal thalamus, preliminary studies (Braford and Northcutt, 1978; Northcutt, 1981a; Echteler and Saidel, 1981; Murakami *et al.*, 1986a) suggest that in *Polypterus* and cyprinids three dorsal thalamic centers might be relays in ascending telencephalic projections, i.e., the anterior, central posterior, and (dorsal) posterior nucleus. The central posterior nucleus is involved in the processing of auditory and/or lateral line information, while Northcutt and Davis (1983) have suggested that the anterior nucleus could serve as a link in a retino-thalamo-telencephalic projection, and that the dorsal posterior nucleus might represent a diencephalic relay in a second (retino-tecto-thalamo-telencephalic) visual pathway.

The ventral thalamic center projecting to the telencephalon is the nucleus ventromedialis thalami. Experimental studies, summarized in Fig. 16A, have shown that this cell mass in acanthopterygian teleosts receives somatosensory, visual, as well as octavolateral input, and projects to Dm, Dd, and Dc.

The third and quantitatively most important source of ascending telencephalic afferents is the region of the posterior tuberculum. In polypteriforms this region is relatively undifferentiated, containing only two different cell masses, but in many teleosts it is differentiated into four or more different nuclei, among which is the remarkable nucleus glomerulosus. The homologization of the individual nuclei within this complex offers serious difficulties. Different subdivisions of this complex receive gustatory, octavolateral, and possibly also visual input, and in particular the octavolateral recipient part projects to the telencephalon, i.e., to Dd and neighboring regions of Dm and/or Dc (Dmd and/or Dcd). In addition to the glomerulosus complex, the posterior tuberculum of acanthopterygian teleosts contains the nucleus prethalamicus, a nucleus relaying visual input from the midbrain tectum to Dl (Murakami *et al.*, 1983; Ito and Vanegas, 1984).

Several of the centers and projections just discussed have been compared with structures and connections of the diencephalon and telencephalon in other vertebrate groups. Thus, Echteler and Saidel (1981) considered the lateral regions (Dd, Dl) of the area dorsalis of the goldfish with regard to their ascending thalamic connections closely comparable to the dorsal and medial pallium of amphibians and reptiles. We think that the existence of the dorsal thalamo-telencephalic projections described by these authors requires confirmation by more critical experiments with more sensitive techniques. It is remarkable that corresponding projections have not been found in detailed studies on the connections of several other teleosts (e.g., Murakami *et al.*, 1983; Ito *et al.*, 1986). Moreover, confirmation of the presence of these dorsal thalamo-telencephalic projections would still not exclude the possibility that the development of these pathways has occurred independently in actinopterygians and terrestrial vertebrates.

The connectivity pattern of the nucleus preglomerulosus or mechanoreceptive thalamic zone (MTZ) has been compared with that of the mammalian medial geniculate nucleus and its homologue in reptiles and amphibians (Finger, 1980; Murakami *et al.*, 1986b) whereas the connections of the nucleus prethalamicus resemble those of the mammalian pulvinar or its homologue in reptiles and amphibians (Ebbesson, 1980; Ito and Vanegas, 1984). However, these similarities offer in our opinion no basis for homologies since both the nucleus preglomerulosus and the nucleus prethalamicus have a position in the ventral diencephalon, i.e., the posterior tuberculum. This region is not homologous with the dorsal thalamus, where the nucleus geniculatus medialis as well as the pulvinar are located.

In spite of the absence of clear homologies with land vertebrates, as just discussed, the organization and distribution of the connections of the area dorsalis (or pallium) shows a remarkable functional resemblance with that of higher vertebrates. Thus, going from lateroventral to medial over the telencephalic surface (Fig. 16) one encounters an olfactory region (Dlp), a region receiving sensory input via the diencephalon with a distinct visual (Dl) and octavolateral area (Dd with Dmd), while the most medial parts of Dmd and Dmv maintain connections with hypothalamic nuclei. Consequently, the connections of Dlp resemble those of the piriform pallium, the connections of Dl, Dd, and lateral parts of Dm those of the general pallium, and those of medial parts of Dm those of the hippocampal pallium of land vertebrates. Surprisingly, this organization is the reverse of that proposed by Holmgren on the basis of topological comparison (see above), but fits a topographic comparison. Dc, located in the center of the area dorsalis, maintains reciprocal connections with all of the superficial fields and is moreover a major source of telencephalic efferents (Fig. 16), connective properties shared with the vertebrate striatum (Murakami *et al.*, 1983), although it should be noted that the striatum, in contrast to Dc, does not project back to superficial (cortical) regions. Accepting the connections of Dc as a basis for comparison with the striatum is not in line with histochemical results (see above), but again fits surprisingly well the topographic comparison of Sheldon (1912; Fig. 18).

In summary, from a functional point of view it is highly interesting that impulses related to different sensory systems may reach the actinopterygian area dorsalis or pallium along three different routes and via three different sets of diencephalic relay stations. However, these various ascending projections yield no clues for securing specific homologies of parts of the actinopterygian pallium with parts of the pallium of terrestrial vertebrates.

6.5. Structure: Presence or Absence of Cortical Equivalents

Similarity in structure has been used as a criterion for the establishment of homologies both in general morphology (see Remane, 1956) and in neuroanatomy. As regards neuroanatomy, Campbell and Hodos (1970) mentioned specifically the importance of the morphology of individual neurons. Needless to say, the presence of such typical neurons as mitral cells or Purkinje cells may help in determining the location and limits of the olfactory bulb and the cerebellum, respectively. As far as we are aware, the noted comparative neuroanatomist J. B. Johnston (1911, 1912) has been the only investigator who has explicitly used data on the structure of areas and individual neurons for the establishment of homologies of parts of the actinopterygian telencephalon. Johnston stated that the cells in the area dorsalis of the present description, as they appear in Golgi preparations, have the following characteristics in common: (1) arrangement in layers with the bases toward the ventricle, (2) the possession of basal dendrites spreading among the laminae and apical dendrites directed outward, (3) the presence on the dendrites of innumerable small spines knobbed at the end, and (4) the axon directed outward among the dendrites and giving off branching collaterals (Johnston, 1911, p. 508). He held that these histological characteristics and the fiber connections of the area dorsalis warrant its interpretation as a primordium hippocampi. Cells with the same characteristics are according to Johnston (1911, 1912) present in the similarly named part of the pallium of the lamprey and of the frog.

Golgi studies of the actinopterygian telencephalon are relatively rare. However, from the papers of Sheldon (1912), Holmgren (1920), and Nieuwenhuys (1959, 1983) we gained the impression that the entire area dorsalis is not characterized by the presence of a single type of cell. Moreover, a laminar arrangement of the perikarya, however characteristic for certain parts of the chondrostean and holostean area dorsalis, is not obvious throughout the area dorsalis of all of the four actinopterygian groups (cf. Figs. 8–14).

An interesting question not related to the institution of homologies is: have certain parts of the actinopterygian telencephalon attained a degree of differentiation that justifies the designation *cerebral cortex?* This question is hard to answer, because we do not know whether the various cytoarchitectonic fields, present in the area dorsalis, represent separate entities or rather form part of larger units. However, we consider it likely that the highly differentiated dorsal area of the telencephalon of osteoglossomorphs is largely occupied by three radially arranged units, each comprising a small-celled superficial zone and a large-celled deeper zone (Fig. 16). If we accept Herrick's (1948) definition of cerebral cortex as "superficial laminated gray matter in the pallial part of the cerebral hemispheres," then the area dorsalis of osteoglossomorphs may be considered to contain true cortical structures. In the evaginated telencephalon of dipnoans and tetrapods the cerebral cortex develops from one or more layers of cells that have migrated out from a smaller ventricular zone toward a generally much larger meningeal zone. However, in osteoglossomorphs and teleosts the cortex comes about in an entirely different way, namely through a combination of eversion and differentiation within a greatly elaborated ventricular zone. This peculiar mode of development has been designated by one of us (Nieuwenhuys, 1962b) as "paradoxical cortex formation." It is true, of course, that the strict matching of the superficial and deeper zones in the area dorsalis of osteoglossomorphs, however remarkable, does not prove a functional coupling by means of reciprocal connections, although it may be noted that such connections do exist in acanthopterygians (Fig. 16C). Golgi and electron microscopic studies will be necessary to establish whether such radial coupling also exists in osteoglossomorphs.

7. Conclusions

1. The telencephalon of actinopterygian fishes develops in a way that differs radically from that in other vertebrate groups. Through an outfolding or eversion process, solid cerebral hemispheres with a remarkably extensive and largely external ventricular surface come about.

2. The extensive comparative neuroembryological studies of Källén (1950a,b, 1951) have shown that the actinopterygian hemispheres can be divided into two principal parts: an area ventralis and an area dorsalis. Moreover, this author has presented evidence strongly suggesting that the area ventralis represents the subpallium and the area dorsalis the pallium of other vertebrates.

3. The area ventralis shows a rather constant histological organization. Most of its parts are in receipt of secondary olfactory fibers. Homologues of the septum and the tuberculum olfactorium of other vertebrate groups can be delineated with reasonable certainty within this area.

4. Whether or not the actinopterygian telencephalon contains a striatum or striatal primordium is uncertain.

5. The area dorsalis shows a distinct progressive differentiation in the series: polypteriforms–chondrosteans + holosteans–primitive teleosts–advanced teleosts.

6. Just like the area ventralis, the area dorsalis receives a strong olfactory input. However, in all actinopterygian groups a large part of the area dorsalis is not under olfactory dominance.

7. The nonolfactory portions of the area dorsalis entertain multifarious connections with each other, with the area ventralis, and with more caudal portions of the CNS. Nonolfactory sensory input may attain the area dorsalis via three different sets of diencephalo-pallial projections, one originating from the dorsal thalamus, another from the ventral thalamus, and a third from the (in many teleosts highly differentiated) posterior tuberculum region.

8. Many attempts have been made to homologize parts of the actinopterygian area dorsalis, with parts of the pallium of other vertebrate groups, but the "touchstone" for the recognition of such homologies has not been found (and may not exist).

8. Abbreviations

bol	Bulbus olfactorius
ca	Commissura anterior
cereb	Cerebellum
D	Area dorsalis telencephali
Dc	Central part of D
Dcd	Part of Dc associated with Dd
Dclb	Part of Dc associated with Dlb
Dcm	Part of Dc associated with Dm
Dc-1,2,3,4	Subregions of Dc
Dd	Dorsal part of D
dienc	Diencephalon

Dl	Lateral part of D
Dlp	Posterior part of Dl
Dm	Medial part of D
Dmd	Dorsal part of Dm
Dmv	Ventral part of Dm
Dm-1,2,3,4	Subregions of Dm
Dp	Posterior part of D according to Northcutt and Braford (1980) (= Dlp)
dpal	Dorsal pallium or primordium pallii dorsalis
E	Nucleus entopeduncularis
ent	Nucleus entopeduncularis
flt	Fasciculus lateralis telencephali
hab	Ganglion habenulae
lc	Locus coeruleus
lih	Lobus inferior hypothalami
lpal	Lateral pallium or primordium piriforme
lvalv	Nucleus lateralis valvulae
mpal	Medial pallium or primordium hippocampi
nI	Nervus olfactorius
NT	Nucleus taeniae
OC	Optic chiasm
P1,2,3	Pallial fields in Polypteriformes
Pl-d	Dorsal part of Pl
Pl-v	Ventral part of Pl
pal	Pallium
paracom	Nucleus paracommissuralis
pom	Area preoptica medialis
PP	Periventricular preoptic nucleus
PPm	Magnocellular part of PP
PPp	Parvocellular part of PP
preglom	Nucleus preglomerulosus
prethal	Nucleus prethalamicus (of Meader)
psv	Pseudoventricle
ptub	Nucleus posterior tuberus
ra	Nucleus raphes anterior
sep	Septum
spcord	Spinal cord
str	Corpus striatum
tc	Tela chorioidea
tect	Tectum mesencephali
tor	Torus semicircularis
tp	Tuberculum posterius
trig	Sensory trigeminal nuclei
troll	Tractus olfactorius lateralis
trolm	Tractus olfactorius medialis
tubol	Tuberculum olfactorium
V	Area ventralis telencephali
Vc	Central part of V
Vd	Dorsal part of V
Vi	Intermediate part of V
Vl	Lateral part of V
vl	Ventriculus lateralis
Vld	Dorsal part of Vl
vmthal	Nucleus ventromedialis thalami
Vn	Cell mass situated in the most dorsal part of V
Vp	Postcommissural part of V
Vs	Supracommissural part of V
Vv	Ventral part of V

ACKNOWLEDGMENTS. The authors are grateful to Mr. J. Russon and Mr. J. de Bekker for the drawings, to Mr. A. Reijnen for the photomicrographs, and to Mrs. M. van de Coevering for secretarial assistance.

9. References

Airhart, M. J., and Kriebel, R. M., 1985, Telencephalic terminals in the major retinal synaptic lamina of the goldfish optic tectum, *Brain Res.* **336:**363–367.

Ariëns Kappers, C. U., 1929, *The Evolution of the Nervous System in Invertebrates, Vertebrates and Man,* Bohn, Haarlem, p. 335.

Ariëns Kappers, C. U., Huber, G. C., and Crosby, E. C., 1936, *The Comparative Anatomy of the Nervous System of Vertebrates, Including Man,* Volume II, Macmillan Co., New York, p. 1845.

Balaban, C. D., and Ulinski, P. S., 1981, Organization of thalamic afferents to anterior dorsal ventricular ridge in turtles. I. Projections of thalamic nuclei, *J. Comp. Neurol.* **200:**95–129.

Bass, A. H., 1981a, Organization of the telencephalon in the channel catfish, Ictalurus punctatus, *J. Morphol.* **169:**71–90.

Bass, A. H., 1981b, Olfactory bulb efferents in the channel catfish, Ictalurus punctatus, *J. Morphol.* **169:**91–111.

Bass, A. H., 1981c, Telencephalic efferents in the channel catfish, Ictalurus punctatus: Projections to the olfactory bulb and optic tectum. *Brain Behav. Evol.* **19:**1–16.

Braford, M. R., Jr., and Northcutt, R. G., 1974, Olfactory bulb projections in the bichir, Polypterus, *J. Comp. Neurol.* **156:**165.

Braford, M. R., Jr., and Northcutt, R. G., 1978, Correlation of telencephalic afferents and SDH distribution in the bony fish Polypterus, *Brain Res.* **152:**157–160.

Braford, M. R., Jr., and Northcutt, R. G., 1983, Organization of the diencephalon and pretectum of the ray-finned fishes, in: *Fish Neurobiology,* Volume 2 (R. E. Davis and R. G. Northcutt, eds.), University of Michigan Press, Ann Arbor, pp. 117–163.

Bruce, L. L., and Butler, A. B., 1984a, Telencephalic connections in lizards. I. Projections to cortex, *J. Comp. Neurol.* **229:**585–601.

Bruce, L. L., and Butler, A. B., 1984b, Telencephalic connections in lizards. II. Projections to anterior dorsal ventricular ridge, *J. Comp. Neurol.* **229:**602–615.

Campbell, C. B. G., and Hodos, W., 1970, The concept of homology and the evolution of the nervous system, *Brain Behav. Evol.* **3:**353–367.

Crosby, E. C., DeJonge, B. R., and Schneider, R. C., 1966, Evidence for some of the trends in the phylogenetic development of the vertebrate telencephalon, in: *Evolution of the Forebrain* (R. Hassler and H. Stephan, eds.), Thieme Verlag, Stuttgart, pp. 117–135.

Daget, J., 1958, Sous-classe des brachiopterygiens, *Trait. Zool.* **13:**2500–2521.

Davis, R. E., Chase, R., Morris, J., and Kaufman, B., 1981, Telencephalon of the teleost Macropodus: Experimental localization of secondary olfactory areas and of components of the lateral forebrain bundle, *Behav. Neural Biol.* **33:**257–279.

Demski, L. S., and Northcutt, R. G., 1983, The terminal nerve: A new chemosensory system in vertebrates? *Science* **220:**435–437.

Droogleever Fortuyn, J., 1961, Topographical relations in the telencephalon of the sunfish, Eupomotis gibbosus, *J. Comp. Neurol.* **116:**249–264.

Ebbesson, S. O. E., 1980, A visual thalamo-telencephalic pathway in a teleost fish (Holocentrus rufus), *Cell Tissue Res.* **213:**505–508.

Ebbesson, S. O. E., and Vanegas, H., 1976, Projections of the optic tectum in two teleost species, *J. Comp. Neurol.* **165:**161–180.

Ebbesson, S. O. E., Meyer, D. L., and Scheich, H., 1981, Connections of the olfactory bulb in the piranha (Serrasalmus nattereri), *Cell Tissue Res.* **216:**167–180.

Echteler, S. M., 1984, Connections of the auditory midbrain in a teleost fish, Cyprinus carpio, *J. Comp. Neurol.* **230:**536–551.

Echteler, S. M., 1985, Organization of central auditory pathways in a teleost fish, Cyprinus carpio, *J. Comp. Physiol.* **156:**267–280.

Echteler, S. M., and Saidel, W. M., 1981, Forebrain connections in the goldfish support telencephalic homologies with land vertebrates, *Science* **212:**683–685.

Fiebig, E., Ebbesson, S. O. E., and Meyer, D. L., 1983, Afferent connections of the optic tectum in the piranha (Serrasalmus nattereri), *Cell Tissue Res.* **231:**55–72.

Finger, T. E., 1980, Nonolfactory sensory pathway to the telencephalon in a teleost fish, *Science* **210:**671–673.

Finger, T. E., and Bullock, T. H., 1982, Thalamic center for the lateral line system in the catfish Ictalurus nebulosus: Evoked potential evidence, *J. Neurobiol.* **13**(1):39–47.

Gage, S. P., 1893, The brain of Diemyctilus viridescens from larval to adult life and comparison with the brain of Amia and of Petromyzon, *The Wilder Quarter Century Book,* Ithaca, pp. 259–314.

Gans, C., 1969, Some questions and problems in morphological comparison, in: *Comparative and Evolutionary Aspects of the Vertebrate Central Nervous System* (J. M. Petras and C. R. Noback, eds.), The New York Academy of Sciences, New York, pp. 506–513.

Goldstein, K., 1905, Untersuchungen über das Vorderhirn und Zwischenhirn einiger Knochenfische, *Arch. Mikrosk. Anat.* **66:**135–219.

Grover, B. G., and Sharma, S. C., 1981, Organization of extrinsic tectal connections in goldfish (Carassius auratus), *J. Comp. Neurol.* **196:**471–488.

Herrick, C. J., 1901, The cranial nerves and cutaneous sense organs of the North American siluroid fishes, *J. Comp. Neurol.* **11:**177–247.

Herrick, C. J., 1921, A sketch of the origin of the cerebral hemispheres, *J. Comp. Neurol.* **32:**429–454.

Herrick, C. J., 1922, *Functional Factors in the Morphology of the Forebrain of Fishes,* Libro en honor de D. Santiago Ramón y Cajal, Madrid, Volume 1, pp. 143–204.

Herrick, C. J., 1933, The amphibian forebrain, *J. Comp. Neurol.* **58:**1–288.

Herrick, C. J., 1948, *The Brain of the Tiger Salamander,* University of Chicago Press, Chicago, p. 470.

Holmgren, N., 1920, Zur Anatomie und Histologie des Vorder- und Zwischenhirns der Knochenfische, *Acta Zool.* **1:**137–315.

Holmgren, N., 1922, Points of view concerning forebrain morphology in lower vertebrates, *J. Comp. Neurol.* **34:**391–459.

Ito, H., and Kishida, R., 1977, Tectal afferent neurons identified by the retrograde HRP method in the carp telencephalon, *Brain Res.* **130:**142–145.

Ito, H., and Kishida, R., 1978, Telencephalic afferent neurons identified by the retrograde HRP method in the carp diencephalon, *Brain Res.* **149:**211–215.

Ito, H., and Vanegas, H., 1983, Cytoarchitecture and ultrastructure of nucleus prethalamicus, with special reference to degenerating afferents from optic tectum and telencephalon, in a teleost (Holocentrus ascensionis), *J. Comp. Neurol.* **221:**401–415.

Ito, H., and Vanegas, H., 1984, Visual receptive thalamopetal neurons in the optic tectum of teleosts (Holocentridae), *Brain Res.* **290:**201–210.

Ito, H., Morita, Y., Sakamoto, N., and Ueda, S., 1980a, Possibility of telencephalic visual projection in teleosts, Holocentrus, *Brain Res.* **197:**219–222.

Ito, H., Butler, A. B., and Ebbesson, S. O. E., 1980b, An ultrastructural study of the normal synaptic organization of the optic tectum and the degenerating tectal afferents from retina, telencephalon, and contralateral tectum in a teleost, Holocentrus rufus, *J. Comp. Neurol.* **191:**639–659.

Ito, H., Murakami, T., and Morita, Y., 1982, An indirect telencephalo-cerebellar pathway and its relay nucleus in teleosts, *Brain Res.* **249:**1–13.

Ito, H., Murakami, T., Fukuoka, T., and Kishida, R., 1986, Thalamic fiber connections in a teleost (Sebastiscus marmoratus): Visual, somatosensory, octavolateral, and cerebellar relay region to the telencephalon, *J. Comp. Neurol.* **250:**215–227.

Jarvik, E., 1968a, *The Systematic Position of the Dipnoi,* Nobel Symposium 4 (T. Orvig, ed.), Almqvist & Wiksell, Stockholm, pp. 223–245.

Jarvik, E., 1968b, *Current Problems of Lower Vertebrate Phylogeny,* Nobel Symposium 4 (T. Orvig, ed.), Almqvist & Wiksell, Stockholm, pp. 497–527.

Johnston, J. B., 1911, The telencephalon of ganoids and teleosts, *J. Comp. Neurol.* **21:**489–591.

Johnston, J. B., 1912, The telencephalon in cyclostomes, *J. Comp. Neurol.* **22:**341–404.

Källén, B., 1950a, A contribution to the ontogenetic development of the nuclei in the forebrain of Lepisosteus, *Acta Anat.* **9:**297–308.

Källén, B., 1950b, Contributions to the ontogeny of the everted forebrain, *K. Fysiogr. Saellsk. Lund Foerh.* **20:**87–110.

Källén, B., 1951, Embryological studies on the nuclei and their homologization in the vertebrate forebrain, *K. Fysiogr. Saellsk. Handl.* **62**(5)**:**34.

Kuhlenbeck, H., 1929, Die Grundbestandteile des Endhirns im Lichte der Bauplanlehre, *Anat. Anz.* **67:**1–51.

Levine, R. L., and Dethier, S., 1985, The connections between the olfactory bulb and the brain in the goldfish, *J. Comp. Neurol.* **237:**427–444.

Luiten, P. G. M., 1981, Afferent and efferent connections of the optic tectum in the carp (Cyprinus carpio L.), *Brain Res.* **220:**51–65.

McCormick, C. A., 1978, Central projections of the lateralis and eight nerves in the bowfin, Amia calva, thesis, Ann Arbor, p. 131.

Matsutani, S., Uchiyama, H., and Ito, H., 1986, Cytoarchitecture, synaptic organization and fiber connections of the nucleus olfacto-retinalis in a teleost (Navodon modestus), *Brain Res.* **373:**126–138.

Meader, R. G., 1939, The forebrain of bony fishes, *K. Ned. Akad. Wet. Proc. Sect. Sci.* **42:**657–670.

Meek, J., Nieuwenhuys, R., and Elsevier, D., 1986a, Afferent and efferent connections of cerebellar lobe C_1 of the mormyrid fish Gnathonemus petersi: An HRP study, *J. Comp. Neurol.* **245:**319–341.

Meek, J., Nieuwenhuys, R., and Elsevier, D., 1986b, Afferent and efferent connections of cerebellar lobe C_3 of the mormyrid fish Gnathonemus petersi: An HRP study, *J. Comp. Neurol.* **245:**342–358.

Miller, R. N., 1940, The telencephalic zonal system of the teleost Corydora paliatus, *J. Comp. Neurol.* **72:**149–176.

Morgan, G. C., Jr., 1975, The telencephalon of the sea catfish Galeichthys felis, *J. Hirnforsch.* **16:**131–150.

Morita, Y., Ito, H., and Masai, H., 1980, Central gustatory paths in the crucian carp, Carassius carassius, *J. Comp. Neurol.* **191:**119–132.

Morita, Y., Murakami, T., and Ito, H., 1983, Cytoarchitecture and topographic projections of the gustatory centers in a teleost, Carassius carassius, *J. Comp. Neurol.* **218:**378–394.

Moy-Thomas, J. A., and Miles, R. S., 1971, *Palaeozoic Fishes,* Chapman & Hall, London, p. 259.

Münz, H., 1982, Centrifugal innervation of the retina by luteinizing hormone releasing hormone

(LHRH)-immunoreactive telencephalic neurons in teleostean fishes, *Cell Tissue Res.* **222:**313–323.

Münz, H., and Claas, B., 1981, Centrifugal innervation of the retina in cichlid and poecilid fishes. A horseradish peroxidase study, *Neurosci. Lett.* **22:**223–226.

Murakami, T., Morita, Y., and Ito, H., 1983, Extrinsic and intrinsic fiber connections of the telencephalon in a teleost, Sebastiscus marmoratus, *J. Comp. Neurol.* **216:**115–131.

Murakami, T., Ito, H., and Morita, Y., 1986a, Telencephalic afferent nuclei in the carp diencephalon, with special reference to fiber connections of the nucleus preglomerulosus pars lateralis, *Brain Res.* **382:**97–103.

Murakami, T., Fukuoka, T., and Ito, H., 1986b, Telencephalic ascending acousticolateral system in a teleost (Sebastiscus marmoratus), with special reference to the fiber connections of the nucleus preglomerulosus, *J. Comp. Neurol.* **247:**383–397.

Nelson, G. J., 1969, Origin and diversification of teleostean fishes, *Ann. N.Y. Acad. Sci.* **167:**18–30.

Nieuwenhuys, R., 1959, The structure of the telencephalon of the teleost Gasterosteurs aculeatus I en II, *K. Akad. Wet. Amsterdam Proc. Sect. Sci.* **62:**341–362.

Nieuwenhuys, R., 1962a, The morphogenesis and the general structure of the actinopterygian forebrain, *Acta Morphol. Neerl. Sci,* **5:**65–78.

Nieuwenhuys, R., 1962b, Trends in the evolution of the actinopterygian forebrain, *J. Morphol.* **111:**69–88.

Nieuwenhuys, R., 1963, The comparative anatomy of the actinopterygian forebrain, *J. Hirnforsch.* **6:**171–192.

Nieuwenhuys, R., 1964, Further studies on the general structure of the actinopterygian forebrain, *Acta Morphol. Neerl. Scand.* **6:**65–79.

Nieuwenhuys, R., 1966, The interpretation of the cell masses in the teleostean forebrain, in: *Evolution of the Forebrain* (R. Hassler and H. Stephan, eds.), Thieme Verlag, Stuttgart, pp. 32–39.

Nieuwenhuys, R., 1967a, Comparative anatomy of olfactory centres and tracts, *Prog. Brain Res.* **23:**1–64.

Nieuwenhuys, R., 1967b, Comparative anatomy of the cerebellum, *Prog. Brain Res.* **25:**1–93.

Nieuwenhuys, R., 1969, A survey of the structure of the forebrain in higher bony fishes (Osteichthyes), *Ann. N.Y. Acad. Sci.* **167:**31–64.

Nieuwenhuys, R., 1982, An overview of the organization of the brain of actinopterygian fishes, *Am. Zool.* **22:**287–310.

Nieuwenhuys, R., 1983, The central nervous system of the brachiopterygian fish, Erpetoichthys calabaricus, *J. Hirnforsch.* **24:**501–533.

Nieuwenhuys, R., and Nicholson, C., 1969, A survey of the general morphology, the fiber connections, and the possible functional significance of the gigantocerebellum of mormyrid fishes, in: *Neurobiology of Cerebellar Evolution and Development* (R. Llinás, ed.), University of Chicago Press, Chicago, pp. 107–134.

Nieuwenhuys, R., and Verrijdt, P. W. Y., 1983, Structure and connections of the telencephalon of the teleost fish Xenomystis nigri. II. The area dorsalis, *Acta Morphol. Neerl. Scand.* **21:**330.

Nieuwenhuys, R., Bauchot, R., and Arnoult, J., 1969, Le dévelopement du télencéphale d'un poisson osseux primitif, Polypterus senegalus Cuvier, *Acta Zool.* **50:**101–125.

Northcutt, R. G., 1981a, Localization of neurons afferent to the telencephalon in a primitive bony fish, Polypterus palmas, *Neurosci. Lett.* **22:**219–222.

Northcutt, R. G., 1981b, Evolution of the telencephalon in nonmammals, *Annu. Rev. Neurosci.* **4:**301–350.

Northcutt, R. G., and Braford, M. R., Jr., 1980, New observations on the organization and evolution of the telencephalon of actinopterygian fishes, in: *Comparative Neurology of the Telencephalon* (S. O. E. Ebbersson, ed.), Plenum Press, New York, pp. 41–97.

Northcutt, R. G., and Davis, R. E., 1983, Telencephalic organization in ray-finned fishes, in: *Fish Neurobiology,* Volume 2 (R. E. Davis and R. G. Northcutt, eds.), University of Michigan Press, Ann Arbor, pp. 203–236.

Oka, Y., and Ueda, K., 1981, Telencephalic afferents in the goldfish: An anterograde degeneration study, *Brain Res. Bull.* **7:**391–394.

Parent, A., 1986, *Comparative Neurobiology on the Basal Ganglia* (R. G. Northcutt, ed.), Wiley, New York.

Parent, A., and Northcutt, R. G., 1982, The monoamine-containing neurons in the brain of the garfish, Lepisosteus osseus, *Brain Res. Bull.* **9:**189–204.

Parent, A., Dube, L., Braford, M. R., Jr., and Northcutt, R. G., 1978, The organization of monoamine-containing neurons in the brain of the sunfish (Lepomis gibbosus) as revealed by fluorescence microscopy, *J. Comp. Neurol.* **182**(3)**:**495–516.

Prasada Rao, P. D., and Finger, T. E., 1984, Asymmetry of the olfactory system in the brain of the winter flounder, Pseudopleuronectes americanus, *J. Comp. Neurol.* **225:**492–510.
Rabl-Rückhard, H., 1882, Zur Deutung und Entwickelung des Gehirns der Knochenfische, *Arch. Anat. Physiol. Anat. Abt.* pp. 111–138.
Remane, A., 1956, *Die Grundlagen des Naturlichen Systems der Vergleichenden Anatomie und der Phylogenetik,* Theoretische Morphologie und Systematiek I, Akad. Verlag. Geest & Portig, Leipzig, p. 634.
Romer, A. S., 1962, *The Vertebrate Body,* Saunders, Philadelphia, p. 627.
Sakamoto, N., and Ito, H., 1982, Fiber connections of the corpus glomerulosum in a teleost, Navodon modestus, *J. Comp. Neurol.* **205:**291–298.
Schnitzlein, H. N., 1968, Introductory remarks on the telencephalon of fish, in: *The Central Nervous System and Fish Behaviour* (D. Ingle, ed.), University of Chicago Press, Chicago, pp. 97–100.
Sheldon, R. E., 1912, The olfactory tracts and centers in teleosts, *J. Comp. Neurol.* **22:**177–338.
Sheldon, R. E., and Brookover, C., 1909, The nervus terminalis in teleosts, *Anat. Rec.* **3:**257–259.
Springer, A. D., 1983, Centrifugal innervation of goldfish retina from ganglion cells of the nervus terminalis, *J. Comp. Neurol.* **214:**404–415.
Springer, A. D., and Gaffney, J. S., 1981, Retinal projections in the goldfish: A study using cobaltous-lysine, *J. Comp. Neurol.* **203:**401–424.
Stensiö, E. A., 1921, *Triassic Fishes from Spitzbergen,* Part 1, Vienna, p. 307.
Story, R. H., 1964, The olfactory bulbar formation and related nuclei of the paddlefish (Polyodon spathula), *J. Comp. Neurol.* **123:**285–298.
Studnicka, F. K., 1896, *Beiträge zur Anatomie und Entwicklungsgeschichte des Vorderhirns der Cranioten,* S.-B. Akad. Boehm. Wiss. Math. -Nat. Kl., Pt. 2.
Vanegas, H., and Ebbesson, S. O. E., 1976, Telencephalic projections in two teleost species, *J. Comp. Neurol.* **165:**(2)**:**181–195.
Vanegas, H., and Ito, H., 1983, Morphological aspects of the teleostean visual system: A review, *Brain Res. Rev.* **6:**117–137.
von Bartheld, C. S., and Meyer, D. L., 1986a, Central projections of the nervus terminalis in the bichir, Polypterus palmas, *Cell Tissue Res.* **244:**181–186.
von Bartheld, C. S., and Meyer, D. L., 1986b, Central connections of the olfactory bulb in the bichir, Polypterus palmas, reexamined, *Cell Tissue Res.* **244:**527–535.
von Bartheld, C. S., and Meyer, D. L., 1986c, Tracing of single fibers of the nervus terminalis in the goldfish brain, *Cell Tissue Res.* **245:**143–158.
von Bartheld, C. S., Meyer, D. L., Fiebig, E., and Ebbesson, S. O. E., 1984, Central connections of the olfactory bulb in the goldfish, Carassius auratus, *Cell Tissue Res.* **238:**475–487.
von Bartheld, C. S., Rickmann, M. J., and Meyer, D. L., 1986, A light- and electron-microscopic study of mesencephalic neurons projecting to the ganglion of the nervus terminalis in the goldfish, *Cell Tissue Res.* **246:**63–70.
Weston, J. K., 1937, Notes on the telencephalon of Mormyrus and Gnathonemus, *K. Akad. Wet. Amsterdam Proc. Sect. Sci.* **40:**894–904.
Wright, J. D., 1967, The telencephalon of the bichir, Polypterus, *Ala. J. Med. Sci.* **4:**252–273.
Wullimann, M. F., 1988, The tertiary gustatory center in sunfishes is not nucleus glomerulosus, *Neurosci. Lett.* **86:**6–10.

3

The Telencephalon of Sarcopterygian Fishes

R. NIEUWENHUYS and J. MEEK

1. Introduction

Crossopterygians are a distinctive and once widespread group of fishes, first known from rocks of Devonian age. They are characterized by the possession of two dorsal fins, the cosmoid structure of their scales and dermal bones, and well-formed fleshy-lobed paired fins. The internal skeleton of the latter is much concentrated so that only a single bone (humerus, femur) articulates proximally with the girdles and distally with two bones (radius and ulna, tibia and fibula). The latter are followed by the more variable distal fin supports. The most important distinguishing feature of these fishes is the division of the neurocranium into two parts, an anterior, ethmo-sphenoid portion and a posterior, otico-occipital portion with an intracranial joint between them. The large, persistent notochord inserts upon the postero-ventral margin of the ethmo-sphenoid portion of the skull (Thomson, 1969; Moy-Thomas and Miles, 1971; Andrews, 1973).

Although several different subdivisions of the crossopterygians have been presented (cf. Jessen, 1967; Thomson, 1969; Andrews, 1973), most taxonomists agree that this group comprises two major kinds of fishes, the Rhipidistia and the Actinistia or Coelacanthini. The former differ from the latter principally in the arrangement and number of the dermal bones of the head, and in having internal nostrils (choanae).

The crossopterygians were long believed to have become extinct by the end of Mesozoic time (some 70 million years ago), but in 1938 a strange fish, caught

---

R. NIEUWENHUYS and J. MEEK • Department of Anatomy and Embryology, University of Nijmegen, 6500 HB Nijmegen, The Netherlands.

off the coast of southeast Africa, proved to be a surviving actinistian crossopterygian, and was named *Latimeria chalumnae.* It lasted until 1952 before a second specimen was secured. Since that time some 80 other specimens have been obtained. All of these have been captured from the waters around the Comores, a small group of islands lying midway between the north tip of Madagascar and the African coast.

The lungfishes or Dipnoi form an extremely ancient group of fish that appeared in the lower Devonian and reached the zenith of its evolution in late Devonian and Carboniferous times (Moy-Thomas and Miles, 1971). A variety of fossil lungfishes have been described from geological formations all over the world, but in the recent fauna this group is represented by only six species. These are distributed over three genera, the African genus *Protopterus* with four species, the South American genus *Lepidosiren* with a single species, and the Australian genus *Neoceratodus,* also with one species. The Australian lungfish *Neoceratodus forsteri,* which closely resembles the dipnoans of early Mesozoic age, is almost certainly the most primitive of the three modern types. Its long, fusiform body is slightly compressed laterally, and is covered with large, rounded scales. The earliest lungfishes possessed leaf-shaped pectoral and pelvic fins with a strong axial skeleton and a proximal fleshy portion that is covered with scales. Similar paired fins are present in *Neoceratodus.*

In the genera *Protopterus* and *Lepidosiren* the body is eel-shaped, with small scales embedded in the skin and the paired fins are reduced to threadlike appendages.

There is a considerable reduction of bone in the internal skeleton, a feature that the dipnoans have in common with most other groups of fishes that have existed for a long time (Moy-Thomas and Miles, 1971).

The reduction of bone and other changes that have taken place during the evolution of the dipnoans, as for instance the replacement of several separate unpaired fins by a single continuous median fin seam, are often interpreted as indicating a return to the larval condition (paedomorphosis).

The crossopterygians or tassel-finned fishes and the dipnoans or lungfishes are frequently included in a common subclass, the Sarcopterygii or fleshy-finned fishes (Romer, 1962, 1969; Schultze, 1986). This classification is based on the fact that they share a number of derived characters such as the structure of their scales and paired fins. The sarcopterygians occupy an important position in the vertebrate kingdom since it is commonly believed that it is from this group of fish that all terrestrial vertebrates evolved. Most early evolutionists believed that dipnoans were on the direct line of evolution leading from fishes to the amphibianlike earliest land-living vertebrates, but at the beginning of the 20th century the view became dominant that most if not all of the resemblances between dipnoans and amphibians are convergent, and that the actual tetrapod ancestor has to be sought among the early rhipidistian crossopterygians. Many structural features, among them being the structure of the dermal skull, the vertebral column, and the paired fin skeleton, have been adduced in favor of this relationship (Schaeffer, 1969). According to this view, *Latimeria,* or as it is also named the coelacanth, although an actinistian rather than a rhipidistian crossopterygian, is among the recent fishes the closest relative of the tetrapods. It is important to note that recently Rosen *et al.* (1981) made a plea for the restoration of the late 19th century view that dipnoans and tetrapods are sister groups, supporting their argument by an impressive list of 20 derived characters (synapomorphies) shared by these two groups. They propose, furthermore, 14 derived characters, for uniting the actinistians, as represented by *Latimeria,* with the dipnoans and the tetrapods, in such a fashion that the former are the sister group of the latter two.

The aim of the present chapter is to provide a survey of the structure and fiber connections of the telencephalon of sarcopterygian fishes. The telencephalon of the crossopterygian *Latimeria* will be considered first and then the telencephalon of the dipnoans will be dealt with. The descriptions of the microstructure will be prefaced by some notes on the gross anatomical relations of the brains treated. Following the descriptions of the crossopterygian and dipnoan telencephala, the discussion section will address the question regarding the extent to which the important position of sarcopterygians at the transition between fishes and terrestrial vertebrates is reflected in the structural organization of their telencephalon.

2. Latimeria chalumnae

2.1. Gross Morphology of the Brain (Figs. 1 and 2)

The gross morphology of the brain of *Latimeria* has been studied by Millot and Anthony (1965), Nieuwenhuys *et al.* (1977), and Northcutt *et al.* (1978). From these studies it appeared that the brain is slender and elongated. There is a considerable discrepancy between the size of the brain and the size of the cranial cavity. Amazingly, the brain occupies less than 1% of the total endocranial volume (Millot and Anthony, 1965; Anthony and Robineau, 1976; Anthony, 1980). As regards its position, the brain confines itself entirely to the most caudal, i.e., the otico-occipital, division of the skull. Due to this caudal position and to the encephalo-cranial discrepancy, most cranial nerves have a very long endo-cranial trajectory and for the same reasons the olfactory peduncles and the hypophyseal duct are also extremely long (Fig. 1). The space between the brain and the inner surface of the skull is filled with a mass of adipose tissue (Millot and Anthony, 1965).

Figure 2 shows that the large subdivisions of the brain, i.e., the rhombencephalon, cerebellum, mesencephalon, diencephalon, and telencephalon, are readily identified in *Latimeria.* The rhombencephalon is well developed. The cerebellum comprises two extremely large auriculae and an unpaired, evaginated corpus cerebelli. The mesencephalon is small and does not show any marked differentiations on its surface. In the diencephalon, ventricular sulci mark the boundaries between the diencephalon, dorsal thalamus, ventral thalamus, and hypothalamus. Remarkably, the infundibular part of the hypothalamus is directed rostrally rather than caudally.

The telencephalon of *Latimeria* consists of four main parts, the telencephalon impar, the cerebral hemispheres, the corpora rostralia, and the olfactory bulbs.

The telencephalon impar is situated dorsal to the optic chiasm. It consists essentially of a pair of massive lateral walls, bounding the ventriculus impar. Dorsally, this cavity is closed by the most caudal part of the tela telencephali which, to a considerable extent, participates in the formation of the velum transversum. The large commissura anterior lies in the rostral part of the telencephalon impar. Immediately rostral to this commissure, the telencephalon impar passes into the cerebral hemispheres, which can be clearly divided into a thickened, dorsal, pallium and a thin-walled, ventral subpallium (Nieuwenhuys, 1965, 1969). Caudally, these two principal parts of the hemispheres are largely separated by a horizontally oriented slitlike extension of the ventricular cavity. More rostrally, these extensions expand into extensive lateral ventricles. The tela telencephali, which is attached dorsally to the pallium, continues medially and forms, with its counterpart, the septum ependymale (Fig. 3). On both sides this ependymal septum attaches below the medial edge of the subpallium. More caudally, however, the ependymal septum loses its connection with the subpallium and becomes a free-hanging median fold, allowing a communication between the ventricular systems of the two forebrain halves.

In front of the cerebral hemispheres the corpora rostralia are found. These structures, representing the most rostral parts of the pallium, are demarcated from the remainder of the pallium by deep transverse grooves. The tela telencephali and the septum ependymale continue bilaterally into a membrane that,

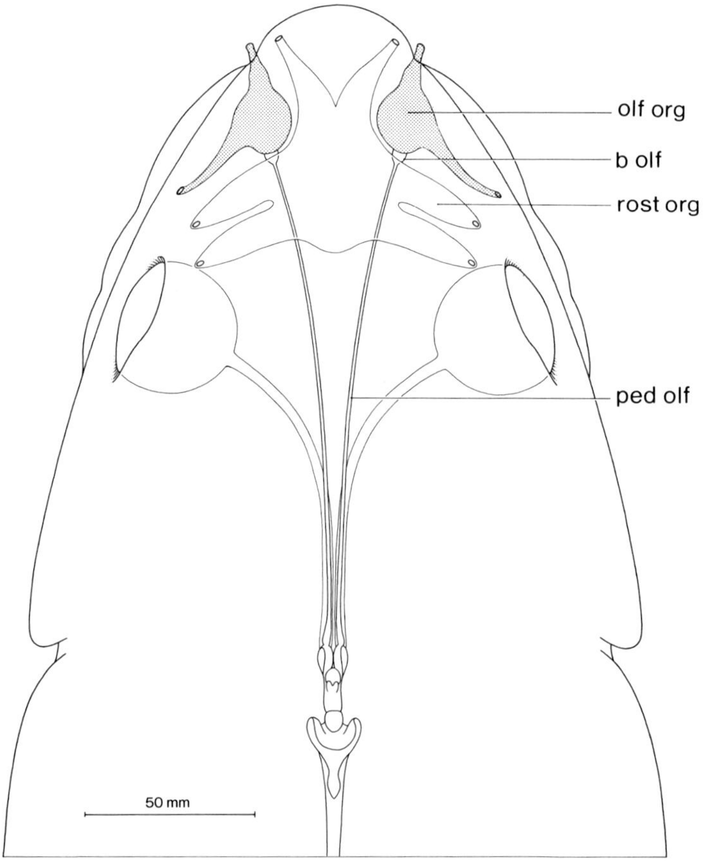

Figure 1. Diagram showing the brain of *Latimeria chalumnae* in position. Dorsal view. Partly based on plate II of Millot and Anthony (1965).

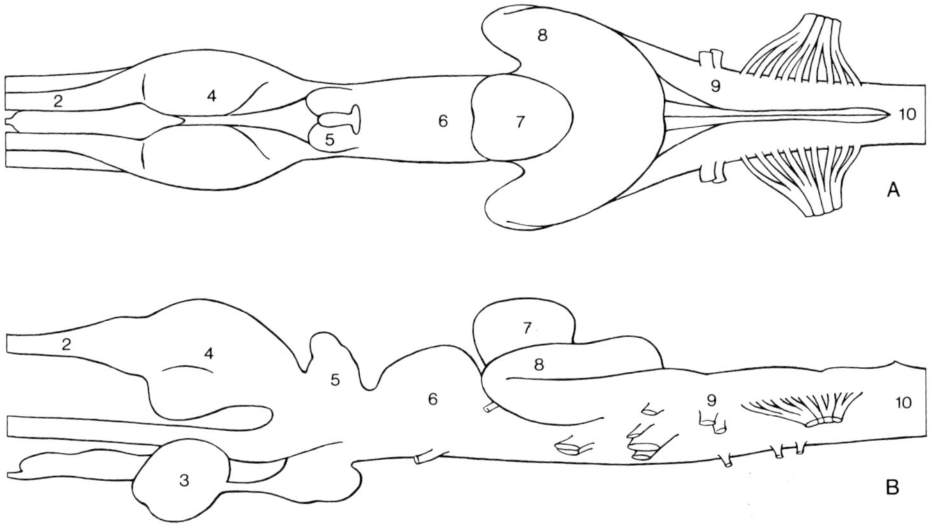

Figure 2. The brain of *Latimeria chalumnae*. (A) Dorsal view; (B) lateral view. Drawn from a Styrofoam model based on a transversely sectioned series. Based on Nieuwenhuys *et al.* (1977). In this figure as well as in Figs. 5 and 6, numbers indicate the following structures: 1, bulbus olfactorius; 2, pedunculus bulbi olfactorii; 3, hypophysis; 4, lobus hemisphericus telencephali; 5, diencephalon; 6, mesencephalon; 7, corpus cerebelli; 8, auricula cerebelli; 9, rhombencephalon; 10, medulla spinalis.

partly surrounding the corpus rostrale, is attached to its dorsolateral and ventrolateral aspects.

The hollow, ellipsoid olfactory bulbs are situated far apart and closely apposed to the olfactory organs. These bulbs are connected with the corpora rostralia by slender olfactory stalks, or peduncles, which may attain a length of 15–20 cm (Millot and Anthony, 1965; Fig. 1). The olfactory peduncles contain a wide canal by means of which the bulbar ventricles communicate with the ventricular cavities of the hemispheres.

2.2. Structure of the Telencephalon (Fig. 3)

The following description of the structure of the telencephalon of *Latimeria* is based on three previous studies by one of us (Nieuwenhuys, 1965, 1967, 1969).

The histological pattern of the olfactory bulb conforms to that found in other fish, with the following layers distinguishable:

1. A layer of olfactory nerve fibers
2. A glomerular layer
3. A rather irregular external cell zone of large, presumably mitral cells
4. An internal, granular cell layer

A sheet of secondary olfactory fibers could not be discerned. In the material studied, these fibers were seen only in the caudal part of the bulb, where they assemble to enter the olfactory peduncle. Passing backward and gradually converging, the olfactory peduncles reach the corpora rostralia which, as already mentioned, represent the most rostral part of the pallium. Histologically, the corpora rostralia appear to contain a poorly differentiated zone of central gray.

Behind the corpora rostralia the hemispheres expand rapidly and constitute a greatly thickened dorsal part and a thin-walled ventral part (Fig. 3). These parts may be considered as pallium and subpallium, respectively (see below). The subpallium, together with the ventral aspect of the pallium, surrounds a narrow but extensive lateral ventricle (Fig. 3). The medial and dorsal surfaces of the pallial body border another ventricular compartment, which is the caudal continuation of the consecutive ventricles of the bulb, the peduncle, and the corpus rostrale. Dorsally and medially, the wall of this compartment is formed by, respectively, the tela and the septum ependymale (Fig. 3). The septum ependymale attaches rostrally to the medial subpallial wall of septal nervous tissue. More caudally, however, the ependymal septum loses its basal attachment, thus allowing a communication between the ventricular systems of the two forebrain halves. The nervous septa become connected ventral to this communication by the lamina terminalis (Fig. 3). If we pass caudally from the level of Fig. 3, the lateral ventricles appear to diminish in size. At the level where the lamina terminalis thickens into the anterior commissure ridge, they are represented only by small lateral depressions and they disappear completely somewhat farther backward. The hemispheres merge at this point with the telencephalon impar.

Surveying the highly remarkable general relations just discussed, and assuming that the telencephalon of *Latimeria*, as that of all other vertebrates, passes through a simple vesicle stage, the following inferences and comparisons may be made:

1. In the ventral (subpallial) part of the forebrain of *Latimeria* a ventrolaterally and rostrally directed "simple" evagination has taken place (Fig. 3). In this respect the subpallium of *Latimeria* conforms to the corresponding region of the chondrichthian, dipnoan, and amphibian forebrains.
2. The dorsal (pallial) part of the telencephalon of *Latimeria* has not participated in the evagination just mentioned. This appears from the fact that the dorsal membranous connection of the nervous walls has widened and is attached laterally. Furthermore, the latimerian pallium is greatly thickened and is expanded intraventricularly (Fig. 3). These are all characteristics reminiscent of the condition found in the everted pallial parts of the telencephalon of actinopterygians.
3. In addition to the processes mentioned, there has been an extensive rostrally and dorsally directed "mixed" evagination in the telencephalon of *Latimeria*, i.e., an evagination in which both the neural and membranous walls have participated. This process has given rise to a complete ependymal septum rostrally, caudally continuing into a free-hanging fold. Although a rather extensive mixed evagination also occurs in the forebrain of the dipnoan *Neoceratodus* (see below), the condition found in *Latimeria* most closely approaches that found in polypteriforms, chondrosteans, and holosteans.

The subdivision of the cerebral hemispheres of *Latimeria* into a dorsal pallium and a ventral subpallium is primarily based on the positional relations of these two parts, but there is also histological evidence for it, as follows:

There is a clear structural difference between the dorsal and ventral areas, the cells in the former being somewhat larger than those in the latter. The two parts are separated by a fairly distinct zona limitans (Fig. 3). This criterion was considered by Johnston (1911a,b) and Holmgren (1922) to be very important in delimiting the pallial from the subpallial region in fish.

The cell bodies in the pallium of *Latimeria* are evenly scattered through the entire width of the thickened wall. In the intermediate and caudal parts of the hemispheres the pallial cells tend to form three separate fields: dorsomedial, ventromedial, and lateral (Fig. 3).

The thin-walled subpallium can be divided into lateral, ventral, and medial areas, which are most likely homologous to the corpus striatum, the tuberculum olfactorium, and the septum, respectively, of chondrichthians, dipnoans, and amphibians. These areas show a very simple histological organization (Fig. 3). The striatum and the olfactory tubercle contain a narrow layer of periventricular cells. These two areas can be clearly delimited from each other only in the caudal part of the telencephalon, where the striatal cell layer is wider than that of the olfactory tubercle. A particularly narrow zone of subependymal cells is found in the septum, but this area contains, in addition, a group of migrated cells that represents the medial septal nucleus. In the more superficial zone of the other subpallial areas, only sparsely scattered cells are present. There is no trace of a cortex tuberculi olfactorii or area superficialis basalis.

The cerebral hemispheres and the telencephalon impar are connected by a transitional area in which the large anterior commissure is situated. In the dor-

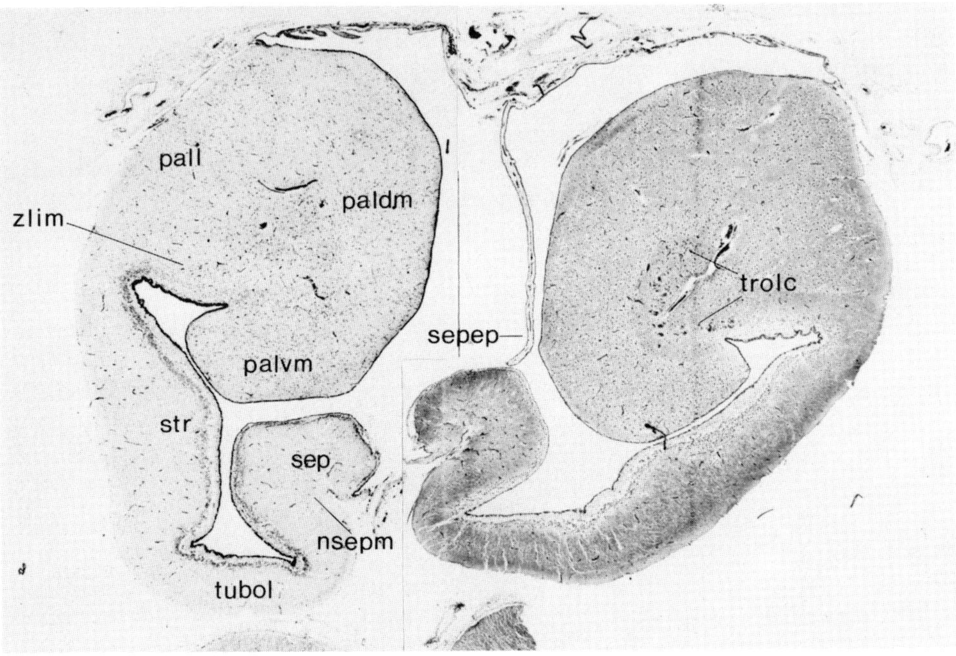

Figure 3. Transverse section through the cerebral hemispheres of *Latimeria chalumnae*. The left half represents a photomicrograph of a Nissl section, the right half that of the adjacent Bodian section.

solateral part of this area the caudal remnants of the lateral and dorsomedial pallial fields are found. The most caudal vestige of the lateral ventricle, which is situated just dorsal to the anterior commissure, is surrounded by the nucleus supracommissuralis, a periventricular cell group that is rostrally continuous with the striatum. A group of fairly large cells, which lies embedded in the decussating fibers of the basal forebrain bundle, constitutes the nucleus commissurae anterioris.

The telencephalon impar consists simply of a pair of massive lateral walls, which are dorsally connected by the caudal part of the wide tela telencephali (for illustrations see Nieuwenhuys, 1965). It is traversed by two large fiber systems, the stria medullaris and the fasciculus basalis telencephali.

The telencephalon impar contains two cell masses, the nucleus preopticus inferior and the nucleus preopticus superior. The nucleus preopticus inferior lines the ventral part of the ventriculus impar. Rostrally and rostroventrally it consists of densely packed, small cells, but dorsocaudally it contains numerous medium-sized elements. Thus, the caudal portion of the nucleus preopticus inferior can be subdivided into a magnocellular and a parvocellular part. The dorsal part of the telencephalon impar contains a wide, diffuse periventricular layer of small cells. More laterally, however, between the stria medullaris and the basal forebrain bundle, a mass of medium-sized cells is found which may be called the nucleus preopticus superior.

The morphological relationships in the dorsal part of the telencephalon impar show a striking similarity to those in the corresponding regions of lungfish. Rudebeck (1945), in his extensive study of the forebrain of *Protopterus*, has presented evidence that the dipnoan nucleus preopticus superior is homologous to the caudal part of the strio-amygdaloid complex of urodeles. A similar interpretation seems to be justified for the nucleus preopticus superior of *Latimeria*. This latter nucleus forms the caudalmost part of a longitudinal zone that rostrally, via the supracommissural nucleus, continues into the striatum. The same holds true for the cell mass in the urodele forebrain that was designated as amygdala by Herrick (1927, 1933).

2.3. Fiber Connections of the Telencephalon

The fiber connections of the telencephalon of *Latimeria* have been studied by Nieuwenhuys (1965, 1967, 1969) and by Kremers (1981). However, it should be emphasized that, due to the rather poor state of preservation of the material available, the analysis of these authors has remained far from complete. They distinguished the following three major fiber systems: the olfactory tracts, the stria medullaris, and the basal forebrain bundle.

Fibers originating in the olfactory bulb run as a compact tract through the olfactory peduncle to the corpus rostrale, where many fibers terminate. Other fibers continue caudally and constitute, together with fibers originating in the corpus rostrale, the tractus olfactorius centralis, a fiber system that, as its name implies, occupies a central position in the pallium (Figs. 3, 4). In the pallium numerous small fascicles leave the tractus olfactorius in a dorsocaudal direction, to constitute the tractus olfactorius habenulae dorsalis (Fig. 4). This tract joins the stria medullaris at the level of transition of the hemispheres into the telencephalon impar. The larger part of the tractus olfactorius centralis runs caudally as a separate bundle through the hemispheres and the telencephalon impar, gradually diminishing in size. During its course through the pallium and the nucleus preopticus superior the central olfactory tract issues fibers into these structures. The caudalmost fibers of the central olfactory tract may be designated as the tractus olfactorius habenulae ventralis. These fibers join the stria medullaris in the rostral part of the diencephalon. Sagittal sections give the impression that by way of the fiber system last mentioned a number of olfactory tract fibers pass uninterruptedly toward the habenula.

The stria medullaris consists of three components. Two of these, the tractus olfactorius habenulae dorsalis and the tractus olfactorius habenulae ventralis, have already been dealt with. The third component arises dorsally in the pallium and runs caudally, slightly ventromedial to the taenia. In the telencephalon impar this tractus pallio-habenularis unites with the tractus olfactorius habenulae dorsalis. After having taken up the tractus olfactorius habenulae ventralis, the stria medullaris passes through the pars dorsalis thalami to the epithalamus.

The large fasciculus basalis telencephali (Fig. 4) connects the various subpallial regions with the dorsal thalamus, the ventral thalamus, the hypothalamus, and the tegmentum mesencephali. It comprises a medial and a lateral component. The thin fibers of the medial component assemble from the septum and presumably also from the tuberculum olfactorium, and pass in the ventromedial part of the bundle to the hypothalamus and the area around the tuberculum posterius. The much coarser fibers of the lateral component originate mainly from the striatum. In the rostral part of the diencephalon this striatofugal component splits up into two bundles: one arching upward to the dorsal thalamus, the other coursing ventrocaudally to the ventral thalamus and the tegmentum mesencephali. Numerous fibers of the basal forebrain bundle decussate through the anterior commissure.

If we survey the fiber relations in the telencephalon of *Latimeria* there appear to be two remarkable points. The first is that olfactory fibers passing to the subpallium were not observed and the second is that, whereas the subpallium by means of the basal forebrain bundle is strongly connected with the thalamus, the hypothalamus, as well as the tegmentum mesencephali, the only detectable caudal connection of the large, massive pallium is with the epithalamus via the

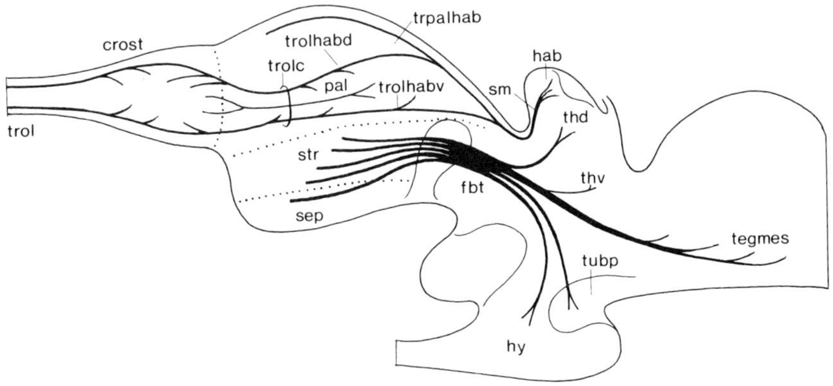

Figure 4. Diagram showing the fiber systems in the telencephalon of *Latimeria chalumnea*.

stria medullaris. The most parsimonious explanation for these peculiar relations seems to be that, due to the poor fixation of the material, the analysis of the fiber connections in the telencephalon of *Latimeria* has remained far from complete. Comparison with sections of the telencephalon of other fishes suggests that the thin-fibered, diffusely arranged pathways were not impregnated. It is conceivable that an olfactory system consisting of such fibers, comparable to e.g. the ventral olfactory tract of lungfishes (Holmgren and van der Horst, 1925; Rudebeck, 1945), or to the experimentally verified subpallial olfactory projection in sharks (Ebbesson and Heimer, 1970; Smeets, 1983), fans out in the subpallium of *Latimeria* and it is likewise conceivable that a thin-fibered bundle, comparable to e.g. the tractus pallii of cartilaginous fishes (Smeets *et al.*, 1983; Smeets and Boord, 1985) and primitive actinopterygian fishes (Johnston, 1911a; Nieuwenhuys, 1983), connects the pallium of *Latimeria* with the hypothalamus.

3. Dipnoans

3.1. Introductory Notes

The CNS of the lungfishes has attracted the attention of a considerable number of authors. The brain as a whole has been dealt with by Burckhardt (1892; *Protopterus annectens*), Bing and Burckhardt (1905; *Neoceratodus*), Holmgren and van der Horst (1925; *Neoceratodus*), Gerlach (1933; *P. annectens*), and Northcutt (1986; *Lepidosiren, Neoceratodus, P. aethiopicus, P. annectens*). The structure of the telencephalon, or parts thereof, has been studied by Elliot Smith (1908; *Lepidosiren*), Holmgren (1922), Rudebeck (1945; *P. annectens*), Nieuwenhuys and Hickey (1965; *Neoceratodus*), Schnitzlein (1966; *Protopterus* sp.), Schnitzlein and Crosby (1967; *Protopterus* sp.), Nieuwenhuys (1967; *P. dolloi, Neoceratodus*), Nieuwenhuys (1969; *P. dolloi*), Capanna and Clairambault (1973), Clairambault and Capanna (1973; *P. dolloi*), Clairambault *et al.* (1974a,b; *P. dolloi*), and Reiner and Northcutt (1987; *P. annectens*).

The following survey of the gross morphology of the dipnoan brain and of the structure and connections of the telencephalon is primarily based on an analysis of histological material of *Lepidosiren, Neoceratodus,* and *P. dolloi.* The literature cited above will be taken into consideration.

3.2. Gross Morphology of the Brain (Figs. 5 and 6)

The brain of *Lepidosiren* is elongated. The width of the rhombencephalon and the telencephalon exceeds that of the very slender intermediate portion of the brain. The rhombencephalon is strongly reminiscent of that of urodele amphibians. However, the alar plate is better developed in dipnoans than in urodeles. Rostrally the rhombencephalic alar plate curves around a lateral recess of the fourth ventricle and continues into a horizontally oriented lamella of tissue, which fuses in the median plane with its counterpart of the opposite side. The rostrolateral wall of the lateral recess represents the auricula cerebelli. The corpus cerebelli is constituted by the fused, horizontally oriented lamellae of tissue and caps the rostral part of the fourth ventricle.

The mesencephalon is small and has largely maintained the early embryonic tubelike condition. A remarkable feature of the midbrain is that its dorsocaudal part extends over the corpus cerebelli.

The diencephalon has only a limited rostrocaudal extent. Its ventricular surface shows three obliquely oriented sulci which in most places define the boundaries between the epithalamus, dorsal thalamus, ventral thalamus, and hypothalamus. The diencephalon is covered by a membranous structure, the tela diencephali, which continues into the tela telencephali, i.e., the membranous roof of the telencephalon impar. Immediately rostral to the foramen interventriculare of Monro the tela telencephali passes over into the lamina terminalis.

The telencephalon consists of three main parts, the telencephalon impar, the cerebral hemispheres, and the olfactory bulbs. The telencephalon impar is situated immediately in front of the optic chiasm. It consists of two relatively thin, slightly diverging, lateral walls which, together with the tela telencephali, surround the ventriculus impar, i.e., the rostral continuation of the third ventricle. The commissura anterior, which represents a thickened portion of the lamina terminalis, is situated in the rostral part of the telencephalon impar. It is at the level of this commissure that the telencephalon impar passes over into a pair of large, rostrally extending, thin-walled cerebral hemispheres. The foramina of Monro which gives access to the telencephalic lateral ventricles are situated directly dorsal to the anterior commissure. The well-developed olfactory bulbs are attached to the rostrodorsal aspect of the telencephalic hemispheres. The fibers of the large olfactory nerves spread in the superficial zone of the olfactory bulbs.

The brains of *Lepidosiren* and *Protopterus* resemble each other closely, but the pattern of that of *Neoceratodus* deviates in several respects. The most salient of these are (cf. Figs. 5 and 6):

1. The auriculae cerebelli are smaller, but the corpus cerebelli is much larger than in *Lepidosiren* and *Protopterus*. The structure last mentioned is not covered by the tectum, and its curved wall caps the rostral part of the

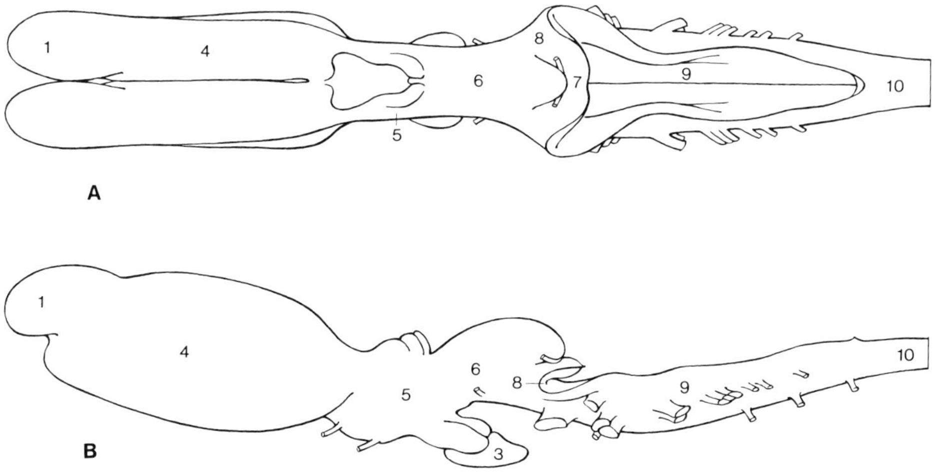

Figure 5. The brain of *Lepidosiren paradoxa*. (A) Dorsal view; (B) lateral view. From Thors and Nieuwenhuys (1979). See Fig. 2 for key.

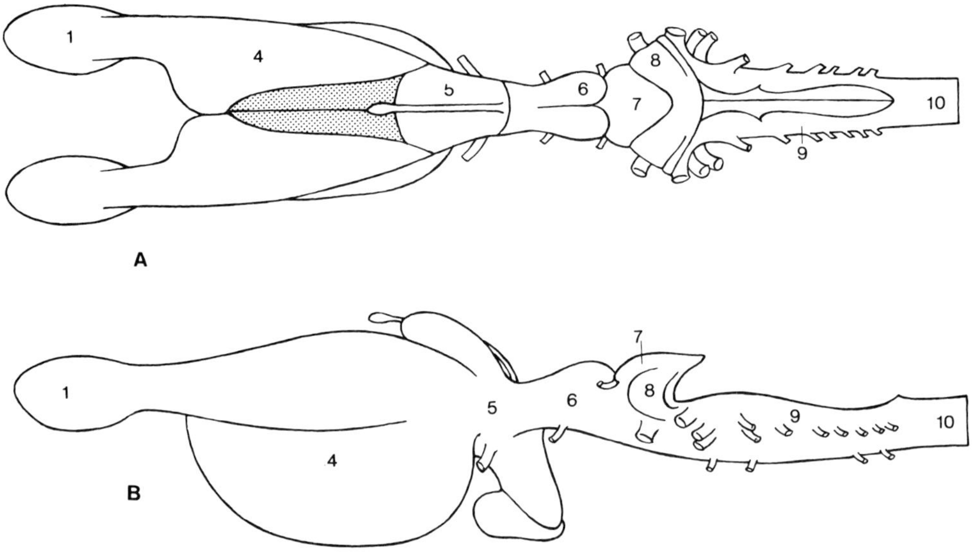

Figure 6. The brain of *Neoceratodus forsteri.* (A) Dorsal view; (B) lateral view. Drawn from a Styrofoam model based on a transversely sectioned series. See Fig. 2 for key.

fourth ventricle. The middle portion of the corpus is thickened and forms a median cerebellar ridge (Holmgren and van der Horst, 1925).

2. The olfactory bulbs are connected to the cerebral hemispheres through short hollow peduncles instead of being sessile.
3. Whereas in *Lepidosiren* and *Protopterus* the cerebral hemispheres consist entirely of nervous tissue, in *Neoceratodus* a part of the medial hemisphere wall is made up of an ependymal membrane (i.e., the stippled area in Fig. 6). Rostrally, these membranous structures bridge only a narrow slit in the nervous walls, but caudally, the distance between their lines of attachment becomes progressively larger. The structures in question together constitute a strongly folded septum ependymale (Fig. 8), which at the level of the foramen of Monro passes over into the, also strongly folded, tela of the telencephalon impar.

3.3. Structure of the Telencephalon (Figs. 7–10)

The telencephalon of dipnoans, like that of most other vertebrates, can be subdivided into a rostral, an intermediate, and a caudal part. The rostral portion consists of two olfactory bulbs, the intermediate region is composed of a pair of large hemispheres, and the relatively small caudal part represents a simply built telencephalon impar. Each of the three structures mentioned surrounds a part of the telencephalic ventricular system.

Histologically, the dipnoan olfactory bulbs show a remarkably definite laminated pattern (Holmgren and van der Horst, 1925; Gerlach, 1933; Rudebeck, 1945; Nieuwenhuys and Hickey, 1965). From outward to inward the following layers can be distinguished: (1) a layer of primary olfactory fibers, (2) a glomerular layer, (3) a layer of mitral cells, (4) a fiber layer, which caudally continues into the secondary olfactory tracts, (5) a wide granular layer, and (6) a subependymal fiber plexus. The inner part of the layer of olfactory nerve fibers and the glomerular zone contain numerous scattered neurons, which Rudebeck (1945) designated as interglomerular cells. He observed that the axons of these elements assemble in the dorsocaudal part of the bulb, forming a separate olfactory pathway: the tractus olfactorius externus.

The dipnoan cerebral hemispheres are clearly of the evaginated type and the evagination has taken place predominantly in a rostral direction. The junctions of the lateral ventricles and the ventricular cavity of the telencephalon impar, i.e., the foramina of Monro, are situated caudally. It has already been mentioned that the hemispheres in *Protopterus* and *Lepidosiren* consist entirely of nervous tissue, whereas in *Neoceratodus* part of the medial hemisphere wall is made up of an ependymal membrane, a structure known as the septum ependymale (cf. Figs. 7 and 8). This peculiar condition is due to the fact that in *Neoceratodus* the evagination process which gives rise to the formation of the hemispheres is not confined to the neural lateral walls of the forebrain anlage,

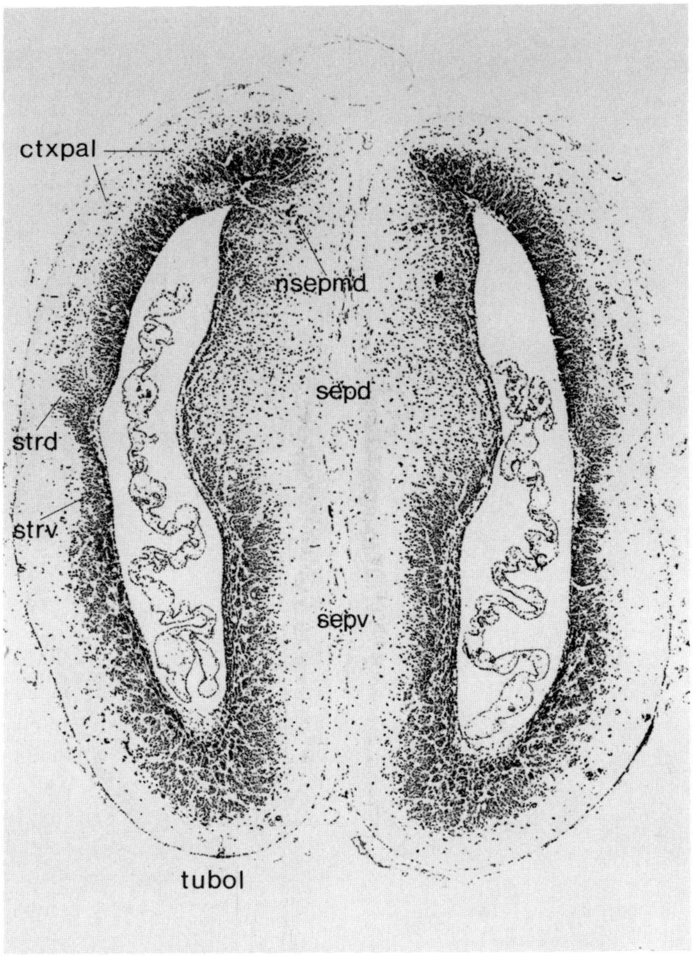

Figure 7. Transverse section through the cerebral hemispheres of *Lepidosiren paradoxa*. Bodian technique.

but also extends over the junction of these lateral walls with the epithelial lamina supraneuroporica (Rudebeck, 1945). Through this "mixed" evagination the membranous roof of the forebrain vesicle becomes in *Neoceratodus* involved in the formation of the medial and dorsal walls of the hemispheres.

The telencephalon of dipnoans is, like the other parts of the brain, characterized by the presence of a continuous zone of periventricular gray (Figs. 6 and 7), which shows only slight cytoarchitectonic differences. In his classical study on the telencephalon of *Lepidosiren*, Elliot Smith (1908) divided the hemispheres into four regions, a dorsolateral pallium, a lateral striatum, a ventral tuberculum olfactorium, and a medial paraterminal body or septum. This subdivision has been adopted by most later students of the dipnoan forebrain (Holmgren, 1922; Holmgren and van der Horst, 1925; Kuhlenbeck, 1929; Gerlach, 1933;

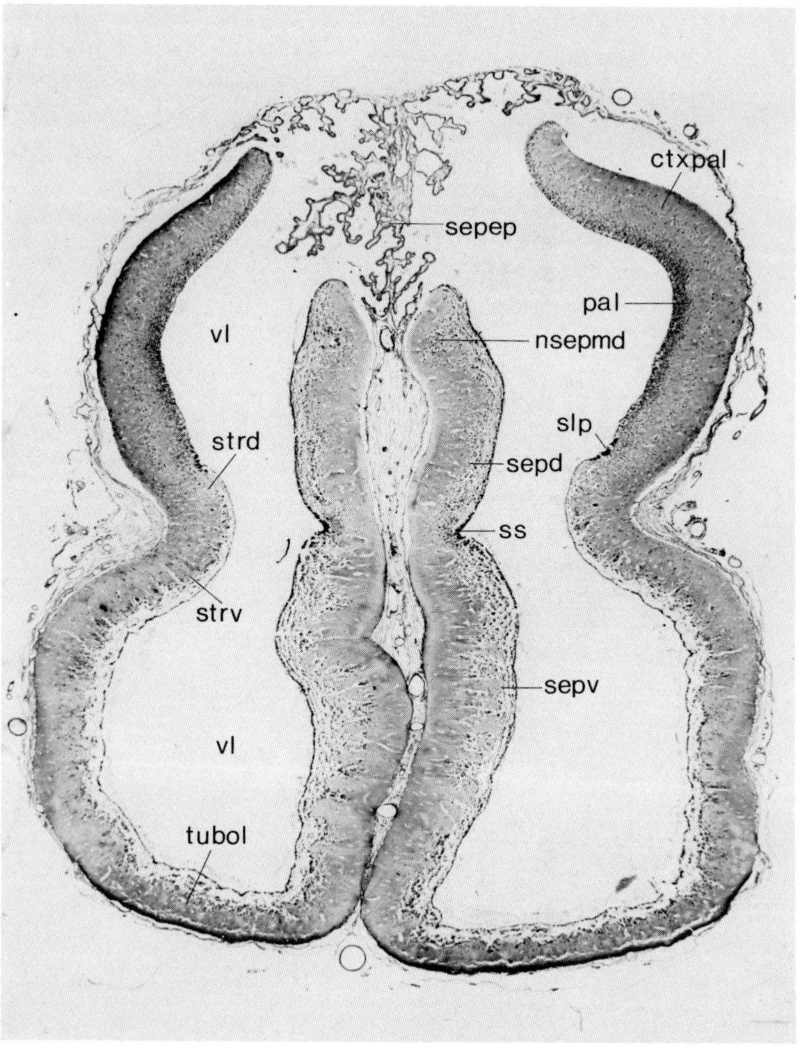

Figure 8. Transverse section through the cerebral hemispheres of *Neoceratodus forsteri*. Bodian technique.

Rudebeck, 1945; Nieuwenhuys and Hickey, 1965) and will also be followed in the present account.

Schnitzlein and Crosby (1967), Clairambault and Capanna (1973), Northcutt (1986), and Reiner and Northcutt (1987) have expressed the opinion that in *Protopterus* the dorsal parts of the medial hemisphere walls do not form part of the septum, but rather represent the medial parts of the pallium. This interpretation will be dealt with below.

The pallium is relatively small in dipnoans. In *Protopterus* and *Lepidosiren* its medial boundary is marked by the angulus dorsalis of the lateral ventricle (Fig. 7), whereas in *Neoceratodus* this boundary is formed by the line of attachment of the septum ependymale (Fig. 8). The boundary between the pallium and the corpus striatum is, throughout most of the hemispheres, marked by a sulcus limitans pallii (Fig. 8) that is heavily ependymated in places. The dipnoan pallium contains two cell layers, an inner zone of small, densely packed, periventricular cells, and an outer layer of more widely spaced, larger elements (Figs. 7–9). The latter is clearly detached from the central gray and, therefore, may be considered to represent a true cerebral cortex. Holmgren and van der Horst (1925) subdivided the pallium of *Neoceratodus* on the basis of local differences in the width of the periventricular cell layer into three fields: a (morphologically) dorsal hippocampal pallium with a thin periventricular layer, an intermediate general pallium with a much thicker layer of ventricular cells, and a ventral pyriform pallium in which the ventricular zone is again thin. We consider it likely that these variations in the width of the periventricular cell layer are caused

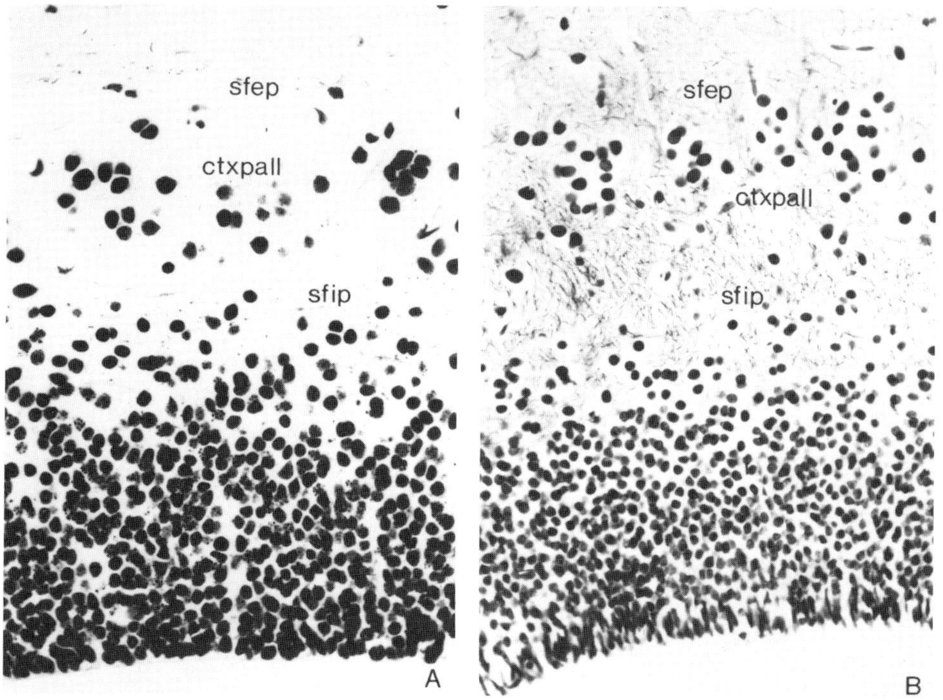

Figure 9. Transverse sections through the pallial part of the cerebral hemispheres of (A) *Lepidosiren paradoxa* (Klüver-Barrera stain) and (B) *Neoceratodus forsteri* (Bodian stain).

simply by the local curvature of the wall and, therefore, do not form a sound basis for such a subdivision.

The Golgi material of Rudebeck (1945) revealed that in *Protopterus* the periventricular pallial zone contains three different kinds of cells, which we designate here as type 1, type 2, and type 3 cells, whereas the migrated cortical layer consists of a single type, type 4 (Fig. 10).

The type 1 cells are very common and have a very short stem dendrite, which gives rise to a number of peripherally extending very spiny branches. Most of these branches extend into the superficial fiber layer of the pallium.

The rare type 2 cells are characterized by a rather long principal dendrite, which arborizes in a fairly widespread area. The main dendrite and its branches are very slender, and the latter end in the superficial pallial layer and among the cortical cells. The axons of these elements could be traced over only a short distance; they are directed toward the ventricular surface.

The type 3 cells are confined to the external parts of the periventricular layer. These elements closely resemble the elements in the cortical layer and represent according to Rudebeck (1945) cortical elements left behind in their matrix layer.

The cortical layer is composed of rather large elements, which in many places are gathered into clusters. Their long, smooth dendrites spread mainly tangentially (Fig. 10). Branches of these dendrites arborize in the superficial pallial fiber zone. The cortical elements contribute axons to both the external and the internal pallial fiber layer.

In *Lepidosiren* and *Protopterus* the rostral one third of the pallium contains a sheet of submeningeally located "supracortical" cells. According to Rudebeck (1945) these cells probably represent a bulbar structure. He observed that their dendrites arborize into glomeruluslike entanglements.

Clairambault and Capanna (1973) interpreted the layers of migrated cells in the pallium in a way that differs radically from the one presented above. They held that the layer designated here as pallial cortex, throughout the length of the

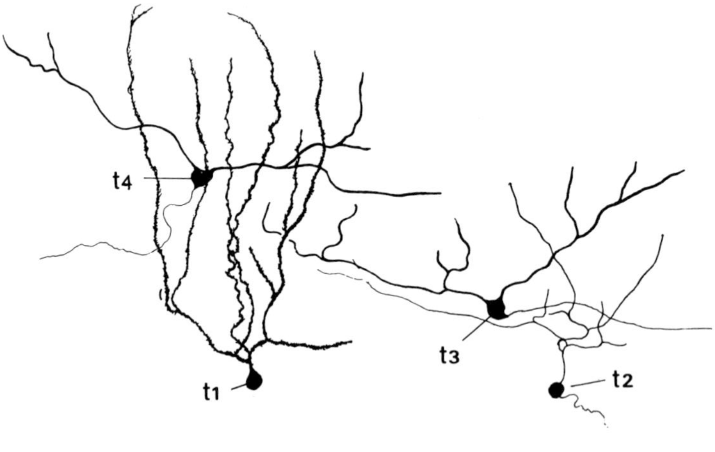

Figure 10. Neurons in the pallium of *Protopterus annectens* as seen in transversely sectioned Golgi preparations. Combined from Fig. 23A and B of Rudebeck (1945). t1–t4, cell types described in the text.

cerebral hemispheres is composed of mitral cells and that the supracortical cells are in fact glial elements.

The *corpus striatum* (i.e., the nucleus olfactorius lateralis of Holmgren, 1922; Holmgren and van der Horst, 1925; and Rudebeck, 1945) forms a part of the lateral hemisphere wall. It is bounded dorsally by the pallium and ventrally by the tuberculum olfactorium. Caudally it grades into the nucleus preopticus superior. Its most rostral part is situated ventral to the caudal portion of the olfactory bulb. Throughout most of its extent the corpus striatum is characterized by a rather thin layer of periventricular gray (Figs. 7 and 8). The boundary with the tuberculum olfactorium is in *Neoceratodus* and in the rostral parts of the hemispheres of *Lepidosiren* and *Protopterus* indistinct. However, in the caudal portion of the hemispheres of the two genera last mentioned, the beginning of the cortex tuberculi olfactorii indicates the ventral boundary of the striatum.

As in urodeles the corpus striatum can be divided into a pars dorsalis and a pars ventralis. Particularly in the rostral and intermediate parts of the hemispheres the pars dorsalis can be easily recognized as a compact, dorsolaterally extending, cell plate (Fig. 7). Reiner and Northcutt (1987) designated what is here called the dorsal part of the striatum as the intercalated nucleus and considered it unclear whether this structure is a pallial or a subpallial cell group. As will be discussed below, Reiner and Northcutt (1987) believe that the striatum extends much more ventrally than indicated in the present account.

Elliot Smith (1908) was the first to comment on the enormous development of the *tuberculum olfactorium* in dipnoans. This structure occupies the entire ventral wall of the hemispheres and extends for some distance into the medial and lateral walls. In *Protopterus* and *Lepidosiren* the rostral portion of the olfactory tubercle projects under the olfactory bulb, whereas its caudal parts expand laterally to the telencephalon impar.

In the tuberculum olfactorium, an inner layer of small cells and an outer layer of somewhat larger elements can be distinguished (Fig. 7). The caudal part of the outer layer is in *Protopterus* and *Lepidosiren* clearly detached from the inner layer, thus forming a cortex tuberculi olfactorii.

In a recent immunohistochemical study, Reiner and Northcutt (1987) considered the areas interpreted here as the ventral part of the striatum and the olfactory tubercle as a single entity, which they designated as the lateral subpallium. They observed that this entity gives rise to a descending substance-P-containing pathway, which terminates in the most ventral part of the tegmentum mesencephali, and that this descending projection is reciprocated by axons originating from tyrosine hydroxylase-positive neurons situated in the same midbrain area. Reiner and Northcutt (1987) considered the descending substance-P-containing fibers and the ascending tyrosine hydroxylase-containing projection comparable to, respectively, the substance-P-containing strionigral pathway and the dopaminergic nigrostriatal projection of amniotes, and arrived at the conclusion that the dipnoan lateral subpallium is homologous to the striatum of land vertebrates. They also produced some immunohistochemical evidence for the presence of a "pallidal" region in the caudal part of the lateral subpallium of their description. In our opinion it is possible that the ventral part of the striatum and the olfactory tubercle of dipnoans together represent the striatum sensu lato of "higher" vertebrates, i.e., include both the dorsal and the ventral striatal moieties as described for mammals (Heimer and Wilson, 1975; Heimer *et al.*, 1982). However, we maintain that the striatum and the olfactory tubercle as delineated by Elliot Smith (1908) and numerous later authors in the dipnoan telencephalon are separate entities. They show distinct structural differences and, although they may well share the two connections described by Reiner and Northcutt, they might well show some differences in their remaining fiber connections (see below).

It has already been mentioned that the dipnoan hemispheres are strongly evaginated in the rostral direction, but it should be added that they also greatly expand ventrally. The latter process results in the enlarged development of the olfactory tubercle but also causes the paraterminal body, or *septal region,* to become exceptionally high (Figs. 7, 8). The septum contains a well-developed layer of periventricular cells, which in most places are more loosely arranged than those in the inner layer of the olfactory tubercle. Cytoarchitectonic differences allow for a subdivision of the septal periventricular gray into a pars dorsalis and a pars ventralis (Figs. 7, 8). A ventricular groove, the sulcus septalis, marks the boundary between these two parts in the caudal portion of the hemispheres (Holmgren and van der Horst, 1925; Rudebeck, 1945). In addition to the zone of periventricular gray, the septal region contains three masses of migrated cells, which occupy dorsal, intermediate, and caudoventral positions, respectively. We consider it probable that these nuclei together represent the medial septal nucleus of other vertebrates, hence we designated them as the dorsal, intermediate, and ventral parts of the medial septal nucleus (Figs. 7, 8). Northcutt (1986; see also Clairambault and Capanna, 1973; Schnitzlein and Crosby, 1967; Reiner and Northcutt, 1987) has expressed the opinion that the region interpreted here as the dorsal part of the septum represents in *Protopterus* the medial part of the pallium, but he left open the possibility that in *Neoceratodus* a medial pallial component is lacking. Our comparative studies (Nieuwenhuys and Hickey, 1965; Nieuwenhuys, 1969) have shown that: (1) all telencephalic parts present in *Protopterus* and *Lepidosiren* can also be recognized in *Neoceratodus,* (2) the dorsal part of the medial hemisphere wall of *Neoceratodus* clearly forms part of the septum, and (3) this pars dorsalis septi of *Neoceratodus* is homologous to the dorsomedial telencephalic quadrant in *Protopterus* and *Lepidosiren.* The highly characteristic nucleus medialis septi, pars dorsalis occupies a corresponding topographical and topological position in the telencephalon of all dipnoans (Figs. 7, 8).

The area directly dorsal to the commissura anterior contains a zone of undifferentiated periventricular gray, which we have termed the nucleus supracommissuralis.

The *telencephalon impar,* i.e., the caudalmost part of the dipnoan forebrain, contains two periventricular cell masses, the nucleus preopticus inferior, or nucleus preopticus proper, and the nucleus preopticus superior.

The nucleus preopticus inferior surrounds the preoptic recess and extends caudally in the suprachiasmatic region. As in other vertebrates it is differentiated into a parvocellular and a magnocellular nucleus. The former consists of densely crowded, small cells; the latter is characterized by the presence of larger elements with relatively pale nuclei (Rudebeck, 1945).

The nucleus preopticus superior passes rostrally, without any distinct boundary, over into the caudal part of the striatum, but its caudal boundary with the ventral thalamus is cytoarchitectonically distinct. Its cells are somewhat less densely packed than those in the nucleus preopticus inferior and show a laminar arrangement. With Rudebeck (1945) we consider it likely that the entity formed by the striatum (Rudebeck, 1945) nucleus olfactorius lateralis and the nucleus preopticus superior is homologous to the strio-amygdaloid complex of urodele amphibians.

Before closing this survey of the cell masses in the telencephalon of lungfishes, it should be noted that Schnitzlein and Crosby (1967) have put forward an interpretation of the parts of the forebrain of *Protopterus* that differs in many respects from the one presented here. In order to give some impression of the degree of disagreement, we mention next a number of the structures claimed by these authors, with (in parentheses) the interpretation given to these structures in the present account: (1) primordial anterior amygdaloid nucleus + primordial basolateral nucleus + primordial corticomedial nucleus + primordial general pallium (cortex pallii), (2) hyperstriatum (corpus striatum, pars dorsalis), (3) neostriatum (corpus striatum, pars ventralis), (4) paleostriatum (lateral part of the tuberculum olfactorium), (5) nucleus lateralis septi (medial part of the tuberculum olfactorium), (6) gyrus dentatus (pars ventralis septi), (7) cornu ammonis (pars dorsalis septi). We are under the impression that the interpretations of Schnitzlein and Crosby are based largely on a forced comparison of this dipnoan forebrain with that of amniotes, rather than with that of other dipnoans, sarcopterygians, or even osteichthians.

3.4. Fiber Connections of the Telencephalon (Fig. 11)

Although recently Reiner and Northcutt (1987) published some immunohistochemical data on the fiber connections of the dipnoan telencephalon (see above), our present-day knowledge of these connections still rests mainly on the analyses of nonexperimental myelin-stained and silver material, carried out by Holmgren and van der Horst (1925), Rudebeck (1945), Nieuwenhuys and Hickey (1965), Schnitzlein and Crosby (1967, 1968), and Nieuwenhuys (1969). The following survey is based on these studies, and on an analysis of normal silver material of *Lepidosiren*.

Secondary olfactory axons pass caudally in the bulbar fiber layer and, on entering the hemisphere, spread over the pallium and the various subpallial regions. The numerous olfactory fibers that enter the pallium constitute an internal and an external layer, the former situated centrally, the latter peripherally to the cortical layer (Fig. 9). These fibers have been termed olfacto-pallial radiations by Holmgren and van der Horst (1925); Rudebeck (1945) summarized them under the name tractus olfacto-dorsolateralis. The internal fiber layer shows a conspicuous concentration in the ventral part of the pallium. This fiber mass, which can be distinguished only in the rostral half of the hemisphere, was named pyriform tract by Holmgren and van der Horst (1925) and tractus olfactorius lateralis internus by Nieuwenhuys and Hickey (1965).

Whereas the internal fiber layer is confined to the pallium, the external layer extends over the basal parts of the hemisphere, reaching the striatum and the lateral parts of the olfactory tubercle. At the level of the striatum the external layer contains a rather compact fiber mass, which may be named the external part of the lateral olfactory tract. A number of fibers of this tract enter the stria medullaris, thus constituting a tractus olfactorius lateralis habenulae.

From the medial parts of the olfactory bulb, fibers stream down into the septal area, forming a medial olfactory tract. According to Holmgren and van der Horst (1925) and Rudebeck (1945), some of the fibers of this tract decussate in the anterior commissure, and form a commissura interbulbaris.

The ventral part of the subpallium, i.e., the large tuberculum olfactorium, also receives numerous secondary olfactory axons. These fibers, which are all unmyelinated, may be included under the name tractus olfactorius ventralis (Rudebeck, 1945).

The fibers of the tractus olfactorius externus, which, as mentioned above, originate from interglomerular cells, spread in the rostral one third of the pallium (Rudebeck, 1945).

In summary, it may be stated that the rostral parts of all regions in the dipnoan hemispheres receive secondary olfactory axons. It should, however, be emphasized that the exact caudal extent of these axons cannot be established in normal material, because during their course through the hemisphere they become inextricably mixed with fibers originating from the various forebrain centers.

The course of the nervus terminalis has been thoroughly studied by Holmgren and van der Horst (1925) in *Neoceratodus*. In this species the terminal nerve has separate anterior and posterior roots (Fig. 11A). The anterior root,

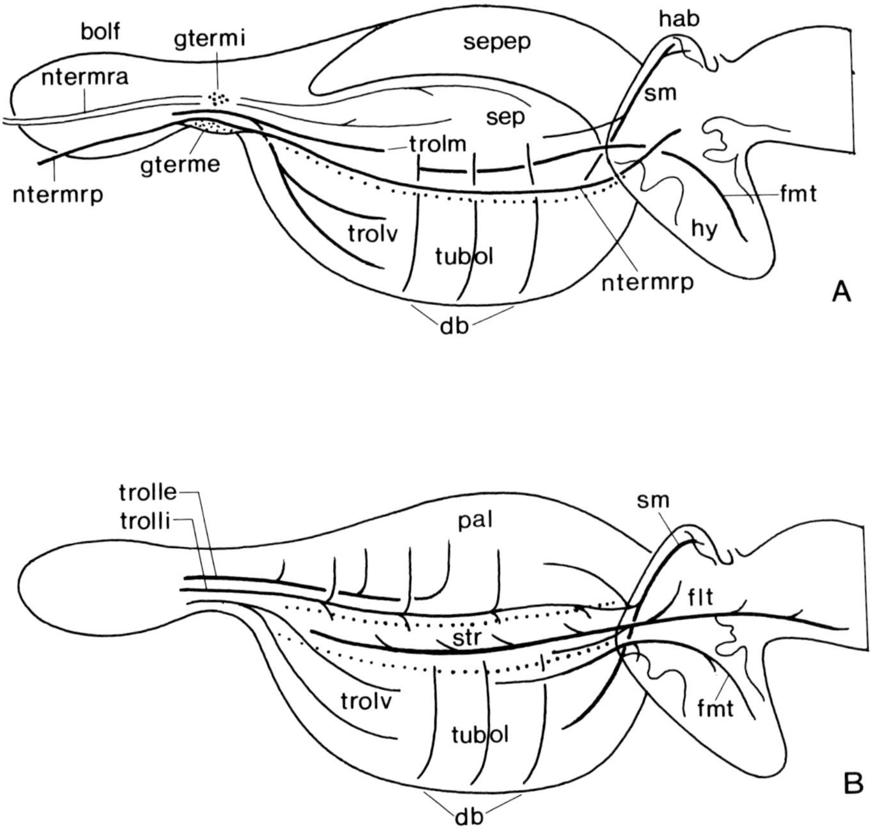

Figure 11. Diagram showing the fiber systems in the telencephalon of *Neoceratodus forsteri.* (A) Medial wall of the hemispheres; (B) lateral wall. Based on Holmgren and van der Horst (1925).

which is composed of three to five small bundles, enters the olfactory peduncle on its medial side, from where it passes caudally in the septum, terminating in the medial and dorsal portions of this part of the medial hemisphere wall. An aggregation of bipolar cells, situated in the most caudal part of the olfactory peduncle, represents, according to Holmgren and van der Horst (1925), an internal terminal ganglion.

The caudal root of the terminal nerve passes caudally between the medial hemisphere walls and enters the brain at the level of the anterior commissure. The fibers of the posterior root decussate partly in this commissure and have been observed to enter the lateral forebrain bundle. The ganglion of the posterior root, i.e., the external terminal ganglion, lies against the medioventral surface of the olfactory peduncle. The posterior root just described has also been observed in *Protopterus.*

Before discussing the connections of the centers situated in the cerebral hemispheres, it should be noted that three fiber systems—the stria medullaris, the fasciculus lateralis telencephali, and the fasciculus medialis telencephali—connect the telencephalon with further caudally situated brain parts.

The *stria medullaris* consists of fibers, which assemble in the most dorsal part of the telencephalon impar, to pass via the most rostral portions of the ventral and dorsal parts of the thalamus to the habenular region. Most of its fibers terminate in the habenular ganglion, but some pass, via the habenular commissure and the contralateral stria, back to the telencephalon.

The *fasciculus lateralis telencephali* or *lateral forebrain bundle* is the largest and the most compact fiber system connecting the telencephalon with further caudally situated structures. The bundle consists of various components but with regard to the origin and termination of these components there is no unanimity in the literature. Holmgren and van der Horst (1925) and Rudebeck (1945) were of the opinion that most of the fibers of this bundle originate from the pallium and pass to the tegmentum mesencephali, hence they termed its main part the tractus corticotegmentalis. Nieuwenhuys and Hickey (1965) and Schnitzlein and Crosby (1967, 1968), on the other hand, held that the principal portion of the bundle connects the striatum with the tegmentum mesencephali. The same groups of authors also reported that the most dorsally situated fibers of the bundle interconnect the striatum with the dorsal thalamus. The polarity of these fibers could not be ascertained. Schnitzlein and Crosby considered it likely that this "dorsal peduncle" of the lateral forebrain bundle has both ascending and descending fascicles. It should be added that the present authors were unable to clearly distinguish this dorsal peduncle in their *Lepidosiren* material. According to our observations the main stream of the bundle passes along the ventral thalamus and the regio tuberculi posterioris to the tegmentum mesencephali. Probably fibers are issued to all of these regions. Its exact caudal extent could not be determined in our normal material, but Bing and Burckhardt (1905) claimed to have traced this bundle to the spinal cord. Interestingly, it has been experimentally shown that the urodelan homologue of the bundle in question does reach the cervical spinal cord (Kokoros and Northcutt, 1977).

The *fasciculus medialis telencephali* or *medial forebrain bundle* consists of thin, mostly unmyelinated fibers that assemble from all parts of the subpallium and gradually converge toward the telencephalon impar. Here they form a loosely arranged fiber mass that, after having passed along the nucleus preopticus inferior, enters the hypothalamus. A large number of its fibers decussate in the anterior commissure. This bundle corresponds to the tractus olfactohypothalamicus of Holmgren and van der Horst (1925) and Rudebeck (1945).

Having discussed the main telencephalic fiber systems, it now remains to present a survey of the connections of the various telencephalic centers. It should be reemphasized that, because experimental hodological studies are lacking, this survey is to be considered as tentative and highly speculative.

The *pallium* contains an internal and an external fiber layer, which are separated by the cortical cell plate (Fig. 9). There can be no doubt that the rostral parts of both of these layers contain numerous secondary olfactory fibers and that throughout the pallium the ventricular as well as the cortical cells contribute fibers to them. The pallium projects via the stria medullaris to the epithalamus. Part of these fibers presumably form a pallial commissure (the commissura pallii posterior of Holmgren and van der Horst, 1925), but others terminate in the ganglia habenulae. Numerous fibers pass from the pallium to the striatum, mainly via a ventral extension of the external pallial fiber layer. With regard to the destination of these fibers there is no unanimity. Holmgren and van der Horst (1925) and Rudebeck (1945) held that the majority of these fibers enter the lateral forebrain bundle (the tractus corticotegmentalis of their terminology), to pass with this system to the tegmentum mesencephali. Nieuwenhuys and Hickey (1965) and the present authors, on the other hand, consider it likely that the fibers that descend from the pallium to the striatum terminate mainly in that structure, which in turn is the source of the bulk of the fibers that descend with the lateral forebrain bundle.

The *septum* receives secondary olfactory fibers as well as fibers from the nervus terminalis anterior and projects via the stria medullaris and the medial forebrain bundle to the epithalamus and hypothalamus, respectively. Moreover, a large, superficially situated associational system connects the septal region with the striatum. Rudebeck (1945), who first described this system, considered it homologous to the diagonal band of Broca, described by Herrick (1921, 1927) for amphibians.

The *tuberculum olfactorium* projects, like the septum, to the habenular ganglion and to the hypothalamus. Although the rostral part of this structure does receive secondary olfactory fibers, this has not been clearly demonstrated for its highly differentiated caudal part. It is also noteworthy that only its most dorsolateral portion receives some myelinated secondary olfactory fibers. These observations and the immunohistochemical findings of Reiner and Northcutt (1987), which have already been discussed, indicate that, contrary to the supposition of Elliot Smith (1908), the dipnoan olfactory tubercle is not exclusively an olfactory center and may well include a homologue of the so-called ventral striatum of mammals.

The connections of the *corpus striatum* are multifarious and intricate. This center receives throughout its extent secondary olfactory fibers via the tractus olfactorius lateralis externus. Other afferents include: (1) projections from the pallium, (2) fibers from the septum via the diagonal band of Broca, (3) fibers ascending from the dorsal thalamus, (4) fibers ascending from the hypothalamus, via a homologue of the olfactory projection tract which Herrick (1921, 1933) observed in amphibians. Herrick (1921) as well as Rudebeck (1945) have suggested that the fiber system in question is comparable to the tractus pallii that Johnston (1911a,b) described for chondrichthian and actinopterygian fishes. It is noteworthy, however, that a tractus pallii in the sense of Johnston (i.e., a massive projection, ascending from the hypothalamus to the pallial parts of the telencephalon) is lacking in both dipnoans and amphibians.

The efferents of the corpus striatum pass with the stria medullaris, the medial forebrain bundle, and the lateral forebrain bundle to the epithalamus, the hypothalamus, and the tegmentum mesencephali, respectively. According to Rudebeck (1945) the caudal part of the striatum (the nucleus olfactorius lateralis in his terminology) constitutes in dipnoans an important center for correlation which, with regard to both position and connections, closely resembles the striatum and adjacent neuropil of urodeles. The dipnoan nucleus preopticus superior represents in Rudebeck's (1945) opinion a direct caudal continuation of this correlation center, homologous to the massa cellularis reuniens or the primordial amygdala of urodeles, as described by Herrick (1927, 1933). On the basis of a thorough analysis of normal material, Rudebeck reported the presence of the following fiber connections of the nucleus preopticus superior in *protopterus:* afferents from the olfactory bulb, the ventral thalamus and the hypothalamus, reciprocal connections with the striatum, the nucleus preopticus superior of the opposite side via the anterior commissure, the nucleus preopticus inferior and the neighboring parts of the thalamus, and efferents to the epithalamus and the hypothalamus.

4. General Discussion and Summary

In the preceding sections of the present chapter a survey has been presented of the structure and connections of the forebrains of the single living crossopterygian fish, *Latimeria,* and of the dipnoans. Crossopterygians and dipnoans constitute together the subclass sarcopterygians and this subclass is of particular evolutionary interest, since it is generally assumed that this group of fish gave rise to the terrestrial vertebrates. In the present, final section of this chapter we will attempt to present an overview of the most important structural and connectional features of the telencephala of the various groups of fishes in comparison to the most primitive group of tetrapods, i.e., the amphibians. Emphasis will be laid on the sarcopterygians, because they are the subject of the present chapter and on the pallial parts of the cerebral hemispheres because they stand central in the present work. For the sake of brevity we will present this overview as a number of statements, to which, wherever necessary, comments will be added.

1. *In the telencephalon of all fishes three main divisions—the olfactory bulbs, the cerebral hemispheres and the telencephalon impar—can be distinguished.*
2. *The olfactory bulbs and the telencephalon impar are relatively constant structural entities in all groups of fishes, but the cerebral hemispheres show a remarkable diversity.*
3. *Grossly speaking it may be said that four different events—simple evagination, mixed evagination, eversion, and local thickening of the wall—are operative during the development of the telencephalon in the various groups of fish.*

During a simple evagination, the side walls of the forebrain anlage bulge out, thus giving rise to ventricular cavities with surrounding nervous tissue. This evagination can take place in lateral, rostral, and caudal directions. By the lateral evagination, the dorsal (pallial) part of the lateral plate is brought into such a position as to cap the ventral (basal) part. This phenomenon is known as the inversion of the pallium.

Mixed evagination differs from simple evagination in that not only the lateral plates, but also the membranous structures of the forebrain anlage are involved. This process leads to the formation of hemispheres that consist partly of nervous and partly of ependymal tissue.

Eversion differs fundamentally from evagination. During this process, the side walls bend outward, and their dorsal connection is transformed into a greatly expanded, membranous tela telencephali. Eversion is thus the antithesis of the pallial inversion in evaginated forebrains.

The last process, local thickening of the hemisphere walls, needs no further comment.

In cartilaginous fishes the cerebral hemispheres are clearly evaginated. However, two characteristic features observed in these fishes deserve mentioning: (1) in most groups there is an extensive fusion between the medial walls of both hemispheres, and (2) in the batoids or rays the lateral ventricles become gradually reduced during development. Due to these two developmental events the batoidean cerebral hemispheres are in the adult stage represented by a single solid mass (Smeets *et al.,* 1983).

In the primitive actinopterygian genera *Polypterus* and *Erpetoichthys* the hemisphere walls show a distinct eversion. The same holds true for the other actinopterygian groups, i.e., the Chondrostei, Holostei, and Teleostei, but here the bending outward is coupled with a marked swelling of the walls. In all actinopterygians there occurs, in addition, a more or less extensive mixed evagination resulting in the formation of an ependymal septum.

In the dipnoans most of the forebrain is evaginated rostrally and ventrally into distinct cerebral hemispheres. Due to this rostral evagination, the hemispheres lie mainly in front of the lamina terminalis. The ventral evagination leads to well-developed subpallial regions. An important structural difference between *Protopterus* and *Lepidosiren* on the one hand, and *Neoceratodus* on the other is that the hemispheres in the former consist entirely of nervous tissue, whereas in the latter, due to a rostral mixed evagination, a part of the medial hemisphere wall is constituted by an ependymal membrane.

The crossopterygian *Latimeria* appears to be the most complicated fish insofar as the general structure of its forebrain is concerned. A simple evagination is found in the ventral part of the hemispheres, whereas the configuration of the dorsal regions may be interpreted as the result of a mixed evagination, a marked thickening, and a slight tendency toward eversion. The forebrain of *Latimeria* thus shows a remarkable combination of the features that exist separately in the dipnoans and in the actinopterygians.

4. *The cerebral hemispheres of all fish groups have been subdivided into a dorsal, pallial part and a basal, subpallial part. However, in most groups there is no general agreement with respect to the extent and the boundaries of these two principal regions* (Fig. 12).

Several previous authors, among them Johnston (1911a,b), Holmgren (1922), and Bäckström (1924), held that in fish cell-free zonae limitantes mark the boundaries between the pallial and the subpallial regions. The presence of such cell-free zones has been clearly demonstrated in cartilaginous fishes (Holmgren, 1922; Bäckström, 1924). However, Smeets (1983; see also Smeets *et al.*, 1983) was of the opinion that in this group the lateral part of the pallium is represented by a cell mass situated ventral to the zona limitans lateralis. He based this interpretation upon the fact that the cell mass in question receives a very substantial olfactory input, just like the lateral or piriform pallium of tetrapods.

As regards the polypteriform fishes there is general agreement concerning

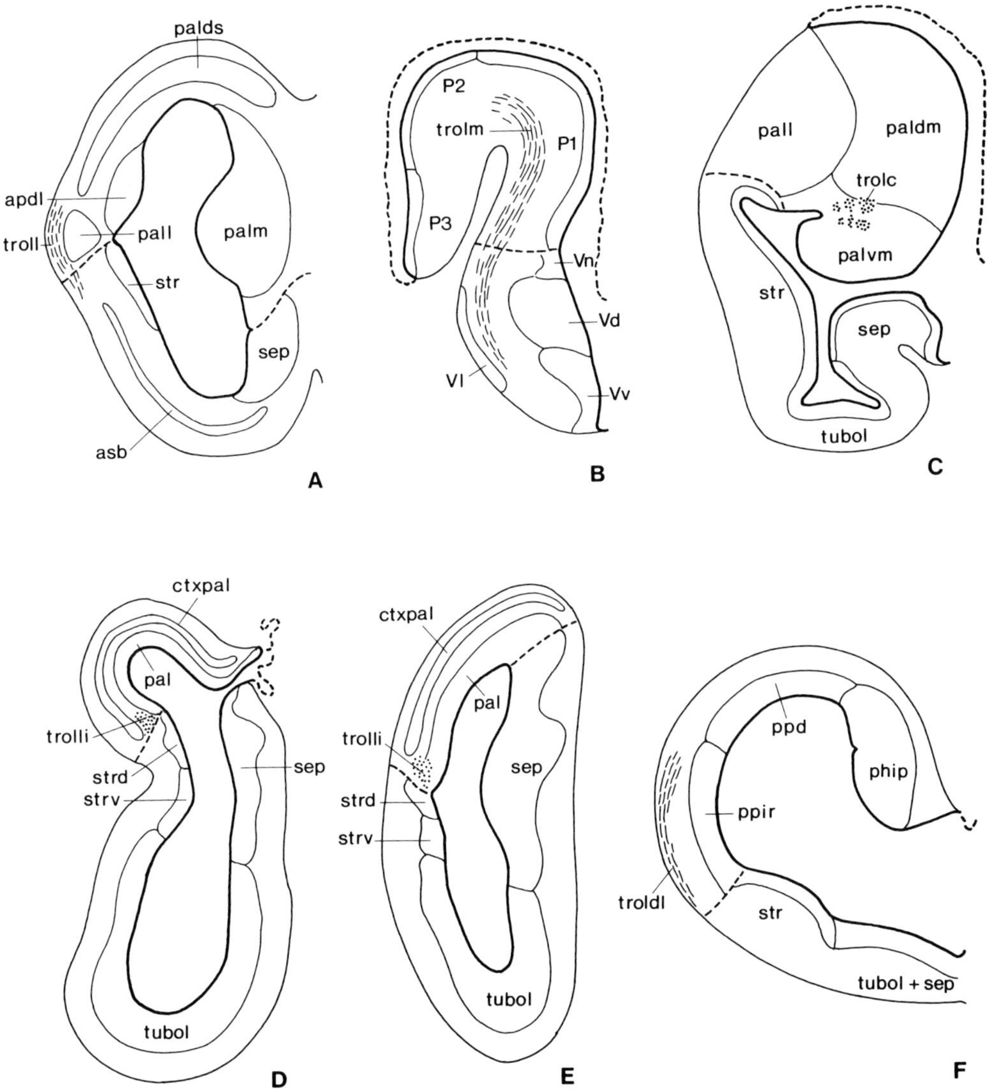

Figure 12. Schematic drawings of sections through the left cerebral hemisphere of a number of anamniotes, showing the position of the pallio-subpallial boundary (heavy interrupted lines), the principal cell masses (thin continuous lines), and the location of the lateral olfactory tract or its homologue. A, *Squalus acanthias;* B, *Erpetoichthys calabaricus;* C, *Latimeria chalumnae;* D, *Neoceratodus forsteri;* E, *Lepidosiren paradoxa;* F, *Ambystoma tigrinum.*

the location of the pallio-subpallial boundary (Holmgren, 1922; Nieuwenhuys, 1962, 1969, 1983; Northcutt and Braford, 1980; Northcutt and Davis, 1983), but this does not hold true for the chondrosteans, holosteans, and teleosts. In the latter groups, Källén (1951a,b) and Nieuwenhuys (1962, 1963) distinguished two principal regions, the area dorsalis telencephali and the area ventralis telencephali, and they considered these regions homologous to the pallium and the subpallium, respectively, of other fish groups and of tetrapods. However, Northcutt and Braford (1980), Northcutt and Davis (1983), and Echteler and Saidel (1981) have argued on the basis of some histochemical and hodological evidence that certain sectors of the actinopterygian area dorsalis (i.e., parts of the fields Dm and Dc in the nomenclature of Nieuwenhuys, 1963) represent the striatum and the globus pallidus of other vertebrate groups and are, hence, of subpallial nature. This interpretation has been discussed in some detail in the previous chapter of the present volume.

As regards dipnoans, all students of the telencephalon of this group have laid the boundary between the pallium and the subpallium in the lateral hemisphere wall at approximately the same place, but there are two distinctly different interpretations of the medial pallio-subpallial border. One group of investigators (Elliot Smith, 1908; Holmgren and van der Horst, 1925; Rudebeck, 1945; Nieuwenhuys and Hickey, 1965; Nieuwenhuys, 1969) is of the opinion that this border is situated dorsally and considers the entire medial hemisphere wall as a subpallial structure. Another group (Schnitzlein and Crosby, 1967; Clairambault and Capanna, 1973; Reiner and Northcutt, 1987), however, interprets the dorsal part of the medial hemisphere wall as a pallial structure, and only the ventral part as subpallial. As mentioned earlier in this chapter, inclusion of the taxonomically highly important Australian lungfish *Neoceratodus* leads in our opinion to the conclusion that in dipnoans the entire medial hemisphere wall is subpallial.

In the sole extant crossopterygian fish *Latimeria,* finally, it is generally agreed that the pallio-subpallial boundary coincides approximately with the transition of the thick-walled dorsal part into the thin-walled ventral part of the hemisphere, and is marked by a fairly distinct zona limitans.

5. *A subdivision of the pallium into medial, dorsal, and lateral formations, the latter of which receives a substantial olfactory projection, has been described for cartilaginous fishes, primitive actinopterygians and tetrapods, but such a tripartitioning is not apparent in sarcopterygians* (Fig. 12).

The studies of Northcutt (1977, 1981b) and Smeets *et al.* (1983) have shown that the pallium of sharks can be subdivided into medial, dorsal, and lateral sectors, and there is experimental evidence indicating that the lateral sector receives a substantial olfactory input (Ebbesson and Heimer, 1970; Smeets, 1983).

The pallium of the most primitive actinopterygians, i.e., the polypteriform fishes, is also divisible into medial, dorsal, and lateral fields, which have been designated as P1, P2, and P3 (Braford and Northcutt, 1974; Northcutt and Braford, 1980; Nieuwenhuys, 1983; Fig. 12B). However, due to the everted condition of the polypteriform telencephalon the medial, dorsal, and lateral pallial fields of this group correspond topologically to the lateral, dorsal, and medial pallial sectors of cartilaginous fishes and tetrapods (Fig. 12A,F). Hence, the fact that the medial pallial field P1 of polypteriform fishes constitutes the main pallial target of secondary olfactory fibers (Braford and Northcutt, 1974; von Bartheld and Meyer, 1986) fits in the general scheme sketched above.

The pallium of amphibians, reptiles, and mammals comprises three principal regions, the medial hippocampal pallium, the dorsal general pallium, and the (olfactory) lateral piriform pallium. Several authors (e.g., Holmgren, 1922; Northcutt, 1981b) have expressed the opinion that homologues of the three main subdivisions of the tetrapod pallium can be distinguished in all gnathostomes. However, we believe that there is no convincing evidence for this tripartitioning in chondrosteans, holosteans, teleosts, and sarcopterygians. For a detailed discussion of the structural relations in the pallium of the actinopterygian groups mentioned we refer to the previous chapter in this volume; here we confine ourselves to the results of our studies of the crossopterygian and the dipnoan pallium.

In the crossopterygian *Latimeria* the large central olfactory tract passes through the pallium in a position roughly corresponding to that of the lateral olfactory tract of cartilaginous fishes and tetrapods and of the main or medial olfactory radiation in polypteriform fishes (Braford and Northcutt, 1974; von Bartheld and Meyer, 1986; Fig. 12C). In the intermediate and caudal parts of the pallium of *Latimeria,* three separate fields, dorsomedial, ventromedial, and lateral, can be distinguished (Nieuwenhuys, 1965). The dorsomedial field surrounds the central olfactory tract and may well be homologous to (part of) the lateral pallium of cartilaginous fishes and tetrapods and to the P1 sector of the polypteriform pallium, but the remaining two pallial fields of *Latimeria* do not correspond topologically with the dorsal or P2 and lateral or P3 areas of the other vertebrate groups mentioned.

Crossopterygians, as represented by *Latimeria,* and dipnoans differ considerably with regard to both the gross appearance and the histological structure of the pallium. However, it is noteworthy that in both groups a large bundle composed of secondary olfactory fibers passes through the pallial region situated adjacent to the pallio-subpallial boundary [i.e., the central olfactory tract in *Latimeria,* as described by Nieuwenhuys (1965), and the pyriform tract of Holmgren and van der Horst (1925) or tractus olfactorius lateralis internus of Nieuwenhuys and Hickey (1965) in dipnoans] (Fig. 12D,E). Thus, it is reasonable to assume that the most ventral part of the dipnoan pallium represents the lateral or piriform pallium of other fish groups and tetrapods (Fig. 12A–C,F). Yet, our nonexperimental material and unpublished results of Northcutt (see Reiner and Northcutt, 1987) suggest that not only the ventral pallial sector, but rather the entire pallium of dipnoans is in receipt of secondary olfactory fibers. Holmgren and van der Horst (1925) indicated that differences in the thickness of the periventricular pallial zone allow a delineation of the ventral, pyriform part of the pallium from the remainder of that structure, whereas Reiner and Northcutt (1987) noted that this boundary is marked by a slight break in the superficial pallial cell layer. We consider neither of these features sufficiently significant to accept this subdivision.

6. *Nonolfactory sensory afferents passing from the dorsal thalamus to the pallium have been described for chondrichthians and actinopterygians, but evidence for the presence of such ascending thalamopallial projections in sarcopterygians is lacking.*

Several previous authors (e.g., Ariëns Kappers *et al.*, 1936) have stressed the importance of nonolfactory thalamophallial projections for the recognition of neopallial pimordia.

Experimental neuroanatomical and physiological studies have shown that in sharks, visual impulses reach the pallium along two different routes, i.e., retino-thalamo-pallial and retino-tecto-thalamo-pallial (Ebbesson and Schroeder, 1971; Schroeder and Ebbesson, 1974; Luiten, 1981), and there is also experimental evidence that in sharks, electrosensory lateral line input reaches the pallium via a dorsal thalamic relay (Bodznick and Northcutt, 1984).

The results of some preliminary studies (Braford and Northcutt, 1978; Northcutt, 1981a; Echteler and Saidel, 1981) suggest the presence of comparable connections in *Polypterus* and in the goldfish.

In dipnoans and in the crossopterygian *Latimeria* the dorsal thalamus is more strongly developed than in chondrichthians and in actinopterygians, and experimental studies may well reveal that this center in sarcopterygians is intercalated in ascending pathways projecting to the pallium. All we can say at present is that our nonexperimental material of *Latimeria, Neoceratodus,* and *Lepidosiren* did not show any trace of direct thalamopallial (or palliothalamic) connections.

7. *The pallium of dipnoans contains a distinct cortical layer. Because corresponding structures are present in the pallium of several anamniote groups, the dipnoan cortex should not be considered as exclusively comparable to the reptilian and mammalian pallial cortices.*

One of the most characteristic features of the dipnoan telencephalon is that the pallial region contains a true cortex throughout its extent and this is the more remarkable, because in almost all regions of the dipnoan brain the neuronal perikarya are confined to a periventricular zone. Elliot Smith (1908), who first described this structure in some detail, compared it directly with the cerebral cortex of reptiles and mammals. However, the investigations of Holmgren and his associates (Holmgren, 1922; Holmgren and van der Horst, 1925; Söderberg, 1922; Bäckström, 1924; Rudebeck, 1945) have revealed that cortical formations, i.e., cell zones that are clearly detached from the periventricular gray, also occur in other anamniotes. Thus, Holmgren (1922) and Bäckström (1924) demonstrated that in the pallium of larval cartilaginous fishes a distinct cortical layer is present, which according to their observations can be subdivided into medial, dorsal, and lateral zones. Later in development these cortical zones mostly lose their character of well-defined cortices, although in the dorsal part of the pallium of adult sharks a diffuse but clearly recognizable cortical formation is still present (Fig. 12A). In adult amphibians no cortical formations are present within the pallium, but Söderberg (1922) and Rudebeck (1945) observed in larval representatives of this group in addition to a periventricular pallial cell layer, a zone of migrated cortical cells. They considered it likely that during further development the cortical layer reunites with the periventricular zone and that, thus, the lack of a differentiated cortex in adult amphibians represents a secondary condition. It is interesting in this context that in dipnoans the neurons that constitute the pallial cortex closely resemble elements situated in the outer zone of the periventricular gray (Rudebeck, 1945; Fig. 10). The lateral pallial cell mass of *Latimeria* corresponds with regard to its topological position to the dipnoan cortex and finally it is noteworthy that in the pallium of the agnathan myxinoids an extensive and well-defined cortical layer is present (Jansen, 1930).

5. Abbreviations

apdl	Area periventricularis dorsolateralis
asb	Area superficialis basalis
b olf	Bulbus olfactorius
crost	Corpus rostrale
ctxpal	Cortex pallii
db	Diagonal band
fbt	Fasciculus basalis telencephali
flt	Fasciculus lateralis telencephali
fmt	Fasciculus medialis telencephali
gterme	Ganglion terminale externum
gtermi	Ganglion terminale internum
hab	Ganglion habenulae
hy	Hypothalamus
nsepm	Nucleus septi medialis
nsepmd	Nucleus septi medialis, pars dorsalis
ntermra	Nervus terminalis, radix anterior
ntermrp	Nervus terminalis, radix posterior
olf org	Olfactory organ
P1, 2, 3	Pallial fields (in polypteriform fishes)
pal	Pallium
paldm	Pallium dorsomediale
palds	Pallium dorsale, pars superficialis
pall	Pallium laterale
palm	Pallium mediale
palvm	Pallium ventromediale
ped olf	Pedunculus bulbi olfactorii
phip	Primordium hippocampi
ppd	Primordium pallii dorsalis
ppir	Primordium piriforme
rost org	Rostral organ
sep	Septum
sepd	Pars dorsalis septi
sepep	Septum ependymale
sepv	Pars ventralis septi
sfep	Stratum fibrosum externum pallii
sfip	Stratum fibrosum internum pallii
slp	Sulcus limitans pallii
sm	Stria medullaris
ss	Sulcus septalis
str	Corpus striatum
strd	Corpus striatum, pars dorsalis
strv	Corpus striatum, pars ventralis
tegmes	Tegmentum mesencephali
thd	Thalamus dorsalis
thv	Thalamus ventralis
trol	Tractus olfactorius
trolc	Tractus olfactorius centralis
troldl	Tractus olfactorius dorsolateralis
trolhabd	Tractus olfactohabenularis dorsalis
trolhabv	Tractus olfactohabenularis ventralis
troll	Tractus olfactorius lateralis
trolle	Tractus olfactorius lateralis externus
trolli	Tractus olfactorius lateralis internus
trolm	Tractus olfactorius medialis
trolv	Tractus olfactorius ventralis
trpalhab	Tractus palliohabenularis
tubol	Tuberculum olfactorium
tubp	Tuberculum posterius
V	Area ventralis telencephali (in actinopterygian fishes)
Vd	Dorsal part of V
Vl	Lateral part of V
vl	Ventriculus lateralis
Vn	Cell mass in the dorsalmost part of V
Vv	Ventral part of V
zlim	Zona limitans

ACKNOWLEDGMENTS. The authors are grateful to Mr. J. Russon and Mr. J. de Bekker for the drawings, to Mr. A. Reijnen for the photomicrographs, and to Mrs. M. van de Coevering for secretarial assistance.

6. References

Andrews, S. M., 1973, Interrelationships of crossopterygians, in: *Interrelationships of Fishes* (P. H. Greenwood, R. S. Miles, and C. Patterson, eds.), Academic Press, New York, pp. 137–177.

Anthony, J., 1980, Evocation des travaux français sur Latimeria notamment depuis 1972, *Proc. R. Soc. London* **208:**349–367.

Anthony, J., and Robineau, D., 1976, Sur quelques caractères juvéniles de Latimeria chalumnae Smith (Pisces, Crossopterygii, Coelacanthidae), *C.R. Acad. Sci.* **283:**1739–1742.

Ariëns Kappers, C. U., Huber, G. C., and Crosby, E. C., 1936, *The Comparative Anatomy of the Nervous System of Vertebrates, Including Man,* Volume II, Macmillan Co., New York, p. 1845.

Bäckström, K., 1924, Contributions to the forebrain morphology in selachians, *Acta Zool.* **5:**123–240.

Bing, R., and Burckhardt, R., 1905, Das Centralnervensystem von Ceratodus forsteri, *Jen. Denkschr.* **4:**513–584.

Bodznick, D., and Northcutt, R. G., 1984, An electrosensory area in the telencephalon of the little skate, *Raja erinacea, Brain Res.* **298:**117–124.

Braford, M. R., Jr., and Northcutt, R. G., 1974, Olfactory bulb projections in the bichir, Polypterus, *J. Comp. Neurol.* **156:**165–178.

Braford, M. R., Jr., and Northcutt, R. G., 1978, Correlation of telencephalic afferents and SDH distribution in the bony fish Polypterus, *Brain Res.* **152:**157–160.

Burckhardt, R., 1892, *Das Centralnervensystem von Protopterus Annectens,* Friedländer, Berlin, p. 64.

Capanna, E., and Clairambault, P., 1973, Some considerations on the forebrain of the bipulmonate Dipnoi, *Rend. Accad. Naz. Lincei* **55:**603–608.

Clairambault, P., and Capanna, E., 1973, Suggestion for a revision of the cytoarchitectonics of the telencephalon of Protopterus, Protopterus annectens (Owen), *Boll. Zool.* **40:**149–171.

Clairambault, P., Capanna, E., Chanconie, M., and Pinganaud, G., 1974a, Tipologia neurale del Septum telencefalico di un Dipnoi lepidosireniforme (Protopterus dolloi Boulenger), *Rend. Accad. Naz. Lincei* **56:**423–431.

Clairambault, P., Capanna, E., Chanconie, M., and Pinganaud, G., 1974b, Typologie neuronique du complexe Strio-amygdaloide de Protopterus dolloi Boulenger, *Rend. Accad. Naz. Lincei* **56:**1017–1025.

Ebbesson, S. O. E., and Heimer, L., 1970, Projections of the olfactory tract fibers in the nurse shark (*Ginglymostoma cirratum*), *Brain Res.* **17:**47–55.

Ebbesson, S. O. E., and Schroeder, D., 1971, Connections of the nurse shark's telencephalon, *Science* **173:**254–256.

Echteler, S. M., and Saidel, W. M., 1981, Forebrain connections in the goldfish support telencephalic homologies with land vertebrates, *Science* **212:**683–685.

Elliot Smith, G., 1908, The cerebral cortex in Lepidosiren, with comparative notes on the interpretation of certain features of the forebrain in other vertebrates, *Anat. Anz.* **33:**513–540.

Gerlach, J., 1933, Über das Gehirn von Protopterus annectens, *Anat. Anz.* **75:**305–448.

Heimer, L., and Wilson, R. D., 1975, The subcortical projections of the allocortex: Similarities in the neural associations of the hippocampus, the piriform cortex, and the neocortex, in: *Golgi Centennial Symposium Proceedings* (M. Santini, ed.), Raven Press, New York, pp. 177–193.

Heimer, L., Switzer, R. D., and Hoesen, G. W. van, 1982, Ventral striatum and ventral pallidum. Components of the motor system? *Trends Neurosci.* **5:**83–87.

Herrick, C. J., 1921, The connections of the vomeronasal nerve, accessory olfactory bulb and amygdala in Amphibia, *J. Comp. Neurol.* **33:**213–280.

Herrick, C. J., 1927, The amphibian forebrain. IV. The cerebral hemispheres of Amblystoma, *J. Comp. Neurol.* **43:**231–325.

Herrick, C. J., 1933, The amphibian forebrain. VI. Necturus, *J. Comp. Neurol.* **58:**1–288.

Holmgren, N., 1922, Points of view concerning forebrain morphology in lower vertebrates, *J. Comp. Neurol.* **34:**391–460.

Holmgren, N., and van der Horst, C. J., 1925, Contribution to the morphology of the brain of Ceratodus, *Acta Zool.* **6:**59–161.

Jansen, J., 1930, The brain of Myxine glutinosa, *J. Comp. Neurol.* **49:**360–507.

Jessen, H., 1967, The position of the Struniiformes, Strunius and Onychodus among the crossopterygians, *Coll. Ctr. Nat. Rech. Sci.* **163:**173–180.

Johnston, J. B., 1911a, The telencephalon of selachians, *J. Comp. Neurol.* **21:**1–113.

Johnston, J. B., 1911b, The telencephalon of ganoids and teleosts, *J. Comp. Neurol.* **21:**489–591.

Källén, B., 1951a, Some remarks on the ontogeny of the telencephalon in some lower vertebrates, *Acta Anat.* **9:**537–548.

Källén, B., 1951b, Embryological studies on the nuclei and their homologization in the vertebrate forebrain, *K. Fysiogr. Saellsk. Lund Handl. N.F.* **47**(5)**:**1–35.

Kokoros, J. J., and Northcutt, R. G., 1977, Telencephalic efferents of the tiger salamander ambystoma tigrinum tigrinum (Green), *J. Comp. Neurol.* **173:**613–628.

Kremers, J. W. P. M., 1981, The brain of the crossopterygian fish *Latimeria chalumnae* Smith, Thesis, Nijmegen, p. 125.

Kuhlenbeck, H., 1929, Die Grundbestandteile des Endhirns im Lichte der Bauplanlehre, *Anat. Anz.* **67:**1–51.

Luiten, P. G. M., 1981, Two visual pathways to the telencephalon in the nurse shark (*Ginglymostoma cirratum*). II. Ascending thalamo-telencephalic connections, *J. Comp. Neurol.* **196:**539–548.

Millot, J., and Anthony, J., 1965, Anatomie de Latimeria chalumnae, in: *Système nerveux et organes de sens* (Ed. de Centre Nat. Rech. Sci.), Volume II, Paris, p. 131.

Moy-Thomas, J. A., and Miles, R. S., 1971, *Palaeozoic Fishes,* Chapman & Hall, London, p. 259.

Nieuwenhuys, R., 1962, Trends in the evolution of the actinopterygian forebrain, *J. Morphol.* **111:**69–88.

Nieuwenhuys, R., 1963, The comparative anatomy of the actinopterygian forebrain, *J. Hirnforsch.* **6:**171–192.

Nieuwenhuys, R., 1965, The forebrain of the crossopterygian Latimeria chalumnae Smith, *J. Morphol.* **117:**1–24.

Nieuwenhuys, R., 1967, Comparative anatomy of olfactory centres and tracts, *Prog. Brain Res.* **23:**1–64.

Nieuwenhuys, R., 1969, A survey of the structure of the forebrain in higher bony fishes, *Ann. N.Y. Acad. Sci.* **167:**31–64.

Nieuwenhuys, R., 1983, The central nervous system of the brachiopterygian fish, Erpetoichthys calabaricus, *J. Hirnforsch.* **24:**501–533.

Nieuwenhuys, R., and Hickey, M., 1965, A survey of the forebrain of the Australian lungfish Neoceratodus forsteri, *J. Hirnforsch.* **7:**433–452.

Nieuwenhuys, R., Kremers, J. P. M., and Huijzen, C. van, 1977, The brain of the crossopterygian fish Latimeria chalumnae: A survey of its gross structure, *Anat. Embryol.* **151:**157–169.

Northcutt, R. G., 1977, Elasmobranch central nervous system organization and its possible evolutionary significance, *Am. Zool.* **17:**411–429.

Northcutt, R. G., 1981a, Localization of neurons afferent to the telencephalon in a primitive bony fish, Polypterus palmas, *Neurosci. Lett.* **22:**219–222.

Northcutt, R. G., 1981b, Evolution of the telencephalon in nonmammals, *Annu. Rev. Neurosci.* **4:**301–350.

Northcutt, R. G., 1986, Lungfish neural characters and their bearing on sarcopterygian phylogeny, *J. Morphol. Suppl.* **1:**277–297.

Northcutt, R. G., and Braford, M. R., Jr., 1980, New observations on the organization and evolution of the telencephalon of actinopterygian fishes, in: *Comparative Neurology of the Telencephalon,* Volume 3 (S. O. E. Ebbesson, ed.), Plenum Press, New York, pp. 41–99.

Northcutt, R. G., and Davis, R. E., 1983, Telencephalic organization in ray-finned fishes, in: *Fish Neurobiology,* Volume 2 (R. E. Davis and R. G. Northcutt, eds.), University of Michigan Press, Ann Arbor, pp. 203–237.

Northcutt, R. G., Neary, T. J., and Senn, D. G., 1978, Observations on the brain of the coelacanth Latimeria chalumnae: External anatomy and quantitative analysis, *J. Morphol.* **155:**181–192.

Reiner, A., and Northcutt, R. G., 1987, An immunohistochemical study of the telencephalon of the African lungfish, Protopterus annectens, *J. Comp. Neurol.* **256:**463–481.

Romer, A. S., 1962, *The Vertebrate Body,* Saunders, Philadelphia, p. 627.

Romer, A. S., 1969, Vertebrate history with special reference to factors related to cerebellar evolution, in: *Neurobiology of Cerebellar Evolution and Development* (R. Llinás, ed.), American Medical Association, Chicago, pp. 1–18.

Rosen, D. E., Forey, P. L., Gardiner, B. G., and Patterson, C., 1981, Lungfishes, tetrapods, paleontology, and plesiomorphy, *Bull. Am. Mus. Nat. Hist.* **167:**163–275.

Rudebeck, B., 1945, Contributions to forebrain morphology in dipnoi, *Acta Zool.* **26:**10–157.

Schaeffer, B., 1969, Adaptive radiation of the fishes and the fish–amphibian transition, *Ann. N.Y. Acad. Sci.* **167:**5–18.

Schnitzlein, H. N., 1966, The olfactory tubercle of the African lungfish, Protopterus, *Ala. J. Med. Sci.* **3:**39–46.

Schnitzlein, H. N., and Crosby, E. C., 1967, The telencephalon of the lungfish, Protopterus, *J. Hirnforsch.* **2**:105–150.

Schnitzlein, H. N., and Crosby, E. C., 1968, The epithalamus and thalamus of the lungfish, Protopterus, *J. Hirnforsch.* **10**:351–371.

Schroeder, D. M., and Ebbesson, S. O. E., 1974, Nonolfactory telencephalic afferents in the nurse shark (*Ginglymostoma cirratum*), *Brain Behav. Evol.* **9**:121–155.

Schultze, H.-P., 1986, Dipnoans as sarcopterygians, *J. Morphol. Suppl.* **1**:39–74.

Smeets, W. J. A. J., 1983, The secondary olfactory connections in two chondrichthians, the shark *Scyliorhinus canicula* and the ray *Raja clavata, J. Comp. Neurol.* **218**:334–344.

Smeets, W. J. A. J., and Boord, R. L., 1985, Connections of the lobus inferior hypothalami of the clearnose skate *Raja eglanteria* (Chondrichthyes), *J. Comp. Neurol.* **234**:380–392.

Smeets, W. J. A. J., Nieuwenhuys, R., and Roberts, B. L., 1983, *The Central Nervous System of Cartilaginous Fishes: Structure and Functional Correlations*, Springer, Berlin, p. 266.

Söderberg, G., 1922, Contributions to the forebrain morphology in amphibians, *Acta Zool.* **3**:65–122.

Thomson, K. S., 1969, The biology of the lobe-finned fishes, *Biol. Rev.* **44**:91–154.

Thors, F., and Nieuwenhuys, R., 1979, Topological analysis of the brain stem of the lungfish *Lepidosiren paradoxa, J. Comp. Neurol.* **187**:589–612.

von Bartheld, C. S., and Meyer, D. L., 1986, Central connections of the olfactory bulb in the bichir, Polypterus palmas, reexamined, *Cell Tissue Res.* **244**:527–535.

4

The Pallium of Anuran Amphibians

TIMOTHY J. NEARY

1. Introduction

This chapter focuses on the pallium (telencephalic roof) of anuran amphibians. It includes descriptions of normal and connectional anatomy based almost entirely on data from ranid frogs, and also offers comparisons with the pallium in reptiles. Only a brief discussion of pallial fields and connections in urodeles is undertaken, largely because of the paucity of new information appearing since the last major review of the amphibian telencephalon (Northcutt and Kicliter, 1980). For much the same reason, no discussion of the pallium in apodans is presented. A description of their pallial areas and references to earlier literature can be found in Northcutt and Kicliter (1980).

The description of normal anatomy relies greatly on previous studies and recognizes three principal pallial fields (medial, dorsal, and lateral). It differs, however, from most other studies in three respects: The lateral amygdala is included in the pallium (after Northcutt, 1974), an expanded definition of the dorsal pallium is used (after Northcutt and Kicliter, 1980), and more subdivisions of the principal fields are delineated. Some of these subdivisions were noted by anatomists in the first quarter of this century, but their descriptions did not gain wide acceptance.

The section on anuran pallial connections summarizes published experimental findings and also reports, for the first time, the results from several preliminary tracing studies, many performed expressly for this chapter. These

TIMOTHY J. NEARY • Department of Anatomy, Creighton University School of Medicine, Omaha, Nebraska 68178.

results generally confirm the work of others, but they also reveal new connections and add credence to many of the field subdivisions seen in normal material.

The comparisons of pallial field connections in anurans and reptiles generally follow traditional lines. Similarly named fields (e.g., medial pallium and medial cortex) are paired for comparison and assumed to be homologous. However, part of the lateral pallium, together with the lateral amygdala, is compared to the reptilian dorsal ventricular ridge. This comparison is based on the homologies proposed by Northcutt and Kicliter (1980).

2. Methods

The analysis of anuran normal pallial anatomy was based primarily on a set of serial sections, taken at 15 μm, through the brain of a bullfrog (*Rana catesbeiana*). The sections were stained by the Bodian method and counterstained with cresyl violet. Nissl-stained sections of the brains from other anuran species were also examined. These were taken from *Ascaphus trueii, Bombina orientalis, Bufo americanus, Centrolenella colymbiphyllum, Dendrobates auratus, Hyla crucifer, Leptodactylus ocellatus,* and *Pseudis paradoxa.*

Tracing studies were performed by applying or injecting wheat germ agglutinin–horseradish peroxidase conjugate (WGA–HRP) into the main olfactory bulb (8 cases), medial pallium (12 cases), dorsal pallium (12 cases), septal area (9 cases), anterior thalamic nucleus (6 cases), or middle thalamic zone (3 cases). Applications in telencephalic areas were made by inserting a sharpened tungsten wire coated with WGA–HRP paste. Iontophoretic injections were made in thalamic areas: Glass pipets (tip diameter: 18–24 μm) were partly filled with 2% WGA–HRP in 0.1 M phosphate buffer (pH 7.4) and, after placement, alternate on–off current (2–5 μA) was passed through the pipets for 4–8 min. Survival times for all studies ranged from 3 to 5 days at 20–25°C. Transported WGA–HRP was visualized using tetramethyl benzidine. For further details, see Neary (1984a).

3. Normal Pallial Anatomy in Anurans

The anuran pallium consists of three principal divisions or fields (medial, dorsal, and lateral pallia) plus a possible fourth population, the lateral amygdala, which may represent a caudoventral extension of the lateral pallium (Northcutt, 1974; Northcutt and Kicliter, 1980). The three principal fields are bounded rostrally by the internal granule cell layer of the olfactory bulb; the transition from bulb to pallium is abrupt, characterized by a thinning of the periventricular cell layers, a reduction in cell density, and an increase in cell size.

3.1. Medial Pallium

The medial pallium in bullfrogs and other anurans is the largest principal field (Figs. 1, 2; see Section 7 for abbreviations). At transverse levels including and rostral to the accessory olfactory bulb, it lies dorsal to the postolfactory eminence (Fig. 1a), a small subpallial region containing partly laminated cells that are smaller than those of the medial pallium. The boundary between the two is marked on the ventricular surface by a short unnamed sulcus that is shallow rostrally and acute caudally. Posterior to the accessory olfactory bulb, the medial pallium is separated from the underlying septal nuclei (Figs. 1b, 2) and diagonal band nucleus (Figs. 1b, 2a) by the cell-poor zona limitans medialis. This border is accompanied by the ventricular sulcus medius.

The medial pallium is bounded ventrolaterally by the dorsal pallium. The transition between the two occurs when the deep cellular laminae of the medial pallium coalesce into the compact inner cell layer of the dorsal pallium (Section 3.2). No obvious change in cell size or Nissl-staining properties occurs in this transition.

Subdivisions of the bullfrog medial pallium cannot be clearly recognized. However, its dorsal and ventral thirds contain generally smaller cells than the middle third, suggesting it can be divided, albeit without great precision, into three parts, as indicated in Fig. 1b. These are: (1) a transitional part (MPt), which lies dorsally and caps the ventricle, (2) a large-celled part (MPl), located centrally, and (3) a small-celled part (MPs), which occupies much of the thickest part of the pallial hemispheric wall. Connectional studies, detailed in Section 4.1, further

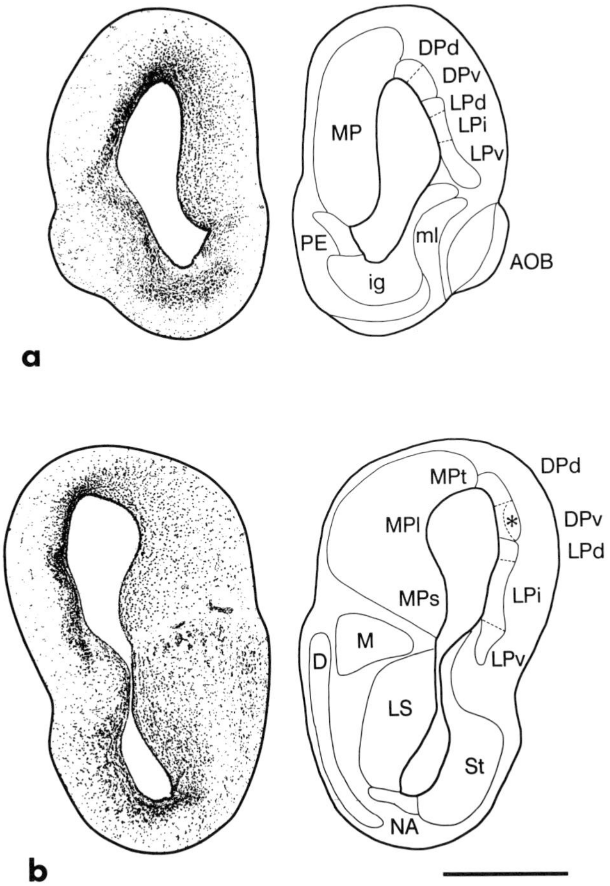

Figure 1. Transverse sections through the bullfrog telencephalon (a) at the level of the accessory olfactory bulb and (b) at a mid-hemispheric level. Bar = 1 mm.

indicate that the medial pallium has adjacent inner and outer cell layers extending through each of its three parts. These layers are almost impossible to delineate with precision in normal material. Comparing normal and experimental material suggests that the outer layer in MPt and MPl is thin and contains fusiform neurons (see below). The outer layer in MPs is slightly wider and contains larger, more scattered neurons than the inner layer, where cells are generally smaller and partly laminated.

Four major cell types are found in Nissl-stained sections of the ranid medial pallium (Clairambault and Derer, 1968): piriform, large and small multipolar, and fusiform. Piriform cells predominate and have basal nuclei covered by an apical cone-shaped cap of cytoplasm directed away from the ventricular surface. The Nissl-positive cytoplasmic caps of multipolar cells are flattened or slighted flared and often extend partly into the base of two or three dendrites. Fusiform neurons generally have two triangular cytoplasmic caps, one frequently larger than the other, on either side of a central, ovoid nucleus. These neurons are

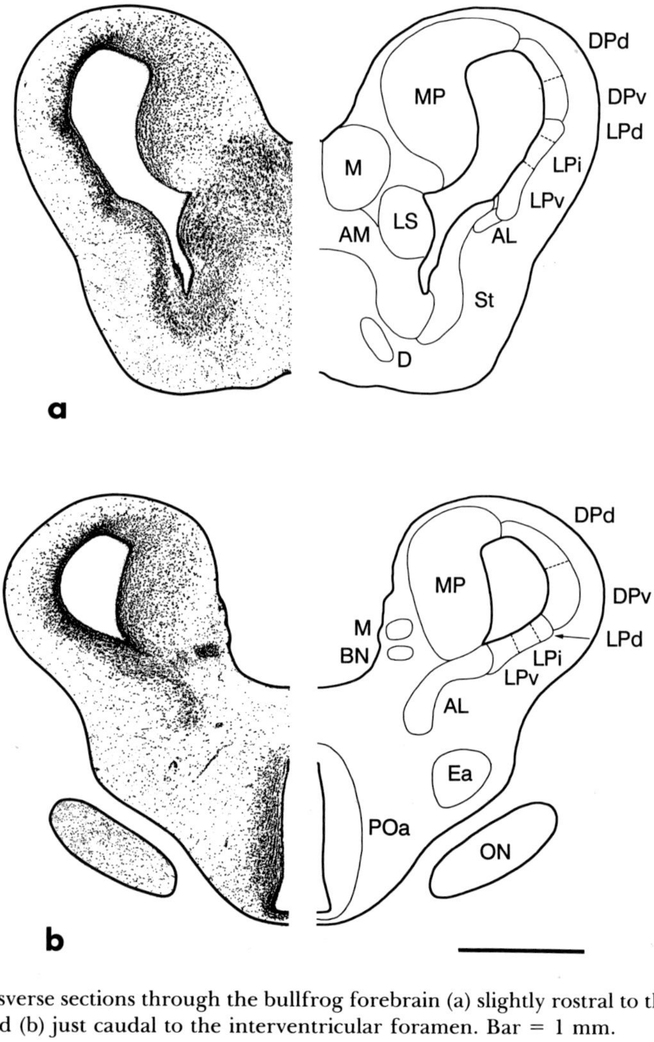

Figure 2. Transverse sections through the bullfrog forebrain (a) slightly rostral to the interventricular foramen and (b) just caudal to the interventricular foramen. Bar = 1 mm.

found in the outer layer of the medial pallium and are oriented with their long axis parallel to the pial surface.

Golgi preparations also reveal several types of cells in the medial pallium. Most frequently reported are pyramidal and piriform cells with dendritic arbors directed toward the outer surface of the hemisphere (Rubaschkin, 1903; Kuhlenbeck, 1921a; Ramón y Cajal, 1946; Hoffman, 1963; Kicliter and Ebbesson, 1976). These probably correspond to the piriform neurons described above. Additionally, Hoffman (1963) described flask-shaped and stellate cells that may represent, respectively, large and small multipolar cells. Finally, fusiform neurons with dendrites paralleling the pial surface are present superficially (Rubaschkin, 1903; Ramón y Cajal, 1946; Kicliter and Ebbesson, 1976) and are almost certainly the same as the fusiform neurons seen in Nissl-stained material.

The medial pallium, as delineated here, corresponds to the medial pallium described by Scalia (1976a) and by Northcutt and Kicliter (1980). It includes the medial pallium and much of the dorsal pallium of several other studies (Hoffman, 1963, 1966; Howell and Hoffman, 1966; Clairambault and Derer, 1968; Northcutt, 1974). The MPl and MPs were first noted in *Bufo* by Röthig (1912), although he did not assign specific names to them. They appear to correspond, for the most part, to Hoffman's (1963) primordial hippocampus *a* and *b*, respectively. A comparison of older terminology for all pallial fields may be found in Clairambault and Derer (1968).

In all of the other anurans studied, the medial pallium has the same basic relations to neighboring neural structures, although the orientation of the neural wall is rotated more to the horizontal in animals with relatively large septal areas. The primary variation in the structure of the medial pallium involves the differentiation of the large- and small-celled parts which, at one extreme, are difficult to discern in *Ascaphus* and *Bombina* and, at the other, are very distinct in *Bufo, Centrolenella, Dendrobates, Hyla, Leptodactylus,* and *Pseudis*. In these latter animals, particularly at transverse levels just in front of the interventricular foramen, the boundary between the large- and small-celled parts of the medial pallium is indicated by a shallow ventricular sulcus, as illustrated in *Bufo* by Röthig (1912).

3.2. Dorsal Pallium

The dorsal pallium (Figs. 1,2) occupies the dorsal third of the lateral hemispheric wall in *Rana*. It is bounded dorsomedially by the medial pallium and ventrally by the dorsal part of the lateral pallium (LPd), which contains uniformly smaller neurons (Neary, 1984a).

Two adjacent cell layers are present in the dorsal pallium: an inner, more compact layer, which begins dorsomedially by the coalescence of deeper cellular laminae of MPt, and an outer, slightly more diffuse layer.

The dorsal pallium can be divided into dorsal and ventral parts (DPd and DPv; Figs. 1, 2). The DPv has a reduced inner layer and a somewhat wider outer layer, indicated by the asterisk in Fig. 1b, compared to the DPd. The inner layer of the DPv often displays a concave outward profile in transverse sections.

In Nissl-stained sections, most dorsal pallial neurons appear to be piriform, although both small multipolar and fusiform neurons are also found in the outer layer. Fusiform neurons are preferentially located at the superficial edge of this layer. Corresponding cell types have been described and illustrated from Golgi material (Rubaschkin, 1903; Ramón y Cajal, 1946; Hoffman, 1963; Kicliter and Ebbesson, 1976). Both piriform and multipolar cells direct their dendrites away from the ventricle, and fusiform neurons have dendrites oriented parallel to the pial surface.

The dorsal pallium recognized here includes all of the dorsal pallium as delimited by Northcutt and Kicliter (1980) and also includes a portion of their LPd. Other studies have included most, if not all, of the dorsal pallium in the lateral (piriform) pallium (Hoffman, 1963, 1966; Howell and Hoffman, 1966; Clairambault and Derer, 1968; Northcutt, 1974; Scalia, 1976a; Kicliter and Ebbesson, 1976). The two parts of the dorsal pallium were not recognized by other authors.

The dorsal pallium and its two parts are present in all anurans examined. In *Ascaphus* and *Hyla,* the DPd and DPv are best differentiated in the caudal half of the hemisphere.

3.3. Lateral Pallium

The bullfrog lateral pallium (Figs. 1,2) is situated in the middle third of the lateral hemispheric wall and is bounded dorsally by the dorsal pallium. Rostrally, its ventral boundary is formed by the internal granule cell layers of the main and accessory olfactory bulbs. Caudal to the accessory bulb, it is separated from the underlying striatum by the zona limitans lateralis, a cell-poor region occupied by the accessory olfactory tract (Scalia, 1972; Northcutt and Royce, 1975). Just forward of the interventricular foramen, this zone gradually disappears as fibers of the accessory olfactory tract pass laterally through the lateral pallium. Slightly caudal to the disappearance of the zona limitans lateralis, the lateral pallium is bounded ventromedially by the lateral amygdala, which contains slightly smaller, more scattered cells. The lateral amygdala remains the ventromedial boundary of the lateral pallium until the separation of the caudal telencephalic pole from the diencephalon. Past this separation, the lateral pallium is bordered medially by the medial pallium.

The lateral pallium may be divided into three parts (Figs. 1, 2): dorsal (LPd), intermediate (LPi), and ventral (LPv). The LPd is the smallest of the three and also contains the smallest neurons. They occupy a compact mass situated over the shallow ventricular sulcus rhinalis and form a fairly distinctive cellular eminence that has been termed the dorsolateral prominence (Röthig, 1912). The LPi consists of an inner layer of closely packed cellular laminae and a thin outer layer of slightly more scattered cells. At midhemispheric levels, it generally presents a lower cell density than either the dorsal or ventral parts.

The LPv appears rostrally as a sheet of scattered, larger cells extending ventrolaterally away from the ventricle. This ventrolateral extension, which continues caudally almost to the interventricular foramen, has been called the lateral prominence (Röthig, 1912). Caudal to the accessory olfactory bulb, the cells of the LPv become smaller and more closely packed and, at transverse levels just rostral to the appearance of the lateral amygdala, they are broken up into loose columns by the laterally coursing fibers of the accessory olfactory tract (Fig. 1b). Immediately rostral to the interventricular foramen, the boundary between the

LPv and LPi is indicated on the ventricular surface by a shallow, as yet unnamed sulcus (Fig. 2a).

The majority of lateral pallial cells appear to be piriform in Nissl-stained material, although some multipolar and fusiform neurons are also present. These are most readily seen in the outer layer of LPi. The picture derived from Golgi studies is similar; most neurons are piriform with rounded bases and dendrites extending laterally or ventrolaterally toward the lateral olfactory tract (Rubaschkin, 1903; Ramón y Cajal, 1946; Hoffman, 1963; Kicliter and Ebbesson, 1976). Multipolar cells with dendrites directed toward the lateral olfactory tract have also been illustrated (Kicliter and Ebbesson, 1976), as have fusiform cells with dendrites paralleling the pial surface (Ramón y Cajal, 1946; Kicliter and Ebbesson, 1976).

The lateral pallium recognized here corresponds most closely to the lateral pallium of Northcutt and Kicliter (1980). It has generally been included in the lateral or piriform pallium of many other authors (Hoffman, 1963; Clairambault and Derer, 1968; Northcutt, 1974; Scalia, 1976a; Kicliter and Ebbesson, 1976), but was considered part of the striatum by Hoffman (1966) and Howell and Hoffman (1966).

The three parts of the lateral pallium were recognized by Kuhlenbeck (1921b) and given similar names. Although Howell and Hoffman (1966) also distinguished these same areas, their nomenclature differed markedly. They termed LPd the "primordial piriform cortex *s*"; LPi, the "striatum, dorsal part"; and LPv, the "striatum, ventral part." LPv corresponds closely to the LPv of Northcutt and Kicliter (1980) and is equivalent to the primordial piriform cortex *s* of Hoffman (1963) and, in part, to the epistriatum of Clairambault and Derer (1968).

The lateral pallium and its component parts are evident in all of the anurans studied, but appear reduced in *Bombina, Centrolenella, Dendrobates,* and *Hyla.*

3.4. Lateral Amygdala

The lateral amygdala (Fig. 2) begins rostrally shortly behind the disappearance of the zona limitans lateralis as a cluster of cells partly wedged between the LPv dorsolaterally and dorsal striatum ventromedially. At these transverse levels, its neurons are slightly smaller and less densely packed than those of the LPv and, conversely, slightly larger and more densely packed than those of the striatum. Caudally, the lateral amygdala expands to form a crescent-shaped lamina partly enveloping, both dorsally and medially, a cell-poor region containing terminals of the accessory olfactory tract—the accessory olfactory neuropil (Scalia, 1972; Northcutt and Royce, 1975). Further caudally, at levels near the rostral end of the magnocellular preoptic nucleus, the lateral amygdala becomes a rounded cluster contiguous with the anterior preoptic area (Neary and Northcutt, 1983). It can be distinguished from the latter by its higher cell density. The lateral amygdala disappears just in front of the optic chiasm. In horizontal sections, the lateral amygdala is also crescentic, enclosing the accessory olfactory neuropil on its rostral, medial, and caudal edges. Thus, in three dimensions, the lateral amygdala is roughly shaped like a cup with its opening directed ventrolaterally.

Lateral amygdalar neurons are primarily piriform (Clairambault and Derer,

1968) and Golgi studies show that they direct their dendrites into the accessory olfactory neuropil (Kicliter and Ebbesson, 1976).

Older descriptions (e.g., Ariëns Kappers *et al.*, 1936) placed the lateral amygdala in the subpallium. Histochemical studies, however, support its inclusion in the pallium (Northcutt, 1974). The lateral amygdala of the present study corresponds to that of Northcutt (1974), Kicliter and Ebbesson (1976), and Northcutt and Kicliter (1980). It represents the corticomedial amygdala of Hoffman (1963, 1966) and Howell and Hoffman (1966), and the amygdala of Clairambault and Derer (1968) and Scalia (1976a).

A lateral amygdala is recognizable in all anuran species examined, although it is relatively small in *Dendrobates.*

3.5. Pallial Fields in Urodeles

Medial, dorsal, and lateral pallial fields are present in urodeles and correspond to the structures of the same name in anurans (Northcutt and Kicliter, 1980). These fields are essentially equivalent to the primordial hippocampus, primordial dorsal pallium, and primordial piriform pallium (dorsolateral olfactory nucleus) delineated by Herrick (1927, 1933) in *Ambystoma* and *Necturus.* Amygdalar nuclei are also present in these animals (Herrick, 1927, 1933) and represent counterparts to the anuran lateral amygdala (Northcutt and Kicliter, 1980).

The medial pallium in salamanders has not been formally divided, although Herrick (1927, 1933) noted "ill-defined" and "obscure" differences between its dorsal and ventral parts. Herrick (1933) suggested that this differentiation might correspond to that seen by Röthig (1912) in anurans (representing the MPl and MPs of this study).

Subdivisions of the dorsal pallium in urodeles have also not been described. However, in the photographs of transverse sections through the telencephalon of *Ambystoma* presented by Northcutt and Kicliter (Figs. 8 and 9, 1980), a slight decrease in cell density is evident in the ventrolateral half of the dorsal pallium. This may indicate the presence of a subdivision comparable to the DPv of frogs.

Several investigators have recognized a lateral (cellular) prominence in urodeles (Röthig, 1912; Herrick, 1927, 1933; Northcutt and Kicliter, 1980) and it appears to represent LPv. The dorsolateral prominence, which caps the sulcus rhinalis in *Cryptobranchus* (Röthig, 1912), may be equivalent to the anuran LPd.

The portion of the telencephalic wall containing the dorsal pallium in salamanders is oriented more to the horizontal than it is in anurans, and forms a cap over the lateral ventricle. This difference in orientation may have led to past confusion in the definition of the anuran dorsal pallium. Until recently, most investigators (e.g., Röthig, 1912; Söderberg, 1922; Ariëns Kappers *et al.,* 1936; Clairambault and Derer, 1968; Kicliter and Ebbesson, 1976) considered the anuran dorsal pallium to be restricted to a very small group of cells that, as in urodeles, caps the lateral ventricle. Northcutt and Kicliter (1980) were the first to observe that the traditional delineation would make the anuran dorsal pallium a highly reduced, tiny transitional region between the medial and lateral pallia, in contrast to the urodele dorsal pallium, which occupies a significant part of the hemispheric wall. Thus, they expanded the area defined as the dorsal pallium in frogs by extending the boundary between the dorsal and lateral pallia ventrolaterally, and their revised definition was essentially followed in the present study.

4. Pallial Connections in Anurans

4.1. Medial Pallium

Most afferents to the anuran medial pallium arise from areas within the telencephalon. Degeneration and peroxidase tracing studies, together, indicate that the medial pallium receives ipsilateral projections from the dorsal and lateral pallia, diagonal band nucleus, medial and lateral septal nuclei, medial amygdala, and the bed nucleus of the pallial commissure (Halpern, 1972; Kokoros, in *Bufo,* 1973; Vesëlkin and Ermakova, 1978; Kicliter, 1979; Ronan and Northcutt, 1979). Vesëlkin and Ermakova (1978) also found evidence for projections to the medial pallium from the lateral amygdala and the nucleus accumbens. The rostral medial pallium receives a restricted input from the ipsilateral olfactory bulb (Scalia *et al.,* 1968; Northcutt and Royce, 1975).

Contralateral telencephalic input to the medial pallium was first shown in anterograde degeneration studies following hemispherectomies, but restricted lesions of the medial pallium did not initially reveal projections to the opposite medial pallium (Halpern *et al.,* 1968; Halpern, 1972). Later degeneration and HRP studies, however, did demonstrate this projection and it appears to be largely symmetrical (Kokoros, 1973; Scalia, 1976a; Vesëlkin and Ermakova, 1978; Ronan and Northcutt, 1979). Other contralateral telencephalic input to the medial pallium comes from the medial septal nucleus, medial amygdala, and bed nucleus of the pallial commissure (Ronan and Northcutt, 1979).

Nontelencephalic afferents to the medial pallium arise bilaterally from the anterior thalamic nucleus (Scalia and Colman, 1975; Vesëlkin and Ermakova, 1978; Ronan and Northcutt, 1979; Neary, 1984a) and from the isthmal raphe (Ronan and Northcutt, 1979). The raphe neurons likely provide a serotonergic input to the medial pallium (Parent, 1973; Yoshida *et al.,* 1983; Ueda *et al.,* 1984).

Results from applications of WGA–HRP in the bullfrog medial pallium, represented by case MP-13 (Figs. 3, 4), confirmed nearly all of these input sources, but differed on a few. Labeled neurons were not found in the nucleus accumbens or lateral amygdala, cell labeling in the bed nucleus of the pallial commissure was indefinite, and labeled cells were present in the ipsilateral suprachiasmatic nucleus (Fig. 4a). Additionally, a few weakly labeled neurons were scattered throughout much of the ipsilateral dorsal isthmus in a loose column beginning ventromedial to nucleus isthmi and extending caudally beneath the cerebellum (Fig. 4d). Catecholaminergic neurons are present in the dorsal isthmal area (Yamamoto *et al.,* 1977; Schmidt, 1980; Inagaki *et al.,* 1981; Yoshida *et al.,* 1983), but it is not known if they provide input to the medial pallium.

Anterograde transport of WGA–HRP to the medial pallium occurred after applications in the olfactory bulb (case OB-2; level not charted), dorsal pallium (case DP-4; Fig. 5a,b), septal area (case S-9; Fig. 6), or anterior thalamic nucleus (case D-61; Fig. 7a,b). These findings support previous evidence for projections to the medial pallium from these populations.

Afferents to the medial pallium do not terminate in a uniform, homoge-

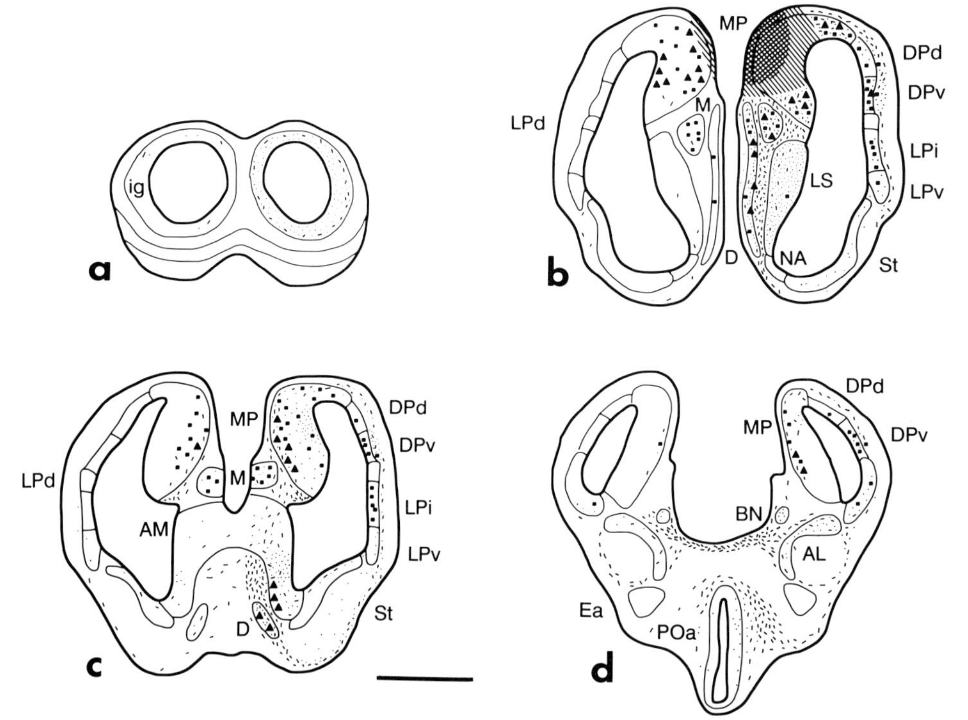

Figure 3. Results from a WGA–HRP application in the medial pallium (case MP-13). See Figure 4 for more caudal levels. In this and the remaining chartings (Figs. 4–9), the transverse levels progress from rostral to caudal, the application or injection site is indicated by hatching, heavily and lightly labeled cells are symbolized by triangles and squares, fibers and terminals are represented by dashes and dots, and, for clarity, only the inner cell layer of the dorsal pallium is outlined. Bar = 1 mm.

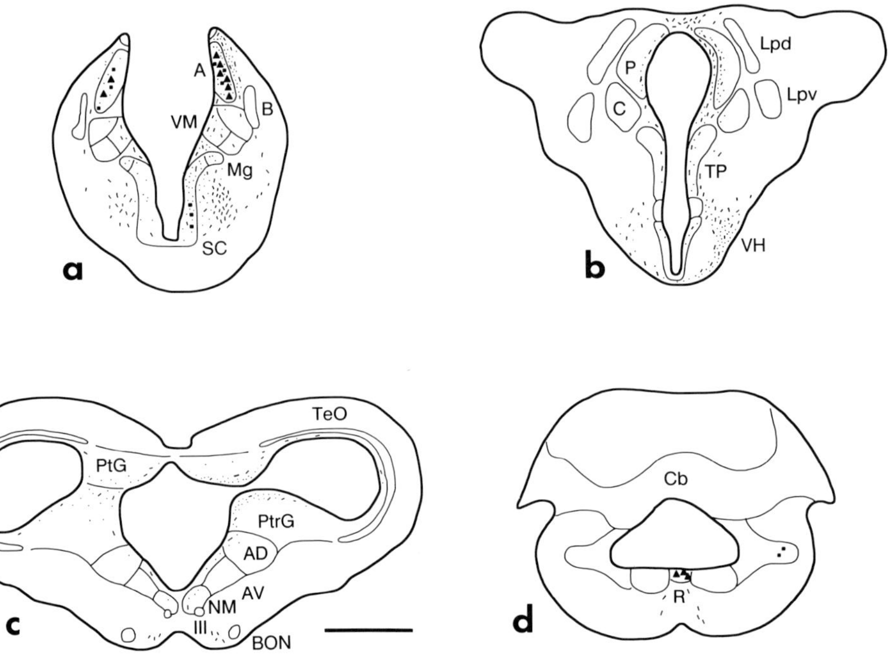

Figure 4. Continuation of results from case MP-13. Bar = 1 mm.

neous manner. Olfactory bulb input is limited to the most rostral and ventral part of the medial pallium, where it is found just beneath the pial surface (Scalia *et al.*, 1968; Northcutt and Royce, 1975). Following hemispherectomies in either *Rana* (Halpern, 1972) or *Bufo* (Kokoros, 1973), degenerating fibers and terminals were present primarily in the ventral two-thirds of the medial pallium. These results indicate that MPs and MPl may receive greater contralateral hemispheric input than MPt. After large diencephalic lesions, ascending degeneration occurred primarily along the medial and dorsal edges of the medial pallium (Kicliter and Northcutt, 1975), suggesting that ascending thalamic input may terminate largely in its outer layer. This terminal pattern was verified after WGA–HRP injections into the anterior thalamic nucleus (case D-61; Fig. 7a,b).

Following WGA–HRP applications in the dorsal pallium or septal area, labeled fibers and terminals in the medial pallium were most heavily concentrated along its medial and dorsal surfaces and in its outer layer (Figs. 5a,b, and 6). Medial pallial applications of WGA–HRP labeled fibers and terminals in the opposite medial pallium, but their distribution could not be accurately determined because, in every case, there was a slight spread of label from the injection site into the opposite hemisphere, followed by uptake and local transport by ependymoglia.

The efferent connections of the medial pallium are more extensive than those of the other pallial divisions. Although initial anterograde degeneration studies (Halpern *et al.*, 1968; Halpern, 1972) indicated that medial pallial projections were, for the most part, limited to a few telencephalic targets (ipsilateral

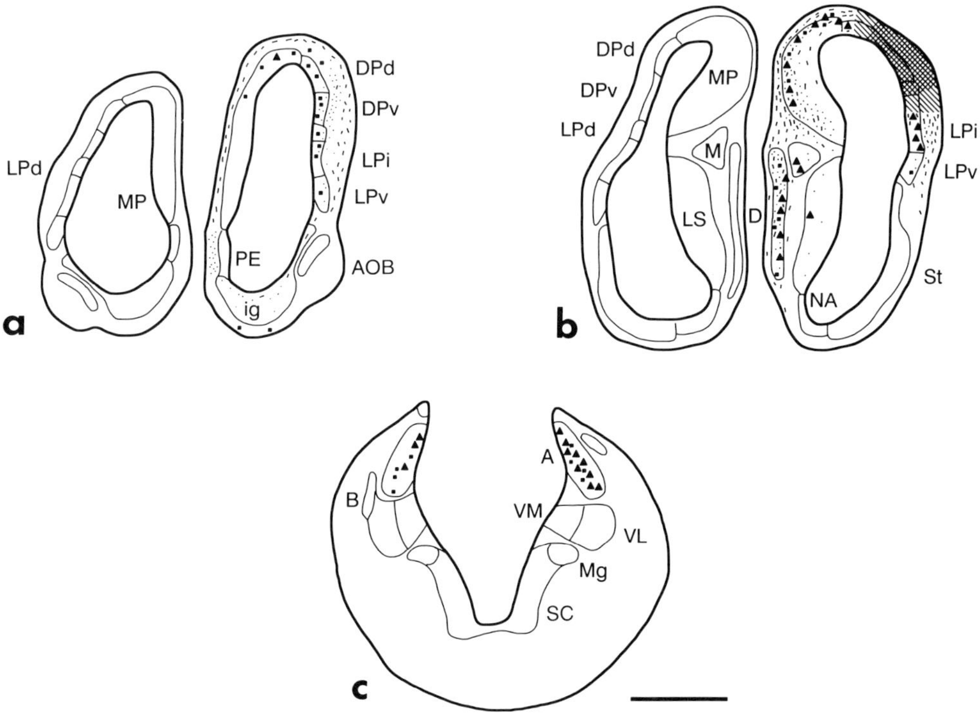

Figure 5. Results from a WGA–HRP application centered in the dorsal pallium (case DP-4). Bar = 1 mm.

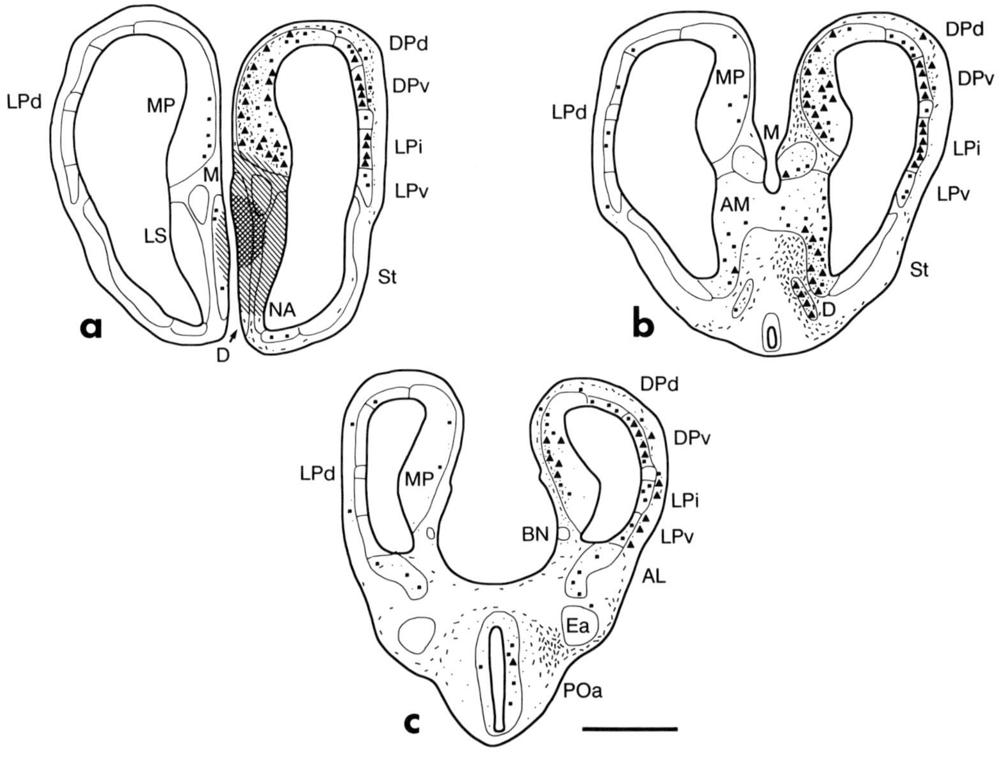

Figure 6. Partial results from a septal area WGA–HRP application centered in the diagonal band nucleus (case S-9). Bar = 1 mm.

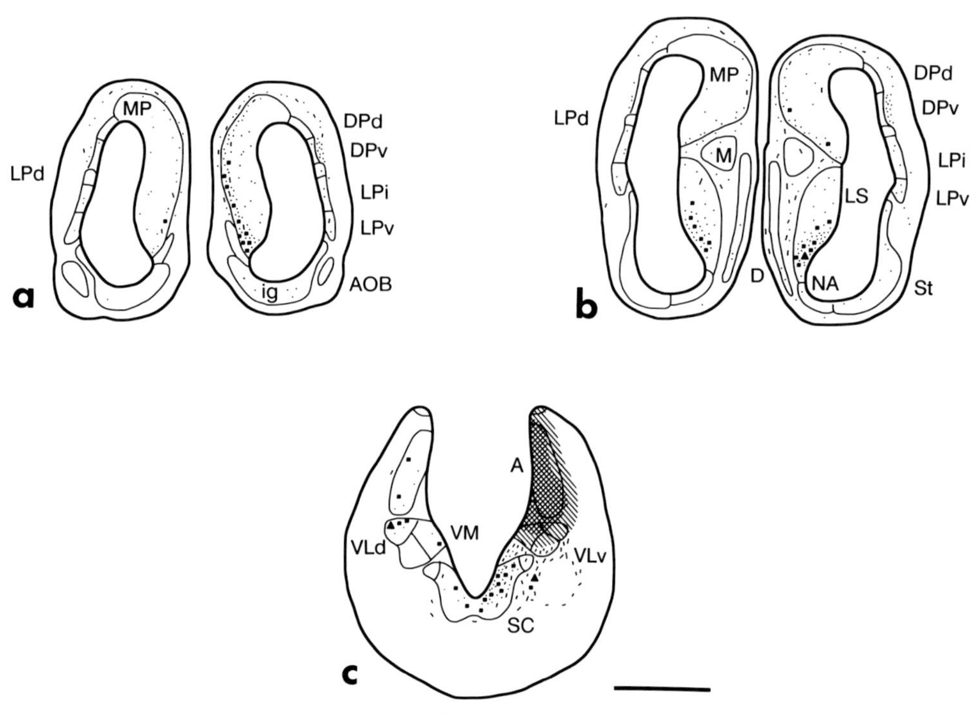

Figure 7. Partial results from a WGA–HRP injection centered in the anterior thalamic nucleus (case D-61). Bar = 1 mm.

medial septal nucleus, postolfactory eminence, and bed nucleus of the pallial commissure), Kokoros (1973) later found evidence for ipsilateral connections to the dorsal and lateral pallia, habenula, and rostral dorsal thalamus, for bilateral projections to both septal nuclei, the diagonal band nucleus, preoptic area, hypothalamus, and ventral thalamus, and for contralateral connections with the opposite medial pallium. Ronan and Northcutt (1979) confirmed most of these efferents and further reported ipsilateral medial pallial input to the internal granule layer of the olfactory bulb, and bilateral input to the bed nucleus of the pallial commissure, both parts of the amygdala, and anterior thalamic nucleus. They did not observe ipsilateral connections with the dorsal pallium or habenula, nor contralateral projections to the lateral septal nucleus. Medial pallial efferents to the olfactory bulb and hypothalamus were previously reported by Scalia (1976a) and have since been confirmed in retrograde tracing studies (Kratskin and Ragimova, 1985; Neary, 1984b).

Placements of WGA–HRP in the medial pallium revealed efferent connections similar to those above (Figs. 3, 4). In the telencephalon, labeled fibers could be traced to the internal granule layer of the olfactory bulb, dorsal pallium (especially DPv), lateral pallium (particularly LPi and LPv), both septal nuclei and the diagonal band nucleus, bed nucleus of the pallial commissure, both parts of the amygdala, striatum, and the opposite medial pallium (Fig. 3). Most of these projections were bilateral, with a stronger ipsilateral representation. A projection to the opposite dorsal pallium, however, was not evident. The most significant differences between these results and those of Ronan and Northcutt

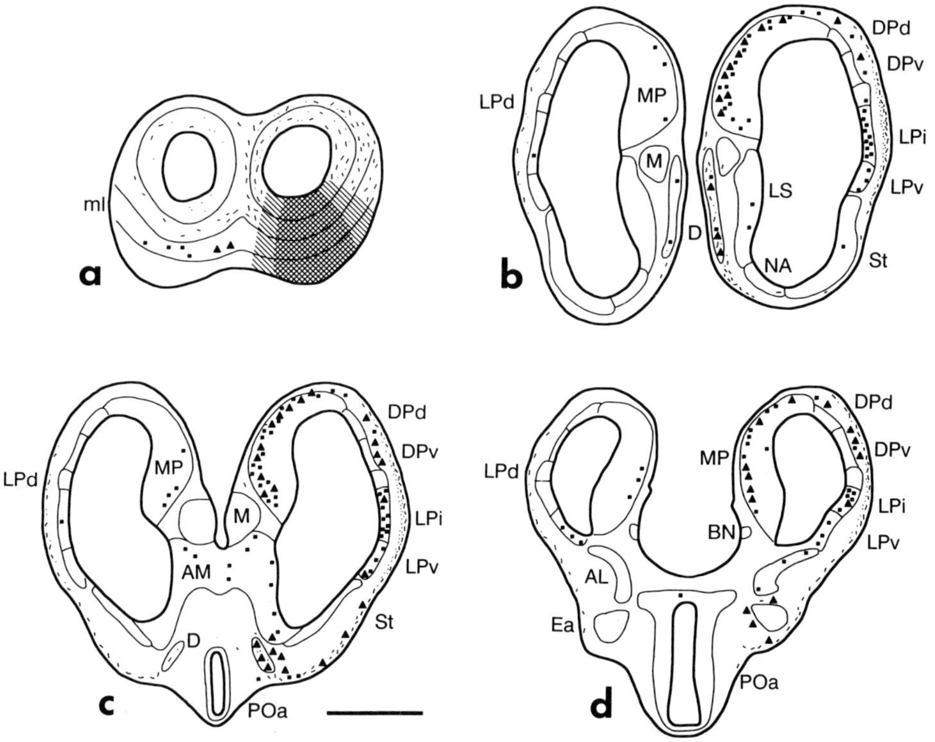

Figure 8. Partial results from a WGA–HRP application in the main olfactory bulb (case OB-2). Bar = 1 mm.

(1979) are the connections to the ipsilateral dorsal pallium and striatum. Evidence for a medial pallial projection to the striatum was also found by Neary (1983).

Descending fibers from the medial pallium exited the telencephalon bilaterally through the medial forebrain bundle (MFB), although some may also have passed into the diencephalon by coursing caudally from the pallial commissure. Major diencephalic targets of the medial pallium were the anterior thalamic nucleus, and structures associated with the course of the MFB—preoptic area, suprachiasmatic nucleus, and infundibular hypothalamus (Figs. 3d, 4a,b). Labeled fibers also left the MFB and formed an extensive diencephalic periventricular plexus that continued into the midbrain. Fibers in this plexus passed through and appeared to contribute terminals to the posterior thalamic nucleus, posterior tuberculum, pretectal and pretoral gray areas, deep layers of the optic tectum, and the periventricular tegmentum (Fig. 4b,c). Although labeled fibers were present in the ventromedial thalamus (Fig. 4a), they appeared to be mostly fibers of passage. Finally, the labeled fibers seen in the isthmal region (Fig. 4d) appeared to be emanating from retrogradely labeled neurons in the raphe.

The outer cell layer of the medial pallium seems to be the predominant source of medial pallial efferents. Following WGA–HRP placements in the dorsal pallium, septal area, anterior thalamic nucleus, or olfactory bulb, most of the labeled medial pallial neurons were present in this layer (Figs. 5a,b, 6, 7a,b, 8b–d). However, a significant number of labeled cells were also present in the inner layer of MPl after tracer placements in the septal area (Fig. 6b). The segregation of neurons afferent to the olfactory bulb was particularly striking and constitutes the best evidence to date for the presence of an outer medial pallial layer. This segregation was not clearly illustrated in Kratskin and Ragimova's (1985) study of olfactory bulb afferents.

Connectional studies suggest that the medial pallium is the major pallial integrative center for olfactory and ascending sensory information. Although it receives only a slight secondary olfactory projection, three other secondary olfactory areas, the lateral and dorsal pallia and the diagonal band nucleus (Scalia *et al.*, 1968; Northcutt and Royce, 1975), provide tertiary olfactory input. Ascending sensory information is most likely relayed from the anterior thalamic nucleus, a polysensory population (Section 4.5.1). Although visual, auditory, and somatosensory stimuli evoke responses in the medial pallium, modality-specific areas have not been found (Karamian *et al.*, 1966; Northcutt, 1970a; Vesselkin et al., 1971; Mudry and Capranica, 1980), and olfactory and nonolfactory responsive areas overlap (Belekhova, 1979).

The medial pallium probably exerts a strong influence on the olfactory system via its projections to the olfactory bulb and secondary olfactory areas. It undoubtedly also has important effects on the limbic system, given its extensive projections to structures associated with the MFB (septal area, preoptic area, suprachiasmatic nucleus, and infundibular hypothalamus). Finally, the medial pallium appears to be the pallial field with the greatest capability of modulating motor functions. It projects both to the striatum and to several populations in the caudal thalamus and pretectum that have descending connections. These populations include the posterior thalamic nucleus, posterior tuberculum, and pretectal gray. All three send efferents to the optic tectum and brain stem (Wilczynski and Northcutt, 1977; Neary and Wilczynski, 1980; Neary, 1982, unpublished observations); the posterior nucleus also projects to the spinal cord (Tóth *et al.*, 1985).

4.2. Dorsal Pallium

Anterograde degeneration experiments have shown that the dorsal pallium receives afferents from the olfactory bulb (Scalia *et al.*, 1968; Northcutt and Royce, 1975), ipsilateral medial pallium (Kokoros, 1973), and thalamus (Kicliter and Northcutt, 1975). Both the olfactory bulb and thalamic inputs are bilateral (Northcutt and Royce, 1975; Kicliter and Northcutt, 1975), and WGA–HRP studies have demonstrated that the thalamic projection arises bilaterally from the anterior thalamic nucleus (Neary, 1984a). Olfactory bulb fibers are located superficially in the dorsal pallium and are most numerous in DPv (Scalia *et al.*, 1968; Northcutt and Royce, 1975). Anterior nucleus afferents ascend in the MFB, pass dorsally from the septal area to supply the medial pallium, and, from there, arch laterally in the dorsum of the hemisphere to course through and terminate in the superficial neuropil of DPd, eventually ending among the outer layer cells of DPv (Kicliter and Northcutt, 1975; Neary, 1984a).

A placement of WGA–HRP centered in DPv (case DP-4, Fig. 5) labeled neurons in the olfactory bulb, ipsilateral medial pallium, and anterior thalamic nucleus. There were also major inputs from the ipsilateral diagonal band nucleus and minor ones from both septal nuclei. Numerous labeled cells were present in LPi, but their proximity to the injection site obscures their significance. Additional evidence for afferents to the dorsal pallium from the medial pallium, septal area, anterior thalamic nucleus, and olfactory bulb was obtained after WGA–HRP applications in these structures (Figs. 3b–d, 6, 7a,b, and 8b–d). Medial pallial, septal, and thalamic afferents terminated primarily in the outer cellular layer of DPv.

Efferents from the dorsal pallium are largely confined to the ipsilateral telencephalon. After a lesion of the dorsal pallium, Kokoros (1973) found degeneration present in the medial pallium, dorsomedial part of the medial septal nucleus (part of the diagonal band nucleus of the present study), and lateral pallium. His findings were corroborated in case DP-4. Labeled fibers could be followed from the placement site ipsilaterally into the outer layer of the medial pallium, diagonal band nucleus, and lateral pallium (Fig. 5a,b). An additional strong projection to the ipsilateral postolfactory eminence was also present (Fig. 5a), as were weak ipsilateral projections to the internal granule cell layer of the olfactory bulb and both septal nuclei (Fig. 5a,b).

Placements of WGA–HRP in the medial pallium, septal area, or olfactory bulb labeled cells in both parts of the dorsal pallium, but most cells were found in DPv (Figs. 3b–d, 6, and 8b–d). Labeled neurons were present only in the outer dorsal pallial layer after olfactory bulb applications, but most were found in the inner layer in both medial pallial and septal cases.

The dorsal pallium, like the medial pallium, appears to be an integrative area for olfactory and ascending sensory activity. Olfactory input arises directly from the main bulb and indirectly from the diagonal band nucleus. The dorsal pallium may respond to visual and somatosensory stimuli (Northcutt, 1970a; Northcutt and Kicliter, 1980). Fibers transmitting polysensory information to the dorsal pallium are likely to arise from the anterior thalamic nucleus (Section

4.5.1). Dorsal pallial efferents pass primarily to structures in the medial hemispheric wall.

4.3. Lateral Pallium

The lateral pallium receives a strong projection from the olfactory bulb over the lateral olfactory tract (Scalia *et al.*, 1968; Northcutt and Royce, 1975), but also receives afferents from the ipsilateral medial pallium (Kokoros, 1973; Ronan and Northcutt, 1979) and from several more caudal sources. Most of the ascending projections to the lateral pallium end in the LPv. Kicliter and Northcutt (1975) traced degenerating fibers to this area after isthmal and tectothalamic hemisections, and Wilczynski and Northcutt (1983) illustrated labeling in or adjacent to the LPv after tritiated proline injections in the lateral amygdala, anterior entopeduncular nucleus, anterior preoptic area, and central thalamic nucleus. The caudal part of the LPv also receives an ascending projection from the infundibular hypothalamus (Neary and Wilczynski, 1977a).

The results from applications of WGA–HRP in the medial pallium or the olfactory bulb confirmed their projections to the lateral pallium (Figs. 3b–d and 8b–d). Tracer placements in the septal area (Fig. 6) also labeled efferents to the lateral pallium. Most of the labeled fibers and terminals present in the lateral pallium after medial pallial and septal applications were found adjacent to LPi and LPv; LPd appeared to receive little input from these areas.

The lateral pallium may also receive afferents from more than one thalamic nucleus. Injections of WGA–HRP in the anterior thalamic nucleus produced sparse anterograde labeling in LPi and LPv (Fig. 7a,b), and injections in the middle thalamic zone (which includes the central nucleus; see Section 4.5) labeled a few fibers and terminals in LPv (Fig. 9a).

The lateral pallium projects primarily to ipsilateral telencephalic areas: olfactory bulb, medial pallium and dorsal pallium, diagonal band nucleus, septal nuclei, and striatum (Scalia, 1976a; Ronan and Northcutt, 1979; Northcutt, 1981; Kratskin and Ragimova, 1985). The caudal part of LPv also projects to a central zone in the infundibular hypothalamus (Neary, 1984b).

Labeled neurons were present in the lateral pallium following WGA–HRP placements in the olfactory bulb (Fig. 8b–d), medial pallium (Fig. 3b–d), and

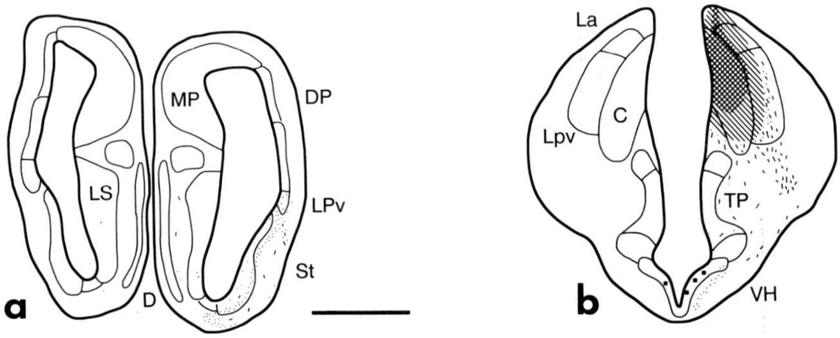

Figure 9. Partial results from a middle thalamic zone WGA–HRP injection placed in the central thalamic nucleus (case D-60). Bar = 1 mm.

septal area (Fig. 6). After olfactory bulb applications, most of the labeled neurons were found in LPi, fewer in LPv, and almost none in LPd. Medial pallial and septal area placements also labeled neurons primarily in LPi.

The lateral pallium appears to be primarily an olfactory area with major outputs directed to the olfactory bulb, medial hemispheric wall, and hypothalamus. It may also receive some ascending sensory afferents from the thalamus, but since these projections are meager, they probably have minor functional significance.

4.4. Lateral Amygdala

The lateral amygdala receives a strong projection to its cell-poor core from the ipsilateral accessory olfactory bulb (Scalia, 1972; Northcutt and Royce, 1975). In addition, it also receives bilateral input from the medial pallium (Ronan and Northcutt, 1979; see also Fig. 3d) and hypothalamus (Neary and Wilczynski, 1977a).

Lateral amygdalar efferents enter the infundibular hypothalamus (Scalia, 1976a) and end in the neuropil surrounding the central zone occupied by LPv terminals (Neary, 1984b). The lateral amygdala may also project to the ventral and dorsal thalamus (Halpern, 1972), septal area (Fig. 6c), and olfactory bulb (Fig. 8d).

4.5. Thalamic Connections

Thalamic afferents to the pallium in anurans are primarily derived from the anterior thalamic nucleus, a population with mixed sensory and limbic affinities. A few pallial afferents may also arise from cells in the middle thalamic zone, which consists of the central thalamic nucleus and the anterior and posteroventral parts of the lateral thalamic nucleus. This zone receives visual, auditory, and somatosensory information from the midbrain roof (optic tectum and torus semicircularis).

4.5.1. Anterior Thalamic Nucleus

Efferents from a variety of sensory and limbic structures, and from others more difficult to classify, converge upon the anterior thalamic nucleus. Retinal axons appear to end on the dendrites of anterior nucleus neurons in a cell-poor terminal region, the neuropil of Bellonci (Scalia and Gergory, 1970). The torus semicircularis projects lightly to the anterior nucleus and its fibers may convey either auditory and/or somatosensory information (Comer and Grobstein, 1981; Neary and Wilczynski, 1986; Neary, 1987, 1988). Somatosensory information reaches the anterior nucleus from a set of cells in the dorsal obex region, the perisolitary band, that is comparable to the dorsal column nuclei and spinal trigeminal nucleus (Neary and Wilczynski, 1977b; Nikundiwe *et al.*, 1982; Urbán and Székely, 1982). Given the variety of sensory afferents to the anterior thalamic nucleus, it is probably a multimodal area with efferents transmitting polysensory information.

Limbic structures afferent to the anterior nucleus include the medial pallium (Ronan and Northcutt, 1979), preoptic area, suprachiasmatic nucleus, and ventral hypothalamic nucleus (Neary and Wilczynski, 1979). Other sources to the anterior nucleus are the posterior entopeduncular nucleus, posterior thalamic nucleus, pretectal gray, and posteroventral tegmental nucleus (Neary and Wilczynski, 1979).

The anterior nucleus has a strong, predominantly ipsilateral projection to the medial pallium and a similar, but smaller projection to the dorsal pallium (Scalia and Colman, 1975; Vesëlkin and Ermakova, 1978; Ronan and Northcutt, 1979; Neary, 1984a). It also sends fibers to the striatum, hypothalamus, and optic tectum (Neary, 1983; Neary and Wilczynski, 1980, 1986).

The results following iontophoretic injections of the anterior thalamic nucleus are represented by those from case D-61 (Fig. 7). Although the ventral thalamus was partly included in this particular injection, similar results, with substantially weaker labeling, were obtained after smaller injections confined to the anterior nucleus. Labeled neurons were found in all the central afferent populations mentioned above and were also present in the posterior tuberculum (not shown), suprachiasmatic nucleus (Fig 7c), and lateral septal nucleus (Fig. 7b). Medial pallial input was derived from outer layer cells (Fig. 7a,b).

Anterior nucleus efferents in case D-61 and others were traced to the targets described earlier, including the medial and dorsal pallia, and striatum (Fig. 7a,b). Additional areas receiving labeled fibers were the lateral septal nucleus (Fig. 7b), lateral pallium (primarily to LPv and LPi; Fig. 7a,b), and the internal granule cell layer of the main olfactory bulb (Fig. 7a).

4.5.2. Middle Thalamic Zone

The optic tectum and torus semicircularis transmit visual, auditory, and somatosensory information to the middle thalamic zone and are its major sources of afferents (Scalia, 1976a,b; Comer and Grobstein, 1981; Neary and Wilczynski, 1986; Hall and Feng, 1987; Neary, 1987, 1988). Other afferents to this zone include the superior olive, secondary isthmal nucleus, and posterior entopeduncular nucleus (Hall and Feng, 1987). The middle thalamic zone projects strongly to the striatum and hypothalamus (Scalia and Colman, 1975; Vesselkin *et al.*, 1980; Wilczynski and Northcutt, 1983; Hall and Feng, 1987; Neary, 1988) and has weaker projections to the torus and secondary isthmal nucleus (Hall and Feng, 1987). A possible weak projection to LPv was illustrated by Wilczynski and Northcutt (1983).

Injections of WGA–HRP into the middle thalamic zone were placed in the central thalamic nucleus, but involved the lateral nucleus to varying extents. These injections labeled cells in the torus semicircularis and secondary isthmal nucleus (not shown), and a few neurons in the ventral hypothalamic nucleus (Fig. 9b).

A strong projection to the infundibular hypothalamus was evident in these cases (Fig. 9b), but anterogradely labeled fibers also coursed forward from the injection sites, through the lateral forebrain bundle, and terminated heavily in the ipsilateral striatum (Fig. 9a). Sparse, but consistent terminal labeling was also present throughout most of the rostrocaudal extent of the ipsilateral LPv (Fig. 9a), confirming the results charted by Wilczynski and Northcutt (1983). Lastly, a sparse projection to the septal area (Fig. 9a) was present in most cases.

4.6. Pallial Connections in Urodeles

The medial pallium in urodeles receives a projection rostrally from the main olfactory bulb (Northcutt and Kicliter, 1980) and a bilateral projection from an anterior thalamic population, the anteroventral zone of Wicht and Himstedt (1986, 1988). The anteroventral zone is very similar to the anterior thalamic nucleus of frogs: it is adjacent to a retinal neuropil, projects back to the optic tectum, and also has a bilateral projection to the dorsal pallium (Wicht and Himstedt, 1986, 1988). The projections of individual pallial fields have not been studied in salamanders. However, pallial connections in general have been documented and appear similar to those in anurans. Kokoros and Northcutt (1977) found that a lesion including the medial and dorsal pallia produced degenerating terminals in the opposite medial pallium, ipsilateral anterior dorsal thalamus, and bilaterally in the septal nuclei, preoptic area, and caudal hypothalamus.

The urodele dorsal pallium receives afferents from the main olfactory bulb (Northcutt and Kicliter, 1980) and, as noted above, from an anterior thalamic group similar to the anuran anterior thalamic nucleus (Wicht and Himstedt, 1986, 1988).

The lateral pallium in salamanders receives input from the main olfactory bulb, and the lateral amygdala receives fibers from the accessory olfactory bulb (Northcutt and Kicliter, 1980). Both of these connections are present in anurans.

5. Pallial Connections in Reptiles

The medial, dorsal, and lateral pallia of amphibians have traditionally been considered homologues of the medial, dorsal (general), and lateral cortices of reptiles. However, in addition to having three pallial cortical areas, reptiles also possess a distinct and prominent structure, the dorsal ventricular ridge (DVR), which has generated controversy both over its pallial or subpallial origin and over the possible relation of its anterior part to mammalian isocortex. Many, if not most, present-day investigators consider the DVR to be largely derived from the pallium, but the question of homology between the anterior DVR and isocortex is far from answered (e.g., Northcutt, 1981; Ulinski, 1983; Bruce and Butler, 1984a).

Northcutt and Kicliter (1980) noted that the amphibian lateral pallium and lateral amygdala, together, have numerous histochemical, topographic, embryologic, and connectional similarities with both the DVR and the lateral cortex of reptiles. Their analysis suggested that the lateral pallium is homologous to both the lateral cortex and part of the DVR, and that the lateral amygdala is homologous to the nucleus sphericus, located in the posterior or basal DVR. More recent evidence lends additional support to their interpretations, and further indicates, as first proposed by Northcutt (1974), that the LPv, in particular, is most closely related to the DVR.

In the discussion that follows, a broad overview of the similarities and differences between reptiles and anurans was attempted. Many conflicting findings in reptiles are not addressed, especially since it is often difficult to assess whether they resulted from technical differences or represent actual variation among species.

5.1. Medial Cortex

The medial cortex of reptiles is generally subdivided into small- and large-celled parts. Röthig (1912) suggested that these may be homologous to the small- and large-celled parts of the anuran medial pallium. However, lack of detailed information on the connections of these subdivisions in frogs precludes such precise comparison at present. Rather, the connections of the medial cortex as a whole will be compared with those of the entire anuran medial pallium.

The medial cortex has several connectional features in common with the medial pallium. It receives a limited, rostral input from the main olfactory bulb (Northcutt, 1970b; Rolon and Skeen, 1980; Reiner and Karten, 1985) and projects to the opposite medial cortex (Voneida and Ebbesson, 1969; Lohman and Mentink, 1972; Bruce andButler, 1984b). The medial cortex has reciprocal connections with the ipsilateral dorsal and lateral cortices, and with the septal area, including the diagonal band nucleus (Lohman and Mentink, 1972; Ulinski, 1975, 1976; Butler, 1976; Lohman and van Woerden-Verkley, 1976; Halpern, 1980; Sligar and Voneida, 1981; Bruce and Butler, 1984b; Desan, 1984). It receives input from an isthmal raphe nucleus (Lohman and van Woerden-Verkley, 1978; Bruce and Butler, 1984b) and from at least one rostral dorsal thalamic group, the nucleus dorsolateralis anterior (DLA; Lohman and van Woerden-Verkley, 1978; Bruce and Butler, 1984b; Desan, 1984). Another rostral dorsal thalamic nucleus, nucleus dorsomedialis (DM) may also project to the medial cortex (Balaban and Ulinski, 1981). (Section 5.2 contains a comparison of these thalamic groups with the anterior thalamic nucleus of anurans.) Lastly, the medial cortex projects to nucleus sphericus (the probable homologue of the lateral amygdala), the hypothalamus, and, possibly, rostral thalamic populations (Northcutt, 1970b; Lohman and Mentink, 1972, Lohman and van Woerden-Verkley, 1976; Halpern, 1980).

Major connectional differences between the reptilian medial cortex and the anuran medial pallium include the apparent absence of projections from the medial cortex to the main olfactory bulb (Rolon and Skeen, 1980), absent or restricted projections to the striatum (Lohman and van Woerden-Verkley, 1976; Bruce and Butler, 1984b; Desan, 1984), efferents to the opposite dorsal cortex (Butler, 1976; Bruce and Butler, 1984b), and receipt of fibers from a mammillary body (Lohman and van Woerden-Verkley, 1978; Bruce and Butler, 1984b), a structure that cannot be identified with certainty in amphibians. Additionally, in contrast to the medial pallium, the medial cortex does not appear to receive input from a retino-recipient thalamic population and it is also not the major pallial target of ascending sensory projections from the thalamus. [However, since part of the medial cortex in lizards is specialized for the receipt of non-olfactory sensory information (Belekhova, 1979), thalamic sensory projections to the medial cortex may be well developed in these animals.] Finally, the medial cortex has not been found to project to the caudal thalamus, pretectum, or midbrain.

5.2. Dorsal Cortex

The cytoarchitecture of the reptilian dorsal cortex is complex and heterogeneous, and has resulted in varying nomenclatures, particularly across orders. In view of recent studies on thalamocortical and intrahemispheric connections in lizards (Bruce, 1982; Bruce and Butler, 1984b) and turtles (Desan, 1984; Heller and Ulinski, 1987), Butler's (1976) suggested subdivision of the dorsal cortex into medial and lateral regions appears the most useful for comparative purposes. The medial region has strong connectional relations with the medial hemispheric wall. In lizards, it corresponds to the dorsal cortex as delimited by Bruce (1982) and Bruce and Butler (1984b) and, in turtles, it is represented by Desan's (1984) area D1. The lateral region receives visual information from a dorsal thalamic, retino-recipient population and is present as the pallial thickening of Bruce (1982) and Bruce and Butler (1984b) in lizards. In turtles, this region corresponds to Desan's (1984) area D2, which includes the pallial thickening and a lateral portion of "general" (dorsal) cortex (Heller and Ulinski, 1987).

The dorsal cortex is similar to the dorsal pallium in having reciprocal connections with the medial cortex and septal nuclei, and in receiving afferents from the diagonal band nucleus (Lohman and Mentink, 1972; Ulinski, 1975; Butler, 1976; Lohman and van Woerden-Verkley, 1976, 1978; Halpern, 1980; Bruce, 1982; Bruce and Butler, 1984b; Desan, 1984). However, it differs significantly from the dorsal pallium with respect to olfactory relationships, commissural connections, thalamic afferents, functional organization, and both striate and extratelencephalic connections:

1. In contrast to the dorsal pallium, the dorsal cortex apparently does not receive olfactory bulb input (Heimer, 1969; Scalia *et al.*, 1969; Northcutt, 1970b; Halpern, 1976; Rolon and Skeen, 1980; Ulinski and Peterson, 1981; Desan, 1984; Reiner and Karten, 1985) nor does it project back to the olfactory bulb (Halpern, 1980; Rolon and Skeen, 1980).

2. The dorsal cortex has commissural connections with the opposite dorsal cortex in some reptiles (Butler, 1976; Bruce and Butler, 1984b), but the dorsal pallium does not connect with the contralateral dorsal pallium.

3. At least three thalamic nuclei project to the dorsal cortex (Hall *et al.*, 1977; Lohman and van Woerden-Verkley, 1978; Balaban and Ulinski, 1981; Bruce and Butler, 1984b; Desan, 1984), but only a single thalamic nucleus projects to the dorsal pallium. All three reptilian nuclei are located in the rostral thalamus and two of these populations, the DM and DLA nuclei, receive inputs from the septum and caudal hypothalamus (Sligar and Voneida, 1981; Hoogland, 1982) and may be considered primarily limbic thalamic nuclei. The third, generally called the dorsal lateral geniculate nucleus (nucleus intercalatus of Bruce and Butler, 1984b), receives afferents from the retina (Hall *et al.*, 1977; Bruce, 1982; Desan, 1984) and is a sensory structure.

It is possible that all three of these reptilian nuclei are differentiated homologues of the anuran anterior thalamic nucleus. The reptilian DM nucleus, like the anterior nucleus, receives inputs from the septum and hypothalamus, noted earlier, and also afferents from a mid-diencephalic entopeduncular nucleus (Hoogland, 1982). Efferent connections of DM include projections to the septum, striatum, dorsal cortex, and, possibly, medial cortex (Parent, 1976; Balaban and Ulinski, 1981; Bruce and Butler, 1984b; Desan, 1984). The anuran anterior nucleus appears to supply comparable areas (Section 4.5.1). The DLA nucleus has the same afferents noted for the DM nucleus (Hoogland, 1982) and projects to the medial, dorsal, and lateral cortices, and striatum (Parent, 1976; Bruce and Butler, 1984b; Desan, 1984). Again, these connections are similar to those of the anterior nucleus. Both DM and DLA may receive some somatosensory information over a spinothalamic projection (Hoogland, 1981). Somatosensory information is also likely to reach the anterior nucleus in frogs, but not over spinothalamic tracts; dorsal column and spinal trigeminal nuclei are more likely sources of this input (Section 4.5.1). Finally, both the lateral geniculate and anterior nucleus receive retinal input and project, respectively, to the dorsal cortex and pallium.

4. The dorsal cortex, as noted at the beginning of this section, can be subdivided, on anatomical grounds, into medial and lateral regions. The medial region has stronger connections with the medial hemispheric wall (including reciprocal connections with the medial cortex) and to the hypothalamus (Lohman and van Woerden-Verkley, 1976; Bruce and Butler, 1984b; Desan, 1984), and has been categorized as a limbic structure (Lohman and Mentink, 1972; Butler, 1976; Bruce, 1982; Desan, 1984). The lateral region receives visual input from the dorsal thalamus and has been considered a sensory structure, possibly homologous to portions of mammalian isocortex (Butler, 1976; Bruce, 1982; Desan, 1984). Although the lateral region may supply the medial hemispheric wall, these projections appear reduced compared to those from the medial region (Lohman and van Woerden-Verkley, 1976; Bruce and Butler, 1984b; Desan, 1984), and its influences on limbic structures may largely be mediated by projections to the medial region (Desan, 1984). Further evidence of the lateral region's isolation from the limbic system is its lack of significant input from the medial cortex (Bruce and Butler, 1984b; Desan, 1984).

The DPd and DPv of frogs may be comparable, on the basis of topography, with the medial and lateral regions of dorsal cortex. The fact that the DPv receives a stronger thalamic projection than DPd (Fig. 7a,b) lends some support to this comparison. However, the DPv also appears to have stronger projections to the medial hemispheric wall (Figs. 3b–d, 6), precisely the opposite of what would be expected if the DPv were equivalent to the lateral dorsal cortical region of reptiles.

The anuran DPv does, however, have another structural feature in common with part of the reptilian lateral dorsal cortex—the pallial thickening. Thalamic afferents to the DPv terminate deeply *among* cells in its outer layer and geniculocortical axons in turtles also appear to synapse among the cell bodies in the cell-rich intermediate layer 2 of the pallial thickening (Heller and Ulinski, 1987). On the other hand, thalamic afferents to the DPd terminate *superficially* in the cell-poor neuropil and geniculocortical afferents to the remainder of D2 and thalamic afferents to D1 also terminate superficially in the cell-poor layer 1 of turtles (Desan, 1984; Heller and Ulinski, 1987).

5. The striatum of reptiles receives a dorsal cortical projection arising primarily from the lateral region (Lohman and van Woerden-Verkley, 1976; Bruce and Butler, 1984b; Desan, 1984). The striatum in frogs, on the other hand, does not receive afferents from the dorsal pallium (Neary, 1983; Wilczynski and Northcutt, 1983; Fig. 5b).

6. The dorsal cortex receives significant ascending input from nonthalamic populations including the hypothalamus and mammillary body, locus coeruleus, superior raphe nucleus, and nucleus parvocellularis medialis in the caudal brain steam (Parent and Poitras, 1974; Bruce and Butler, 1984b; Ouimet *et al.*, 1985). Similar projections have not been demonstrated in anurans.

The dorsal cortex also sends fibers, arising primarily from its medial region, to a number of extratelencephalic targets including the rostral thalamus, hypothalamus, and optic tectum (Lohman and Mentink, 1972; Lohman and van Woerden-Verkley, 1976; Hall *et al.*, 1977; Halpern, 1980; Bruce and Butler, 1984b). The dorsal pallium in frogs, however, does not appear to send fibers outside the telencephalon.

5.3. Lateral Cortex

The lateral cortex is comparable to the anuran lateral pallium in several respects. It receives fibers from the main olfactory bulb (Heimer, 1969; Scalia *et al.*, 1969; Northcutt, 1970b; Halpern, 1976; Rolon and Skeen, 1980; Ulinski and Peterson, 1981; Desan, 1984; Reiner and Karten, 1985) and reciprocates this projection (Halpern, 1980; Rolon and Skeen, 1980). The lateral cortex also has reciprocal connections with the medial cortex (Lohman and Mentink, 1972; Ulinski, 1976; Halpern, 1980; Bruce and Butler, 1984b; Desan, 1984) and receives input from the DLA thalamic nucleus (Bruce and Butler, 1984b). This last projection may be similar to a projection of the anterior thalamic nucleus to the lateral pallium in amphibians.

The lateral cortex also has several other connections that are difficult to compare at present. These include afferents from the diagonal band nucleus, nucleus sphericus, contralateral lateral cortex, and nucleus paraventricularis of the hypothalamus (Bruce and Butler, 1984b). The diagonal band nucleus in frogs may provide input to the lateral pallium, but information regarding afferents from sources comparable to the remaining reptilian structures is not available.

5.4. Dorsal Ventricular Ridge

The DVR is generally subdivided into anterior and posterior or basal portions (see Ulinski, 1983, for review of structure and connections; see also Bruce and Butler, 1984a, for a study of connections). The anterior DVR contains separate, largely modality-specific regions that receive visual, auditory, or somatosensory information relayed from dorsal thalamic nuclei. Two of these thalamic populations, nucleus rotundus and nucleus reuniens (medialis), appear comparable to middle thalamic zone groups in frogs: nucleus rotundus receives afferents from the optic tectum and nucleus reuniens receives afferents from the torus semicircularis. The thalamic nuclei relaying somatosensory information from the spinal cord and dorsal column nuclei to the anterior DVR do not have clear counterparts in the amphibian diencephalon. Although studies indicate that the middle thalamic zone in frogs may receive somatosensory information, this input appears to arise from the torus semicircularis (Section 4.5.2). Other afferents to the anterior DVR arise from the DM thalamic nucleus, the dorsal cortex, and populations in the midbrain tegmentum and isthmus that probably provide monoaminergic input. The anterior DVR has two major efferent projections: to the striatum and posterior DVR. It also has a limited input to the dorsal cortex and may send fibers to the core of the ventromedial hypothalamic nucleus (Voneida and Sligar, 1979).

In lizards and snakes, the posterior DVR receives accessory olfactory bulb input (to nucleus sphericus), and in turtles and crocodilians, a portion of the posterior DVR receives fibers from the main olfactory bulb. The anterior DVR also projects to the posterior DVR. The major output of the posterior DVR is directed to the hypothalamus: In snakes, nucleus sphericus may project to the core of the ventromedial hypothalamic nucleus (Halpern, 1980), but in lizards, it may supply the area surrounding the ventromedial nucleus (Voneida and Sligar, 1979). Other efferents pass to the opposite posterior DVR and internal granule layer of the accessory olfactory bulb.

Amphibians lack an obvious structure that might be compared to the DVR as a whole. Nevertheless, Northcutt and Kicliter (1980) have summarized evidence suggesting that part of the lateral pallium may be homologous both to lateral cortex and to portions of the DVR and, further, that the lateral amygdala is homologous to the nucleus sphericus in the posterior DVR of snakes and lizards. In particular, they noted that neither the lateral pallium nor the DVR displays high cholinesterase and monoamine levels, that both the lateral amygdala and nucleus sphericus are the principal targets of the accessory olfactory bulb, and, lastly, that both the DVR and lateral cortex are derived embryologically from the same matrix zone.

In addition to these points, several other similarities between the LPv and lateral amygdala, on one hand, and the DVR, on the other, may exist:

1. Both LPv and the DVR apparently receive input from dorsal thalamic populations relaying information from the midbrain roof. In anurans, these populations are found in the middle thalamic zone and, in reptiles, they are the reuniens and rotundal nuclei.
2. Both the LPv and the DVR seem to receive input from a rostral thalamic nucleus that, in turn, receives afferents from the hypothalamus and septum. In frogs, this nucleus is the anterior thalamic nucleus. In reptiles, it is the DM nucleus.
3. The caudal LPv and lateral amygdala both project to the caudal hypothalamus and their fibers appear to terminate in separate regions. Similarly, the DVR appears to contain at least two populations that project to the caudal hypothalamus and their fibers also have differing termination patterns.

6. Summary and Conclusions

The pallium in anurans contains three principal fields, the medial, dorsal, and lateral pallia, and a fourth associated field, the lateral amygdala, which is closely related to the lateral pallium. Comparable pallial divisions can be recognized in salamanders, and connectional data, although quite limited, suggest that they have afferents and efferents similar to those in frogs.

The anuran medial pallium is the largest principal field. Three subdivisions can be recognized (small-celled, large-celled, and transitional), but these are poorly differentiated in *Rana* compred, for example, to *Bufo* or *Hyla*. The medial pallium can also be divided into adjacent inner and outer cell layers that extend through all of the subdivisions. The outer layer appears to receive the majority of medial pallial afferents and to be the primary source of medial pallial efferents.

The medial pallium has the most extensive connections of any pallial field. These connections include reciprocal connections with the olfactory bulb, dorsal and lateral pallia, septal area, and anterior thalamic nucleus. The medial pallium projects strongly to diencephalic limbic structures, such as the suprachiasmatic nucleus and hypothalamus, and also has somewhat weaker projections to the striatum, caudal thalamus, pretectum, and midbrain.

In general, the medial pallium appears to be an important integrative area for olfactory and ascending polysensory information. Its efferents to the olfactory bulb are likely to have significant effects on the olfactory system, and its projections to the basal diencephalon directly affect the limbic system. The medial pallium probably also influences motor functions via its projections both to the striatum and to various caudal diencephalic and pretectal populations that, in turn, project to the optic tectum, brain stem, or spinal cord.

The medial pallium has traditionally been considered homologous to the medial cortex of reptiles. Both areas have basically similar connections with other pallial fields, the septal area, and hypothalamus. However, the medial cortex does not appear to project to the olfactory bulb, has more restricted striatal projections, and apparently has no efferents to the caudal thalamus, pretectum, or midbrain. Furthermore, it does not receive afferents from a retino-recipient thalamic population nor does it appear to be the major pallial recipient of ascending sensory thalamic projections. Medial cortical connections absent from the anuran medial pallium include projections to the opposite dorsal cortex and afferents from the mammillary body.

The medial cortex projects to fewer structures than the medial pallium. This may indicate that several types of medial pallial efferents, present in ancestral amphibians, were lost in the evolution of modern reptiles. Alternatively, several medial pallial connections may have developed independently in anurans.

The anuran dorsal pallium has dorsal and ventral parts, DPd and DPv, with adjacent inner and outer cell layers. Compared to the DPd, the DPv receives the majority of afferents from the olfactory bulb, medial telencephalic wall, and thalamus, and has somewhat stronger projections to the olfactory bulb and medial telencephalic wall. Differences between the inner and outer layers are most obvious in DPv: anterior nucleus afferents terminate in its outer layer, cells in this layer project to the olfactory bulb, and its inner layer has stronger projections to the medial telencephalic wall.

The connections of the dorsal pallium are, for the most part, confined to the telencephalon. It has reciprocal connections with the olfactory bulb, medial pallium, and septal area and it projects strongly to the postolfactory eminence. The dorsal pallium also receives afferents from the anterior thalamic nucleus. These fibers probably transmit polysensory information and represent the only known extratelencephalic connections of the dorsal pallium.

The dorsal pallium seems to be an integrative area for olfactory and ascending sensory activity and, in this regard, it is similar to the medial pallium. However, the dorsal pallium receives a smaller thalamic projection and has a much more limited set of efferents. Therefore, its importance as an integrative area is probably secondary to that of the medial pallium.

The dorsal cortex of reptiles has a few connections in common with the dorsal pallium. These include reciprocal connections with medial hemispheric structures and afferents from a retino-recipient thalamic nucleus. Most dorsal cortical connections, however, are not shared by the dorsal pallium. Among these are commissural relations with the opposite dorsal cortex, afferents from multiple rostral thalamic nuclei and several brain-stem populations, and projections to the striatum, thalamus, hypothalamus, and optic tectum. The only connections possessed by the dorsal pallium, but not dorsal cortex, are olfactory bulb afferents and efferents. The dorsal cortex apparently differs from the dorsal pallium in having separate lateral sensory and medial limbic regions. Such a functional segregation is not evident in the dorsal pallium, although the DPv may be related to the pallial thickening portion of the lateral region.

The dorsal cortex of reptiles has far more extensive connections than the anuran dorsal pallium. This suggests either that the dorsal cortex has acquired several new connections or, conversely, that the dorsal pallium has lost many connections. Although the dorsal cortex and medial pallium have some common connections not possessed by the dorsal pallium (e.g., projections to the striatum), these similarities may be instances of homoplasy, perhaps in part related to the fact that both the dorsal cortex and medial pallium receive significant ascending sensory input (see below).

The anuran lateral pallium has three parts (dorsal, intermediate, and ventral) and is the major target of the lateral olfactory tract. The olfactory bulb projection is reciprocated, especially by its intermediate part, LPi, and, to a lesser extent, by its ventral part, LPv. The lateral pallium also has reciprocal connections with the medial pallium and septal area. It appears to receive afferents from the anterior thalamic nucleus and projects to the striatum. Its efferents to the medial wall arise primarily from the LPi. The LPv, however, also has distinctive connections. These include reciprocal connections with the hypothalamus and, possibly, afferents from the middle thalamic zone. The lateral pallium, as a whole, appears to be an olfactory structure; it receives only slight and limited ascending projections from the thalamus.

The anuran lateral pallium and reptilian lateral cortex have similar connections and are likely homologues. However, a portion of the anuran lateral pallium, the LPv, together with the lateral amygdala, may also be partly homologous with another reptilian pallial region, the DVR. These two anuran fields and the DVR appear to have a similar set of thalamic afferents, possess dual projections to hypothalamus, and send efferents to the striatum. The major difference between the LPv and the DVR involves the representation of olfactory and nonolfactory sensory systems. The LPv is primarily an olfactory structure, receives little ascending sensory input, and has no specialized areas dedicated to that input. The anterior DVR, on the other hand, is relatively isolated from the olfactory system, receives abundant ascending sensory input, and has specialized areas for receipt of visual, auditory, and somatosensory information relayed from separate thalamic nuclei.

Since it is doubtful that ancestral amphibians possessed a DVR (Northcutt and Kicliter, 1980), the sparse projection from the anuran middle thalamic zone to the LPv is probably not the remnant of a once robust pathway. However, since the middle thalamic zone has efferents that may be specialized (Neary, 1988), its projection to the LPv may also be derived and have no evolutionary relationship to thalamo-DVR pathways in reptiles.

The fourth anuran pallial field, the lateral amygdala, appears to be a caudoventral extension of the LPv. It is the primary target of the accessory olfactory bulb, receives medial pallial fibers, and has reciprocal connections with the hypothalamus. Its likely reptilian homologue, nucleus sphericus, has most of these connections, although hypothalamic afferents to nucleus sphericus remain to be demonstrated.

In conclusion, anurans and reptiles display significant differences in pallial organization and connections. If these differences reflect changes that took place in reptilian evolution, then the following shifts in pallial relations to sensory, limbic, and motor systems may have occurred:

1. Pallial involvement with the olfactory system became reduced in reptiles. The reciprocal connections between the dorsal cortex and olfactory bulb were lost and medial cortical projections to the olfactory bulb disappeared.

2. Reptiles developed pallial sensory areas that received specific ascending projections from the thalamus and these areas also became relatively isolated from the limbic and olfactory systems. The lateral region of dorsal cortex developed into a visual area receiving ascending input from a specialized, retinorecipient thalamic nucleus, the lateral geniculate, which, in turn, developed from an amphibian anterior thalamic nucleus. Lateral dorsal cortical connections with the medial hemispheric wall were reduced, and direct relations with the olfactory bulb, as noted above, were lost. The anterior part of the DVR differentiated from the lateral pallium, lost its olfactory bulb connections, and developed specific auditory, somatosensory, and visual regions, each receiving input from nuclei derived from the middle thalamic zone.

3. Thalamic sensory projections to reptilian pallial limbic areas decreased and the pallial sensory areas became the primary sources for visual, auditory, and somatosensory input to the limbic pallium. Specifically, ascending sensory projections to the medial cortex (and possibly medial dorsal cortex) were lost or reduced, and the anterior DVR projection to the posterior DVR became a major route for sensory input to the limbic pallium. A secondary route, from the lateral dorsal cortex to the medial dorsal cortex, may also have emerged.

4. The development of specific sensory areas in the reptilian pallium was paralleled by the development of new areas affecting motor function. The decrease in ascending sensory projections to medial cortex was accompanied by a reduction or loss of medial cortical projections to the striatum, caudal thalamus, pretectum, and midbrain. The two major sensory areas, lateral dorsal cortex and the anterior DVR, developed strong projections to the striatum and have become the pallial areas most likely to affect motor function.

These proposals are based largely on the assumption that amphibian pallial evolution has been static. However, both the degree and the nature of possible specializations shown by the pallium in anurans are, for the most part, unknown.

For these questions to be answered, detailed studies of pallial connections in closely related anamniotes, particularly salamanders and lungfish, are still needed.

7. Abbreviations

A	Anterior thalamic nucleus	M	Medial septal nucleus
AD	Anterodorsal tegmental nucleus	Mg	Magnocellular preoptic nucleus
AM	Medial amygdala	ml	Mitral cell layer of the olfactory bulbs
AL	Lateral amygdala	MP	Medial pallium
AOB	Accessory olfactory bulb	MPl	Medial pallium, large-celled part
AV	Anteroventral tegmental nucleus	MPs	Medial pallium, small-celled part
B	Nucleus of Bellonci	MPt	Medial pallium, transitional part
BN	Bed nucleus of the pallial commissure	NA	Nucleus accumbens
BON	Basal optic nucleus	NM	Nucleus of the medial longitudinal fasciculus
C	Central thalamic nucleus	ON	Optic nerve
Cb	Cerebellum	P	Posterior thalamic nucleus
D	Diagonal band nucleus	PE	Postolfactory eminence
DP	Dorsal pallium	POa	Anterior preoptic area
DPd	Dorsal pallium, dorsal part	PtG	Pretectal gray
DPv	Dorsal pallium, ventral part	PtrG	Pretoral gray
Ea	Anterior entopeduncular nucleus	R	Isthmal raphe nucleus
ig	Internal granule cell layer of the olfactory bulbs	SC	Suprachiasmatic nucleus
La	Lateral thalamic nucleus, anterior division	St	Striatum
Lpd	Lateral thalamic nucleus, posterodorsal division	TeO	Optic tectum
Lpv	Lateral thalamic nucleus, posteroventral division	TP	Posterior tuberculum
LPd	Lateral pallium, dorsal part	VH	Ventral hypothalamic nucleus
LPi	Lateral pallium, intermediate part	VL	Ventrolateral thalamic nucleus
LPv	Lateral pallium, ventral part	VLd	Ventrolateral thalamic nucleus, dorsal part
LS	Lateral septal nucleus	VLv	Ventrolateral thalamic nucleus, ventral part
		VM	Ventromedial thalamic nucleus

Acknowledgments. I thank Marilee Webb for her excellent histological and photographic assistance, Jerry Schneider for developing the graphics software used in making the figures, and Laura Bruce for her numerous perceptive comments on the manuscript. These studies were supported by NSF grant BNS-7924699 and BRS grants to Creighton University.

8. References

Ariëns Kappers, C. U., Huber, G. C., and Crosby, E. C., 1936, *The Comparative Anatomy of the Nervous System of Vertebrates, Including Man,* reprinted 1960, Hafner, New York.

Balaban, C. D., and Ulinski, P. S., 1981, Organization of thalamic afferents to anterior dorsal ventricular ridge in turtles. I. Projections of thalamic nuclei, *J. Comp. Neurol.* **200:**95–129.

Belekhova, M. G., 1979, Neurophysiology of the forebrain, in: *Biology of the Reptilia,* Volume 10 (C. Gans, R. G. Northcutt, and P. Ulinski, eds.), Academic Press, New York, pp. 287–359.

Bruce, L. L., 1982, Organization and evolution of the reptilian forebrain: Experimental studies of

forebrain connections in lizards, Ph.D. dissertation, Department of Anatomy, Georgetown University.
Bruce, L. L., and Butler, A. B., 1984a, Telencephalic connections in lizards. II. Projections to anterior dorsal ventricular ridge, *J. Comp. Neurol.* **229:**602–615.
Bruce, L. L., and Butler, A. B., 1984b, Telencephalic connections in lizards. I. Projections to cortex, *J. Comp. Neurol.* **229:**585–601.
Butler, A. B., 1976, Telencephalon of the lizard *Gekko gecko* (Linnaeus): Some connections of the cortex and dorsal ventricular ridge, *Brain Behav. Evol.* **13:**396–417.
Clairambault, P., and Derer, P., 1968, Contributions à l'étude architectonique du Télencéphale des Ranidés, *J. Hirnforsch.* **10:**123–172.
Comer, C., and Grobstein, P., 1981, Organization of sensory inputs to the midbrain of the frog, *Rana pipiens, J. Comp. Physiol.* **142:**161–168.
Desan, P. H., 1984, The organization of the cerebral cortex of the pond turtle, *Pseudemys scripta elegans,* Ph.D. dissertation, Department of Neurobiology, Harvard University.
Hall, J. A., Foster, R. E., Ebner, F. F., and Hall, W. C., 1977, Visual cortex in a reptile, the turtle (*Pseudemys scripta* and *Chrysemys picta*), *Brain Res.* **130:**197–216.
Hall, J. C., and Feng, A. S., 1987, Evidence for parallel processing the frog's auditory thalamus, *J. Comp. Neurol.* **258:**407–419.
Halpern, M., 1972, Some connections of the telencephalon of the frog, *Rana pipiens, Brain Behav. Evol.* **6:**42–68.
Halpern, M., 1976, The efferent connections of the olfactory bulb and accessory olfactory bulb in the snakes, *Thamnophis sirtalis* and *Thamnophis radix, J. Morphol.* **150:**553–578.
Halpern, M., 1980, The telencephalon of snakes, in: *Comparative Neurology of the Telencephalon* (S. O. E. Ebbesson, ed.), Plenum Press, New York, pp. 257–295.
Halpern, M., Scalia, F., and Riss, W., 1968, On the nature of the hippocampal commissure in the frog, *Brain Behav. Evol.* **1:**155–174.
Heimer, L., 1969, The secondary olfactory connections in mammals, reptiles, and sharks, *Ann. N.Y. Acad. Sci.* **167:**129–146.
Heller, S. B., and Ulinski, P. S., 1987, Morphology of geniculocortical axons in turtles of the genera *Pseudemys* and *Chrysemys, Anat. Embryol.* **175:**505–515.
Herrick, C. J., 1927, The amphibian forebrain. IV. The cerebral hemispheres of *Amblystoma, J. Comp. Neurol.* **43:**231–325.
Herrick, C. J., 1933, The amphibian forebrain. VI. *Necturus, J. Comp. Neurol.* **58:**1–288.
Hoffman, H. H., 1963, The olfactory bulb, accessory olfactory bulb, and hemisphere of some anurans, *J. Comp. Neurol.* **120:**317–368.
Hoffman, H. H., 1966, The structure of the forebrain of *Bufo marinus, Ala. J. Med. Sci.* **3:**286–298.
Hoogland, P. V., 1981, Spinothalamic projections in a lizard, *Varanus exanthematicus:* An HRP study, *J. Comp. Neurol.* **198:**7–12.
Hoogland, P. V., 1982, Brainstem afferents to the thalamus in a lizard, *Varanus exanthematicus, J. Comp. Neurol.* **210:**152–162.
Howell, E. M., and Hoffman, H. H., 1966, The forebrain of *Hyla cinerea, Ala. J. Med. Sci.* **3:**270–285.
Inagaki, S., Shiosaka, S., Takatsuki, K., Sakanaka, M., Tagaki, H., Senba, E., Matsuzaki, T., and Tohyama, M., 1981, Distribution of somatostatin in the frog brain, *Rana catesbeiana,* in relation to location of catecholamine-containing neuron system, *J. Comp. Neurol.* **202:**89–101.
Karamian, A. I., Vesselkin, N. P., Belekhova, M. G., and Zagorulko, T. M., 1966, Electrophysiological characteristics of tectal and thalamo-cortical divisions of the visual system in lower vertebrates, *J. Comp. Neurol.* **127:**559–576.
Kicliter, E., 1979, Some telencephalic connections in the frog, *Rana pipiens, J. Comp. Neurol.* **185:**75–86.
Kicliter, E., and Ebbesson, S. O. E., 1975, Organization of the "nonolfactory" telencephalon, in: *Frog Neurobiology* (R. Llinás and W. Precht, eds.), Springer-Verlag, Berlin, pp. 946–972.
Kicliter, E., and Northcutt, R. G., 1975, Ascending afferents to the telencephalon of ranid frogs: An anterograde degeneration study, *J. Comp. Neurol.* **161:**239–254.
Kokoros, J. J., 1973, Efferent projections of the telencephalon in the toad (*Bufo marinus*) and tiger salamander (*Ambystoma tigrinum*), Ph.D. dissertation, Department of Anatomy, Case Western Reserve University.
Kokoros, J. J., and Northcutt, R. G., 1977, Telencephalic efferents of the tiger salamander *Ambystoma tigrinum tigrinum* (Green), *J. Comp. Neurol.* **173:**613–628.
Kratskin, I. L., and Ragimova, N. G., 1985, Origin of the main olfactory bulb centrifugal innervation in the frog, *Rana temporaria:* Combined histochemical investigation, *Dokl. Akad. Nauk SSSR* **282:**747–749.

Kuhlenbeck, H., 1921a, Zur Histologie des Anurenpalliums, *Anat. Anz.* **54:**280–285.

Kuhlenbeck, H., 1921b, Die Regionen des Anurenvorderhirns, *Anat. Anz.* **54:**304–316.

Lohman, A. H. M., and Mentink, G. M., 1972, Some cortical connections of the tegu lizard (*Tupinambis teguxin*), *Brain Res.* **45:**325–344.

Lohman, A. H. M., and van Woerden-Verkley, I., 1976, Further studies on the cortical connections of the tegu lizard, *Brain Res.* **103:**9–28.

Lohman, A. H. M., and van Woerden-Verkley, I., 1978, Ascending connections to the forebrain in the tegu lizard, *J. Comp. Neurol.* **182:**555–574.

Mudry, K. M., and Capranica, R. R., 1980, Evoked auditory activity within the telencephalon of the bullfrog (*Rana catesbeiana*), *Brain Res.* **182:**303–311.

Neary, T. J., 1982, Forebrain projections to the obex region in ranid frogs, *Neurosci. Abstr.* **8:**765.

Neary, T. J., 1983, A caudal striatal division in ranid frogs, *Neurosci. Abstr.* **9:**1064.

Neary, T. J., 1984a, Anterior thalamic nucleus projections to the dorsal pallium in ranid frogs, *Neurosci. Lett.* **51:**213–218.

Neary, T. J., 1984b, A lectin-peroxidase study of hypothalamic afferents in ranid frogs, *Anat. Rec.* **208:**125.

Neary, T. J., 1987, Ascending projections of the torus semicircularis in ranid frogs, *Neurosci. Abstr.* **13:**305.

Neary, T. J., 1988, Forebrain auditory pathways in ranid frogs, in: *The Evolution of the Amphibian Auditory System* (B. Fritzsch, T. Hetherington, M. Ryan, W. Wolkowiak, and W. Wilczynski, eds.), Wiley, New York, pp. 233–252.

Neary, T. J., and Northcutt, R. G., 1983, Nuclear organization of the bullfrog diencephalon, *J. Comp. Neurol.* **213:**262–278.

Neary, T. J., and Wilczynski, W., 1977a, Autoradiographic demonstration of hypothalamic efferents in the bullfrog, *Rana catesbeiana, Anat. Rec.* **187:**665.

Neary, T. J., and Wilczynski, W., 1977b, Ascending thalamic projections from the obex region in ranid frogs, *Brain Res.* **138:**529–533.

Neary, T. J., and Wilczynski, W., 1979, Anterior and posterior thalamic afferents in the bullfrog, *Rana catesbeiana, Neurosci. Abstr.* **5:**144.

Neary, T. J., and Wilczynski, W., 1980, Descending inputs to the optic tectum in ranid frogs, *Neurosci. Abstr.* **6:**629.

Neary, T. J., and Wilczynski, W., 1986, Auditory pathways to the hypothalamus in ranid frogs, *Neurosci. Lett.* **71:**142–146.

Nikundiwe, A. M., de Boer-van Huizen, R., and ten Donkelaar, H. J., 1982, Dorsal root projections in the clawed toad (*Xenopus laevis*) as demonstrated by anterograde labeling with horseradish peroxidase, *Neuroscience* **7:**2089–2103.

Northcutt, R. G., 1970a, Pallial projection of sciatic, ulnar, and mandibular branch of the trigeminal afferents in the frog (*Rana catesbeiana*), *Anat. Rec.* **166:**356.

Northcutt, R. G., 1970b, The telencephalon of the western painted turtle *Pseudemys scripta, Ill. Biol. Monogr.* No. 43.

Northcutt, R. G., 1974, Some histochemical observations on the telencephalon of the bullfrog, *Rana catesbeiana* Shaw, *J. Comp. Neurol.* **157:**379–390.

Northcutt, R. G., 1981, Evolution of the telencephalon in nonmammals, *Annu. Rev. Neurosci.* **4:**301–350.

Northcutt, R. G., and Kicliter, E., 1980, Organization of the amphibian telencephalon, in: *Comparative Neurology of the Telencephalon* (S. O. E. Ebbesson, ed.), Plenum Press, New York, pp. 203–255.

Northcutt, R. G., and Royce, G. J., 1975, Olfactory bulb projections in the bullfrog *Rana catesbeiana, J. Morphol.* **145:**251–268.

Ouimet, C. C., Patrick, R. L., and Ebner, F. F., 1985, The projection of three extrathalamic cell groups to the cerebral cortex of the turtle *Pseudemys, J. Comp. Neurol.* **237:**77–84.

Parent, A., 1973, Distribution of monoamine-containing neurons in the brain stem of the frog, *Rana temporaria, J. Morphol.* **139:**67–78.

Parent, A., 1976, Striatal afferent connections in the turtle (*Chrysemys picta*) as revealed by retrograde axonal transport of horseradish peroxidase, *Brain Res.* **108:**25–36.

Parent, A., and Poitras, D., 1974, The origin and distribution of catecholaminergic axon terminals in the cerebral cortex of the turtle (*Chrysemys picta*), *Brain Res.* **78:**345–358.

Ramón y Cajal, P., 1946, El Cerebro de los Batracios, *Trab. Inst. Cajal de Invest. Biol.* **38:**41–111.

Reiner, A., and Karten, H. J., 1985, Comparison of olfactory bulb projections in pigeons and turtles, *Brain Behav. Evol.* **27:**11–27.

Rolon, R. R., and Skeen, L. C., 1980, Afferent and efferent connections of the olfactory bulb in the soft-shelled turtle (*Trionyx spinifer spinifer*), *Neurosci. Abstr.* **6**:305.

Ronan, M. C., and Northcutt, R. G., 1979, Afferent and efferent connections of the bullfrog medial pallium, *Neurosci. Abstr.* **5**:146.

Röthig, P., 1912, Beiträge zur Studium des Centralnervensystems der Wirbelthiere. 5. Die Zellanordnungen im Voderhirn der Amphibien, mit besonderer Berücksichtigung der Septumkerne und ihr Vergleich mit den Verhältnissen bei Testudo und Lacerta, *Verh. K. Akad. Wet. Amsterdam* (Ser. 2) **17**:1–73.

Rubaschkin, W., 1903, Zur Morphologie des Gehirns der Amphibien, *Arch. Mikrosk. Anat.* **62**:207–243.

Scalia, F., 1972, The projection of the accessory olfactory bulb in the frog, *Brain Res.* **36**:409–411.

Scalia, F., 1976a, Structure of the olfactory and accessory olfactory systems, in: *Frog Neurobiology* (R. Llinás and W. Precht, eds.), Springer-Verlag, Berlin, pp. 213–233.

Scalia, F., 1976b, The optic pathway of the frog: Nuclear organization and connections, in: *Frog Neurobiology* (R. Llinás and W. Precht, eds.), Springer-Verlag, Berlin, pp. 386–406.

Scalia, F., and Colman, D. R., 1975, Identification of telencephalic afferent thalamic nuclei associated with the visual system of the frog, *Neurosci. Abstr.* **1**:46.

Scalia, F., and Gregory, K., 1970, Retinofugal projections in the frog: Location of the postsynaptic neurons, *Brain Behav. Evol.* **3**:16–29.

Scalia, F., Halpern, M., Knapp, H., and Riss, W., 1968, The efferent connexions of the olfactory bulb in the frog: A study of degenerating unmyelinated fibres, *J. Anat.* **103**:245–262.

Scalia, F., Halpern, M., and Riss, W., 1969, Olfactory bulb projections in the South American caiman, *Brain Behav. Evol.* **2**:238–262.

Schmidt, F. S., 1980, Catecholaminergic neurons of treefrog isthmotrigeminal tegmentum, *Exp. Brain Res.* **39**:235–237.

Sligar, C. M., and Voneida, T. J., 1981, Efferent projections of the septum in the tegu lizard, *Tupinambis nigropunctatus, Anat. Rec.* **210**:197–202.

Söderberg, G., 1922, Contributions to the forebrain morphology in amphibians, *Acta Zool.* **3**: 1–57.

Tóth, P., Csank, G., and Lázár, G., 1985, Morphology of the cells of origin of descending pathways to the spinal cord in *Rana esculenta.* A tracing study using cobaltic-lysine complex, *J. Hirnforsch.* **26**:365–383.

Ueda, S., Nojyo, Y., and Sano, Y., 1984, Immunohistochemical demonstration of the serotonin neuron system in the central nervous system of the bullfrog, *Rana catesbeiana, Anat. Embryol.* **169**:219–229.

Ulinski, P. S., 1975, Corticoseptal projections in the snakes *Natrix sipedon* and *Thamnophis sirtalis, J. Comp. Neurol.* **164**:375–388.

Ulinski, P. S., 1976, Intracortical connections in the snakes *Natrix sipedon* and *Thamnophis sirtalis, J. Morphol.* **150**:463–484.

Ulinski, P. S., 1983, *Dosal Ventricular Ridge: A Treatise on Forebrain Organization in Reptiles and Birds,* Wiley, New York.

Ulinski, P. S., and Peterson, E. H., 1981, Patterns of olfactory projections in the desert iguana, *Dipsosaurus dorsalis, J. Morphol.* **168**:189–227.

Urbán, L., and Székely, G., 1982, The dorsal column nuclei of the frog, *Neuroscience* **7**:1187–1196.

Vesëlkin, N. P., and Ermakova, T. V., 1978, Studies on the thalamic connections in the frog *Rana temporaria* by peroxidase technique, *J. Evol. Biochem. Physiol.* **14**:117–122.

Vesselkin, N. P., Agayan, A. L., and Nomokonova, L. M., 1971, A study of thalamotelencephalic afferent systems in frogs, *Brain Behav. Evol.* **4**:195–206.

Vesselkin, N. P., Ermakova, T. V., Kenigfest, N. B., and Goikovic, M., 1980, The striatal connections in the frog *Rana temporaria:* An HRP study, *J. Hirnforsch.* **21**:381–392.

Voneida, T. J., and Ebbesson, S. O. E., 1969, On the origin and distribution of axons in the pallial commissures in the tegu lizard (*Tupinambis nigropunctatus*), *Brain Behav. Evol.* **2**:467–481.

Voneida, T. J., and Sligar, C. M., 1979, Efferent projections of the dorsal ventricular ridge and the striatum in the tegu lizard, *Tupinambis nigropunctatus, J. Comp. Neurol.* **186**:43–64.

Wicht, H., and Himstedt, W., 1986, Two thalamo-telencephalic pathways in a urodele, *Triturus alpestris, Neurosci. Lett.* **68**:90–94.

Wicht, H., and Himstedt, W., 1988, Topologic and connectional analysis of the dorsal thalamus of *Triturus alpestris* (Amphibia, Urodela, Salamandridae), *J. Comp. Neurol.* **267**:545–561.

Wilczynski, W., and Northcutt, R. G., 1977, Afferents to the optic tectum of the leopard frog: An HRP study, *J. Comp. Neurol.* **173**:219–230.

Wilczynski, W., and Northcutt, R. G., 1983, Connections of the bullfrog striatum: Afferent organization, *J. Comp. Neurol.* **214**:321–332.

Yamamoto, K., Tohyama, M., and Shimuzu, N., 1977, Comparative anatomy of the topography of catecholamine containing neuron system in the brain stem from birds to teleosts, *J. Hirnforsch.* **18**:229–240.

Yoshida, M., Nagatsu, I., Kondo, Y., Karasawa, N., Ohno, T., Spatz, M., and Nagatsu, T., 1983, Immunohistocytochemical localization of the neurons containing catecholamine-synthesizing enzymes and serotonin in the brain of bullfrog (*Rana catesbeiana*), *Acta Histochem. Cytochem.* **16**:245–258.

5

The Cerebral Cortex of Reptiles

PHILIP S. ULINSKI

1. Introduction

Reptiles and mammals are the two groups of vertebrates with well-developed cerebral cortices. Ray-finned fishes have forebrains that develop by an eversion of the rostral neural tube that reduces the roof of the telencephalon to a thin membrane (Nieuwenhuys, 1982; Northcutt and Davis, 1983). They have olfactory cortices on the ventrolateral walls of their cerebral hemispheres (e.g., Braford and Northcutt, 1974), but the eversion process seemingly precludes the formation of a cortical roof to the telencephalon. Cartilaginous fishes (such as sharks), fleshy-finned fishes (lungfishes and coelacanths), amphibians, reptiles, birds, and mammals all have forebrains that develop via a fundamentally different process. This involves an evagination of the rostral neural tube resulting in paired lateral ventricles, interventricular foramina, and a neuronal roof to the telencephalon that can form an extensive cerebral cortex. However, the telencephalic roof of sharks forms a solid mass of neurons lacking the lamination usually associated with the cerebral cortex (Smeets *et al.*, 1983). Although amphibians (Northcutt and Kicliter, 1980) and lungfishes (Northcutt, 1986) have laminated cortices, they show little migration of neurons away from the ependyma. Birds have a small cerebral cortex, ostensibly resulting from a secondary reduction of the reptilian pattern (Benowitz, 1980). It is only in reptiles and mammals that the telencephalic roof develops into extensive and multilayered cortices.

A fundamental understanding of the nature and significance of such cortices clearly requires an understanding of the commonality and differences of

PHILIP S. ULINSKI • Department of Organismal Biology and Anatomy, and Committee on Neurobiology, University of Chicago, Chicago, Illinois 60637.

design seen in the cerebral cortex of reptiles and mammals. However, the brains of reptiles are largely unmyelinated, which made it difficult to trace out even major fiber systems until the introduction of Nauta degeneration methods in the 1960s. The relatively small brain of reptiles made physiological studies using evoked potential techniques or large stimulation electrodes difficult, so there was little neurophysiological work on reptiles until the advent of microelectrode methods. Factual information about the central nervous systems of reptiles was consequently limited principally to cytoarchitectonic descriptions of the brain until roughly 1970 (reviews: Ariëns Kappers, 1921; Ariëns Kappers *et al.,* 1936; Goldby and Gamble, 1957; Northcutt, 1967).

Since 1970 there has been a dramatic increase in the information available about the brains of reptiles due to the application of modern neuroanatomical and behavioral methods. Some of this information is summarized in earlier reviews of forebrain or cortical organization of reptiles: Northcutt (1970, 1978, 1981), Butler (1978, 1980), Belekhova (1979), Halpern (1980), Peterson (1980), Ulinski (1983), Parent (1986), and Schwerdtfeger and Smeets (1988). Most recently, *in vitro* preparations of reptilian eye cup (Baylor and Fettiplace, 1976), ear (Crawford and Fettiplace, 1980, 1981a,b), and whole brain (Mori and Shepherd, 1979; Mori *et al.,* 1981; Kriegstein, 1985) have been developed and used to greatly extend the body of information on the cellular physiology and pharmacology of the cerebral cortex of reptiles. It is, therefore, now possible for the first time to provide an adequate outline of how the cerebral cortex is organized in reptiles.

After providing some brief, background information on reptiles and their forebrains, this chapter reviews information on the structure and function of the cerebral cortex in reptiles with emphasis placed on work done since 1970. Wherever possible, an attempt has been made to place results within a common nomenclatorial framework. Thus, a uniform nomenclature for cortical areas and cell types is provided and information is presented using this nomenclature wherever such a synthesis seems warranted. The chapter concludes with a consideration of the differences and similarities in the cortices of reptiles and mammals.

2. Background Information

2.1. Phylogenetic Relationships

Reptiles living in Recent geological times are placed into four orders within the class Reptilia (e.g., Romer, 1956; Schmidt and Inger, 1957). The order Chelonia contains the turtles and tortoises which are characterized by their unique shells (Pritchard, 1967). It is a small order of about 200 species that live in both marine and freshwater environments as well as terrestrial habitats ranging from forests to deserts. Most neurobiological work has been done on turtles in the family Emydidae, which includes the closely related genera *Pseudemys* and *Chrysemys* in North America and the genus *Emys* in Europe.

The order Crocodylia includes crocodiles, alligators, and gavials, all characterized by the elongated rostrum or snout present in their skulls (e.g., Langston, 1973). There are about 20 species of crocodilians distributed in fresh and brackish waters in the Americas, Africa, Asia, and Australia. Essentially all neurobiological information has been obtained on the American alligator, *Alligator mississippiensis*, and its close relatives, the caimans (e.g., *Caiman sklerops* and *C. crocodilus*) of South America.

The order Squamata includes about 5000 species of lizards, snakes, and amphisbaenians. Lizards occur throughout most of the world and typically have four legs, although several groups of lizards have elongate bodies and show limb reduction. Neurobiological work has been carried out principally on the green iguana (*Iguana iguana*), a herbivorous American lizard; on the tegu lizards (*Tupinambis teguixin* and *T. nigropunctatus*), moderate sized and omnivorous South American forms; on monitor lizards (*Varanus exanthematicus* and *V. niloticus*), large, carnivorous Asian lizards; and on the medium-sized European lizards of the genus *Lacerta*. Snakes are characterized by absence of all but vestigial hind limbs in some forms and a highly modified skull affording it the ability to swallow prey larger than its head. Work has been done on boa constrictors (*Constrictor constrictor*), garter snakes (*Thamnophis sirtalis*), water snakes [*Nerodia* (= *Natrix*) *sipedon*] and several species of rattlesnakes. The rattlesnakes and pythons have attracted particular interest (e.g., Berson and Hartline, 1988) because of their infrared detection systems used for prey capture. Amphisbaenians (Gans, 1978) are a small group of burrowing reptiles that some workers consider as closely related to lizards.

The fourth order, Rhynchocephalia, includes only a single living species. This is the sphenodon or tuatara (*Sphenodon punctatum*) that lives on a group of small islands off the coast of New Zealand. Sphenodons bear a superficial resemblance to medium-sized lizards. However, they demonstrate a number of technical characteristics that place them in a group that appeared in the fossil record in the Triassic period, but declined in numbers by the Cretaceous period.

Although a complete discussion of the evolutionary history of reptiles is beyond the scope of this chapter (see Carroll, 1969, 1987), a few points about the relationships of reptiles (Fig. 1) are immediately relevant to neurobiological issues. The first is that the species of Recent reptiles represent the surviving endpoints of a very large and successful radiation of vertebrates that flourished principally in the Mesozoic era, from 225 to 70 million years ago. It included species living in all conceivable habitats, from subterranian burrowers, to desert dwellers, to inhabitants of tropical rain forests to free-flying forms (the pterosaurs), as well as the highly diverse dinosaurs. It also included a lineage of reptiles that diverged early in reptilian evolution from the stem reptiles, and eventually gave rise to mammals midway through the Mesozoic (Hotton *et al.*, 1986). Second, it is important to bear in mind that these synapsid reptiles were well on their evolutionary way prior to appearance in the fossil record of any of the four groups of reptiles represented in the Recent fauna. Mammals and lizards, for example, appear at roughly the same point in geological time, the results of long and independent lines of evolution. No evolutionary priority is accorded to being a mammal, and lizards and mammals must be viewed as being as equally advanced and equally specialized. There is no reason to view mammals as in any sense having evolved from or been derived from any of the Recent groups of reptiles.

A related point that has caused particular confusion in the neurobiological literature concerns the relationship between turtles and mammals. Several authors have taken the view that the brains of turtles are of particular value in understanding the evolution of mammalian cortex because turtles appear early in the fossil record and have changed little during their evolution. It is true that, with the rhynchocephalians, turtles appear in the fossil record earlier than representatives of the other Recent orders. However, reptilian evolution was well underway by the time turtles appeared, and the lines leading to mammals had separated off at least several million years before the first turtles appeared in the fossil record. Thus, turtles (unfortunately) provide no privileged view of the stem reptiles or the ancestors of mammals. The idea that turtles have not changed since their appearance is also misleading. There have been significant changes in the structure of the shell (Zangerl, 1969) and the first turtles could not retract their necks while modern turtles either fold their necks to the side (side-necked or pleurodiran turtles) or pull their necks into their shells (cryp-

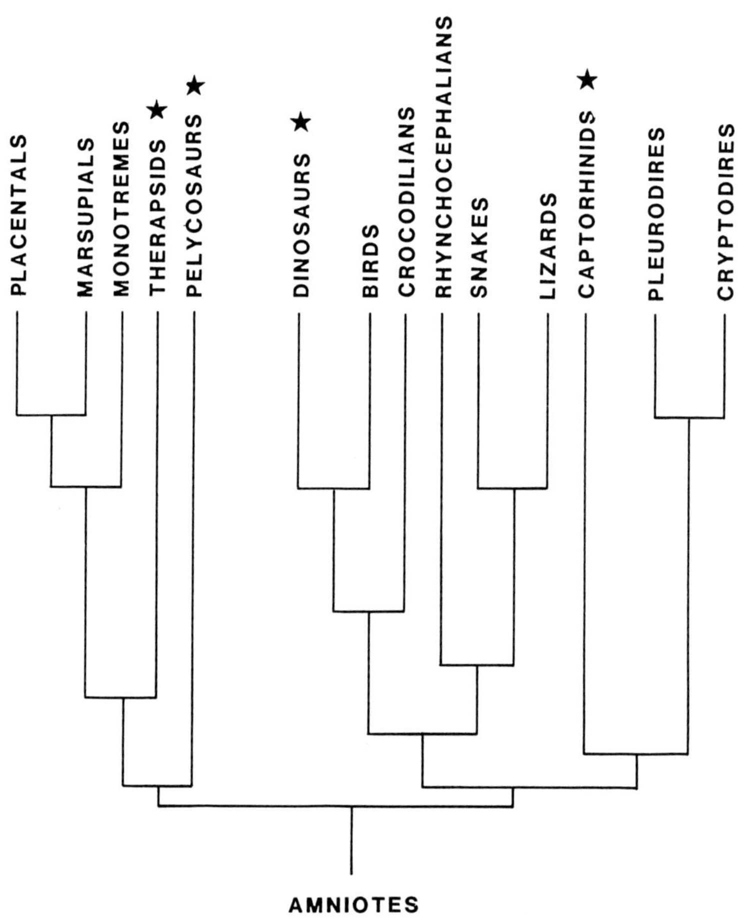

Figure 1. Phylogenetic relationships of major groups of reptiles. This is a cladogram summarizing the phylogenetic relationships of several major groups of reptiles. The relationship of birds and mammals to various groups of reptiles is also shown. The four orders of recent reptiles are the turtles (represented by two clades, the pleurodire or side-necked turtles and the cryptodires or turtles that withdraw their heads into their shells), lizards, snakes, rhynchocephalians (the sphenodon or tuatara), and crocodilians. Two extinct groups of reptiles (captorhinids and dinosaurs) are marked by stars. Notice that birds are the sister group of the dinosaurs. The cluster of clades on the left is composed of groups that diverged very early in evolutionary time from the stem reptiles and are represented in recent times by the placental, marsupial, and monotreme mammals. Two extinct groups (the therapsids and pelycosaurs) are marked by stars.

todiran turtles). Relevant here, the cortex of cryptodiran turtles is a distinctly laminated structure throughout while that of pleurodiran turtles contains a nuclear structure (Riss *et al.*, 1969). Thus, observations on the cortex of cryptodiran turtles will not necessarily hold generally for turtles, let alone for all of the reptiles or mammals.

The idea of a close relationship between turtles and mammals has attracted attention over the years principally because of a general interest in reconstructing the evolutionary history of mammalian brains. Accepting a close relationship between turtles and mammals held out the promise of deducing the structure of the brain in an animal close to the phylogenetic origin of mammals, but it is now apparent that this approach is based upon a misunderstanding of evolutionary interrelationships of reptiles and mammals. A more profitable strategy is to avoid selecting any one species of reptile as particularly important in understanding the evolution of mammalian cortex, and to recognize that the members of the four orders of living reptiles are essentially random samples from the reptilian adaptive radiation. Thus, features of cortical organization that are common to the living reptiles are likely to be those that evolved early in the evolution of reptiles and have been retained throughout the evolution of each, independent lineage. An important goal of studies of cortical organization in reptiles is, then, to understand the basic pattern of cortical organization and determine how it resembles and differs from the pattern of cortical organization seen in mammals. This point is considered in more detail in Section 9.

2.2. General Structure of Forebrain

The forebrain is reptiles (Fig. 2) develops from the prosencephalon of the neural tube, which, as in all vertebrates, is divided into a rostral telecephalon and a caudal diencephalon (Senn, 1979). The walls of the diencephalon develop ventricular sulci that divide it into four basic components (Fig. 3). The epithalamus includes the roof organs such as the pineal and parapineal (see Quay, 1979) and the habenula (Fig. 3, HB). The structure and connections of the habenular nuclei in reptiles generally resemble those of mammals. In particular, the habenular nuclei in the monitor lizard, *Varanus benegalensis,* project to the interpeduncular nucleus and the superior raphe (Distel and Ebbesson, 1981).

Ventral to the epithalamus is the dorsal thalamus, which consists of several discrete nuclear groups (Papez, 1935; Cruce, 1974; Butler and Northcutt, 1973; Balaban and Ulinski, 1981a). There are considerable interspecific differences in the cytoarchitecture of the thalamus, and comparisons have been hampered because different nomenclature schemes have often been used for different

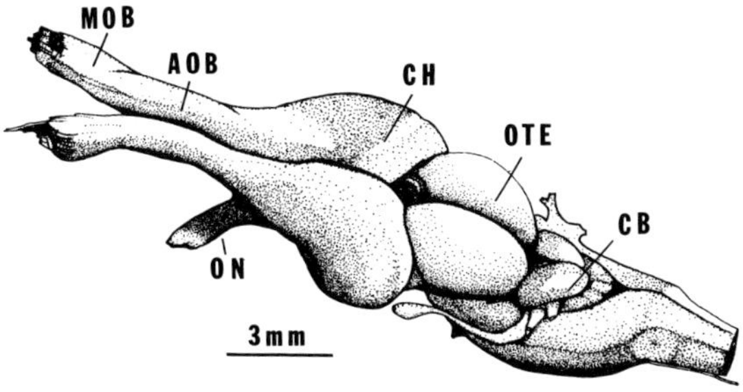

Figure 2. Brain of a reptile. This drawing of the brain of the eastern garter snake, *Thamnophis sirtalis,* illustrates the general features of the brains of reptiles. It is shown in a dorsolateral view. AOB, accessory olfactory bulb; CB, cerebellum; CH, cerebral hemisphere; MOB, main olfactory bulb; ON, optic nerve; OTE, optic tectum. Drawing by Dr. D. M. Dacey.

species. One common feature is a large, ellipsoidal nucleus situated in the center of the thalamus (e.g., Rainey, 1979; Rainey and Ulinski, 1982a,b) called nucleus rotundus (Fig. 3, RO), which receives bilateral projections from the optic tectum (Hall and Ebner, 1970a) and is responsive to visual stimuli. A dorsal lateral geniculate complex (Fig. 3, DLGN) that receives direct projections from retinal ganglion cells has been identified in all of the groups of reptiles (see Ulinski, 1983, for references). There are considerable interspecific differences in the cytoarchitecture and neuronal organization of the dorsal lateral geniculate complex, some of which will be discussed in Section 7.1.1b. Nucleus rotundus is surrounded dorsally, medially, and ventrorostrally by a group of nuclei that can be called collectively the perirotundal nuclei. They are interconnected with com-

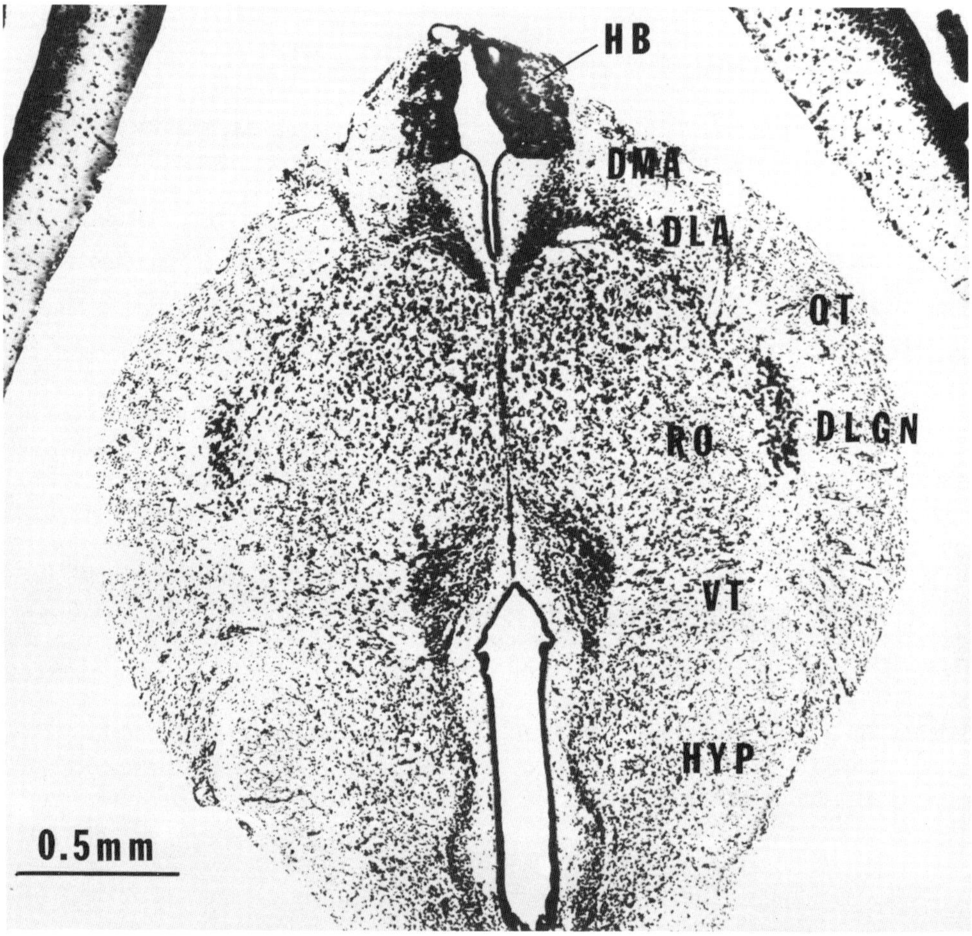

Figure 3. General feature of the diencephalon. This transverse Nissl section through the diencephalon of the red-eared turtle, *Pseudemys scripta elegans,* shows the general organization of the diencephalon in reptiles. The epithalamus is represented by the habenula (HB). The dorsal thalamus includes the dorsal lateral geniculate nucleus (DLGN) situated internal to the optic tract (OT), nucleus rotundus (RO), and two "perirotundal nuclei," the dorsomedial anterior (DMA) and dorsolateral anterior (DLA) nuclei. The ventral thalamus (VT) is a narrow band of tissue ventral to the dorsal thalamus. The hypothalamus (HYP) comprises the ventral third of the diencephalon. It consists of both scattered cells laterally and a vertical plate of cells medially.

ponents of the limbic system. A nucleus that receives auditory input from the torus semicircularis (the midbrain homologue of the inferior colliculus) is present in all groups of reptiles, but has been called by different names in different species. The caudal thalamus also includes a nucleus that receives somatosensory information from the dorsal column nuclei and spinal cord.

The ventral thalamus (Fig. 3, VT) is a relatively small wedge of tissue situated ventral to the dorsal thalamus and dorsal to the forebrain bundle system, the major avenue by which the brain stem is interconnected with the telencephalon. The ventral lateral geniculate nucleus (not present in Fig. 3) is situated at the lateral edge of the ventral thalamus, medial to the optic tract. It receives direct retinal projections, has reciprocal connections with the optic tectum, and projects to the brain-stem reticular formation. The suprapeduncular nucleus is positioned dorsal to the forebrain bundles and receives descending projections from the striatum. The anterior entopeduncular nucleus is embedded in the forebrain bundle. It appears to be an analogue to the thalamic reticular or perigeniculate nucleus of mammals that contains GABAergic neurons and may be a source of inhibitory inputs to dorsal thalamic nuclei (Ulinski, 1987c).

The hypothalamus forms (Fig. 3, HYP) the floor of the diencephalon and contains several cytoarchitecturally distinct nuclei (Crosby and Showers, 1969; Cruce, 1974; Butler and Northcutt, 1973). Many neurons in the dorsal hypothalamus are arranged in plates of cells that parallel the third ventricle. A ventromedial nucleus is typically present as a discrete mass of neurons in the caudal, ventral hypothalamus. A cluster of nuclei in the caudal hypothalamus are called the mammillary nuclei. Hypothalamic neurons generally have multipolar, isodendritic dendritic fields (Franzoni and Fasolo, 1982; Subhedar and Rama Krishna, 1984; Prasada Rao *et al.*, 1981; Prasada Rao and Subhedar, 1977). Some have one or two processes that extend into the ependymal lining of the third ventricle and contact the cerebrospinal fluid. The hypothalamus is interconnected with limbic components of the forebrain.

The telencephalon (Fig. 4) contains lateral ventricles interconnected with the third ventricle through interventricular foramina (Fig. 4, asterisk). The anterior commissure is situated rostral and dorsal to these foramina, and the forebrain bundle system passes ventral to the anterior commissure. Its medial component, the medial forebrain bundle, courses between the ventral diencephalon and the ventromedial components of the telencephalon. Its lateral component, the lateral forebrain bundle (Fig. 4, LFB), is equivalent to the internal capsule of mammals and courses between the dorsal diencephalon and the dorsal components of the telecephalon, which it reaches by turning dorsally once it has passed rostral to the anterior commissure.

The paired cerebral hemispheres contain three major components. The pallium forms the roof of the hemispheres and lies dorsal to the lateral ventricles. In mammals, the entire pallium forms a cerebral cortex. In reptiles (and birds) the pallium has two quite distinct parts, although both contain antigens that cross-react with antisera prepared against mammalian pallium (Kriegstein *et al.*, 1986). The first is the cerebral cortex, which extends from the dorsal edge of the septum, over the lateral ventricles and down onto the ventrolateral surface of the hemispheres. The cortex (Fig. 4, M, DM, D, and L) is a laminated structure, and a consideration of its organization is the major focus of this chapter. As in mammals, the cerebral cortex of reptiles receives ascending projections from nuclei in the dorsal thalamus. Particular regions of the cortex give rise to efferent

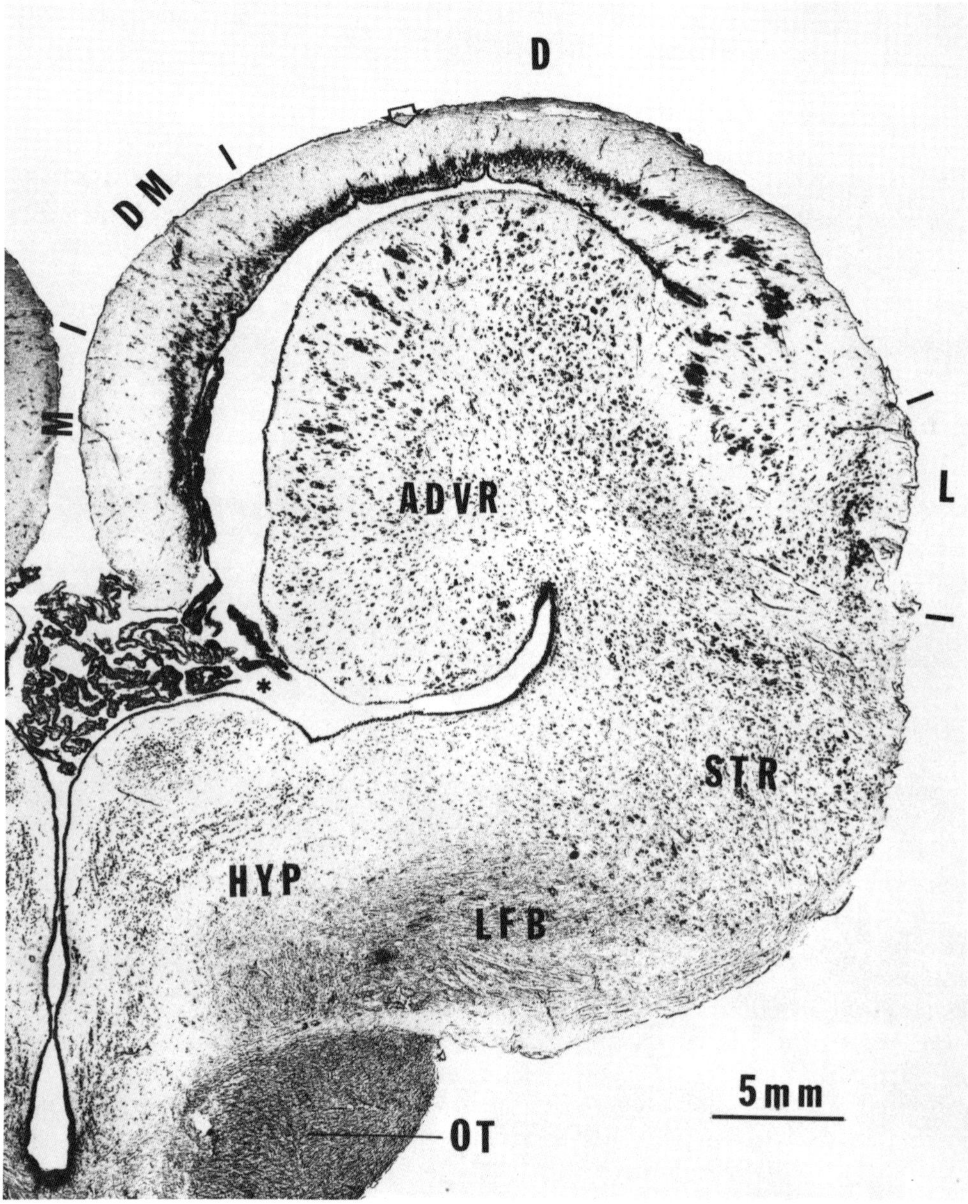

Figure 4. Cerebral hemisphere of a reptile. This transverse Nissl section through the cerebral hemisphere of the red-eared turtle, *Pseudemys scripta elegans,* illustrates the general features of the telencephalon of reptiles. The section passes through the interventricular foramen (marked by the asterisk), which connects the third and lateral ventricles. The dorsal telencephalon or pallium has two components. The anterior dorsal ventricular ridge (ADVR) protrudes into the lateral ventricle. The cerebral cortex lies dorsal to the lateral ventricle. It contains four cytoarchitectonic areas: medial cortex (M), dorsomedial cortex (DM), dorsal cortex (D), and lateral cortex (L). The ventral lateral wall of the telencephalon is formed by the striatum (STR), which blends caudally with the hypothalamus (HYP). Fibers in the lateral forebrain bundle (LFB) system are seen coursing through the striatum. OT, optic tract.

projections to the septum and hypothalamus that generally resemble the fornix system of mammals. Also, the visual cortex has efferent projections to the dorsal lateral geniculate nucleus, nucleus rotundus, and the optic tectum (at least in some species). However, there is no indication that reptiles have efferent projections to the rhombencephalon or spinal cord that might resemble the corticospinal or corticobulbar tracts of mammals.

The second part of a pallium is unique to reptiles and birds. It is a ridge of tissue that protrudes into the lateral ventricles from the ventrolateral wall of the hemisphere and is therefore called the dorsal ventricular ridge (DVR). DVR is divided into two parts at about the level of the anterior commissure. The anterior dorsal ventricular ridge (Fig. 4, ADVR) receives ascending projections from several of the sensory relay nuclei of the dorsal thalamus. These projections are arranged in a pattern that is the same for the three orders of reptiles that have been studied (there are no data for *Sphenodon*). Nucleus rotundus projects to the lateral edge of ADVR, which is known to be responsive to visual stimuli in at least turtles and lizards. The somatosensory nuclei of the caudal thalamus project to a central region of ADVR in crocodilians and lizards. The auditory nucleus of the caudal thalamus projects to the medial region of ADVR in crocodilians and lizards. The organization of ADVR has been discussed extensively by Ulinski (1983). The basal dorsal ventricular ridge (BDVR; not present in Fig. 4) is situated caudal and ventral to ADVR. Its connections bear a general resemblance to the mammalian amygdala. BDVR in snakes, lizards, and *Sphenodon,* for example, contains a discrete structure called nucleus sphericus (Fig. 11B, NS). It receives inputs from the accessory olfactory bulb and projects to the ventromedial nucleus of the hypothalamus, and thus resembles the corticomedial nuclear group of the amygdala of mammals. There is less information on BDVR than there is on ADVR, but what little is known is reviewed by Ulinski (1983).

The second major component of the cerebral hemispheres is the striatum (Fig. 4, STR), which makes up the ventral lateral walls of the hemispheres. The striatum of reptiles bears a general resemblance in its cytology, connections, and histochemistry to the caudate, putamen, and globus pallidus of mammals (Reiner *et al.,* 1984; Parent, 1986). Some attempts have been made to relate specific components of the striatum of reptiles to elements of the basal ganglia in mammals. However, these have not yielded clear results, and will not be pursued here.

The third major component of the cerebral hemispheres is the septum (Figs. 6 and 7, S). The precommissural part of the septum forms the ventromedial wall of the cerebral hemisphere, while its postcommissural part extends caudally over the anterior commissure. The septum contains dorsolateral, ventromedial, and ventrolateral nuclear groups. The majority of septal neurons are multipolar and isodendritic, but those lying adjacent to the midline are typically bipolar with the long axes of the dendritic fields oriented parallel to the surface of the brain (Ulinski, 1977a; Martin-Perez *et al.,* 1981). The septum contains neurons that stain positively for GABA and acetylcholinesterase (Schwerdtfeger *et al.,* 1986). The septum in tegu lizards projects caudally to the periventricular and mammillary nuclei of the hypothalamus and to the perirotundal thalamic nuclei (Hoogland *et al.,* 1978; Sligar and Voneida, 1981).

Using this overview of the forebrain of reptiles as a background, we can now turn to a more detailed consideration of the cortical component of the pallium.

3. Cytoarchitecture of Cortex

The cerebral cortex is a sheet of tissue between 0.5 and 1.0 mm thick that forms a fissureless roof for the cerebral hemisphere (Fig. 5). It contains three layers defined by variations in the density of neurons. The outer layer 1 is bounded dorsally by the end feet of ependymal tanycytes that extend radially through the cortex and by the pia (see Stark, 1979, for a review of the meninges of reptiles). It contains relatively few, scattered somata. The intermediate layer 2 contains a large number of more densely packed somata. It forms a continuous sheet of cells extending from the medial to the lateral edge of the cortex in many species. In others, breaks occur in layer 2 during the course of embryonic development and the edges of the intermediate region of layer 2 come to lie deep in the medial and lateral regions. The cortex thus has five layers in these regions, which are called the medial and lateral superpositions and marked by stars in Fig. 5. The configuration of the cortex within the superpositions resembles that present at the point where the dentate gyrus meets area CA3 within the hippocampal formation of mammals. The inner layer 3 contains a moderate number of loosely packed somata. A layer of fibers called the alveus (Fig. 5, AL) is situated deep in layer 3. A distinct, sheetlike aggregation of cells known as the cell plate (Fig. 5, CP) occurs in layer 3 in some reptiles (Unger, 1906; de Lange, 1911; Rose, 1923; Ulinski, 1974). The inner surface of layer 3 is bounded by ependymal cell somata (Fig. 5, E).

The cortex is divided into several cytoarchitectonic areas based on variations in cell density and the configuration of the three layers. The same areas can be recognized in representatives of all four orders (Figs. 6–9), but there are variations between taxa in the detailed cytoarchitecture of each area (see Table I for references). The areas are rostrocaudally elongate strips that extend from the caudal edge of the olfactory bulbs to the caudal pole of the cortex (Fig. 10). They constitute most of the cortex, but the cortex is reduced to thin membranes at some points. This typically occurs along the caudoventral surface of each cerebral hemisphere where the cortex becomes a noncellular membrane that forms the medial face of the lateral ventricle. The cortex is also reduced to a membrane in the olfactory cortex of anoline lizards that have a highly reduced olfactory system (Armstrong *et al.,* 1953). Several nomenclature systems have been used for the cortical areas over the years, but most workers now recognize four areas named (from medial to lateral): medial cortex, dorsomedial cortex, dorsal cortex, and lateral cortex.*

Medial cortex (also called small-celled medial cortex and ventral medial cortex) lies on the dorsal edge of the septum (Figs. 6–9, M). It is narrow in width rostrally where it blends with the anterior olfactory nucleus (Fig. 10, AON), and expands caudally so that it forms a relatively large percentage of the caudal pole of the hemisphere. Caudal to the septum it thus lies dorsal to the thin membrane mentioned in the previous paragraph. It is characterized in all forms by the

*Many authors recognize only three cortical areas in reptiles. These would correspond to the lateral area, dorsal area, and the combination of the dorsomedial and medial areas. However, most of the authors recognize two components in the medial of these three areas. Since the cytoarchitecture, cytology, and connections of these two components vary in all groups of reptiles, it seems preferable to distinguish them as distinct cytoarchitectonic areas.

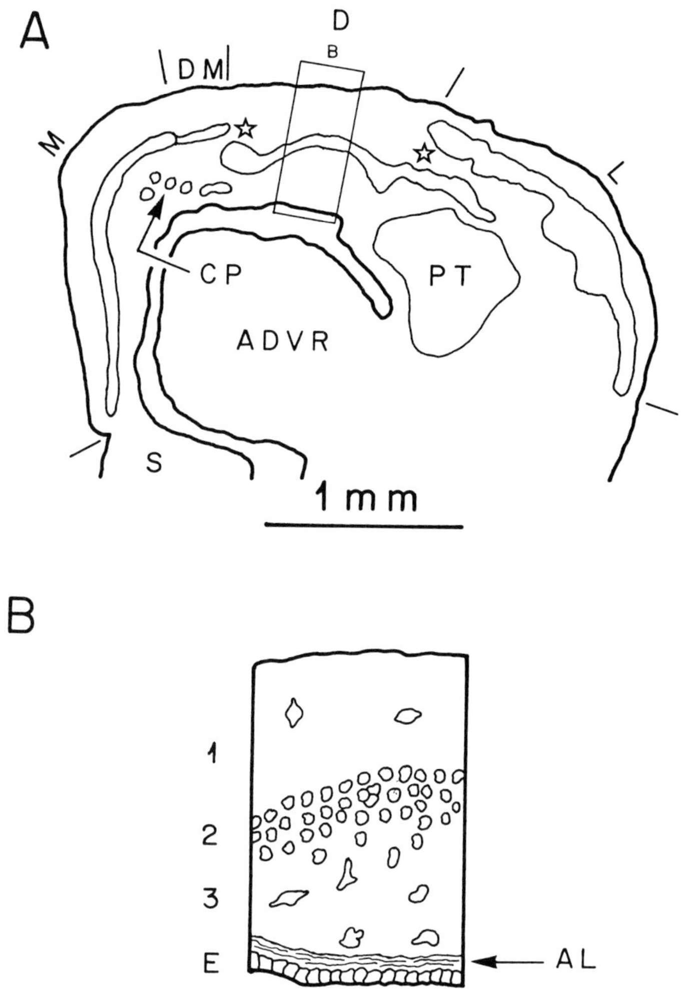

Figure 5. General features of cerebral cortex. (A) Drawing of a transverse section through the pallium of a tegu lizard to illustrate the nomenclature of the major components of cerebral cortex in reptiles. The cortex is divided into four major cytoarchitectonic areas: medial cortex (M), dorsomedial cortex (DM), dorsal cortex (D), and lateral cortex (L). In some forms, a distinct cluster, or several clusters of cells, are seen in the lateral region of dorsal cortex. This component of dorsal cortex is sometimes called the pallial thickening (PT). The cortex in reptiles is fundamentally a three-layered cortex, but in some species the second layer is ruptured at the medial and lateral borders of dorsal cortex, resulting in the medial and lateral superpositions (marked by stars). In some species, clusters of cells occur in the deep layer of medial, dorsomedial, and dorsal cortex. These are called the cell plate (CP). The rectangle marked "B" in dorsal cortex is magnified in panel B. (B) Transverse section through dorsal cortex to illustrate the nomenclature of cortical layers. The cortex contains three layers. The outer layer 1 contains a few scattered cells (open profiles). The intermediate layer 2 contains many, densely packed somata. The deep layer 3 contains a moderate number of scattered somata. A distinct bundle of unmyelinated fibers, the alveus (AL), runs through the deep part of layer 3 of medial, dorsomedial, and dorsal cortex. The inner surface of the cortex is formed by the ependyma (E). ADVR, anterior dorsal ventricular ridge; S, septum.

presence of relatively few neurons in layers 1 and 3. Somata of neurons in layer 2 are small, densely packed, and stain intensely in Nissl stains.

Dorsomedial cortex (also called large-celled medial cortex and dorsal medial cortex) forms the dorsomedial convexity of the cerebral hemispheres. Like the medial cortex, it contains relatively few neurons in layers 1 and 3. The cells in dorsomedial cortex are always larger than those in medial cortex, but there are some marked interspecific differences in the cytoarchitecture of dorsomedial cortex. Dorsomedial cortex in squamates (Figs. 8 and 9, DM) has a densely packed layer 2, which is relatively narrow in lizards but wider in snakes. By contrast, neurons in layer 2 of dorsomedial cortex of turtles (Fig. 7, DM) and crocodilians (Fig. 6, DM) are loosely packed so that layer 2 is relatively thick in species in these taxa.

Dorsal cortex (also called general cortex) forms the dorsal surface of the cerebral hemispheres (Figs. 6–9, D). It has few neurons in layer 1, many densely packed neurons in layer 2, and a moderate number of loosely packed neurons in layer 3. It is usually possible to divide dorsal cortex into three subareas based on subtle differences in cytoarchitecture. These subareas have not been explicitly recognized by all authors, and—when recognized—have been subjected to a variety of nomenclatures. The simplest nomenclature recognizes medial, inter-

Table I. Studies of the Cytoarchitecture of Cortex in Reptiles[a]

Turtles and tortoises	
Chelydra serpentina	Riss *et al.* (1969)
Chrysemys picta	Riss *et al.* (1969), Northcutt (1970)
Podocnemis unifilis	Riss *et al.* (1969)
Pseudemys scripta	Desan (1984)
Testudo hermanni	Platel *et al.* (1973)
Crocodilians	
Alligator mississippiensis	Crosby (1917)
Alligator lucius	Rose (1923)
Caiman sklerops	Riss *et al.* (1969)
Crocodylus palustris	Rose (1923)
Lizards	
Anolis carloinensis	Greenberg (1982)
Chamaeleo lateralis, C. vulgaris	Beckers *et al.* (1971/1972), Shanklin (1930)
Gekko gecko	Smeets *et al.* (1986)
Iguana iguana	Northcutt (1967)
Lacerta viridis	Beckers *et al.* (1971/1972), Goldby (1934)
Scincus scincus	Platel (1969)
Tupinambis sp.	Curwen (1937), Ebbesson and Voneida (1969)
Snakes	
Coluber constrictor	Carey (1967)
Constrictor (= Boa) constrictor	Ulinski (1974)
Crotalus adamanteus, C. atrox	Warner (1931, 1946)
Eryx johni	Ulinski (1974)
Natrix (= Nerodia) sipedon	Ulinski (1974)
Amphisbaenians	
Monopeltis guentheri	Beckers *et al.* (1971/1972)
Sphenodon	
Sphenodon punctatum	Cairney (1926), Durward (1930)

[a]This table is a selected list of references on the cytoarchitecture of cortex in reptiles, arranged by taxonomic group. Other, especially older, references are contained in the bibliographies of the papers listed.

mediate, and lateral subdivisions of dorsal cortex. Layer 2 of the medial and lateral subdivisions is sometimes slanted in adult forms and juts ventral to layer 2 of medial cortex and lateral cortex, respectively, resulting in the medial and lateral superpositions. Layer 2 retains its embryonic condition in other adult forms and is continuous with layer 2 of medial and lateral cortex. The lateral subdivision of dorsal cortex in cryptodiran turtles merits special comment (Fig. 7). It is greatly expanded and its lateral edge forms a scroll-like convolution that was named the pallial thickening by Johnston (1915). The pallial thickening is characterized by the presence of many large clusters of neurons with closely apposed somata.

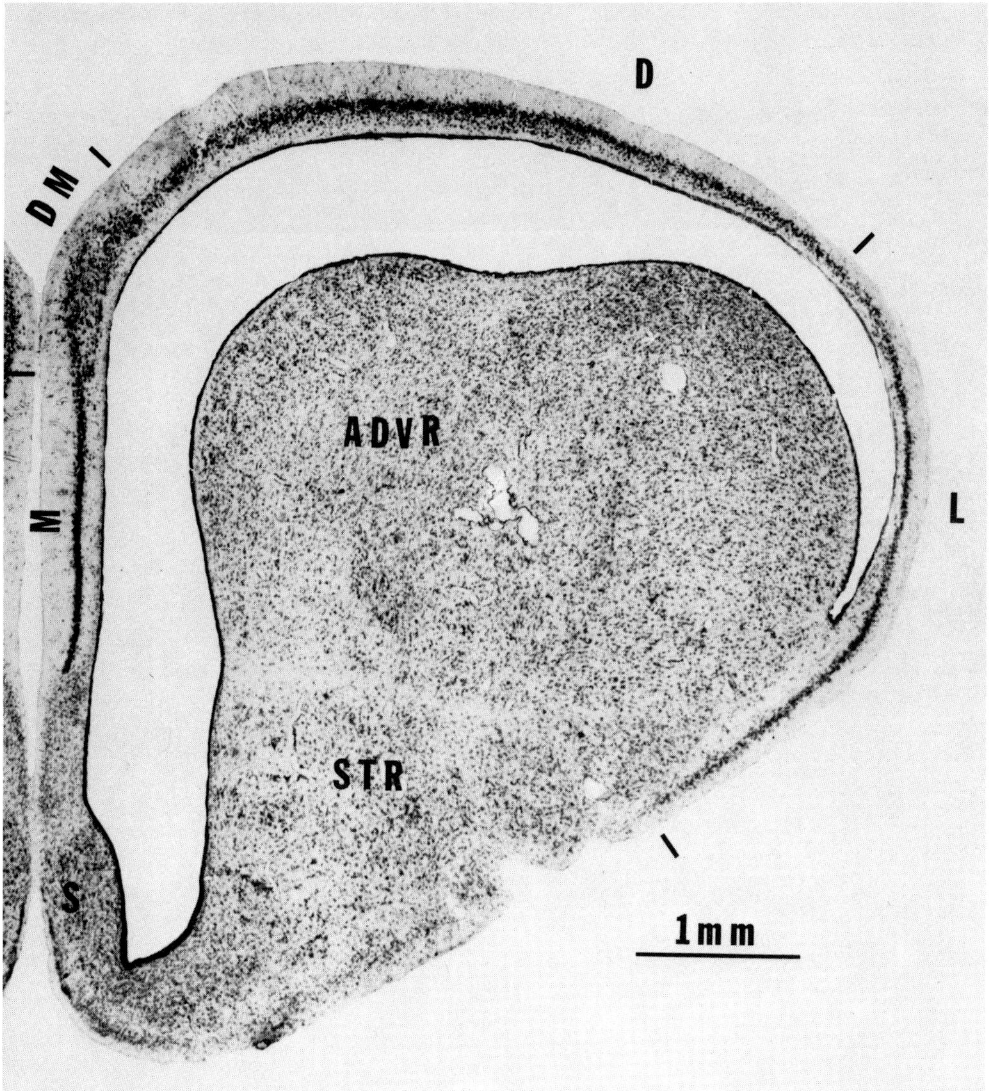

Figure 6. Telencephalon of a crocodilian. Figures 6 through 9 show examples of transverse sections through the telencephalons of representatives of each of the major groups of Recent reptiles. This is a Nissl section from the telencephalon of an alligator, *Alligator mississippiensis*. The cortex is divided into medial cortex (M), dorsomedial cortex (DM), dorsal cortex (D), and lateral cortex (L). ADVR, anterior dorsal ventricular ridge; S, septum; STR, striatum.

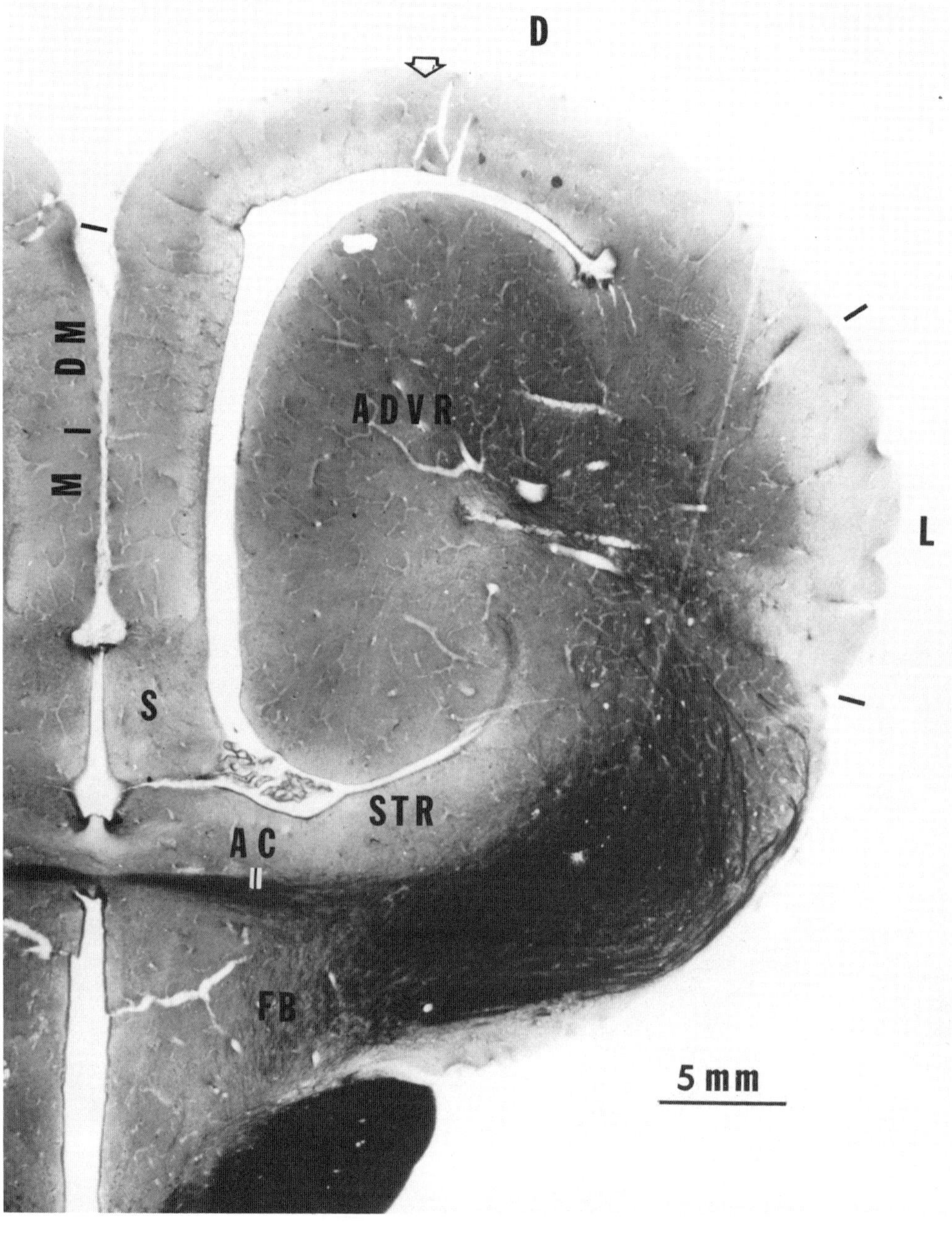

Figure 7. Telencephalon of a turtle. This is a section from the telencephalon of the red-eared turtle, *Pseudemys scripta elegans,* stained with a myelin stain. It is slightly rostral to the section shown in Fig. 3. The cortex is situated dorsal to the postcommissural septum (S). It contains medial cortex (M), dorsomedial cortex (DM), dorsal cortex (D), and lateral cortex (L). The open arrow marks the divisions of dorsal cortex that are designated D1 (laterally) and D2 (medially) by Desan (1984). The section passes through the anterior commissure (AC) and the forebrain bundles (FB) as they turn laterally into the striatum (STR). ADVR, anterior dorsal ventricular ridge.

Pallial thickening has often been regarded as a separate entity. However, Desan (1984) recently divided dorsal cortex of cryptodiran turtles into two divisions. His area D1 corresponds to the medial subarea of dorsal cortex; his area D2 includes the intermediate subarea and the pallial thickening. In *Sphenodon,* the lateral subarea of dorsal cortex curves ventrally and fuses during the course of ontogenesis with the cell-cluster zone of ADVR, so that ADVR and cortex appear continuous in adult *Sphenodon* (Hines, 1923).

Lateral cortex forms the ventrolateral wall of the hemisphere. It is situated dorsal to the lateral olfactory tract, which carries the axons of mitral cells from the main olfactory bulb to lateral cortex. This cortical area is typically divided

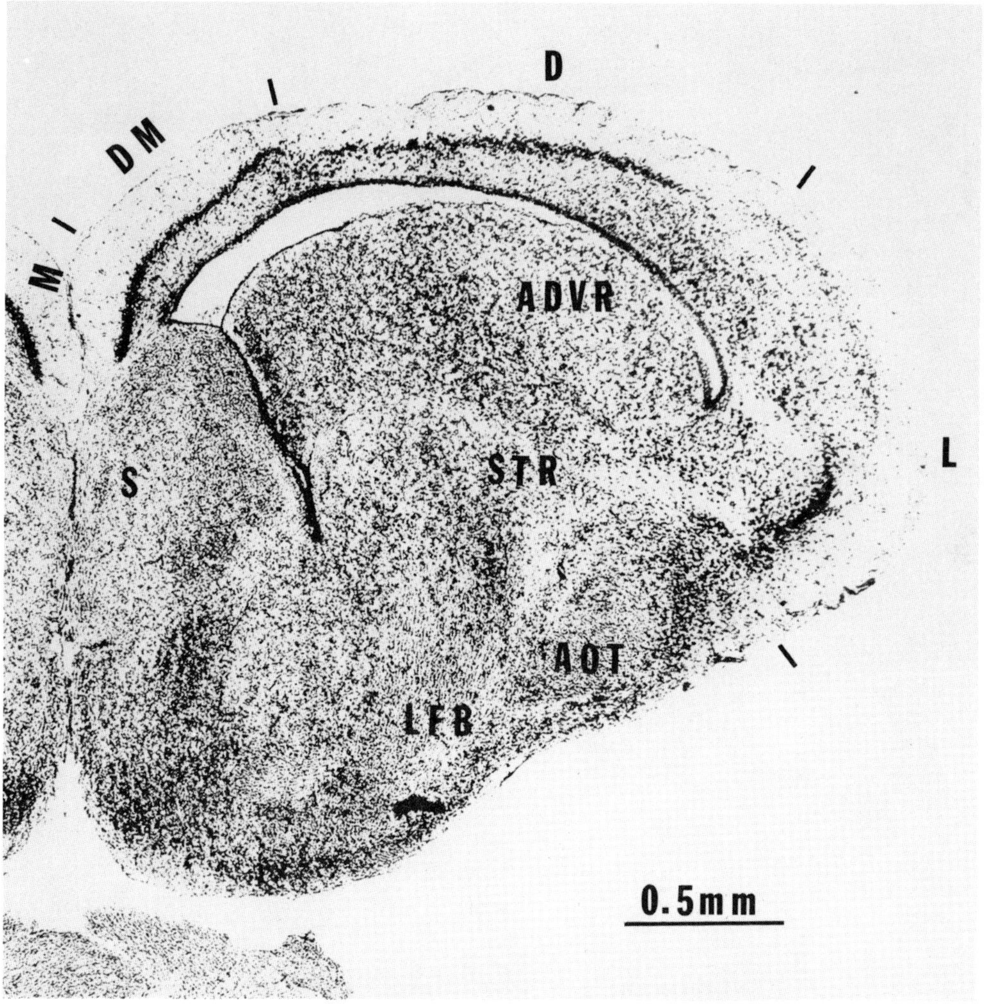

Figure 8. Telencephalon of a snake. This is a Nissl section from the telencephalon of a water snake, *Natrix sipedon.* At this level, more rostral than the section shown in Fig. 2, the cortex is situated dorsal to the precommissural septum (S). It contains medial cortex (M), dorsomedial cortex (DM), dorsal cortex (D), and lateral cortex (L). The accessory olfactory tract (AOT) carries fibers from the accessory olfactory bulb (see Fig. 2) to nucleus sphericus in the caudal telencephalon (see Fig. 11B). This tract, and the perifascicular nuclei that surround it, are best developed in snakes and some lizards. ADVR, anterior dorsal ventricular ridge; LFB, lateral forebrain bundle; STR, striatum.

into four subareas (Fig. 11). Two rostral areas (Fig. 11, drL, vrL) occupy the part of the lateral cortex that is situated rostral to the anterior commissure. Caudal areas (Fig. 11, dcL, vcL) are situated caudal to the anterior commissure. The rostral and caudal parts of lateral cortex are each divided into dorsal and ventral parts on cytoarchitectonic criteria. Thus, the dorsal rostral cortex has somata relatively densely packed in layer 2, while the ventral rostral cortex has somata

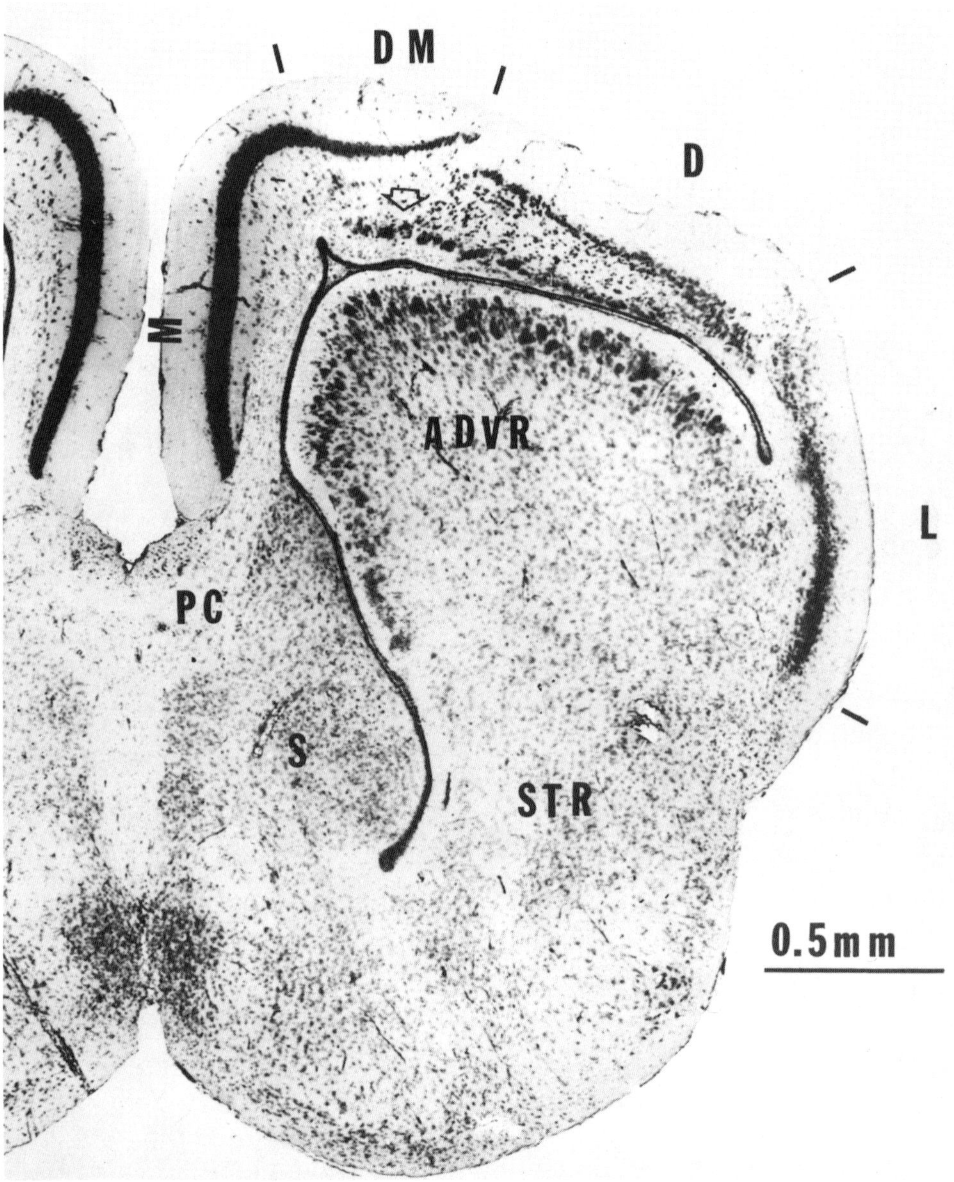

Figure 9. Telencephalon of a lizard. This is a Nissl section from the telencephalon of the tokay gecko, *Gekko gecko.* At this level, the pallial commissure (PC) connects the two cortices through the dorsal part of the precommissural septum (S). The cortex contains medial cortex (M), dorsomedial cortex (DM), dorsal cortex (D), and lateral cortex (L). Notice the distinct cell plate in dorsomedial and dorsal cortex. ADVR, anterior dorsal ventricular ridge; STR, striatum.

more loosely packed in layer 2. Layer 3 of rostral cortex usually has a distinct accumulation of loosely packed somata.

4. Development

The trilaminar pattern of cortex arises ontogenetically as a result of a radial migration of neuroblasts from a ventricular layer lining the inner surface of the cerebral hemispheres of embryos. General features of this process can be identified by studying sections through the forebrains of selected developmental stages of embryos (Herrick, 1892; Tandler and Kantor, 1907; Johnston, 1916; Kirsche, 1972; Hines, 1923; Reese, 1910; Hetzel, 1974; Warner, 1946; Kallen, 1951, 1956; Berquist and Kallen, 1954; Goffinet, 1983). The roof of the hemisphere consists of a neuroepithelium of actively dividing cells in early embryos. As neuroblasts begin to migrate from the neuroepithelium, an inner layer of scattered cells and a cell-free outer (or marginal) layer can be recognized. The trilaminar pattern of the adult cortex is evident in later embryonic stages in which layer 1 contains a few neurons, usually with horizontally aligned processes, layer 2 contains many neurons with bipolar, vertically aligned processes, and layer 3 contains a moderate number of scattered neurons. Layer 2 is initially

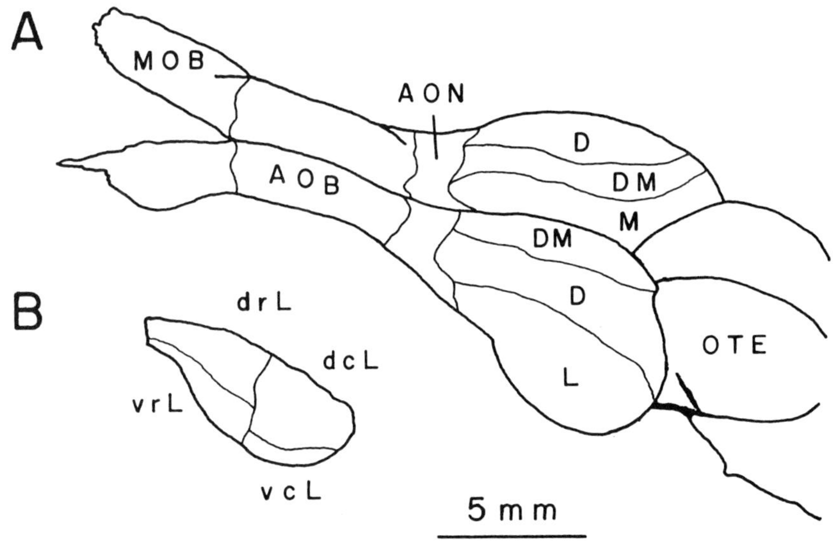

Figure 10. (A) Distribution of cortical areas. This is a drawing of the brain of a garter snake, *Thamnophis sirtalis,* similar to that shown in Fig. 2 to illustrate the position of the cortical areas upon the cerebral hemisphere. Because of the orientation of the brain, the drawing shows areas on the lateral surface of the left hemisphere and the medial surface of the right hemisphere. Medial (M), dorsomedial (DM), dorsal (D), and lateral (L) cortex have the form of rostrocaudally elongate strips. They abut rostrally with the anterior olfactory nucleus (AON), which encircles the peduncle of the olfactory bulb. Dorsal cortex is sometimes divided into three subareas (now shown). Lateral cortex is divided into four subareas. (B) Subareas of lateral cortex. This is a magnification of lateral cortex from panel A to show the position of the dorsal rostral lateral (drL), ventral rostral lateral (vrL), dorsal caudal lateral (dcL), and ventral caudal lateral (vcL) subareas. The cytoarchitecture of these areas is illustrated in Fig. 11.

a continuous stratum of cells in all species, but is ruptured medially and laterally to form the two superpositions in those species demonstrating this cytoarchitectonic feature.

Details of the migratory process have been investigated by tritiated thymidine labeling in the turtle *Emys orbicularis* and the lizard *Lacerta trilineata* (Goffinet *et al.*, 1986). Cortical neurons are born over a period of about 9 days late in the first half of embryogenesis. There are lateral-to-medial and rostral-to-caudal gradients in the timing of cortical migrations, so that, for example, neuroblasts migrate from the ventricular zone into lateral cortex earlier in development than they do in medial cortex. Migration occurs by means of an outside-in pattern within each cortical area with the first born neuroblasts migrating into layer 1 while those neuroblasts generated later in development migrate into layer 3. Glial cells continue to be born after the major wave of neuronal generation has subsided. Radial glial cells are present in embryos (Goffinet, 1983). Instances of neurons that are closely apposed to radial glial cells are sometimes seen in Golgi preparations of embryos, suggesting that radial glia play a role in guiding the migration of cortical neuroblasts in reptiles as they do in mammals.

Although the major fraction of cortical neurogenesis occurs during embryogenesis, two bits of evidence suggest some neurogenesis can occur in adult lizards of the genus *Lacerta*. The first is that there is a progressive increase in the

Figure 11. Cytoarchitecture of cortical areas. Shown are transverse sections through the brain of a water snake, *Nerodia sipedon*, to illustrate the cytoarchitecture of the cortical areas identified in Fig. 10. (A) Section through the rostral hemisphere; (B) section through the caudal pole of the hemisphere. Abbreviations as in Fig. 10.

number of neurons in medial cortex of *Lacerta* with age (Lopez-Garcia *et al.*, 1984). The second is that spindle-shaped cells in layer 3 of medial cortex that appear to be immature, migrating neurons have been identified at both the light and electron microscope levels (Garcia-Verdugo *et al.*, 1986). These cells are in direct contact with the radial processes of ependymal tanycytes. They may originate from a subventricular or germinative zone that lines the lateral part of medial cortex (Kirsche, 1972; Garcia-Verdugo *et al.*, 1981).

5. Cortical Cells

5.1. Glial Cells

Cerebral cortex of reptiles contains two types of glial cells. First are ependymal cells or tanycytes (Fig. 12) that are the adult derivatives of the embryonic radial glial cells. The somata of these cells form a pseudostratified epithelium (the ependyma) that lines the ventricular surface of the cortex and the anterior

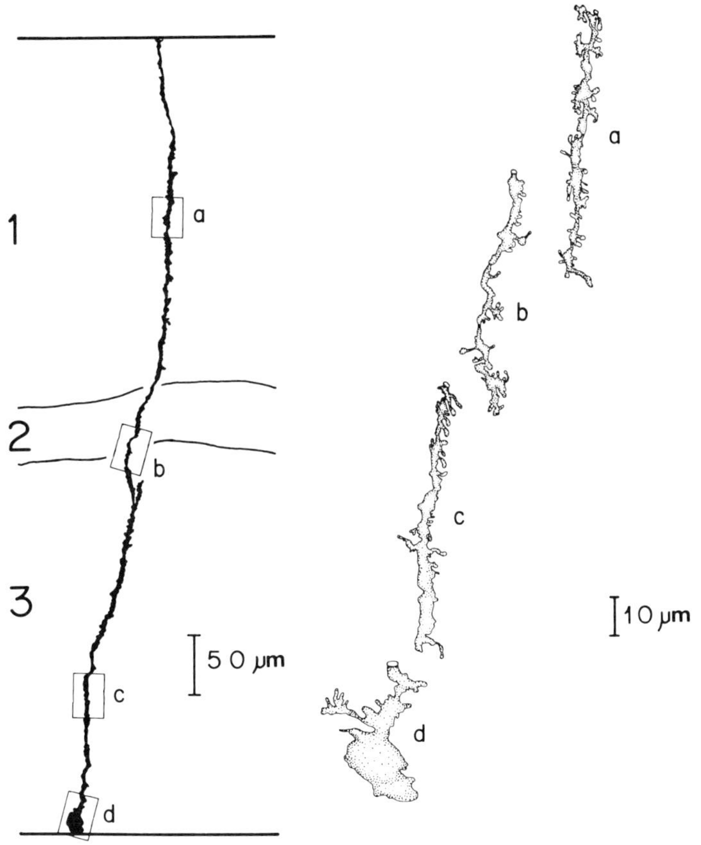

Figure 12. Ependymal cell. On the left is a camera lucida drawing of an ependymal cell from a Golgi preparation of the brain of a water snake, *Natrix sipedon.* The parts of the cell enclosed in rectangles are shown at higher magnification on the right. Most of the cell is covered by velate processes, which decrease in density as the cell passes through layer 2 of medial cortex. From Ulinski (1977a).

dorsal ventricular ridge. The ventricular surface of each tanycyte is heavily ciliated. The cell body itself is elongated, with its sides joined to those of its neighbors via junctional complexes (Ulinski, 1977a, 1979; Wouterlood *et al.*, 1981; Davila *et al.*, 1985; Weiss and Ulinski, 1985). The cell body contains a nucleus and a well-developed Golgi apparatus. A short neck region extends from the cell body and then continues into a long radial process that extends the full width of the cortex. These can span the cortex unbranched, or divide into daughter processes that continue toward the pia. The processes contain a core of microtubules and filaments surrounded by a rim of cytoplasm with many glycogen particles (Ulinski, 1977a, 1979; Garcia-Verdugo *et al.*, 1981; Wouterlood *et al.*, 1981; Weiss and Ulinski, 1985). The radial processes bear velate processes that extend into the neuropil. The velate processes can be unevenly distributed along the shaft of the radial process. In medial cortex, tufts of velate processes occur in layers 3 and 1, but there are fewer velate processes in layer 2, which is composed of many densely packed somata (Lacey, 1978). The velate processes contain glycogen and insinuate themselves between neuronal somata and dendrites (see Weiss and Ulinski, 1985). The radial processes end in footlike expansions just below the pial surface (Wouterlood *et al.*, 1981; Davila *et al.*, 1985). These contain many mitochondria and expand into horizontally elongate and thin processes. They overlap each other and are joined by gap junctions to form a glial limiting membrane, positioned internal to the basal lamina, that separates the brain from the pia.

Both the cell bodies and the radial processes of tanycytes are postsynaptic to axon terminals involved in chemical synapses (Ebner and Colonnier, 1975; Weiss and Ulinski, 1985). These involve asymmetric active zones and round synaptic vesicles. Intracellular recordings have been achieved from tanycytes with *in vitro* preparations of turtle dorsal cortex (Connors and Ransom, 1982). Recordings made from tanycytes while stimulating fiber systems within the cortex failed to produce any sign of postsynaptic potentials, so the function of neuron–tanycyte synapses is unclear. Weiss and Ulinski (1985) discuss several possibilities, including the modulation of transport of substances into or out of the ventricular system or a role in controlling the extracellular ionic environment by mediating movements of the velate processes.

The remaining glial cells are oligodendrocytes and astrocytes that are scattered throughout the cortex. They have not been extensively studied, but seem to resemble the glial cells of mammals both in Golgi preparations (Ramón, 1891, 1896; Conty, 1953; King, 1966; Kruger and Maxwell, 1967; Stensaas and Stensaas, 1968; Inoue *et al.*, 1974; Wouterlood *et al.*, 1981; Guirado *et al.*, 1984) and in electron microscopic material. Both fibrous and protoplasmic astrocytes have been described.

5.2. Cytology of Neurons

Cortical neurons are comparable in their cytology to small- and medium-sized neurons in other regions of vertebrate nervous systems (Davydova and Smirnov, 1969; Ulinski, 1977a, 1979; Ulinski and Rainey, 1980; Davila *et al.*, 1986). Small neurons have a modicum of cytoplasm that contains scattered ribosomes, but few organized stacks of rough endoplasmic reticulum. Larger neurons have more cytoplasm and some stacks of rough endoplasmic reticulum that appear as a diffuse basophilia at the light microscope level in Nissl stains. The organelles found in cortical neurons are generally unremarkable. However, neurons in the cortex of reptiles, and elsewhere in their central nervous system, have more glycogen than is commonly found in neurons in mammals. It appears as granules scattered generally throughout the cytoplasm and dendrites of most neurons. This abundance of glycogen may reflect differences between reptiles and mammals in their dependence upon aerobic versus anaerobic metabolism (e.g., Bennett and Dawson, 1976). Nuclei are frequently indented by fingers of cytoplasm.

Dendrites of cortical neurons can be divided into two types based on their cytology. Some have a cytoplasm filled with neurotubules and neurofilaments running within the core of the dendrite (Fig. 13). Others seemingly lack an extensive cytoskeleton (Fig. 14) and have been labeled as "pale" dendrites by some authors (Davydova and Smirnov, 1969; Ebner and Colonnier, 1975; Ulinski, 1977a). Some dendrites are smooth, or bear a series of swellings or dilatations. Others are covered with excrescences or dendritic spines (Fig. 15). These vary in their morphology from simple, filiform appendages, to spines with a knoblike swelling distally to those with a relatively large head that end in a fingerlike extension invaginated into a presynaptic element (e.g., Ulinski, 1977a; Davydova and Goncharova, 1979; Martinez-Guijarro *et al.*, 1984). Spines are invariably postsynaptic to profiles that resemble axon terminals (Ebner and Colonnier, 1975; Ulinski, 1977a). The synaptic contacts involve asymmetric active zones and presynaptic elements with clear, round synaptic vesicles in the great majority of cases, but some instances of spinelike profiles postsynaptic to terminals with clear, pleomorphic vesicles have been reported (Ebner and Colon-

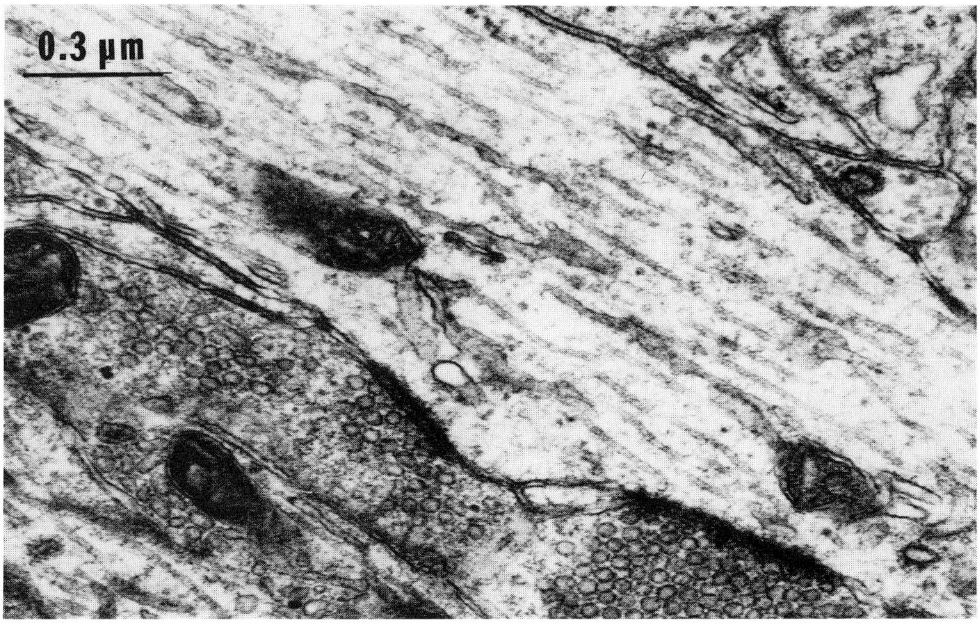

Figure 13. Dendrite. This is a longitudinal section through a dendrite from layer 1 of medial cortex of a water snake, *Natrix sipedon*. Like most dendrites, it contains many neurotubules. It is postsynaptic to two axon terminals.

nier, 1975). Spine apparatuses have not been described in reptiles. Dendrites frequently show structures called vacuolated invaginations (Ebner and Colonnier, 1975; Ulinski, 1979; Wouterlood *et al.*, 1981). These consist of regions of plasma membrane that invaginate into the dendrite forming a membrane-bound space confluent with the extracellular space (Fig. 16, VI). Several large, membrane-bound vesicles are enclosed in this space. Vacuolated invaginations are often found adjacent to an axon terminal (Fig. 16, AT) that is presynaptic to the dendrite in which they occur, and in some sections it can be determined that the axon terminal gives rise to a flap of its plasma membrane that extends into the associated dendrite. Thus, the vesicles in the invagination are actually inside the axon terminal. The function of vacuolated invaginations is not known, but Wouterlood *et al.* (1982) found a fivefold increase in the number of vacuolated invaginations in the medial cortex of the lizard *Agama* following lesions of the adjacent dorsal and lateral cortical areas. These areas contain neurons that project to medial cortex (see Section 7.2), so the increase in vacuolated invaginations is correlated with the presence of degenerating axon terminals in medial cortex.

Cortical neurons show a marked tendency to form groups or clusters of cells with closely apposed somata. This is dramatically the case in layer 2 of medial cortex, which consists of many, very densely packed somata (Fig. 17). Electron microscopic observation shows that the plasma membranes of neighboring cells are closely apposed. The somata of neurons in other regions of the cortex are usually not as densely packed as in medial cortex, but the presence of clusters of neurons that are apposed to each other but isolated from neighboring cells is

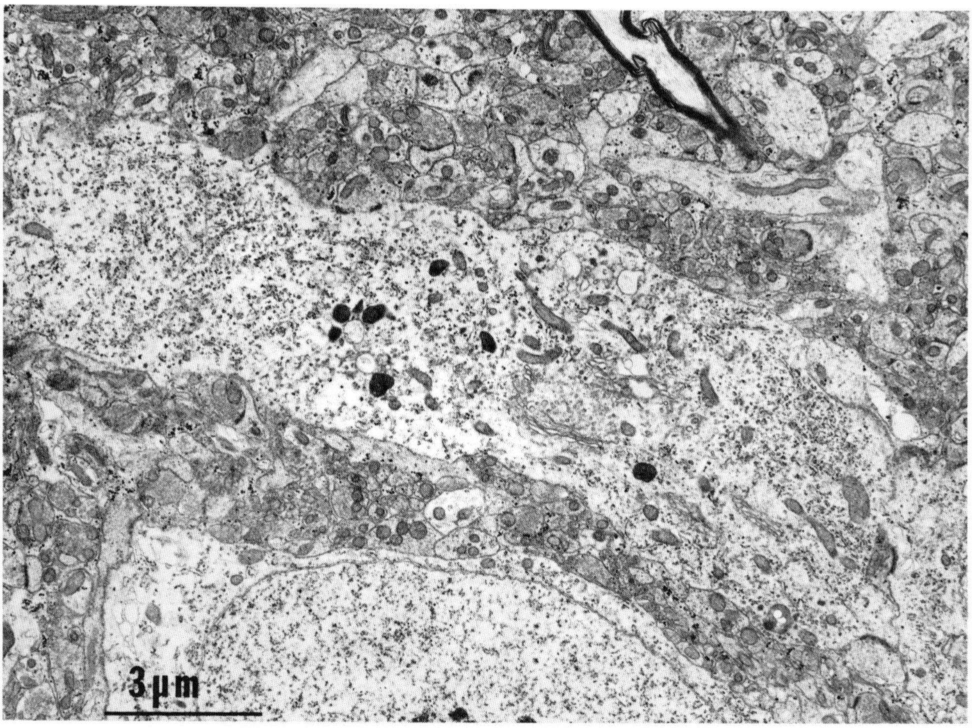

Figure 14. Pale dendrite. This is an example of a dendrite that lacks obvious neurotubules. It is from layer 3 of medial cortex of a water snake, *Natrix sipedon.*

common. Very large and distinct clusters are seen in the pallial thickening region of the dorsal cortex of cryptodiran turtles (Fig. 7). Smaller and less distinct clusters of neurons are seen throughout the cortex (e.g., Fig. 9) in many species. Three-dimensional reconstruction of layer 2 of turtle dorsal cortex indicates that clusters are present in medial dorsal cortex and that they become larger and more isolated in the pallial thickening. Such clusters are not unique to the cortex, but are actually a very dominant feature of the forebrains of reptiles and birds (see Ulinski, 1983). In some regions, such as the medial cortex of snakes (Fig. 17), apposed neurons are linked via membrane specializations that, according to Ulinski (1977a), resemble gap junctions. In other regions, membrane specializations have not been observed and the regions of contact resemble what Sotelo *et al.,* (1974) called "casual appositions." Ulinski (1977a) suggests that somatic appositions constitute a substrate for electrical interactions between cortical neurons, either via electrical synapses or via ephaptic interactions. However, Wouterlood *et al.* (1981) argue that the somatic appositions are more reminiscent of adhesive contacts between neurons in other regions of central nervous systems.

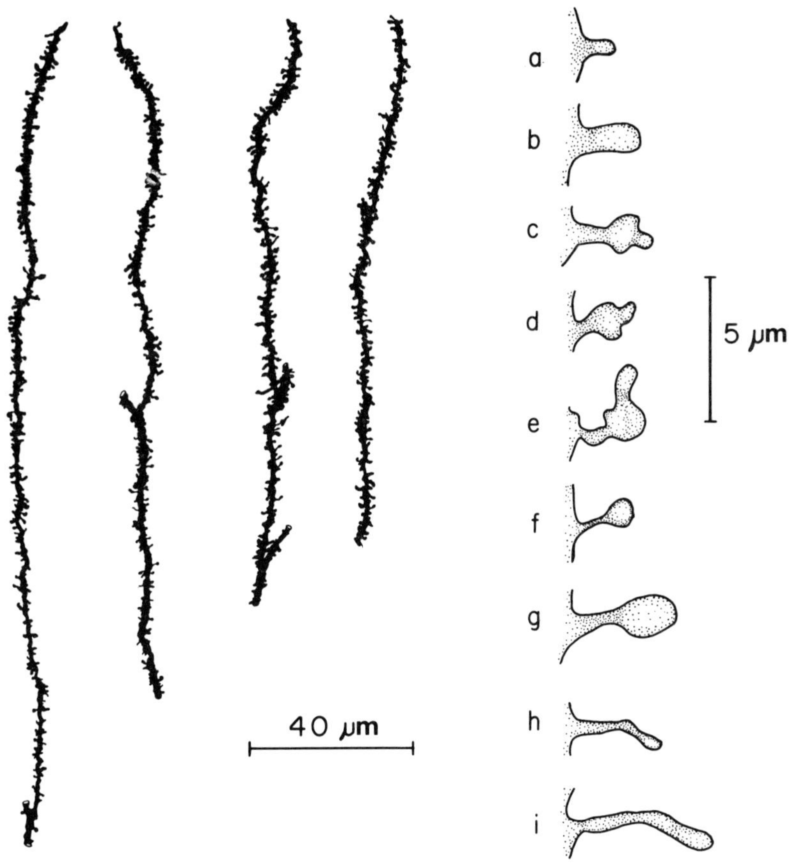

Figure 15. Dendritic spines. Shown are examples of spiny dendrites from a Golgi preparation of medial cortex of a water snake, *Natrix sipedon.* Four examples of dendritic segments are shown on the left. On the right are the range of morphologies seen in dendritic spines. They range from small stubs (a, b), through pedunculated spines (c–g), through long, filiform appendages (h, i). From Ulinski (1977a).

The overall morphology of cortical neurons has been studied in Golgi preparations of several different species of reptiles. There have been a few recent studies that used intracellular injection of HRP to demonstrate the morphology of cortical neurons, but this method has not yet been employed systematically in reptiles. There is clearly a considerable variation in the morphology of cortical neurons. Most authors have used descriptive names to designate different groups of neurons, and there has been no systematic classification of cortical neurons in reptiles. The sketches in Fig. 18 summarize the major varieties of neurons that have been described.

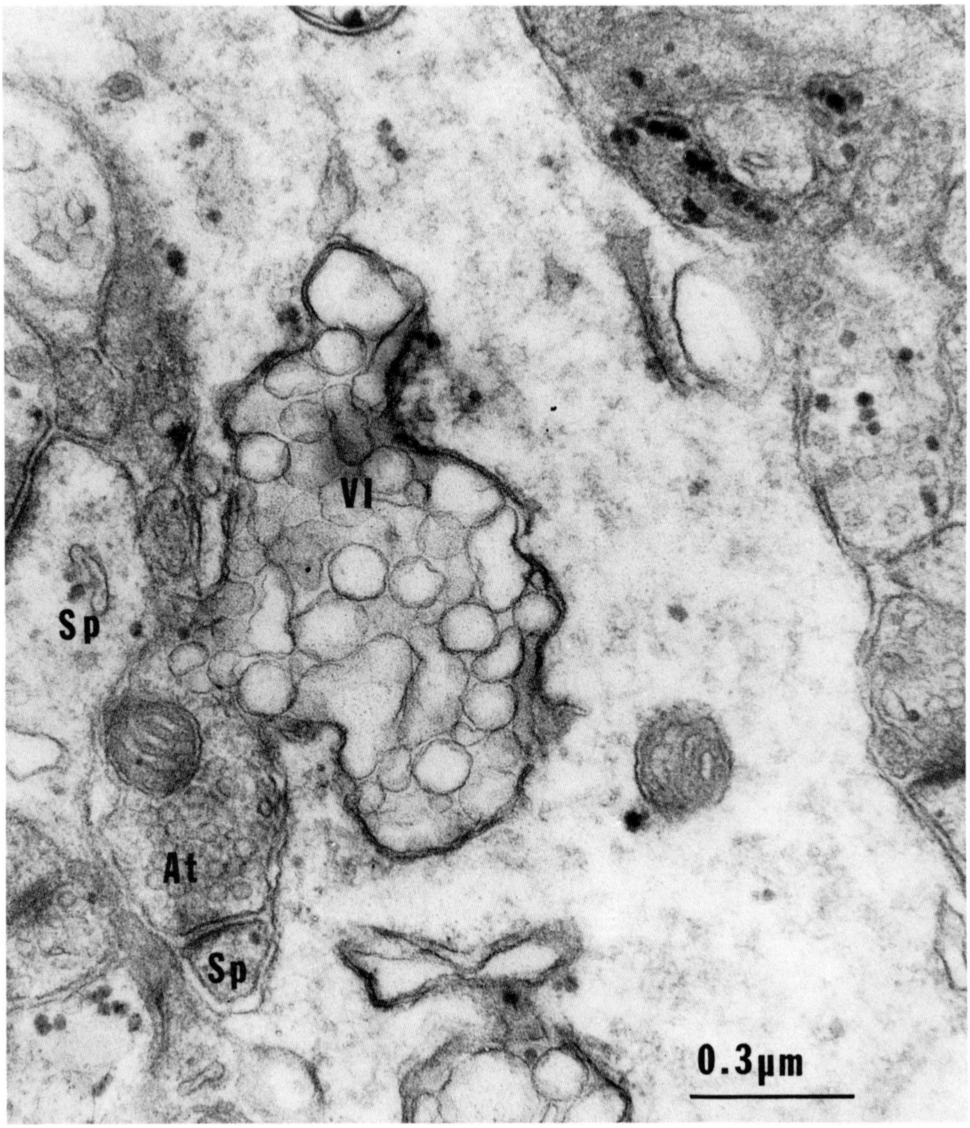

Figure 16. Vacuolated invagination. This is a vacuolated invagination in a dendrite from medial cortex of a water snake, *Natrix sipedon.* The axon terminal to the left (At) contacts a small dendritic spine (Sp), and also invaginates into the large dendrite on the right. This vacuolated invagination (VI) contains many, large membrane-bound vesicles.

Stellate neurons (Fig. 18A) have relatively small, oval or round somata and bear dendrites that radiate in a spherical dendritic field confined within a given cortical layer. They vary in the extent to which they bear dendritic spines. The axons of stellate cells have not been well characterized, but Desan (1984) did obtain good impregnations of stellate cells that had axon systems arborizing extensively within the vicinity of the cells' dendritic fields.

A second group of cortical neurons (Fig. 18B) have medium-sized somata that are fusiform or oval in shape and frequently have their long axes oriented vertically. These neurons have a single apical dendrite that extends vertically into adjacent cortical layers and several basal dendrites that extend toward the ependyma. The dendrites have a moderate cover of dendritic spines. Again, the axons have not been well characterized, but it has been established that some have collaterals that extend vertically into adjacent cortical layers. These neurons can be reasonably compared to the pyramidal neurons of mammalian cortex, and this term will be used here. However, two caveats are in order. The first is that the term *pyramidal* has been used promiscuously within the literature on reptile cortex and is often applied to cells that bear little real resemblance to

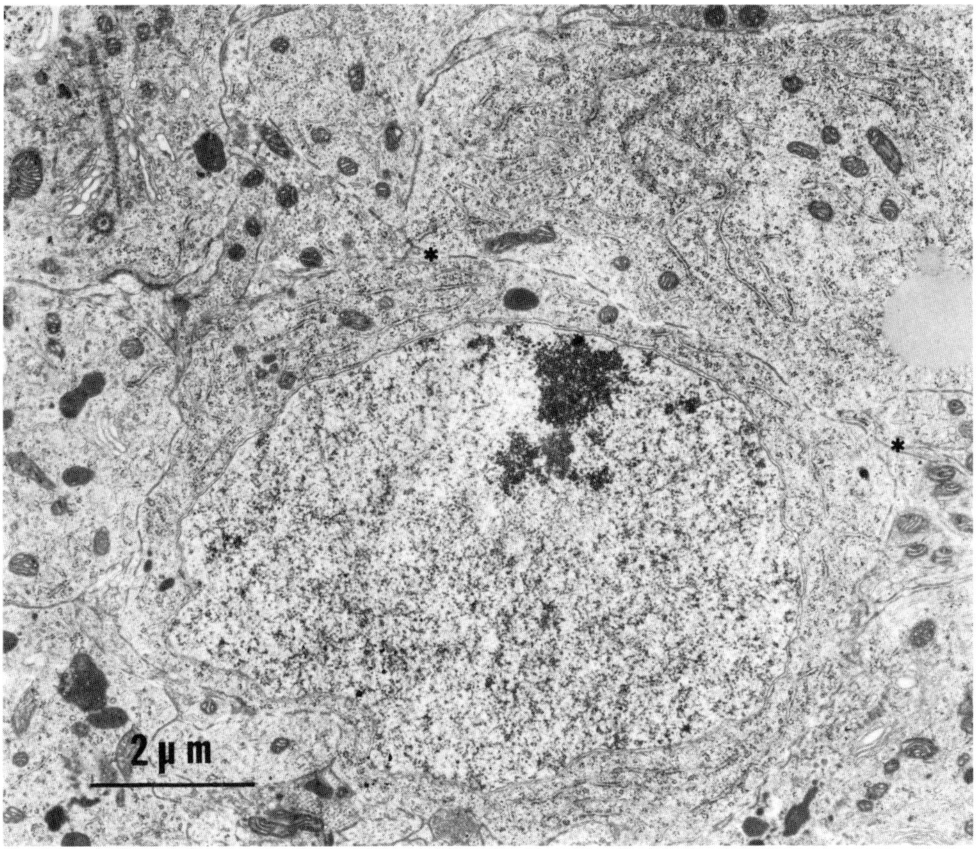

Figure 17. Neurons with apposed somata. This is an example of cortical neurons with apposed somata from medial cortex of a water snake, *Natrix sipedon*. The soma that occupies the bottom of the figure is closely apposed to a second soma at the upper right. The line of apposition is marked by several specialized junctions (asterisks).

mammalian pyramidal cells. The second is that there remains some discussion as to what constitutes a pyramidal cell within the mammalian literature. In general, it seems appropriate to use the term *pyramidal* conservatively in describing reptilian neurons, and imprudent to attach too great a significance to the term.

A third, and quite distinct, group of cells (Fig. 18C) can be designated as double pyramidal cells or bitufted cells. They have moderate-sized, oval or fusiform somata, but are distinguished by their possession of two tufts of vertically oriented dendrites, one from each pole of the soma. Each tuft of dendrites hence forms a pyramidal configuration, so the cell has two pyramids or tufts of

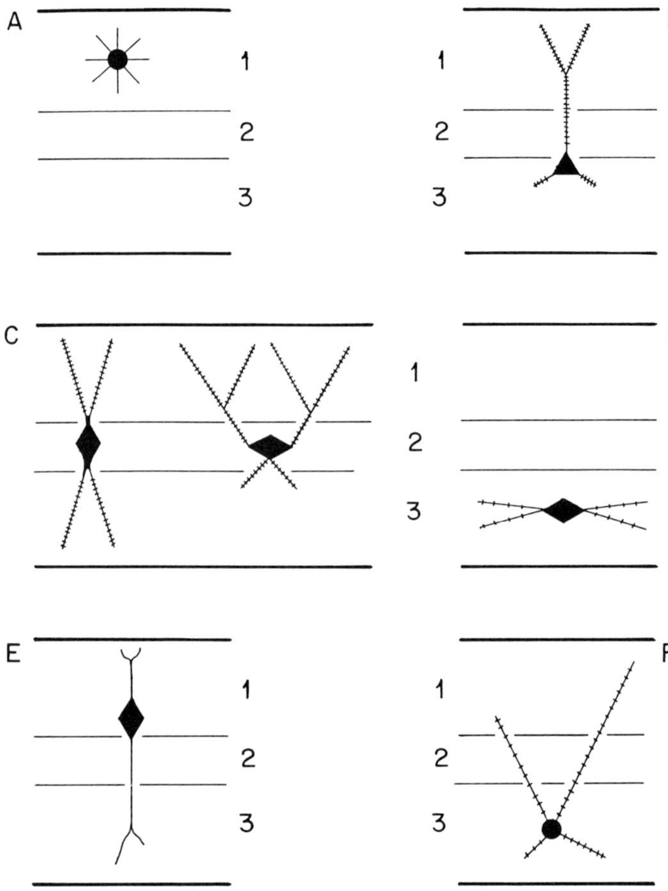

Figure 18. Cortical cell types. These drawings show, in cartoon form, the principal types of cells seen in cerebral cortex of reptiles. Each cell is positioned in a segment of cortex, with layers 1, 2, and 3 indicated. (A) Stellate cells have oval or round somata that give rise to several dendrites, extending in all directions. They are most frequent in layer 1. (B) Pyramidal cells have triangular-shaped somata with a single ascending dendrite and several descending dendrites. They are found in layers 2 and 3. (C) Double pyramidal cells typically (like the cell shown on the left) have vertically fusiform somata that bear two conical tufts of dendrites. They occur in layers 2 and 3. However, double pyramidal cells frequently occur with reduced descending dendritic trees, like the cell shown on the right. (D) Horizontal cells have horizontally elongated somata and bear dendrites that extend considerable distances from each pole. They occur in layer 3. (E) Bipolar cells have vertically fusiform somata and bear only a single, relatively unbranched dendrite from each pole. They occur in all three layers. (F) Multipolar cells have round or oval somata and give rise to several, long dendrites that may cross laminar boundaries. They occur in all three layers.

dendrites. The dendrites are invested heavily with dendritic spines. Axons of double pyramidal cells can bear collaterals that either ascend or descend into adjacent cortical layers. There is clearly a range in the morphology of double pyramidal cells. All have a dendritic field extending vertically toward the pial surface. However, the extent to which the descending dendritic field is developed varies considerably.

Horizontal cells (Fig. 18D) are characteristic of the third layer of the cortex. They have fusiform somata whose long axes are oriented parallel to the ependymal surface. The somata bear dendrites from their poles. The dendrites usually extend for considerable distances within layer 3, but may eventually turn and ascend through layer 2 before terminating in layer 1.

Bipolar cells (Fig. 18E) have vertically elongate somata and a single dendrite that leaves each pole of the soma.

Multipolar cells (Fig. 18F) have oval or round somata that bear several dendrites in all directions. The dendrites extend throughout the depth of the cortex, branching several times and bearing relatively few spines.

5.3. Physiology of Neurons

Information on the cellular physiology of cortical neurons is limited to a few *in vivo* studies (e.g., Pivavarov and Trepakov, 1972) and, more recently, extensive, *in vitro* studies in turtles (Connors and Kriegstein, 1986; Kriegstein and Connors, 1986; Kriegstein, 1987; Larson-Prior and Slater, 1988). Membrane properties of cortical neurons are generally unremarkable, except that some pyramidal or double pyramidal cells have relatively long membrane time constants, i.e., 140 msec (Connors and Kriegstein, 1986). The cause or significance of this is not known.

A striking feature of cortical neurons is that they can be divided into two categories based on differences in their firing patterns (Fig. 19). Experiments in which cells were studied with intracellular recording techniques and then filled with markers such as Lucifer yellow or HRP suggest these physiological groupings correlate with morphological differences. Intracellular injection of current in neurons classified as pyramidal or double pyramidal cells on morphological grounds (Fig. 19A) produces a train of action potentials that have a relatively slow firing frequency and accommodate rapidly (Connors and Kriegstein, 1986; Shen and Kriegstein, 1987). Action potentials with two amplitudes are seen. Large-amplitude potentials have long (1–3 sec), hyperpolarizing afterpotentials and result principally from an inward sodium conductance, with some involvement of a calcium conductance. Action potentials with small amplitudes typically precede the large-amplitude potentials and may trigger them. They are distinct from action potentials triggered by antidromic activation of axons, and may represent dendritic spikes. A second group of neurons includes layer 1 stellate cells as well as layer 3 horizontal cells (Connors and Kriegstein, 1986; Shen and Kriegstein, 1987). Intracellular current injection in these cells produces a relatively high frequency of firing that accommodates slowly. Nothing is known of the ionic basis of the firing pattern of these cells.

Double pyramidal cells in dorsomedial cortex frequently discharge in periodic bursts of multiple action potentials (Shen and Kriegstein, 1987). In some instances, hyperpolarizing the cells reveals the presence of underlying EPSPs, suggesting the cells burst as a result of synaptic activation. However, other cells show no sign of such EPSPs and appear to burst endogenously. Both types of bursting cells accelerate their firing as a result of orthodromic synaptic activation. Bursting also increases when the cortex is treated with bicuculline methiodide, either by focal application or by submerging the cortex in a bath containing the agent. This indicates bursting is normally suppressed by intrinsic GABAergic mechanisms. Similarly, intracellular studies of neurons in turtle medial cortex in *in vitro* preparations (Larson-Prior and Slater, 1988) show they are normally under the influence of large, spontaneous IPSPs. Application of bicuculline or picrotoxin blocks the IPSPs and leads to a large depolarizing shift and epileptiform discharge that can be blocked, in turn, by NMDA antagonists.

Cortical neurons generally show various mixtures of EPSPs and IPSPs when activated by either electrical (Kriegstein and Connors, 1986; Shen and Kriegstein, 1987; Kriegstein, 1987) or natural (Pivavarov and Trepakov, 1972; Kriegstein, 1987) stimulation. EPSPs can be elicited by focal application of L-glutamate. IPSPs can involve two components. The first is a short-latency hyperpolarizing event with a reversal potential of −70 mV. It is blocked by bicuculline methiodide and reversed by chloride injection. These responses are probably chloride conductances mediated via GABA A receptors. Long-latency hyperpolarizing events are insensitive to bicuculline and are not reversed by chloride injection. They have reversal potentials near the potassium reversal potential as calculated by the Nernst equation. They can be tentatively attributed to a potassium conductance mediated via a GABA B receptor.

Cortical neurons show both habituation and potentiation of their activity as the result of prolonged activation. Habituation is seen *in vivo* in turtle visual cortex (Gusel'nikov and Pivavarov, 1978). A neuron that responds to a moving stimulus shows decreased activity in response to repeated presentation of the stimulus. Consistent with this, paired application of cortical shocks with short interstimulus intervals results in a suppression of the activity of pyramidal cells in intracellular recordings (Kriegstein and Connors, 1986). On the other hand, pairs of pulses with long interstimulus intervals, up to several hundreds of

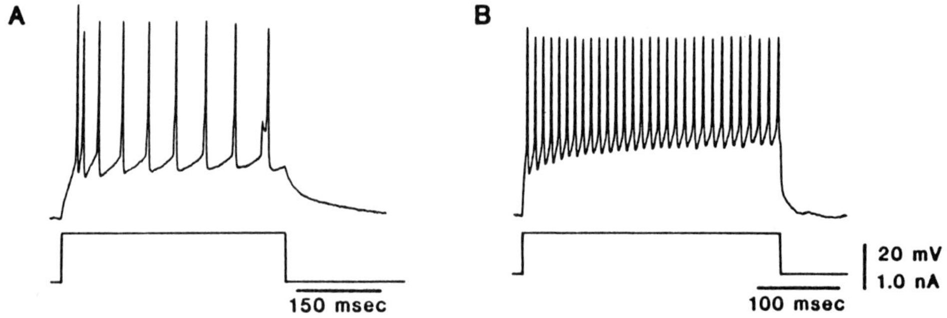

Figure 19. Physiological properties of cortical cells. Shown are the two firing patterns seen in cortical cells. Each panel represents an intracellular recording from a cortical cell during a depolarizing current pulse. (A) This firing pattern is typical of double pyramidal and pyramidal cells. It consists of a number of relatively broad action potentials and smaller "spikelets" which may be dendritic spikes. There is strong accommodation. (B) This firing pattern is typical of stellate and horizontal cells. Action potentials are relatively narrow. There are no spikelets and little accommodation. From Connors and Kriegstein (1986).

milliseconds, lead to an increased response in pyramidal cells that is apparently mediated by a greatly enhanced EPSP.

5.4. Neurotransmitters

There is considerable evidence for neurons in cerebral cortex that use GABA as a neurotransmitter. Immunohistochemical techniques have been used to localize GABA and related marker substances to cerebral cortex. Schwerdtfeger and Lorente (1988a,b) used monoclonal antibodies against GABA to study the distribution of GABA-like immunoreactivity in the lizard *Podarcis* (= *Lacerta*) *hispanica* at both the light and electron microscopic levels. GABA-positive somata are found in layers 1 and 3 of the medial and dorsomedial cortices and are typically associated with aspiny or sparsely spiny dendrites. GABA-positive axon terminals are plentiful, particularly in layer 1, and form symmetric active zones. They are typically presynaptic to unlabeled somata or dendrites. The distribution of GABA-positive neurons is similar in the medial and dorsomedial cortex of turtles (Shen and Kriegstein, 1987). Blanton *et al.* (1987) used antibodies against GABA and its synthetic enzyme glutamic acid decarboxylase (GAD) as well as histochemical procedures for the GABA degradative enzyme, GABA-transaminase, to study potential GABAergic neurons in the dorsal cortex of turtles. These markers label many neurons in layers 1 and 3 as well as a few neurons in layer 2. Labeled neurons include stellate, multipolar, and horizontal cells, but not pyramidal and double pyramidal cells. *In vitro* receptor binding studies using tritiated fluonitrazepam indicate that benzoidazepine receptors (which are part of a complex containing GABA A receptors) are present in turtle cortex (Schlegel and Kriegstein, 1987). They are found in all layers and areas, but are relatively rare in dorsomedial cortex and plentiful in the outer third of layer 1 of the other areas. It is likely that stellate and horizontal cells throughout the cortex are GABAergic and responsible for generating IPSPs mediated by GABA A receptors and chloride channels in pyramidal and double pyramidal cells.

Both *in vivo* and *in vitro* studies on turtle embryos show that these GABAergic responses develop early in cortical differentiation (Blanton and Kriegstein, 1987; Shen *et al.*, 1987; Kriegstein and Shen, 1987; Kriegstein *et al.*, 1988). Experiments in which cells were labeled with HRP *in vitro* show that horizontal cells are the first relatively differentiated cells to develop in turtle cortex; stellate cells develop slightly later. Both are immunoreactive for GABA soon after they differentiate. Whole-cell patch-clamp studies of cells dissociated from embryonic cortex show that GABA-mediated chloride currents are also present early in development, indicating that GABA A receptors are present and functional before the formation of layer 2 or of synapses.

Neurons with somata in layer 2 of the cortex of turtle embryos stain positively with antibodies against the excitatory amino acid glutamate (Blanton and Kriegstein, 1987). Local application of L-glutamate in intracellular recording experiments results in EPSPs in several classes of cortical cells (Kriegstein and Connors, 1986). It appears likely then that some intracortical connections of pyramidal and double pyramidal cells may be mediated via glutaminergic mechanisms.

Immunohistochemical techniques have been used to localize the peptide substance P in the forebrains of both turtles and caiman (Brauth *et al.*, 1983;

Reiner *et al.*, 1984). Somata that stain positively for substance P are found in layers 1 and 2 of medial, dorsomedial, and dorsal cortical areas. Terminal labeling is found in lateral cortex, principally in layers 2 and 1. These observations raise the possibility that projections from medial cortex to lateral cortex use substance P as a neurotransmitter.

Members of the enkephalin family of peptides are found in neurons throughout the cortex of *Caiman* (Brauth, 1984), turtles (Reiner, 1983), and lizards (Naik *et al.*, 1981), principally in layer 2. A small number of neurons that stain positively for somatostatin have been reported in dorsal cortex of turtles (Bear and Ebner, 1983) and the lizard *Ctenosauria pectinata* (Goossens *et al.*, 1980).

Although neurons that stain positively with antibodies against the acetylcholine synthetic enzyme, choline acetyltransferase, are found in the basal forebrain of turtles (Desan, 1984; Hohmann *et al.*, 1983), there is no indication of cholinergic neurons within the cortex.

6. Neuronal Organization of Cortical Areas

Representatives of the morphological groups of neurons described in Section 5.2 are found in all four cytoarchitectonic areas of cortex, but each area contains a particular mixture of cell types and morphological varieties, giving it a distinct neuronal organization.

6.1. Medial Cortex

Layer 1 of medial cortex contains a small number of cells (Minelli, 1966; Lacey, 1978; Davydova and Goncharova, 1979; Ulinski, 1977a; Davila *et al.*, 1985; Berbel *et al.*, 1987). Some are stellate in configuration; others resemble small horizontal cells. In snakes (Ulinski, 1977a), they have oval or round somata, 5 to 10 μm in diameter, that bear two to five dendrites (Fig. 20). The dendrites contain a flocculent cytoplasm with few neurotubules and are relatively smooth, bearing few dendritic spines. They form arbors 200 to 400 μm in diameter. Axons of these stellate cells have not been well characterized, but seem to extend some distance horizontally in layer 1 before branching. Both the somata and dendrites of stellate cells receive many synaptic contacts, predominantly from axon terminals with clear, round synaptic vesicles and forming asymmetric contacts.

Layer 2 is characterized by a large number of small double pyramidal cells with densely packed somata that frequently form casual appositions (Garcia-Verdugo *et al.*, 1981; Davila *et al.*, 1985) or are linked by specialized junctions (Ulinski, 1977a; Wouterlood *et al.*, 1981). They are morphologically quite heterogeneous (Minelli, 1966; Ramón, 1896; Davydova and Goncharova, 1979; Ulinski, 1977a; Lacey, 1978; Davila *et al.*, 1985; Berbel *et al.*, 1987; Guirado *et al.*, 1984; Ebbesson and Voneida, 1969), and we do not know how many cell types are actually present. One clear type is the candelabra cell described by Ulinski (1977a) in the snake *Natrix* (Fig. 21). These neurons have vertically oriented, fusiform somata that bear an extensive array of dendrites in layer 1. The dendrites branch once or twice and ultimately reach the pial surface. They are heavily covered by dendritic spines that range from slender, filiform appendages to mushroom-like in appearance. The spines and shafts of these dendrites are postsynaptic to axon terminals, the great majority of which contain clear, round synaptic vesicles and form asymmetric contacts. The lower pole of each candelabra cell soma gives rise to a single dendrite that is initially smooth, but soon divides into several spinous branches that descend throughout the upper part of layer 3. Proximal shafts of the descending dendrites give rise to one or sometimes two axons that invariably descend into layer 3. Each axon has collateral branches containing many varicosities *en passant,* superficially in layer 3, and then enters the alveus deep in layer 3 where it bifurcates into a medial branch that enters the septum and a lateral branch that courses toward dorsomedial cortex. Both branches bear varicosities *en passant.* Electron microscopic studies of the alveus (Ulinski, 1977a) show it consists of fascicles of tightly packed, unmyelinated axons that dilate into varicosities at intervals. These contain clear, round synaptic vesicles that form asymmetric contacts on dendrites and somata in layer 3. The candelabra cells are the source of major efferent projections from the cortex, which will be described in Section 7.3.

Other layer 2 neurons can be viewed as variants of candelabra cells. One type that has been described frequently in lizards are the superficial fusiform cells first described by Pedro Ramón (1896) in *Lacerta* (Fig. 22), but since reported in several lizards (Lacey, 1978; Davila *et al.,* 1985; Berbel *et al.,* 1987). They generally resemble candelabra cells, but have ascending dendrites that run for a considerable distance along the inner border of layer 2 before turning into layer 1.

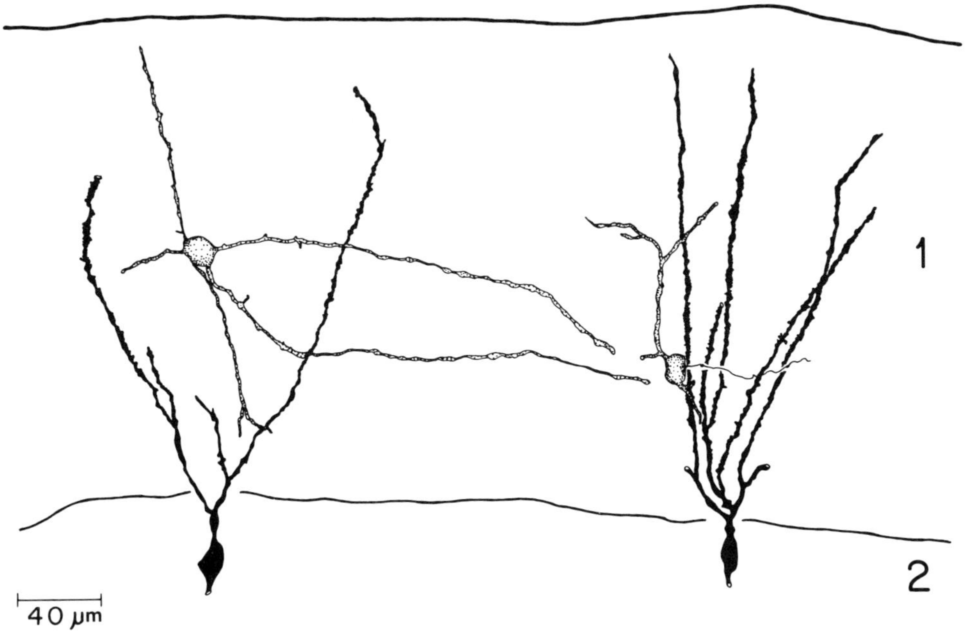

Figure 20. Stellate cells from medial cortex. This is a camera lucida drawing of two stellate cells from Golgi preparations of medial cortex of a water snake, *Natrix sipedon.* The stellate cells are situated in layer 1 and bear several dendrites that extend in all directions. The cell on the right bears an axon that runs horizontally, intersecting the dendrites of candelabra cells. From Ulinski (1977a).

There is considerable reason to believe that the axons of cells in layer 2 of medial cortex contain zinc. Zinc can be detected histochemically using the method of Timm in several areas of cerebral cortex of lizards (Lopez-Garcia *et al.*, 1983; Perez-Clausell, 1988), including layer 3 of medial cortex and layers 1 and 3 of dorsomedial cortex, as well as the dorsal septum. These contain abundant Timm-reactive presynaptic boutons which make asymmetric contacts with dendrites and dendritic spines. The only known fiber system with a distribution that correlates precisely with the Timm-positive regions of the telencephalon are the efferent fibers from medial cortex to the septum and dorsomedial cortex (Olucha *et al.*, 1988). Recent work in mammals suggests zinc may be released by synaptic terminals and can differentially modulate the activity of different classes of excitatory amino acid receptors (Westbrook and Mayer, 1987; Peters *et al.*, 1987). Since both afferents to the cortex and intracortical axons may be glutaminergic, cells in layer 2 of medial cortex could play a complicated role in

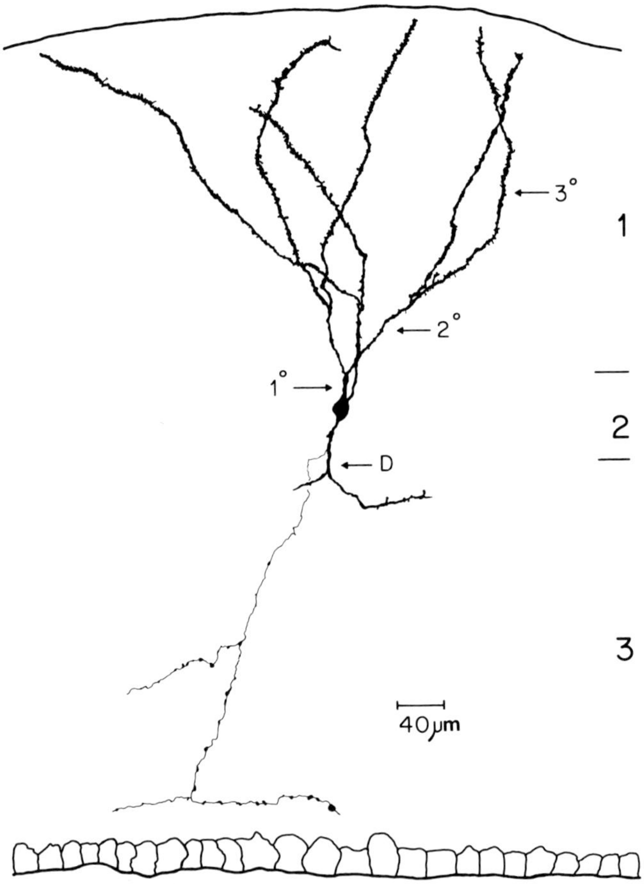

Figure 21. Candelabra cell. This is a camera lucida drawing showing the morphology of a candelabra cell, a type of modified double pyramidal cell characteristic of medial cortex. It is from Golgi preparations of a water snake, *Natrix sipedon*. Ascending dendrites extend in layer 1 and consist of proximal portions (1°) and distal portions (2° and 3°). Descending dendrites (D) branch in the upper half of layer 3. The axon descends through layer 3 and bifurcates into medial and lateral branches in the alveus. From Ulinski (1977a).

controlling the overall activity of neurons in dorsomedial cortex and the septum through the release of both a traditional neurotransmitter as well as an ion with a neuromodulatory function.

Layer 3 typically contains a moderate number of scattered neurons, but there is a tendency in some species for cells deep in layer 3 to form clusters or a more extensive cell plate. In snakes (Ulinski, 1977a), these periventricular cells are pyramidal cells with fusiform somata (Fig. 23), oriented horizontally in layer 3, and joined to their neighbors by rows of appositions resembling those seen on candelabra cells. Descending, spiny dendrites spread tangentially from the somata within the alveus. A single ascending dendrite extends radially from each soma through layer 3 and into layer 2. It is initially 2 μm thick and lacks dendritic spines, but eventually bifurcates into spinous branches as it leaves layer 2. In lizards, layer 3 contains neurons that range from spiny multipolar cells to pyramidal cells to horizontal cells (Davila *et al.*, 1985; Berbel *et al.*, 1987; Guirado *et al.*, 1984). The horizontal cells have thick dendrites that are presynaptic to Timm-positive boutons, which suggests they are postsynaptic to layer 2 cells. A population of multipolar cells with distinctive, long dendritic spines are also postsynaptic to Timm-positive boutons (Lopez-Garcia *et al.*, 1988). They are GABA-positive and bear axons that ascend to layer 1. Thus, they are probably inhibitory interneurons.

Figure 22. Lizard cortex. This is a drawing of a Golgi preparation through medial, dorsomedial, and dorsal cortex in a lizard, *Lacerta.* It shows several cell types. Note in particular the candelabra cells and the "superficial fusiform cell" (a) in medial cortex. From Ramón (1896).

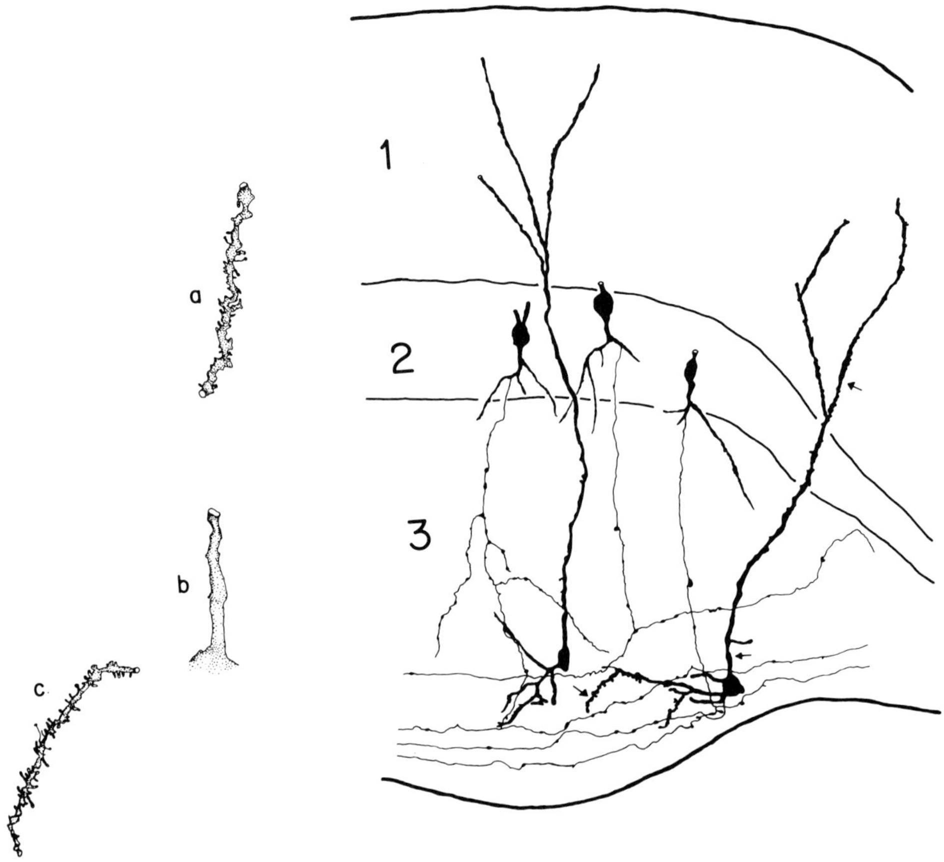

Figure 23. Pyramidal cells. This is a camera lucida drawing of pyramidal cells from the cell plate of medial cortex. It is from Golgi preparations of a water snake, *Natrix sipedon.* the somata and descending dendrites of the cells lie in the alveus; their ascending dendrites extend through layer 1. An axon originates from the soma on the right and runs into the alveus. The insets show details of the descending dendrite (c), proximal (b) and distal ascending (a) dendrites from the cell on the right at the levels indicated by the arrows. Note that descending axons of candelabra cells intersect the pyramidal cells. From Ulinski (1977a).

6.2. Dorsomedial Cortex

Bipolar cells in layer 1 of dorsomedial cortex have fusiform, vertically oriented somata with three to five dendrites that extend up to 150 μm from the soma (Minelli, 1966; Ulinski, 1979; Lacy, 1978). Some are devoid of spines or excrescences, while others bear a moderate number of spines. The morphology of their axons is poorly known.

Layer 2 is characterized by double pyramidal cells (Fig. 24) distinctly larger than those present in layer 2 of medial cortex (Ramón, 1891; Minelli, 1966; Ebbesson and Voneida, 1969; Davydova and Goncharova, 1979; Lacey, 1978; Ulinski, 1979). They have fusiform, vertically oriented somata that are distinguished ultrastructurally by a relatively large number of lysosomes and lipofuchsin granules (Ulinski, 1979). They are relatively loosely packed, so contacts with neighboring somata are rarer than in medial cortex and the appositions characteristic of candelabra cells are not found in dorsomedial cortex. Each pole of the soma gives rise to a tuft of dendrites, consisting of a smooth proximal shaft that branches into several spinous daughter branches. The shafts are contacted by axon terminals containing clear, round vesicles and forming symmetric contacts (Ulinski, 1979). The spines are contacted by axons with clear, round vesicles and forming asymmetric contacts (Ulinski, 1979; Guirado *et al.*, 1984). The ascending dendrites extend through layer 1 almost to the pial surface. The descending dendrites extend through the upper half of layer 3. There are some variations in th size and shape of the dendritic fields (see Ulinski, 1979). Axons originate either from the somata of double pyramidal cells or from proximal dendrites near the soma. They descend into layer 3 and arborize into a series of branches forming a conical field. Some branches extend medially and participate in the commissural connections described in Sections 7.2.

Layer 3 contains several distinct types of neurons (Ramón, 1891; Minelli, 1966; Ulinski, 1979). First is a population of "displaced" double pyramidal cells that resemble the layer 2 cells. Second is a population of multipolar cells with round or oval somata (Fig. 26). Dendrites originate from all parts of their somata. They are initially smooth, but eventually are covered by dendritic spines. The ascending dendrites are long and can extend into layer 2; the descending

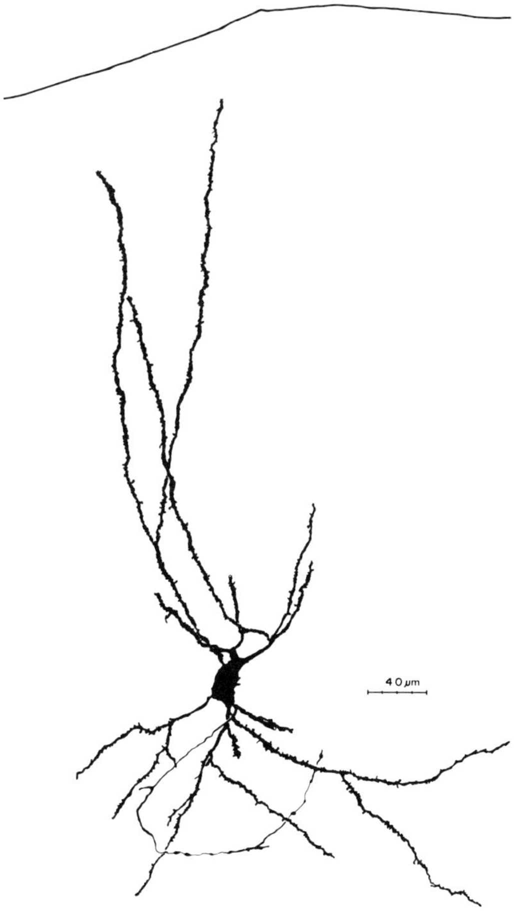

Figure 24. Double pyramidal cell. A camera lucida drawing of a double pyramidal cell from a Golgi preparation of dorsomedial cortex of a water snake, *Natrix sipedon*. The soma of the cell is located in layer 2. From Ulinski (1979).

dendrites are shorter and extend through layer 3 to the ependyma. Finally, horizontal cells have somata in the inner half of layer 3 (Fig. 25). These are fusiform and oriented horizontally. Each pole of the soma gives rise to one or two dendrites that eventually branch and extend for considerable distances in layer 3 parallel to the ventricular surface. The dendrites bear relatively few spines, but sometimes have dilatations.

6.3. Dorsal Cortex

The cytoarchitecture of dorsal cortex varies considerably between species, but the only detailed investigations of its cellular structure have been carried out in cryptodiran turtles (Davydova and Smirnov, 1969; Northcutt, 1970; Davydova and Goncharova, 1979; Desan, 1984) and, to a lesser extent, in lizards (Ramón, 1891; Ebbesson and Voneida, 1969; Guirado *et al.*, 1987).

Layer 1 contains scattered stellate cells. Stellate cells have their somata situated deeper in layer 1 and dendrites that extend in all direction from their somata. The dendrites are generally confined to layer 1 and can vary in morphology from smooth to beaded to sparsely spinous to those covered with a moderate number of spines. The axons of these cells arborize extensively within layer 1 bearing many varicose collaterals. Horizontal cells are situated in the upper part of layer 1, nearer the pia, and have dendritic fields that are more dorsoventrally flattened with individual dendrites extending roughly parallel to the pial surface. The axons of these cells also extend horizontally. The stellate cells are GABAergic and demonstrate a tonic firing pattern (Section 5).

Layer 2 of turtles contains two distinct cell types (Northcutt, 1970; Davydova and Goncharova, 1979; Minelli, 1966; Desan, 1984). The majority are double

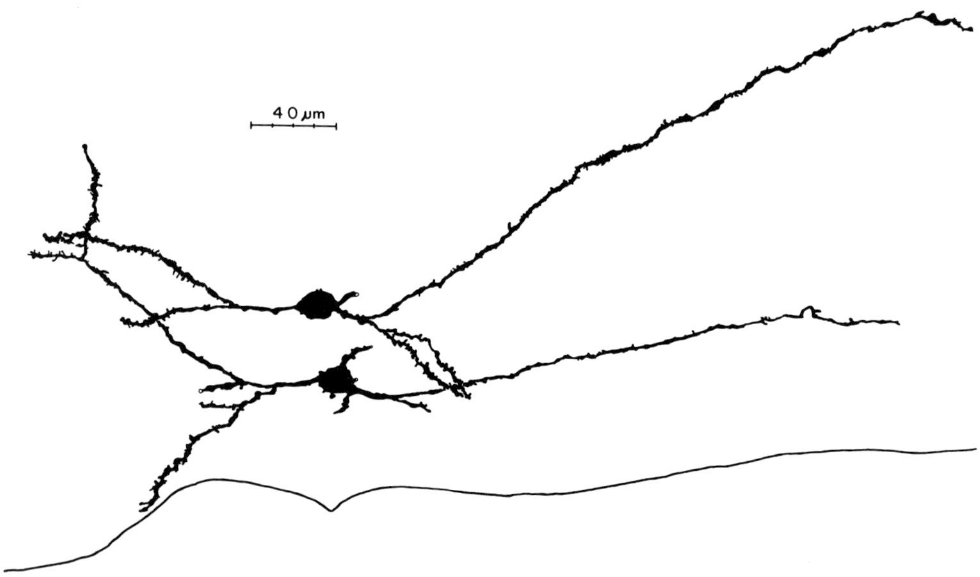

Figure 25. Horizontal cells. Camera lucida drawing of two horizontal cells from Golgi preparations of dorsomedial cortex of a water snake, *Natrix sipedon*. The ventricular surface of dorsomedial cortex is shown at the bottom of the drawing. From Ulinski (1979).

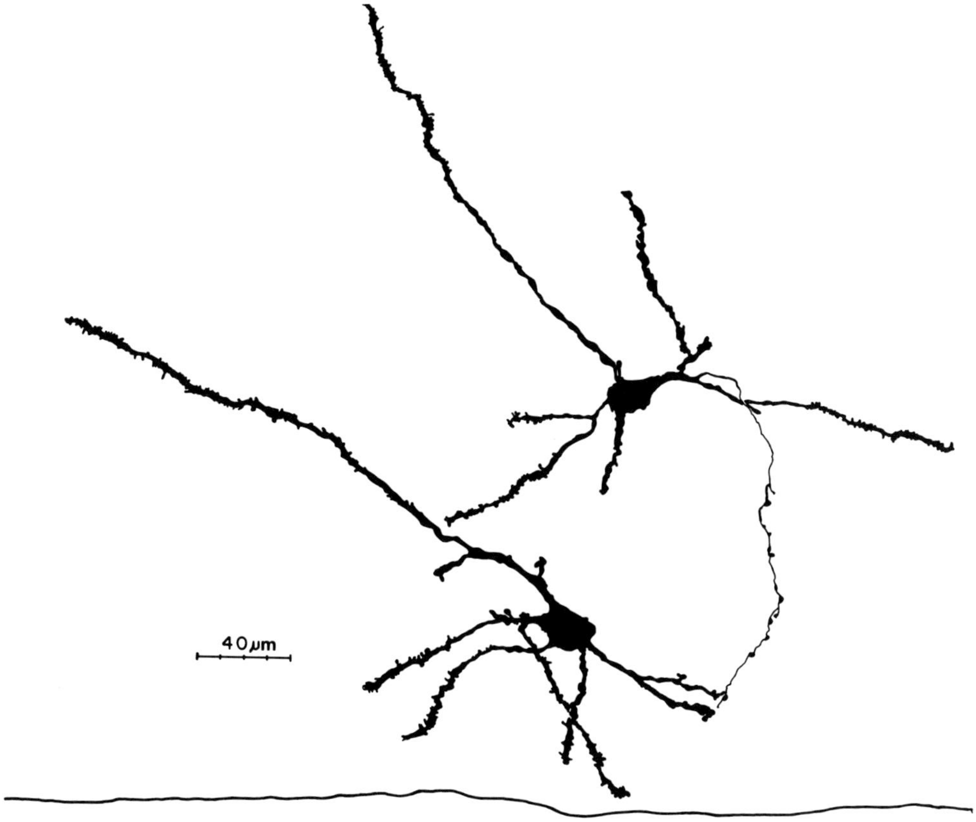

Figure 26. Multipolar cells. Camera lucida drawing of two multipolar cells from a Golgi preparation of dorsomedial cortex of a water snake, *Natrix sipedon*. The somata of the cells are situated in layer 3. The ventricular surface is shown at the bottom of the drawing. From Ulinski (1979).

pyramidal cells, each having two conical dendritic fields, one ascending into layer 1 and the other descending into layer 3 (Fig. 27). The proximal dendrites are free of dendritic spines, but spines gradually appear as the dendrites are followed out from the somata and then decrease in number toward the distal tips of the dendrites. Some of the spines are long or filiform, others are pedunculated, and many end in complex heads that invaginate into presynaptic axon terminals. The spines are postsynaptic to axon terminals that almost always contain clear, round synaptic vesicles and form asymmetric junctional contacts. The morphology of the axons of double pyramidal cells has not been systematically studied, but many descend into layer 3 where they collateralize. Some collaterals ascend obliquely into layer 1; others run in layer 3 into adjacent cortical layers. Although most double pyramidal cells have two dendritic fields, they vary systematically in their development according to the position of the cell in dorsal area. Layer 2 is situated close to the ependyma at the medial edge of dorsal area, and the double pyramidal cells in this region have very reduced basal dendritic fields. The size of the basal dendritic fields increases in cells situated at progressively more lateral regions in dorsal cortex, so those in the pallial thickening (the convoluted region adjacent to lateral cortex) have ascending and descending fields that are of equal size. The double pyramidal cells

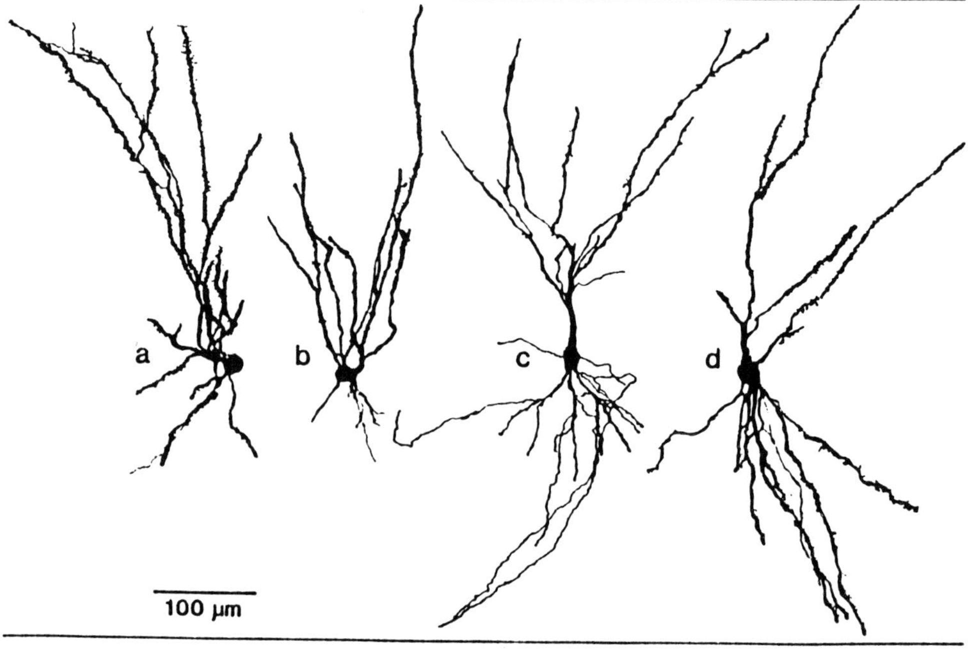

Figure 27. Double pyramidal cells. Drawings of double pyramidal cells from dorsal cortex of the red-eared turtle, *Pseudemys scripta elegans*. These cells are all from area D2 of Desan (1984) and illustrate the way in which the morphology of double pyramidal cells in this area varies as a function of mediolateral position. Cell a is situated toward the medial border of area D2; other cells are situated at progressively more lateral positions. Note the way in which the length of the descending tuft of dendrites increases as cells are situated more laterally. From Connors and Kriegstein (1986).

exhibit a slow firing pattern, a tendency to accommodate, and, possibly, dendritic spikes. The second type of cell present in layer 2 of dorsal cortex is a multipolar cell that has a round soma and several smooth dendrites that extend into layers 1 and 3. Layer 2 of lizards contains double pyramidal and multipolar cells, but also contains distinct populations of pyramidal cells (Ramón, 1891; Ebbesson and Voneida, 1969; Guirado *et al.*, 1987).

Layer 3 contains principally horizontal cells with fusiform somata and dendrites that extend from their poles, running concentric with the ependymal surface. Little is known about the axons of the horizontal cells, except that at least some project out of the cortex to the brain stem (Ulinski, 1986b). Many of the horizontal cells stain positively for markers associated with GABA (Shen and Kriegstein, 1987). In addition, layer 3 may contain bipolar cells.

6.4. Lateral Cortex

The cellular composition of lateral cortex has been most extensively studied in the snakes *Constrictor, Thamnophis,* and *Natrix* (Ulinski and Rainey, 1980) with a few observations available in lizards (Ebbesson and Voneida, 1969) and turtles (Minelli, 1966; Davydova and Goncharova, 1979). Layer 1 of all four cytoarchitectonic subareas of lateral cortex is characterized by stellate cells with oval or round somata and dendrites confined to layer 1. The dendrites vary from smooth to heavily decorated with spines. The axons of stellate cells appear to collateralize in layer 1, within the dendritic field of the parent cell. However, each of the four cytoarchitectonic areas has a distinct mixture of cell types in layers 2 and 3.

The rostral and caudal parts of lateral cortex are distinguished principally by differences in layer 2. Rostral lateral cortex is characterized by a population of distinct layer 2 cells, called bowl cells by Ulinski and Rainey (1980) (Fig. 28). These neurons have fusiform somata that give rise to a dendritic arbor, about 600 μm in diameter, extending into layer 1 with an overall bowl-shaped configuration. The somata are densely packed, with their plasma membranes touching in casual appositions (Ulinski and Rainey, 1980; Garcia-Verdugo *et al.*, 1986), and receive relatively few axonal contacts. The proximal shafts of bowl cell dendrites are smooth. The more distal primary and secondary branches are heavily covered by dendritic spines that are postsynaptic to axon terminals with clear, round synaptic vesicles and forming asymmetric junctions. The axons of bowl cells descend into layer 3 and collateralize extensively. One branch of each cell is efferent from lateral cortex. Bowl cells in ventral rostral lateral cortex have axons that course ventrally. Bowl cells in dorsal rostral lateral cortex have axons that course dorsally and form a fascicle of axons, each bearing varicosities *en passant,* that courses over the dorsal surface of the brain to end in medial cortex (Fig. 29). Rostral lateral cortex also is characterized by a cell plate or "secondary lamina" of somata in layer 2 which seems to be composed principally of stellate cells. These have relatively large somata and dendrites that extend both within layer 3, and reach some distance through layer 2 and into layer 1.

Caudal lateral cortex is characterized by pyramidal cells in layers 2 and 3 (Fig. 30). Some have a distinctly pyramidal-shaped somata and an apical dendrite

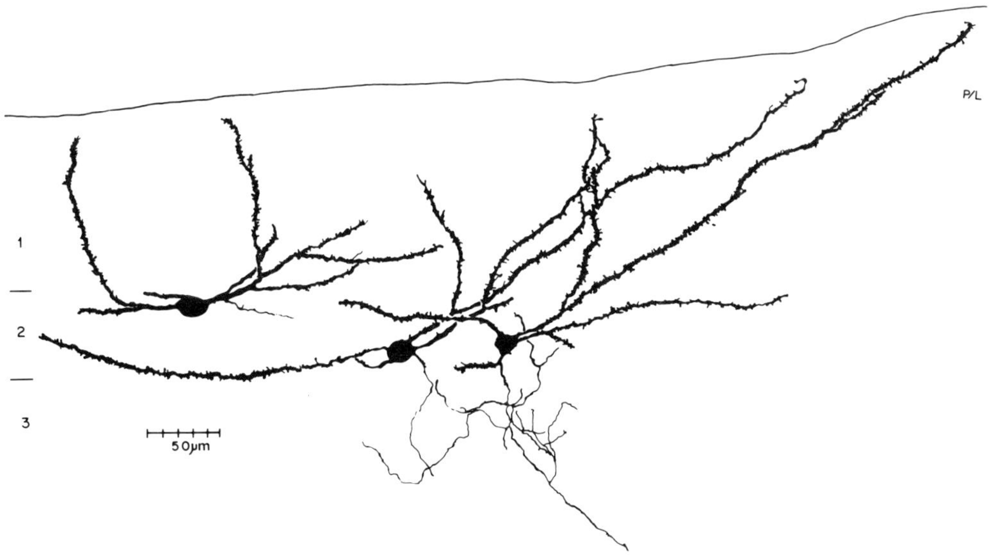

Figure 28. Bowl cells. Camera lucida drawing of bowl cells, a form of modified double pyramidal cell characteristic of lateral cortex. These are from Golgi preparations of lateral cortex of a water snake, *Natrix sipedon.* All of the cells are from dorsal lateral cortex. The pial surface is represented by the line at the top of the drawing and the borders of layer 2 are indicated by horizontal lines on the left. From Rainey and Ulinski (1980).

that ascends vertically and unbranched through layers 3 and 2. Others, more often found in layer 3, have horizontally oriented fusiform somata that bear an apical dendrite from one pole. This dendrite then curves sharply and ascends through layer 2. The proximal shafts of the apical dendrites are relatively smooth, but branch within layer 1 into a conical arbor of spiny dendrites. The basal dendrites of pyramidal cells are moderately spiny and form a conical arbor in layer 3.

Horizontal and double pyramidal cells are found in lateral cortex, but constitute a relatively small percentage of the total neuronal population as compared to other cortical areas.

7. Cortical Connections

7.1. Afferents to Cortex

The overall pattern of afferents to the cortex from subcortical structures has been studied in turtles (Desan, 1984) and lizards (Bruce and Butler, 1984a) using the retrograde transport of HRP. These studies indicate cortex receives inputs

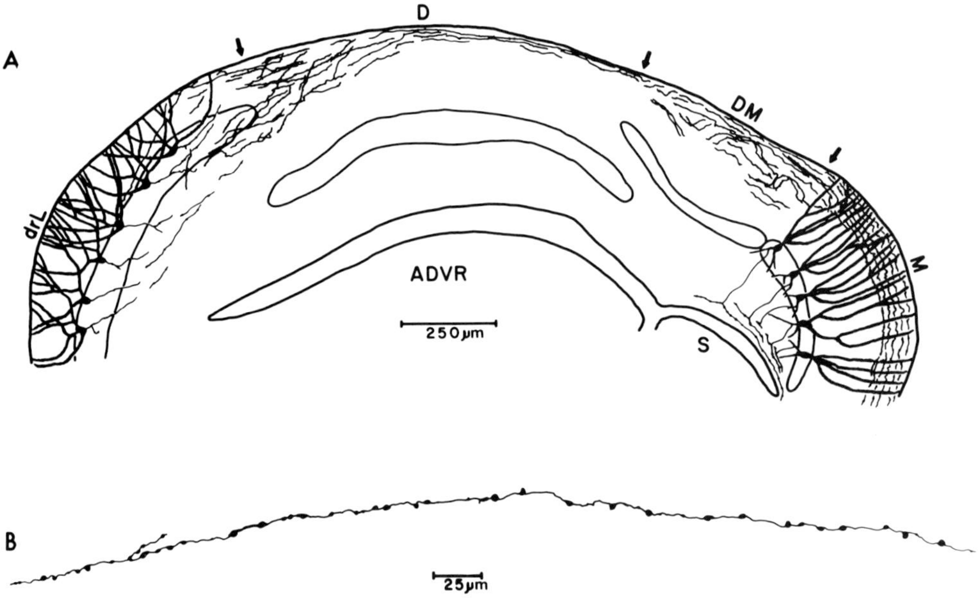

Figure 29. Bowl cell projections to medial cortex. (A) This is a drawing of the dorsal half of a coronal section through dorsal lateral cortex in a water snake, *Natrix sipedon*. Dorsal rostral lateral cortex (drL) is at the left. Examples of several bowl cells are depicted to illustrate the course of bowl cell axons. Notice that the axons of cells whose somata are situated ventrally in drL proceed medially through layer 3 and then curve dorsally. The axons of cells whose somata are situated ventrally course through layer 2 and dorsally into layer 1. Both sets of axons course together at the border of drL and dorsal cortex (D), which is marked by the arrow on the left. The bowl cell axons course across dorsal cortex just below the pial surface and enter dorsomedial cortex (DM). At this point the fascicle expands so that the fibers occupy the outer third of layer 1 in medial cortex (M). They intersect the distal ascending dendrites of the candelabra cells. (B) Camera lucida drawing of a segment of a bowl cell axon. It could be traced out of drL, across the full width of dorsal area and into dorsomedial area. Note the varicosities, which can be seen along its trajectory. From Rainey and Ulinski (1980).

from a considerable variety of subcortical structures, which can be grouped into three categories for the purposes of discussion: (1) sources of sensory information, (2) structures related to the limbic system, and (3) brain-stem and basal forebrain systems.

7.1.1. Sources of Sensory Input

7.1.1a. Olfactory System. All reptiles have a main olfactory bulb that receives input from the olfactory mucosa of the nasal cavity. In addition, snakes, *Sphenodon,* and many lizards have accessory olfactory bulbs that receive input from the vomeronasal or Jacoboson's organs. These are particularly well developed in forms with bifid tongues used to transport substrate-bound odorants to the vomeronasal cavities. Crocodilians lack vomeronasal organs; it is still uncertain whether or not turtles have vomeronasal organs (see Parsons, 1970). The accessory olfactory bulbs, when present, project to a distinctive nucleus in the amygdalar complex known as nucleus sphericus (Halpern, 1976; Ulinski and

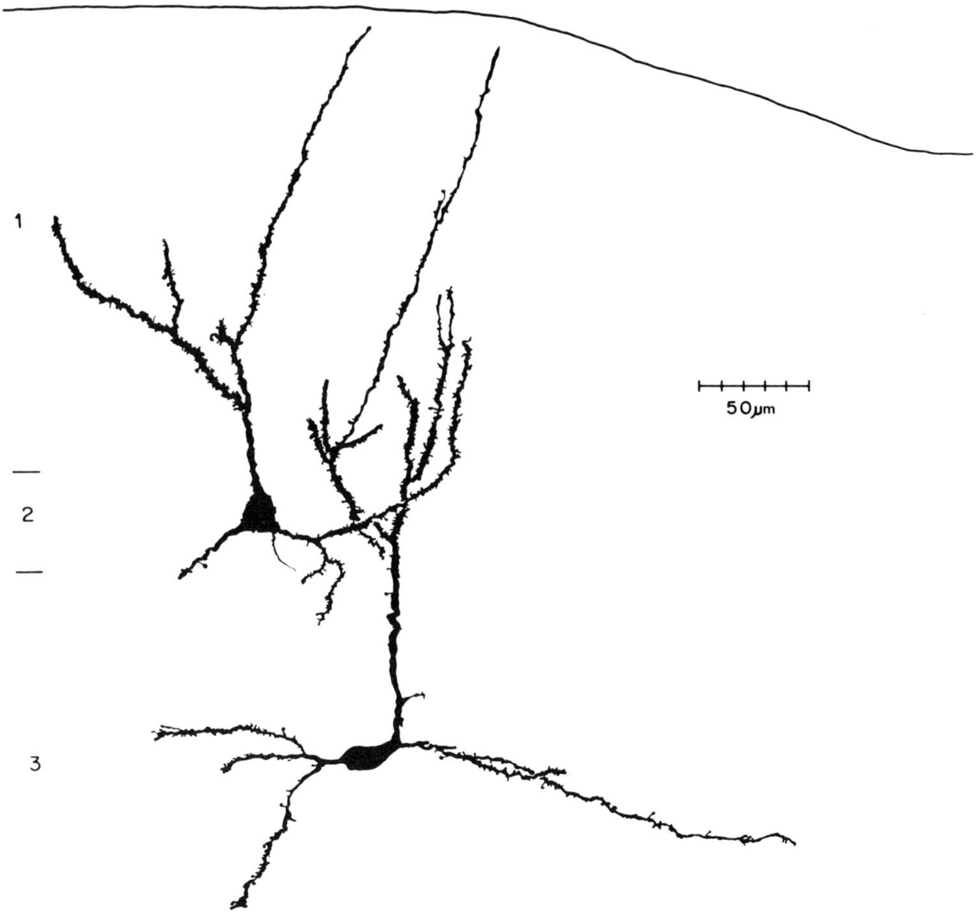

Figure 30. Pyramidal cells. Camera lucida drawing of pyramidal cells from Golgi preparation of lateral cortex of a water snake, *Natrix sipedon.* The pial surface is shown at the top of the drawing and the boundaries of layer 2 are demarcated by horizontal lines on the left. Note that the cell whose soma is positioned in layer 2 has a basal dendrite that recurves into layer 1, while the layer 3 pyramidal cell has its basal dendrites confined to layer 3. From Rainey and Ulinski (1980).

Kanarek, 1973; Heimer, 1969; Ulinski and Peterson, 1981). The main olfactory bulbs project to several structures situated around the rostral telencephalon (e.g., Halpern, 1976; Ulinski and Peterson, 1981; Scalia *et al.*, 1969; Reiner and Karten, 1985). These include the olfactory tubercle, diagonal band nuclei, and lateral cortex (Fig. 31).

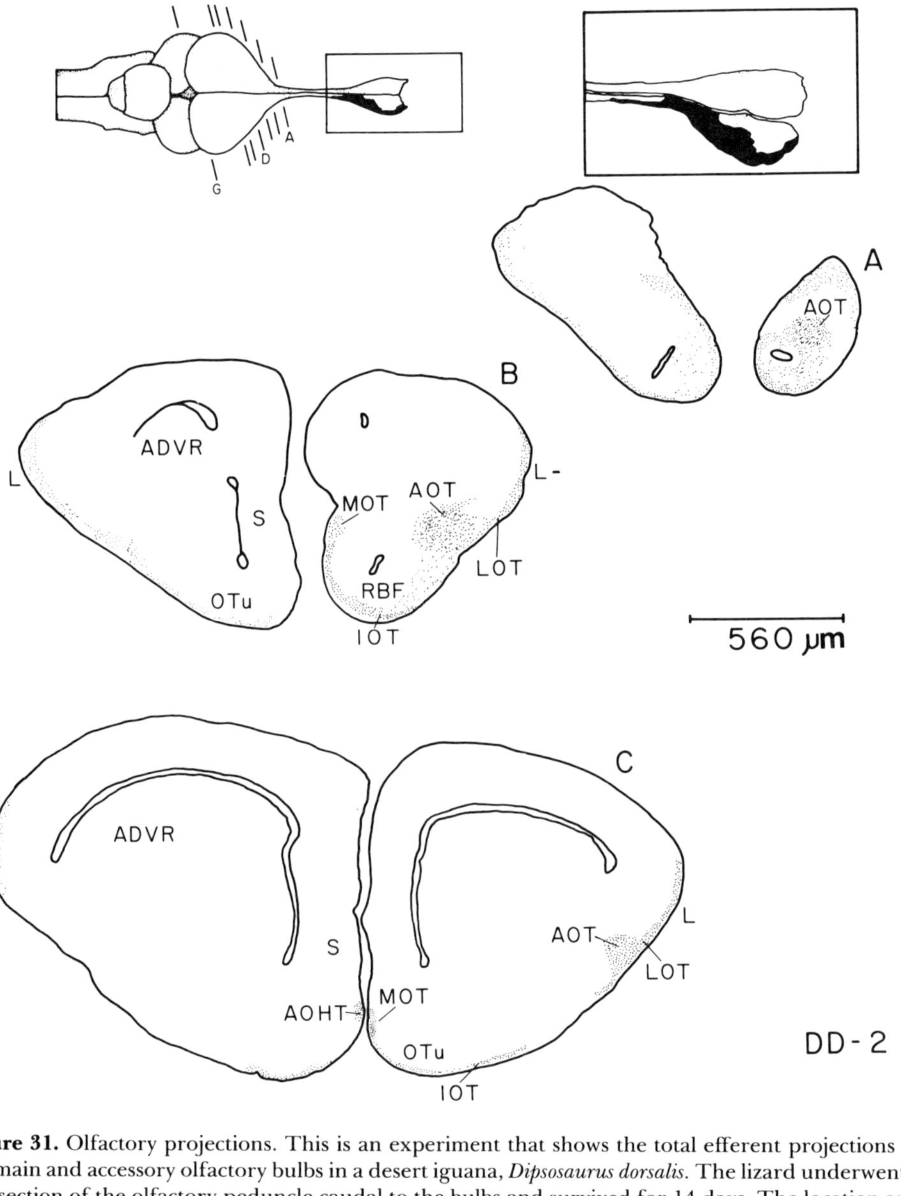

Figure 31. Olfactory projections. This is an experiment that shows the total efferent projections of the main and accessory olfactory bulbs in a desert iguana, *Dipsosaurus dorsalis.* The lizard underwent a transection of the olfactory peduncle caudal to the bulbs and survived for 14 days. The location and extent of the lesion are shown in the inset on the right. The planes of the transverse sections A through G are shown on the dorsal view of the brain on the left. Note that the sections are oriented obliquely relative to the brain's long axis so that the right half of the brain in each section is rostral to the left half. The appearance of degenerated elements in Fink–Hemier preparations is charted on sections A through G. Axons from mitral cells in the main olfactory bulb run through the lateral olfactory tract (LOT) to the rostral subareas of the ipsilateral lateral cortex (L). They also cross the

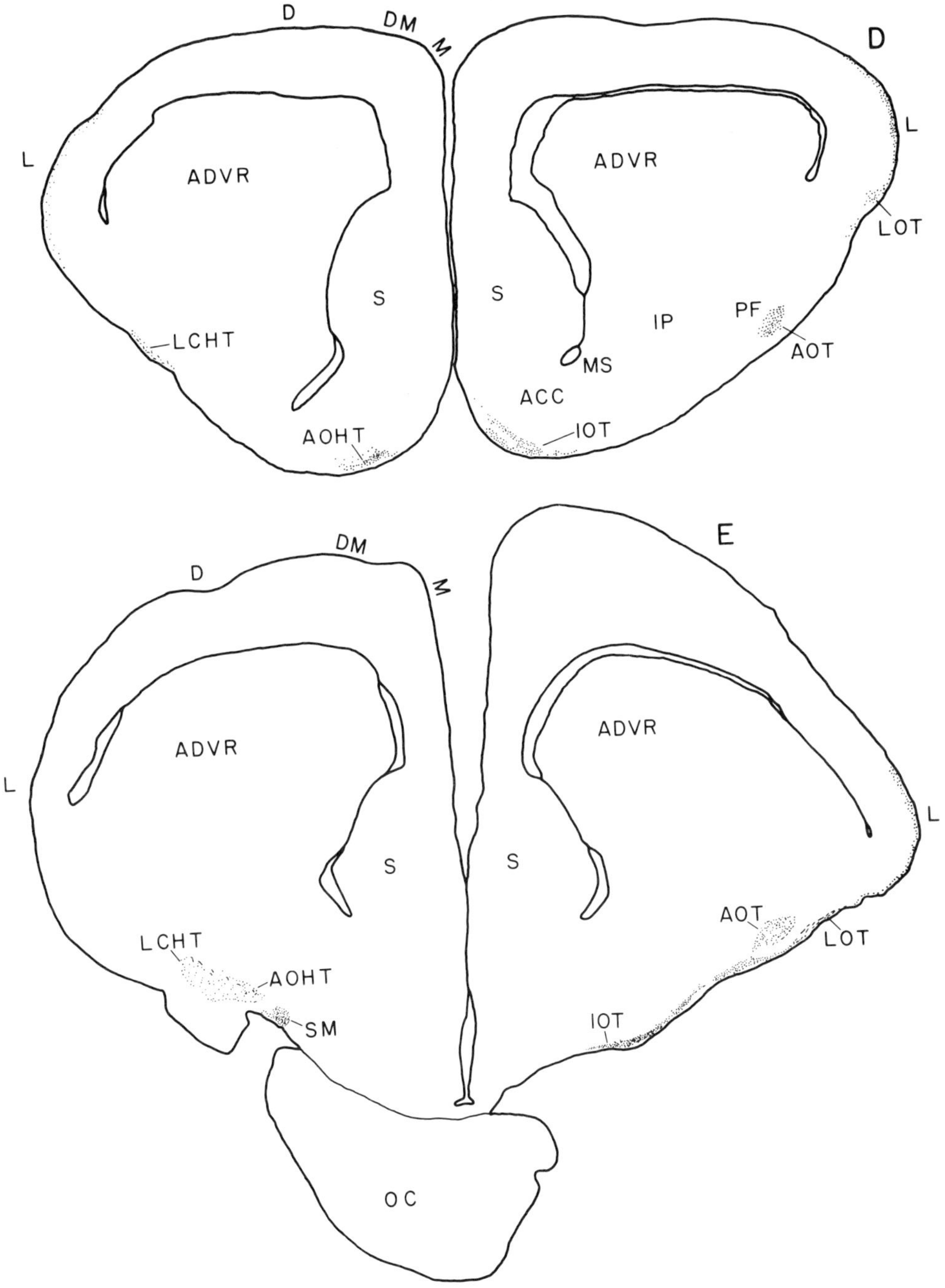

Figure 31. (*Continued*)

midline through the stria medullaris system (SM) and habenular commissure (HC) to reach the contralateral lateral cortex. Axons from the accessory olfactory bulb run through the accessory olfactory tract (AOT) to the nucleus sphericus. AC, anterior commissure; ACC, nucleus accumbens; ADVR, anterior dorsal ventricular ridge; AOHT, anterior olfactohabenular tract; D, dorsal cortex; DM, dorsomedial cortex; IOT, intermediate olfactory tract; IP, intrapeduncular nucleus of the striatum; LCHT, lateral corticohabenular tract; M, medial cortex; MOT, medial olfactory tract; MS, medial striatal nucleus; OC, optic chiasm; OT, optic tract; OTu, olfactory tubercle; PF, perifascicular complex; RBF, retrobulbar formation; S, septum; T, optic tectum. From Ulinski and Peterson (1981).

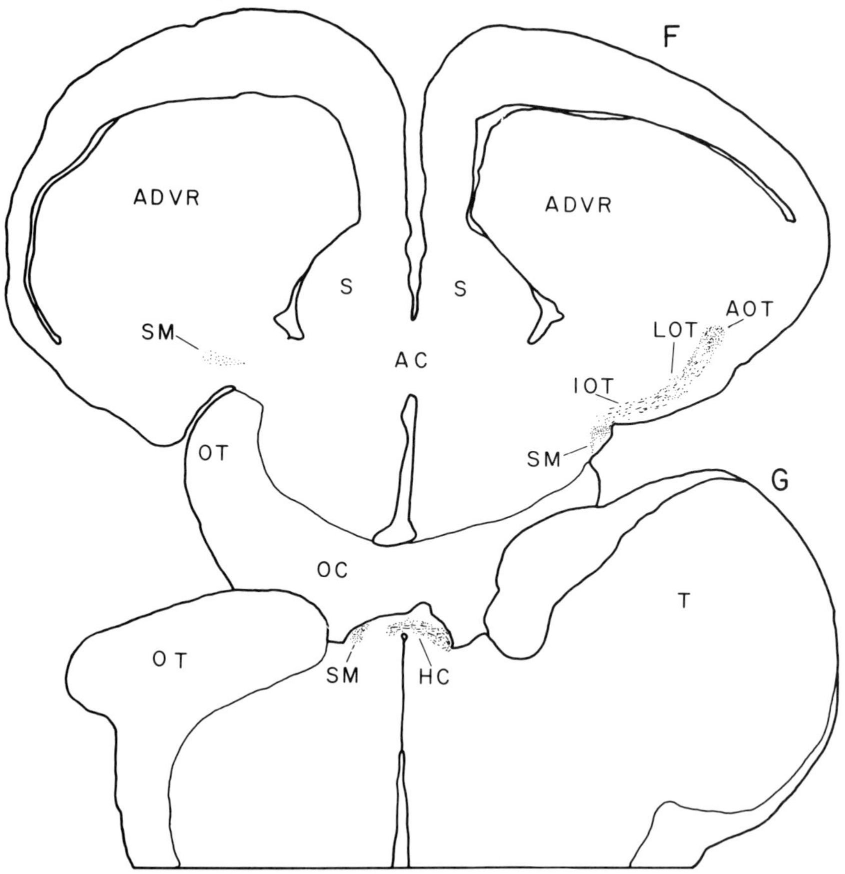

Figure 31. (*Continued*)

Efferent fibers from the main olfactory bulb exit its lateral surface and run into the ipsilateral lateral cortex via the lateral olfactory tract. The pattern of main olfactory bulb input to lateral cortex varies between species. In snakes, it terminates throughout the dorsoventral extent of the rostral lateral cortex, but ends precisely at its border with the caudal lateral cortex (Halpern, 1976; Ulinski and Rainey, 1980). The same situation is present in the fence lizard *Lacerta viridis* and the desert iguana *Dipsosaurus dorsalis* (Ulinski and Peterson, 1981). Anoles of the genus *Anolis* have relatively reduced olfactory systems (Armstrong *et al.*, 1953). The olfactory bulb complex is small and the rostral lateral cortex is reduced to a pallial membrane and receives only a sparse olfactory bulb input. Heimer (1969) reported a rather different situation in the tegu lizard *Tupinambis*. He found main olfactory bulb inputs to the full rostrocaudal extent of lateral cortex. This resembles the pattern of olfactory bulb projections in turtles (Gamble, 1956; Balaban, 1977; Skeen *et al.*, 1984; Reiner and Karten, 1985) and *Caiman* (Scalia *et al.*, 1969) in which the olfactory bulbs project all along the ventrolateral surface of the cortex.

In contrast to mammals, there are direct projections to the contralateral lateral cortex in reptiles. These are effected via the habenular commissure by fibers that run caudally in the lateral olfactory tract and then cross the base of the hemisphere, join the stria medullaris and thereby gain access to the habenular commissure. They turn rostrally and run through the contralateral stria medullaris and lateral corticohabenular tracts to reach the caudal edge of the contralateral cortex. Orrego (1961, 1962) found that electrical stimulation of the olfactory bulb produced a surface negative evoked potential in the contralateral olfactory cortex with a latency of 50–100 msec.

Main olfactory bulb projections are limited to the outer one-quarter of layer 1 of lateral cortex (Ulinski and Rainey, 1980; Ulinski and Peterson, 1981; Skeen *et al.* 1984). Individual fibers run caudally in the lateral olfactory tract (Peterson and Ulinski, 1981). They initially have dorsoventral trajectories, but gradually turn to run rostrocaudally in layer 1. They consist of thin axons bearing many varicosities *en passant* and terminate principally upon dendritic spines via axon terminals that form asymmetric active zones and contain clear, round synaptic vesicles (Rainey and Ulinski, 1980). There have been no intracellular studies of their synaptic activity, but Orrego's (1961, 1962) finding that electrical stimulation of the olfactory bulbs of turtles elicits surface negative evoked potentials in lateral cortex suggests they give rise to EPSPs in cortical neurons. The orientation of the dendrites of bowl cells and their numerical predominance in the rostral lateral cortex suggest they are the principal postsynaptic targets of olfactory bulb afferents.

HRP experiments in lizards (Bruce and Butler, 1984a; Martinez-Garcia *et al.*, 1986) show the rostral lateral cortex receives olfactory information indirectly via structures in the basal telencephalon as well as directly from the main olfactory bulbs. HRP injections retrogradely label somata in the diagonal band nuclei and the nucleus of the lateral olfactory tract. These structures receive inputs from the main olfactory bulbs and could serve as an indirect relay of olfactory information to lateral cortex. The relationship of the direct and indirect olfactory inputs within lateral cortex is not known.

7.1.1b Dorsal Lateral Geniculate Nucleus. Early theories of telencephalic evolution held that the cerebral hemispheres of nonmammals were dominated by olfactory inputs and received little or no nonolfactory sensory input. This was consistent with the view that there are no direct thalamic projections to the dorsal telencephalon in reptiles (e.g., Powell and Kruger, 1960; Kruger and Berkowitz, 1960). However, in 1970 Hall and Ebner (Hall and Ebner, 1970b) used Nauta degeneration techniques to convincingly demonstrate a projection from the dorsal lateral geniculate nucleus to the cortex in turtles. Geniculocortical projections have since been studied extensively in turtles and demonstrated in several lizards. Their presence is problematic in snakes and crocodilians, and unexplored in *Sphenodon.*

The dorsal lateral geniculate complex in turtles is a rostrocaudally elongate structure situated internal to the optic tract. It consists of a cell plate of densely packed neurons that makes up its medial face and a neuropil of scattered cells situated between the cell plate and the optic tract (Rainey and Ulinski, 1986). The cell plate cells have dendrites that extend radially into the neuropil. The neuropil cells have dendrites arranged parallel to the optic tract (Rainey and Ulinsky, 1986). The retina projects bilaterally and topographically to the geniculate complex (Hall and Ebner, 1970a; Knapp and Kang, 1968a,b; Bass and Northcutt, 1981a,b; Ulinski and Nautiyal, 1988). The retinogeniculate axons form restricted terminal arbors (Sjostrom and Ulinski, 1985) that synapse on geniculate neurons via synapses with clear, round synaptic vesicles and asymmetric junctional complexes (Ulinski and Nautiyal, 1988). These terminals form triadic relationships with pairs of cell plate dendrites as well as making relatively isolated contacts along the more proximal shafts of the cell plate cells. The geniculate neurons have axons that exit the medial face of the geniculate complex (Heller and Ulinski, 1987), course ventrally and join the dorsal peduncle of the lateral forebrain bundle (Fig. 32). They proceed rostrally into the striatum where they initially lie at the very superficial edge of the dorsal peduncle as it passes beneath the floor of the lateral ventricle (Hall and Ebner, 1970b; Heller and Ulinski, 1987). However, they eventually turn laterally and extend dorsally between the inner edge of the lateral cortex and the anterior dorsal ventricular ridge. Both the Hall and Ebner and Heller and Ulinski studies are subject to some ambiguity because of the difficulty of restricting lesions or injections of tracer substances solely to the geniculate complex. However, the trajectory suggested for the geniculocortical fibers by these studies is consistent with the pattern of labelling seen in the cortex following the transynaptic transport of tritiated amino acids (Desan, 1984).

Geniculate neurons can be retrogradely labeled following cortical HRP injections (Hall *et al.*, 1977; Belekhova *et al.*, 1979), and Ulinski (1987b) was able to label neurons in the neuropil layer as well as both sublayers of the cell plate by cortical injections. It therefore appears that all of the cell populations in the geniculate project to the cortex. It is less certain whether all of the individual cells in each population project to the cortex, but it seems likely from the density of retrograde labeling that at least a sizable majority do. This implies that the geniculate may lack true interneurons, although interactions between geniculate neurons can occur via dendrodendritic synapses (Ulinski, 1987a) and there may be a population of neurons in the lateral forebrain bundle that effectively serve as thalamic interneurons (Ulinski, 1987c).

Both the earlier orthograde degeneration study of Hall and Ebner (1970b) and the most recent orthograde HRP study of Heller and Ulinski (1987) indicate that once the geniculocortical axons reach the dorsal telencephalon, they run through the pallial thickening and then course dorsally into dorsal cortex. They are situated around the somata of layer 2 neurons in the pallial thickening, but occupy the outer one-third of layer 1 within dorsal cortex where they terminate precisely at the boundary between areas D1 and D2 of Desan (1984). Hall and Ebner interpreted degeneration in the pallial thickening as fibers passing through the pallial thickening without synapse. However, the Golgi-like HRP fills of geniculocortical axons obtained by Heller and Ulinski show that the axons have varicosities *en passant* as they pass through the pallial thickening. Individual axons could synapse upon the proximal dendrites or somata of neurons in the pallial thickening and then pass medially, approximately in the transverse plane, to synapse upon the distal dendrites of neurons in the medial part of D2. Because it is not possible to determine the precise arrangement of geniculocortical afferents in traditional sections, Mulligan and Ulinski (1990) used orthograde transport of HRP in an *in vitro* whole brain preparation to determine their trajectories. They folded back the cerebral cortex and made pressure or iontophoretic HRP injections in the superficial part of the lateral forebrain bundle as it passes beneath the lateral ventricle. Following sacrifice, the cortex was removed and processed as a whole-mount preparation. Orthogradely labeled axons could be traced out of the lateral forebrain bundle and into the cortex.

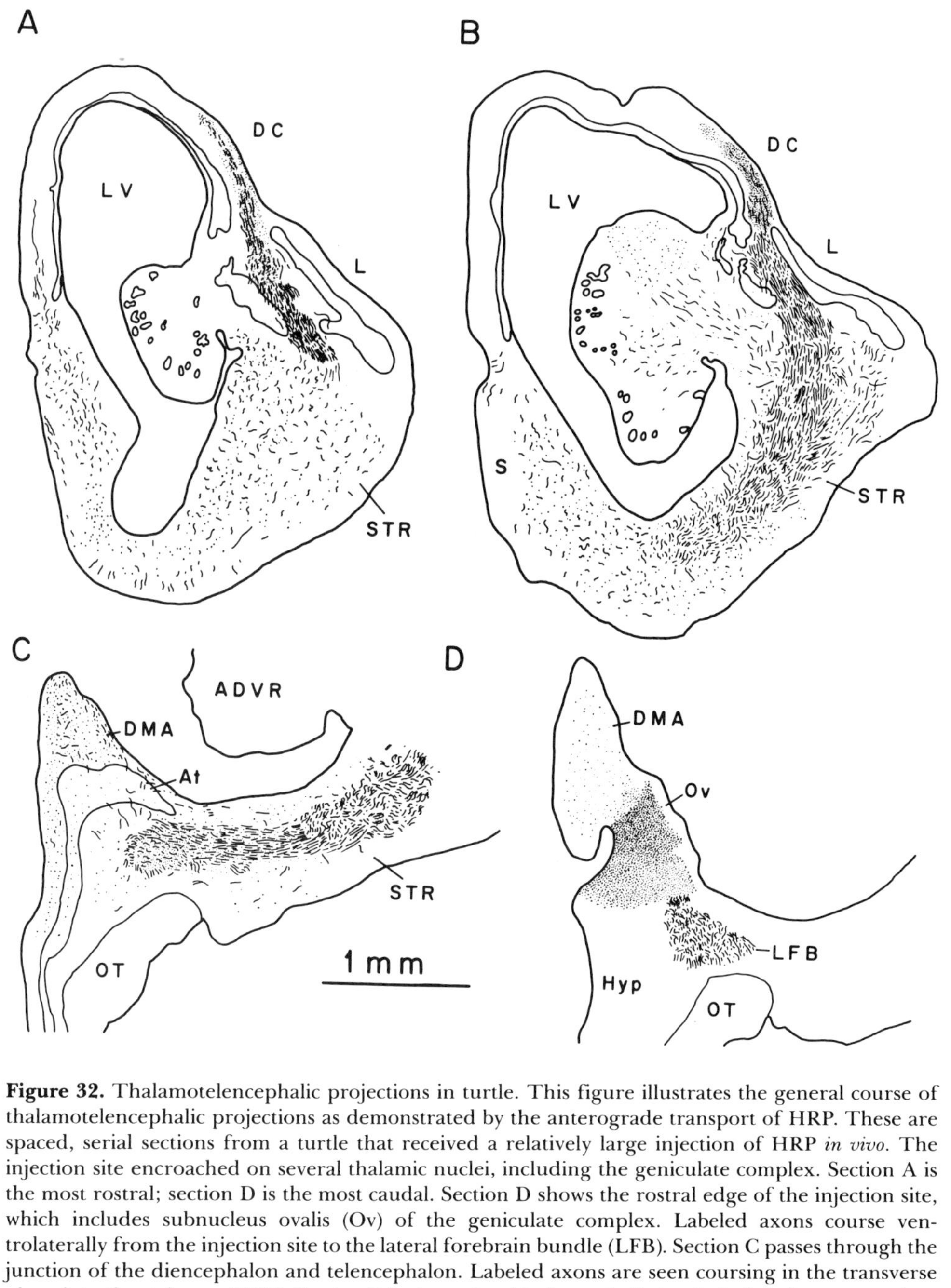

Figure 32. Thalamotelencephalic projections in turtle. This figure illustrates the general course of thalamotelencephalic projections as demonstrated by the anterograde transport of HRP. These are spaced, serial sections from a turtle that received a relatively large injection of HRP *in vivo*. The injection site encroached on several thalamic nuclei, including the geniculate complex. Section A is the most rostral; section D is the most caudal. Section D shows the rostral edge of the injection site, which includes subnucleus ovalis (Ov) of the geniculate complex. Labeled axons course ventrolaterally from the injection site to the lateral forebrain bundle (LFB). Section C passes through the junction of the diencephalon and telencephalon. Labeled axons are seen coursing in the transverse plane into the striatum (STR). Section B is through midtelencephalic levels. Labeled axons run into the septum (S) and through the striatum (STR) into the anterior dorsal ventricular ridge (ADVR) and medial to the lateral cortex (L). This latter set of fibers pass through the pallial thickening and into the dorsal cortex (DC). At, area triangularis of the ventral thalamus; DMA, dorsomedial anterior nucleus; Hyp, hypothalamus; LV, lateral ventricle; OT, optic tract. From Heller and Ulinski (1987).

Individual axons fan out as they enter the cortex and run radially from lateral to medial across dorsal cortex, bearing varicosities *en passant* at irregular intervals. Thus, an individual geniculocortical axon can synapse upon a band of cortical neurons that extends from lateral to medial across dorsal cortex. The functional implications of this arrangement will be considered in Section 8.2.

The ultrastructure of geniculocortical afferents has been studied by Ebner and Colonnier (1975, 1978) and Smith *et al.* (1980). They made large thalamic lesions and sampled degenerating synapses along a strip of tissue situated at the boundary of D2 and D1. Thalamic synapses—presumably including the geniculocortical synapses—involve presynaptic elements that contain clear, round synaptic vesicles and form asymmetric junctional complexes. They occur upon the dendritic spines of cortical neurons and the smooth dendritic shafts of layer 1 stellate cells. Smith *et al.* estimate that each double pyramidal cell receives 300 thalamic synapses and each stellate cell receives 1800 thalamic synapses.

Pivovarov and Trepakov (1972) and Kriegstein (1987) recorded intracellularly from layer 1 and layer 2 neurons following electrical stimulation of the optic nerve or photic stimulation of the retina. Such recordings show EPSPs that are often followed by IPSPs. The EPSPs can be attributed to geniculocortical afferents. The EPSPs recorded in cortical cells following activation of the geniculocortical path resemble those obtained following intracortical application of L-glutamate (Kriegstein and Connors, 1986), and they are blocked by antagonists of the quisqualate and kainate types of excitatory amino acid receptors (Larson-Prior *et al.,* 1989).

The organization of the dorsal lateral geniculate complex in reptiles other than turtles is less completely understood. The diencephalon of lizards and snakes contains several cytoarchitecturally distinct structures that receive retinal inputs (see Ulinski, 1983, for references), and it is not certain how to relate these to the dorsal lateral geniculate nuclei of mammals. Most workers would agree that a structure named the dorsal lateral geniculate nucleus should receive direct retinal input and project to the telencephalon, while the ventral lateral geniculate nucleus should receive direct retinal input and have reciprocal projections with the optic tectum. Using these criteria, it appears that only the very rostralmost of the several retinorecipient structures in the diencephalon of squamates qualifies as the dorsal lateral geniculate nucleus; the several more caudal structures should be considered the ventral lateral geniculate nucleus.

A candidate for the dorsal lateral geniculate nucleus has been identified in three genera of lizards. Lohman and van Woerden-Verkely (1978) used orthograde degeneration and retrograde HRP transport to show that a plate of cells in the rostral diencephalon of the tegu lizard, *Tupinambis,* known to receive retinal input, projects to the lateral telencephalic wall. Bruce and Butler (1984b) used the retrograde transport of HRP following telencephalic injections in the tokay gecko, *Gekko gecko,* and the green iguana, *Iguana iguana,* to show that a similarly placed structure (named nucleus intercalatus) projects to the lateral telencephalic wall. The region of the telencephalon that receives geniculate inputs is situated between the anterior dorsal ventricular ridge and lateral cortex, ventral to the edge of dorsal cortex, in each species and corresponds to the area that has often been designated as the pallial thickening. It consists of rather distinct clusters of cells in *Gekko,* but is less clearly defined in *Iguana* (Bruce and Butler, 1984a,b) and *Tupinambis* (Ebbesson and Voneida, 1969). Nothing is known of the organization of the geniculotelencephalic projections in lizards.

Until quite recently it appeared that the organization of the thalamic visual nuclei of snakes was quite different from that of other reptiles. Snakes seemed to lack a nucleus rotundus while having a relatively large dorsal lateral geniculate complex (e.g., Ulinski, 1977b). This situation was related to theories about the phylogenetic history of snakes (Northcutt and Butler, 1974; Senn and Northcutt, 1973) and to general patterns of brain evolution (Ebbesson, 1980). However, Dacey and Ulinski (1983) reexamined the problem and were able to use HRP techniques to show that snakes have a nucleus rotundus which is prominent, but situated more caudally in the thalamus than in other reptiles. Also, du Lac and Dacey (1981) analyzed the geniculate complex in snakes in some detail and argued that only the most rostral part of the complex, the nucleus ovalis of Warner (1931) and other authors (Halpern and Frumin, 1973), represents the dorsal lateral geniculate complex. The nucleus ovalis receives bilateral retinal projections and possibly gives rise to a small projection to the telencephalic wall between the anterior dorsal ventricular ridge and lateral cortex (Ulinski, unpublished observations). The remainder of the geniculate complex contains several cytoarchitecturally distinct regions that jointly correspond to the ventral lateral geniculate complex. These have reciprocal, topographically organized projections with the optic tectum (du Lac and Dacey, 1981; Dacey and Ulinski, 1983, 1986a,b) and project to the brain-stem reticular formation (Ulinski *et al.*, 1983).

7.1.1c. Dorsal Ventricular Ridge. In addition to demonstrating a source of visual input to cerebral cortex in turtles, Hall and Ebner (1970b) showed that nucleus rotundus—which receives inputs from the optic tectum—projects to the anterior dorsal ventricular ridge. Sensory inputs to ADVR have since been extensively studied in both reptiles and birds (see Ulinski, 1983), and it appears there is a general pattern in which auditory information projects to the medial, caudal region of ADVR, somatosensory information projects via the spinal cord and dorsal column nuclei to the central region of ADVR, and visual information projects via the optic tectum to a lateral region of ADVR. It is possible that the trigeminal sensory nuclei project directly to a ventrolateral region of ADVR (see Clark and Ulinski, 1984; Bruce and Butler, 1984b) and that the vestibular nuclei project to a region of ADVR immediately internal to lateral cortex (Kunzle, 1985).

There is some indication that ADVR provides a second route by which visual information reaches visual cortex. HRP injections in visual cortex retrogradely label neurons in the dorsal area of ADVR of turtles (Balaban, 1978, 1979; Desan, 1984). This is part of ADVR that receives projections from nucleus rotundus in turtles (e.g., Balaban and Ulinski, 1981a,b) and is responsive to visual stimuli (Dunser *et al.*, 1981). Belekhova *et al.* (1979) provided physiological evidence for this projection by demonstrating a cortical response to visual stimuli in turtles that had sustained lesions of the dorsal lateral geniculate complex. These responses could reach the cortex via a relay in ADVR, and have a latency that is 20 msec longer than the early cortical response due to the geniculocortical pathway. Similarly, HRP injections in the lateral part of dorsal cortex retrogradely label neurons in that lateral region of ADVR in *Gekko* (Bruce and Butler, 1984b).

Whether or not there are projections to the cortex from the auditory and somatosensory areas of ADVR is unknown. However, Orrego and Lissenby

(1962) recorded evoked potentials in the cortex of turtles following somatosensory stimulation and Moore and Tschirigi (1962) found auditory, somatosensory, and visual multimodal responses in dorsal cortex of alligator. There is no indication of direct projections from the auditory thalamic nuclei to the cortex in any reptile. The organization of the somatosensory nuclei of the thalamus is not clearly established, but there is some evidence for spinal cord projections to the rostral thalamus in turtles (Kunzle and Schnyder, 1984). Since some of the perirotundal nuclei in this region project to the cortex, they may participate in a route by which somatosensory information can reach the cortex.

7.1.2. Limbic System Structures

7.1.2a. Hypothalamus. Each of the four cortical areas receives projections from one or more hypothalamic nuclei (Desan, 1984; Bruce and Butler, 1984a). The medial area receives projections principally from the mammillary nuclei, bilaterally, and a sparser projection from the periventricular region of the ipsilateral hypothalamus. Dorsomedial cortex receives a sparse projection from the ipsilateral mammillary nuclei. Dorsal cortex receives a bilateral projection from the mammillary nuclei, as well as projections from the lateral, preoptic, and paraventricular areas of the hypothalamus. Lateral cortex receives projections from the ipsilateral paraventricular nucleus.

The laminar organization of hypothalamic-cortical projections within the various areas is not known, but clues as to which neurotransmitters may be used are provided by Reiner *et al.* (1985) who employed immunohistochemical techniques to show that axons within layer 2 of turtle cortex that show substance P- or cholecystokinin-like immunoreactivity are absent from animals that have undergone hypothalamic lesions. Also, the periventricular hypothalamus contains somata that stain positively with antibodies against dopamine (Smeets *et al.*, 1986, 1987; Smeets, 1988). Fibers that stain with dopamine-like immunoreactivity are present in all of the cortical areas, although the details of their distribution vary between species. In *Pseudemys,* scattered fibers are found in layers 1 and 2 of both medial and dorsomedial areas, in layer 1 of D1, layer 2 of the pallial thickening, and layers 2 and 3 of both the dorsal and ventral parts of lateral cortex (Smeets *et al.,* 1987). In *Gekko,* labeled fibers are found in layers 1 and 3 of medial cortex, the outer part of layer 1 of dorsomedial cortex, layers 1 and 2 of dorsal cortex, and densely in layer 2 of the ventral rostral lateral cortex (Smeets *et al.,* 1986). In *Python,* fibers are found in layers 1 and 3 of medial and dorsomedial cortex, throughout dorsal cortex (but particularly in layer 3), and throughout lateral cortex (but particularly in layer 2 of ventral rostral lateral cortex) (Smeets, 1988). The significance of these differences is not known.

7.1.2b. Septum. The septum projects to one degree or another to medial, dorsomedial, and lateral cortical areas. These projections have been studied using the retrograde transport of HRP in snakes (Reperant, 1976), lizards (Lohman and van Woerden-Verkely, 1978; Bruce and Butler, 1984a), and turtles (Desan, 1984). Orthograde tracing studies in tegu lizards (Lohman and van Woerden-Verkely, 1976; Hoogland *et al.,* 1978; Sligar and Voneida, 1981) indicate that the precommissural septum projects to layer 1 of the medial and intermediate parts of dorsal cortex.

7.1.2c. Perirotundal Nuclei. These are a set of nuclei that partially cover the rostral face of nucleus rotundus. They vary in character and extent between species, but dorsomedial and dorsolateral anterior nuclei (Fig. 3) are generally distinguished. The dorsomedial anterior nucleus lies between nucleus rotundus and the third ventricle and continues dorsally to lie between nucleus rotundus and the habenula. The dorsolateral anterior nucleus is situated between the dorsomedial anterior nucleus and the lateral geniculate complex. An anterior nucleus is recognized lying ventral to the rostral pole of nucleus rotundus in turtles. Nothing is known of the neuronal organization or transmitter profile of the perirotundal nuclei, except that cortical injections in *Caiman* can label more than 99% of the neurons in the dorsolateral anterior nucleus, which suggests this nucleus has few intrinsic neurons (Pritz and Stritzel, 1987).

There have been no systematic studies of the afferents of the perirotundal nuclei. However, orthograde tracing studies (Hoogland *et al.*, 1978; Sligar and Voneida, 1981) show the septal nuclei project to the ipsilateral perirotundal nuclei. Since the septum receives projections from the medial and dorsal areas of the cortex, it then forms a link between the cortex and perirotundal nuclei. Also, Halpern (1974) reports a direct projection from the medial area of the cortex to the anterior thalamus in garter snakes, *Thamnophis sirtalis*.

Retrograde HRP experiments establish that the perirotundal nuclei project widely to the telencephalon. They have been retrogradely labeled following injections of HRP into the anterior dorsal ventricular ridge of turtles (Belekhova *et al.*, 1979; Balaban and Ulinski, 1981a) and lizards (Lohman and van Woerden-Verkely, 1978; Bruce and Butler, 1984a). They are also labeled following HRP injections restricted to specific cortical areas in turtles (Hall *et al.*, 1977; Ouimet *et al.*, 1985; Desan, 1984), lizards (Lohman and van Woerden-Verkely, 1978; Bruce and Butler, 1984a), and *Caiman* (Pritz and Stritzel, 1987). The projection appears to be a diffuse one, so the results of injections are sometimes difficult to interpret. However, there seems to be a pattern to the projection of the perirotundal nuclei upon the cortex. For example, Bruce and Butler recognized three perirotundal nuclei: the dorsomedial nucleus, the dorsolateral anterior nucleus, pars magnocellularis (DLAm), and the dorsolateral anterior nucleus, pars parvocellularis (DLAp). DLAm projects bilaterally to the medial area of cortex. DLAm and DLAp project bilaterally to the dorsomedial area, with cortical injections retrogradely labeling more cells in DLAp than in DLAm. DLAp also projects bilaterally to the dorsal area and sparsely to the ipsilateral lateral cortex.

Experiments designed to study the laminar organization of perirotundal projections to cortex have yielded contradictory results. On the one hand, Balaban and Ulinski (1981a) used autoradiographic tracing methods to suggest that the perirotundal nuclei project diffusely throughout the depth of medial cortex in turtles. On the other hand, Lohman and van Woerden-Verkely (1978) used orthograde degeneration methods to suggest that the perirotundal nuclei terminate in a fairly discrete band within the medial third of layer 1 in tegu lizards. Desan (1984) used autoradiographic methods on turtles and suggested that perirotundal projections to cortex involve both diffuse and specific systems. Thus, a discrete band of label is present in the middle third of layer 1 of medial and dorsomedial cortex, but some diffuse label above background is also present.

7.1.3. Brain Stem and Basal Forebrain

7.1.3a. Locus Coeruleus. Noradrenergic neurons in reptiles have been studied using the Falck–Hillarp histofluorescence methods (Baumgarten and Braak, 1968; Parent and Poirer, 1971; Yamamoto and Shimizu, 1977; see Parent, 1979, 1986, for reviews) and immunohistochemical methods employing antibodies against the synthetic enzyme tyrosine hydroxylase (Wolters *et al.*, 1984; Brauth, 1988). These studies demonstrate a cluster of catecholaminergic neurons in the dorsolateral isthmus region of the brain that have been designated the locus coeruleus (Cruce and Nieuwenhuys, 1974; ten Donkelaar and Nieuwenhuys, 1979). Transections of the brain stem rostral to the locus coeruleus result in a reduction of fluorescence in the forebrain (Parent and Poitras, 1974), and HRP injections in both the dorsal ventricular ridge (Balaban and Ulinski, 1981a) and dorsal cortex (Bruce and Butler, 1984a; Ouimet *et al.*, 1985) retrogradely label neurons in the locus coeruleus. The existence of a coeruleocortical pathway to the pallium is therefore well established in at least turtles and lizards. Both Falck–Hillarp and glyoxylic acid fluorescence methods demonstrate a plexus of noradrenergic fibers in dorsal cortex (Parent, 1973; Parent and Poitras, 1974; Ouimet *et al.*, 1981, 1985). Fibers are varicose and present predominantly in the outer 100 μm of layer 1. Fluorescence is depleted by pretreatment with reserpine (Ouimet *et al.*, 1985), and biochemical analysis shows this region of cortex to have a high norepinephrine concentration (Ouimet *et al.*, 1981). Ultrastructural analysis using tissue fixed with potassium permanganate (Ouimet *et al.*, 1981) shows that noradrenergic terminals contain small granular vesicles and are presynaptic to dendritic spines. The cytology of their active zones varies, with some being symmetric and others asymmetric.

7.1.3b. Raphe Nuclei. The distribution of indoleaminergic neurons in reptiles has been studied with histofluorescence techniques in lizards (Braak *et al.*, 1968) and turtles (Parent, 1973, 1979; Parent and Poitras, 1974), and with immunohistochemical techniques in lizards (Wolters *et al.*, 1985; Smeets and Steinbusch, 1988) and turtles (Umeda *et al.*, 1983). Serotonergic neurons are distributed in the several raphe nuclei spread throughout the rostrocaudal extent of the brain stem (for cytoarchitectonic descriptions of the raphe nuclei see Cruce and Nieuwenhuys, 1974; Newman and Cruce, 1982; ten Donkelaar and Nieuwenhuys, 1979; Wolters *et al.*, 1985; ten Donkelaar *et al.*, 1987). Neurons in the medial superior raphe nucleus are retrogradely labeled following injections of HRP in the dorsomedial and dorsal cortical areas (Lohman and van Woerden-Verkely, 1978; Bruce and Butler, 1984a; Ouimet *et al.*, 1985). Serotonergic fibers are present in the inner half of layer 1 of dorsal cortex of the turtle *Pseudemys* (Ouimet *et al.*, 1985). In *Gekko* (Smeets and Steinbusch, 1988) serotonergic fibers are found in layers 1 and 3 of medial and dorsomedial cortex, layer 2 of dorsal cortex, and throughout lateral cortex.

7.1.3c. Reticular Formation. The midbrain reticular formation gives rise to a bilateral projection to the cortex in the snakes *Thamnophis sirtalis* and *Natrix sipedon* (Ulinski, 1981). The projection arises from isodendritic, multipolar neurons situated in the dorsolateral midbrain tegmentum. Their axons ascend in the medial forebrain bundle and some cross the midline in the anterior pallial commissure. Fibers turn dorsally on both sides of the brain and pass through the septum to enter the cortex. They run laterally in the alveus and turn to terminate in layer 1 of medial cortex and layers 1, 2, and 3 of dorsomedial cortex. They synapse upon dendritic spines as boutons with round synaptic vesicles and forming asymmetric active zones. Khatchaturian *et al.* (1984) report the presence of a group of neurons in the dorsolateral midbrain tegmentum of the lizard *Anolis* that stain with antibodies against β-endorphin, ACTH, and α-melanocyte stimulating hormone (all peptides in the proopiomelanocorticotropin family). Fibers that show immunoreactivity against these antisera are also found in the cortex of *Anolis,* so these peptides may be associated with the reticulocortical system.

7.1.3d. Basal Forebrain. It is now generally agreed that the basal forebrain of mammals contains several nuclear groups that give rise to cholinergic projections to the cortex; a similar system seems to be present in reptiles. Injections of HRP in both the dorsal and lateral cortical areas of geckos (Bruce and Butler, 1984a) and turtles (Desan, 1984) retrogradely label neurons in the nuclei of the diagonal band and in a cell group situated dorsal to the diagonal band, but ventral to the septal complex. These groups also contain neurons that stain positively with antibodies against choline acetyltransferase in turtles (Desan, 1984; Mufson *et al.,* 1984). Studies combining choline acetyltransferase immunocytochemistry with retrograde HRP labeling have not been done on reptiles, but it seems likely these two groups are equivalent to the cholinergic groups in the basal telencephalon. Thus, as in mammals, the diagonal band complex contains cholinergic neurons that project to the cortex. The more dorsal group in reptiles generally resembles the basal nucleus of Meynert of mammals.

The outer part of layer 1 of dorsal cortex of turtles stains positively for acetylcholinesterase and contains high levels of choline acetyltransferase and ACh (Hohmann *et al.,* 1983). Also, receptor binding techniques show that the entire dorsal cortex, including the pallial thickening, contains a moderate concentration of muscarinic cholinergic receptors (Schlegel and Kriegstein, 1987). Both extracellular and intracellular recordings in turtle dorsal cortex following focal application of ACh demonstrate a biphasic response of cortical neurons to ACh (Blanton *et al.,* 1986). A hyperpolarization, lasting from 250 msec to 15 sec, that may be associated with a potassium current, is followed by a voltage-sensitive depolarization that lasts up to 60 sec and is associated with a decrease in membrane conductance. These responses can be mimicked by electrical stimulation of the nucleus basalis, so there is evidence that the basal forebrain can modulate cortical activity via a cholinergic projection to at least the dorsal and lateral cortical areas.

7.2. Intracortical Connections

The four cytoarchitectonic areas of the cortex are interconnected via an extensive set of association and commissural connections, which have been studied in lizards (Voneida and Ebbesson, 1969; Lohman and Mentink, 1972; Lohman and van Woerden-Verkely, 1976), snakes (Ulinski, 1976), and turtles (Desan, 1984). It should be noted that experiments on the intracortical connections are technically difficult because of the problems involved in localizing lesions or injections of tracer substances to individual areas, which are very small, and ambiguities caused by fiber-of-passage problems. Nonetheless, some general patterns have been established and the introduction of new tracing techniques, such as the use of *Phaseolus vulgaris* leukoagglutinin as a tracer substance, promises that future studies will have greater resolution than has been possible so far. The intracortical connections can be divided into several sets for the purposes of discussion.

First are projections from the lateral, dorsal, and dorsomedial areas to the ipsilateral medial area (Fig. 33). These have a laminar organization such that axons from each of the three areas terminate in a discrete band within medial cortex (Fig. 34). Since the candelabra cells of medial cortex have their somata densely packed in layer 2, both the ascending and descending components of their dendritic trees are in register. Consequently, each segment of the dendritic tree of a candelabra cell receives inputs from a different cortical area. Axons from the bowl cells of layer 2 of the dorsal part (Desan, 1984) of lateral cortex run as a distinct fascicle in the superficial part of layer 1 of dorsal and dorsomedial cortex (Rainey and Ulinski, 1980). When the fascicle reaches medial cortex, it expands to occupy the outer third of layer 1 (Ulinski, 1976; Martinez-Garcia *et al.*, 1986). Each individual axon bears varicosities *en passant* and extends across the full medial-to-lateral extent of medial cortex, intersecting, and presumably synapsing upon, the dendrites of many candelabra cells. Since the outer third of layer 1 contains predominantly axon terminals with clear, round synaptic vesicles synapsing upon dendritic spines via asymmetric junctions, it is likely that the bowl cell terminals have these characteristics.

Fink–Heimer degeneration (Lohman and Mentink, 1972; Ulinski, 1976) and orthograde HRP studies (Desan, 1984) show dorsal cortex projects to the middle third of layer 1 of medial cortex. It is not entirely certain which of the three subareas of dorsal area give rise to this projection. However, there is some indication (Desan, 1984) that the more lateral subareas of dorsal cortex (e.g., area D2 of Desan in turtles) project to the medial subarea of dorsal cortex (i.e.,

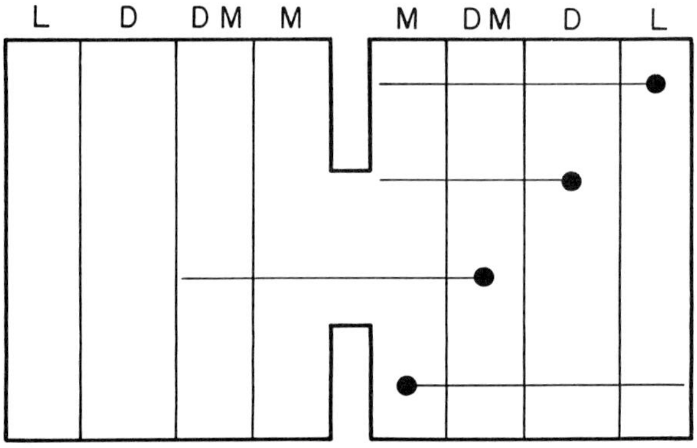

Figure 33. Commissural and association projections. This figure illustrates the organization of the major commissural and association projections that interconnect the four cortical areas. The cortical areas of the two cerebral hemispheres are diagrammed as if they have been flattened into rostrocaudally elongate strips. The two cortices are connected across the midline by the pallial commissure. Commissural and association projections tend to run in transverse planes. Three association projections are shown. First, lateral cortex (L) projects medially through dorsal (D) and dorsomedial (DM) cortex to medial cortex (M). Second, dorsal cortex projects medially to medial cortex. Third, medial cortex projects lateral to dorsomedial, dorsal, and lateral cortices. The principle association projection involves projections from dorsomedial cortex, through the ipsilateral and contralateral medial cortices, to the contralateral dorsomedial cortex. Lizards have commissural projections linking the two lateral cortices through the habenular commissure; these are now shown.

area D1 of Desan), which projects in turn to medial area. Electron microscopic degeneration studies in the lizard *Agama* (Wouterlood *et al.*, 1982) indicate that axons originating in dorsal cortex terminate as boutons with clear, round synaptic vesicles, predominantly upon dendritic spines in medial cortex.

The dorsomedial area (Lohman and Mentink, 1972; Ulinski, 1976) gives rise to a projection to the inner third of layer 1 and the outer half of layer 3 of medial cortex. Thus, lateral and dorsal cortex project to the ascending dendrites of the candelabra cells, but dorsomedial cortex projects to both the ascending and descending dendrites.

Medial cortex of snakes and lizards gives rise to a laterally directed projection that reciprocates the projections just described (Ulinski, 1976; Desan, 1984). It arises from the candelabra cells, whose axons descend to layer 3 and bifurcate within the alveus. The medially directed branches give rise to an efferent projection to the septum that is described below. The laterally directed branches project back to dorsomedial, dorsal, and lateral areas. The laminar organization of these projections is not as well documented as are the projections to medial area. However, the projection to dorsomedial area appears to involve collaterals of candelabra cell axons that ascend from the alveus and extend through most of the depth of dorsomedial cortex, sparing only the outer part of layer 1 which remains unstained in Timm-stained material (Perez-Clausell, 1988). The projection to dorsal cortex involves all but the most superficial part of layer 1. The

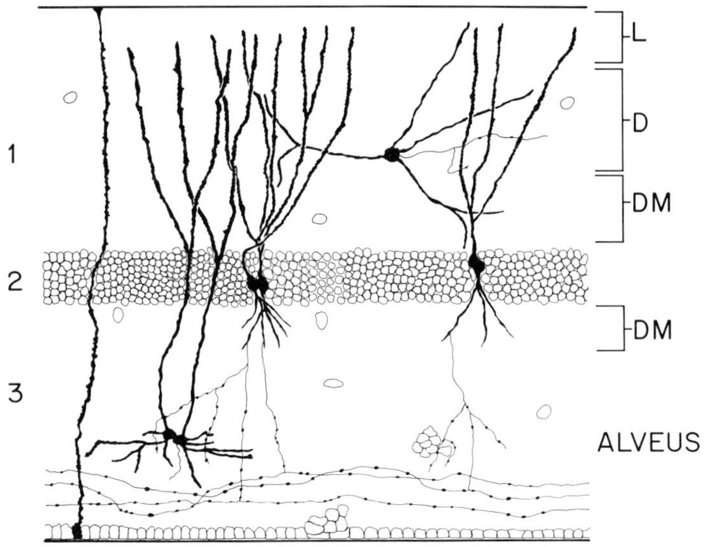

Figure 34. Laminar organization of medial cortex. This is a semidiagrammatic depiction of snake medial cortex. The somata of candelabra cells are densely packed in layer 2, so that their ascending and descending dendrites are in register in layers 1 and 3, respectively. Afferent systems tend to run through medial cortex in the transverse plane and therefore intersect the dendrites of the candelabra cells. Afferents from lateral cortex (L) run in the outer part of layer 1. Afferents from dorsal cortex (D) run in the intermediate part of layer 1. Afferents from the dorsomedial cortex (DM) run in the inner part of layer 1 and the upper part of layer 3. The axons of the candelabra cells descend in layer 3 and bifurcate to run in the alveus. The laterally directed branch gives rise to association projections to dorsal, dorsomedial, and lateral cortices; the medially directed branch gives rise to efferent projects to the septum and hypothalamus. From Ulinski (1977a).

projection to lateral cortex involves the inner part of layer 1, thus complementing the projection of the ipsilateral and contralateral olfactory bulbs which terminate superficially in lateral cortex (Section 7.1.1a). Desan (1984) provides evidence that the analogous feedback projections in turtles arise from cells in both medial and dorsomedial cortex. Injections of lateral cortex with HRP retrogradely label neurons at the border of dorsal and dorsomedial cortex, while injections in D1 retrogradely label neurons at the border of medial and dorsomedial cortex.

Both the medially and laterally directed association connections conform to a lamellar organization such that both sets of axons run in approximately transverse planes (Ulinski, 1976; Ulinski and Rainey, 1980). Thus, bowl cells in a rostrocaudally compact band of lateral cortex project to a row of candelabra cells situted at roughly the same rostrocaudal level, and those candelabra cells project back to the same set of bowl cells. Because the axons that effect these projections bear varicosities *en passant,* it is likely that each candelabra cell in such a band is contacted by the axons of many bowl cells in the corresponding band.

In addition to these reciprocal projections in the transverse plane, there is evidence for association projections running rostrocaudally within the cortex. A rostrocaudally directed fascicle of fibers was described in normal sections of lizard cortex (Lohman and Mentink, 1972). Desan (1984) reported, on the basis of both anterograde HRP and tritiated proline injections, that axons run considerable rostrocaudal distances within specific cortical areas in turtles. Also, the projection from D2 to D1 involves a fascicle of fibers that course from D2 and run rostrocaudally along the medial border of D1.

There have been some claims that dorsal cortex (Ware, 1974; Gaidenko, 1978) participates in commissural connections with homotypic areas in the contralateral hemisphere, but these have not been generally substantiated and their existence should be regarded as provisional. There is evidence for commissural connections between lateral cortex on the two sides of the brain in lizards (Martinez-Garcia *et al.,* 1986; Butler, 1976) in which neurons in lateral cortex on one side of the brain project to the contralateral lateral cortex via the habenular commissure. However, commissural connections between the two lateral cortices seem to be absent in turtles.

There is clear evidence for commissural projections between the dorsomedial areas of the two hemispheres (Lohman and Mentink, 1972; Ulinski, 1976)(Fig. 35). These undoubtedly involve axons of layer 2 double pyramidal cells whose axons run medially through medial cortex. It is not known if single double pyramidal cells both project to medial cortex and participate in the commissural projections. Axons from dorsomedial cortex cross the midline through the pallial commisure (a U-shaped band of axons situated dorsal to the anterior commissure), run through the contralateral septum, and terminate in the contralateral medial and dorsomedial areas. The terminal zone in the medial area overlaps that originating from the ipsilateral dorsomedial area. The terminal zone in dorsomedial cortex is principally in layer 1. It is positioned deep in layer 1 at the medial edge of dorsomedial cortex, but then slants dorsally so that it extends superficially in layer 1 at the lateral edge of dorsomedial cortex. This terminal zone complements the projection from medial cortex to dorsomedial cortex and corresponds to the clear zone in Timm-stained material.

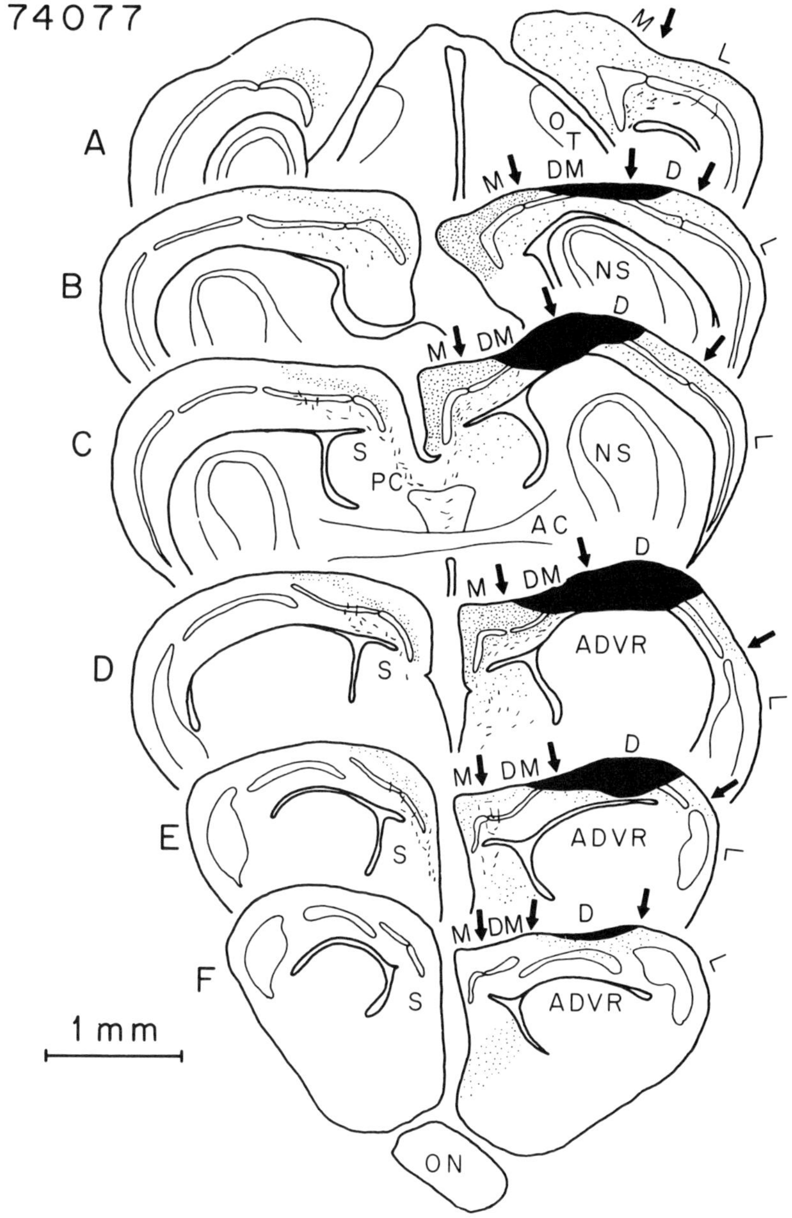

Figure 35. Commissural projections from dorsomedial cortex. The projections of dorsomedial cortex are shown in an anterograde degeneration experiment in which a lesion was made involving dorsomedial cortex (DM). Since dorsal cortex (D) and lateral cortex (L) do not project to the contralateral hemisphere in snakes, all of the commissural projections can be attributed to damage to dorsomedial cortex. Commissural axons cross the midline in the pallial commissure (PC) and run in layer 1 and 3 of the contralateral medial cortex (M). They terminate in layer 1 of the contralateral dorsomedial cortex. AC, anterior commissure; ADVR, anterior dorsal ventricular ridge; NS, nucleus sphericus; ON, optic nerve; OT, optic tract; S, septum. From Ulinski (1976).

7.3. Cortical Efferents

No projections from dorsomedial cortex to noncortical structures are known. However, each of the other cortical areas has projections efferent from the cortex. These have been studied in lizards (Butler, 1976; Lohman and Mentink, 1972; Lohman and van Woerden-Verkely, 1976), snakes (Halpern, 1974; Ulinski, 1975), and turtles (Gaidenko, 1978; Desan, 1984).

Lateral cortex receives bilateral projections from the main olfactory bulb as well as several olfactorecipient structures (Section 7.1.1a). Its dorsal rostral and dorsal caudal parts give rise to the projection that courses over the dorsal surface of the cortex to medial cortex, as described in Section 7.2.

Dorsal cortex gives rise to three sets of efferent projections. First are projections to the ipsilateral septum (Lohman and Mentink, 1972; Lohman and van Woerden-Verkely, 1976) involving axons that run medially through layer 3 and turn ventrally to enter the septum (Fig. 36). They terminate in the ventrolateral part of the septum along the full rostrocaudal length of both the parts of the septum situated rostral to the anterior commissure (precommissural septum) and those caudal to the anterior commissure (postcommissural septum). In tegu lizards, this projection originates from dorsal cortex (Lohman and van Woerden-Verkely, 1976). Second, dorsal cortex projects to the anterior dorsal ventricular ridge (Butler, 1976). Neurons in D2 project throughout the dorsal area of the anterior dorsal ventricular ridge in turtles (Balaban, 1978; Desan, 1984). Since D2 receives projections from the dorsal lateral geniculate complex and dorsal

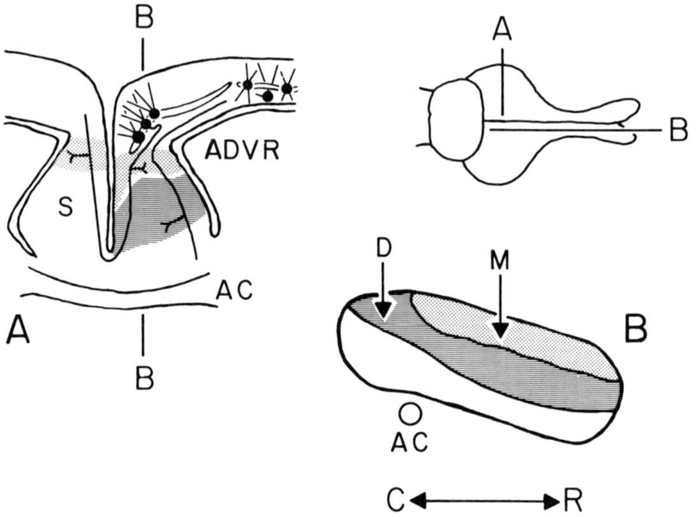

Figure 36. Corticoseptal projections. This figure illustrates the general organization of corticoseptal projections in snakes. The sketch of the snake brain shows the levels of the diagrammatic sections depicted in A and B. (A) Transverse section showing the course and termination (stipple) of axons of candelabra cells in medial cortex to the dorsal septum bilaterally and the course of the axons of unknown cells in dorsal cortex to their termination (stripes) in ventral septum ipsilaterally. (B) Parasagittal section of the septum showing the positions of the terminal fields of projections from medial cortex (stipple) and dorsal cortex (stripes) relative to the anterior commissure (AC). ADVR, anterior dorsal ventricular ridge; D, dorsal cortex; M, medial cortex; S, septum. From Ulinski (1975).

area of ADVR receives projections from nucleus rotundus, the projection from D2 to dorsal area serves to interconnect the telencephalic visual areas. Finally, dorsal cortex gives rise to efferent projections to visual structures in the brain stem. These have been best documented in turtles in which area D2 of Desan projects to the dorsal lateral geniculate complex, nucleus rotundus, and, perhaps, the optic tectum. Lesions of dorsal cortex (Gaidenko, 1978; Hall *et al.*, 1977) and HRP injections in D2 (Ulinski, 1986b; Desan, 1984) demonstrate that fibers course ventrally from dorsal cortex, proceed medial to lateral cortex and through the striatum to reach into the brain stem through the ventral peduncle of the lateral forebrain bundle. Axons turn dorsally and run into either nucleus rotundus or the dorsal lateral geniculate complex. These axons are rectilinear, of fine caliber, and bear varicosities *en passant.* Those entering nucleus rotundus extend from ventral to dorsal through the nucleus and can, therefore, intersect the dendritic fields of neurons scattered throughout rotundus. Those entering the dorsal lateral geniculate complex run from ventral to dorsal in the inner part of the structure's neuropil layer. These axons are also of fine caliber and bear varicosities *en passant.* They are positioned so as to intersect the smooth, proximal shafts of cell plate cells. Some axons can be traced caudally in the forebrain bundle into the midbrain tegmentum, where they turn dorsally and apparently terminate in the deep layers of the optic tectum. The existence of corticotectal cells has not been demonstrated with full certainty in turtles, but has been shown in the lizard *Agama* (Elprana *et al.,* 1980).

One issue surrounding the projections from dorsal cortex to the brain stem concerns their precise origin from within dorsal cortex. Desan (1984) used HRP injections to show that area D1 gives rise to intracortical projections to medial cortex, while area D2 gives rise to projections to the brain stem. This latter point has been confirmed using the retrograde transport following injections of HRP in the thalamus (Ulinski, 1986b). Such injections produce retrogradely labeled cells scattered throughout layers 2 and 3 of area D2, which include pyramidal, multipolar, or stellate cells and horizontal cells. A second question is whether a single cell projects to several structures (such as the geniculate complex, rotundus, and tectum) via collateral branches, or whether a single cell projects to only one structure. Definitive resolution of this issue requires double labeling experiments, but the available evidence favors a single cell projecting to multiple targets (Ulinski, 1986b).

Like dorsal cortex, medial cortex gives rise to efferent projections to the septum (Ulinski, 1975; Halpern, 1980). Axons run from medial cortex medially in the alveus to terminate in the dorsal part of the ipsilateral precommissural septum, complementing the projections of dorsal cortex more ventrally in the septum (Fig. 32). In addition, the axons continue ventrally, cross the midline in the pallial commissure, and proceed dorsally to give rise to a less dense terminal zone in the contralateral precommissural septum. Halpern (1974, 1980) reported that lesions of medial cortex in snakes also produce a bundle of degenerated axons that continues through the septum, passes caudal to the anterior commissure, and continues medially through the thalamus to terminate in the caudal hypothalamus. Degeneration was also found in the anterior thalamus, but could not be unequivocally attributed to an origin in the medial cortex because lesions in dorsal cortex also produce degeneration in the anterior thalamus.

8. Functional Organization of Cortical Areas

Previous sections have reviewed the cellular, neurochemical, and connectional properties of the four cortical areas. This section considers what little information is available on the function of each of the areas.

8.1. Lateral Cortex

Lateral cortex receives information from the main olfactory bulb both directly via the lateral olfactory tract and indirectly via olfactorecipient structures in the base of the forebrain. It receives a feedback projection from medial cortex and ascending projections from the perirotundal nuclei of the thalamus, the median raphe, dopaminergic cell groups in the hypothalamus, and the cholinergic cells of the diagonal band-basal forebrain system. It gives rise to a major efferent projection to medial cortex, providing it with potential influence over the septum and hypothalamus. Finally—at least in lizards—there are commissural connections between the dorsal rostral lateral cortices.

Orrego (1961, 1962) studied the physiology of lateral cortex in turtles using electrical stimulation of the olfactory bulbs and showed that neurons in lateral cortex are activated by stimulation of the olfactory bulb. These findings have been confirmed by other workers (see Belekhova, 1979), although the precise role of lateral cortex in olfactory behaviors is not known (see Ulinski and Peterson, 1981, for a review).

8.2. Dorsal Cortex

It is clear that dorsal cortex of turtles is involved in visual information processing. It receives an organized, although not topographic, projection from the dorsal lateral geniculate complex and projects back to visual structures in the brain stem such as the dorsal lateral geniculate complex, nucleus rotundus, and the optic tectum. Also, presentation of visual stimuli alters the electrical activity of dorsal cortex in both extracellular and intracellular recording experiments (Orrego, 1962; Bass *et al.*, 1983; Pivovarov and Trepakov, 1972; Kriegstein, 1987). Nonetheless, the precise role of dorsal cortex in the visual behavior of turtles has been problematic. Neurophysiological experiments have failed to find neurons with easily interpretable receptive field properties, and lesions of dorsal cortex do not produce obvious deficits in visual tasks such as pattern recognition (Bass *et al.*, 1973; Cranny and Powers, 1983).

However, the receptive field properties of cortical neurons suggest involvement in behavioral responses to moving objects in the turtle's environment. Neurons respond well to small, moving stimuli anywhere in the turtle's visual space (Mazurskaya, 1973; Mazurskaya *et al.*, 1976). They habituate rapidly (Gusel'nikov and Pivovarov, 1978), and the habituation shows some directional tuning so that repeated presentation of the stimulus moving in one direction causes habituation in a cortical neuron, whereas presentation of a stimulus moving in another direction elicits a brisk response. In addition to responding to a continuous moving stimulus, cortical cells also respond well to two small stimuli presented at two disjunct points in visual space (Mazurskaya, 1973). Units located superficially in the cortex (presumably stellate cells) have response properties that differ from those located deep in the cortex (presumably double pyramidal cells). Superficial units respond preferentially to stimuli presented simultaneously, while deep units respond preferentially to stimuli presented successively, with some temporal delay, at two disjunct points. The optimum interstimulus interval varies as a function of the positional disparity of the two points, being—for example—200 msec for a disparity of 20° of visual arc. From a behavioral viewpoint, these properties are consistent with the idea that dorsal cortex is involved in analyzing moving stimuli as they move through the turtle's visual world. Thus, individual cells can respond to a give stimulus at different points in visual space and over time spans that would occur as turtles encounter moving objects.

Some of these properties can be explained in terms of the known anatomy and cellular physiology of neurons in dorsal cortex. Cortical cells have wide-field receptive fields that require a convergence of information from all points of visual space onto neurons situated throughout dorsal cortex. The dorsal lateral geniculate complex receives a topographically organized representation of the retinal surface (Ulinski and Nautiyal, 1988), and although a few geniculate neurons have wide-field receptive fields, most have receptive fields that subtend about 30° of visual arc (Boiko, 1980). Some convergence occurs in the projections from the retina to the geniculate, but a further convergence must occur deeper into the geniculocortical pathway. This convergence occurs in two steps (Mulligan and Ulinski, 1990; Cosans and Ulinski, 1990; Ulinski, 1988). Information about points along the vertical meridians of visual space apparently converges upon cortical neurons as a result of the point-to-line character of geniculocortical projections, while information about points along the horizontal meridians converges upon cortical neurons as a result of intracortical connections.

The tendency for cortical neurons to habituate strongly is expected of a motion analysis system and is consistent with the presence of inhibitory neurons in the cortex. Both the layer 1 stellate cells and the double pyramidal cells of layer 2 receive thalamic inputs (Smith *et al.*, 1980) which are apparently excitatory (Pivovarov and Trepakov, 1972; Kriegstein, 1987). The stellate cells make GABAergic synapses upon double pyramidal cells. Thus, activation of the geniculocortical system evokes an EPSP–IPSP sequence in double pyramidal cells. Since the stellate cells make extensive synaptic contacts in layer 1 (Desan, 1984) and have a tonic firing pattern (Connors and Kriegstein, 1986), activation of geniculate neurons would lead to initial activation of double pyramidal cells followed by a strong inhibition. Since stellate cells respond preferentially to stimuli that occur simultaneously in different regions of visual space, they will tend to bias double pyramidal cells toward transitory responses to local, moving stimuli. At this point, we have only the most rudimentary idea of how dorsal cortex contributes to the turtle's ability to respond to moving stimuli. However, what is known suggests that it is involved in a smooth pursuit system functionally similar to the motor systems in primates that allow them to follow a moving target with their eyes (Lissberger *et al.*, 1987).

Although these results on turtles seem promising, they shed little light on the variations in the cytoarchitecture and connections of dorsal cortex that are seen between members of different groups of reptiles. Indeed, dorsal cortex is the cortical area that shows the greatest interspecific variation. It is particularly interesting that cryptodiran turtles have a large input to dorsal cortex from the dorsal lateral geniculate complex while this system is relatively modest in lizards and perhaps absent in crocodilians. Members of each of the major orders of reptiles vary in the relative development of their eyes and the extent to which they use visual behavior in their day-to-day lives. However, many lizards—including those species that have been studied using anatomical techniques—use visual cues extensively both in feeding behaviors and in social communication. Why the geniculocortical system should be significantly smaller in size in these animals than in turtles is not at all obvious. The lateral regions of dorsal cortex, which receive geniculate input, also vary in their cytoarchitecture between species. Some forms have a large pallial thickening area adjacent to lateral cortex, while in others the geniculate input terminates in only a small cluster of neurons. Thus, the fraction of dorsal cortex that corresponds to a visual area is quite large in turtles but quite small in lizards.

It seems clear that dorsal cortex as classically defined is not a unitary structure, but it is not certain how to relate the subdivisions of dorsal cortex in one species to its subdivisions in another species. One suggestion (e.g., Butler, 1976) is that dorsal cortex contains two major components. A lateral part receives geniculate input and is a primary visual cortex. The medial part receives inputs from the lateral part, but not directly from the geniculate. It thus serves to relay visual information to the dorsomedial and medial cortical areas. Desan's (1984) subdivision of dorsal cortex into a lateral area (D2) which includes pallial thickening and a medial area (D1) reflects this idea in that there is a sequential projection from the geniculate to D2 to D1 and then to medial cortex. The medial part may also receive inputs from other sensory modalities (e.g., Moore and Tschirigi, 1962; Orrego and Lissenby, 1962; Belekhova, 1979).

8.3. Dorsomedial Cortex

Dorsomedial cortex lacks efferent projections that leave the cortices of the two hemispheres, so it seems to play a role in interconnecting the cortical areas. Its most obvious connections form the largest commissural system in the cortex, linking the dorsomedial area of one side of the brain with its opposite member, and a reciprocal connection with medial cortex. In at least turtles, the double pyramidal cells that are the major cell type found in layer 2 of dorsomedial cortex tend to fire in bursts of action potentials, either as a result of endogenous bursting properties or as a result of a rhythmic synaptic drive (Shen and Kriegstein, 1987).

Electroencephalographic studies on crocodilians (Pyrethon and Dusan-Pyrethon, 1969; Flannigan *et al.,* 1973), turtles (Herman *et al.,* 1964; Pyrethon and Dusan-Pyrethon, 1969; Karmanova and Churnosov, 1972; Vasilescu, 1970; Flannigan, 1974b; Flannigan *et al.,* 1974; Walker and Berger, 1973), lizards (Flannigan, 1974a) and snakes (Pyrethon and Dusan-Pyrethon, 1969) show that rhythmic discharges can be recorded from the cortex, olfactory bulbs, and optic tectum of reptiles. Reptiles generally demonstrate a variety of behavioral arousal states, including periods of quiescence that resemble mammalian sleep periods in that the animal assumes a species-specific, stereotypic rest posture, shows an elevated behavioral response threshold and decreased respiration. Changes in behavioral state are correlated with EEG changes in that alert and active states are correlated with low-amplitude EEG patterns and quiescent periods are correlated with the presence of high-amplitude spikes in the EEG record. Whether these spikes correspond to the slow waves seen in mammalian sleep or to "limbic system spikes" recorded from the hippocampus of mammals during sleep (Hartse and Rechtschaffen, 1974) is not clear (see Walker and Berger, 1973). However, Hartse and Rechtschaffen (1974) showed that systemic administration in the tortoise *Geochelone carbonaria* of atropine sulfate, a drug that acts on central muscarinic cholinergic receptors, increases the frequency of high-voltage spikes in the EEG record. Since ACh hyperpolarizes neurons in dorsal cortex via muscarinic receptors, it is possible that atropine sulfate disinhibits neurons in dorsal cortex, thereby increasing the synaptic drive to endogenously bursting neurons in dorsomedial cortex. Thus, dorsomedial cortex may play a role in organizing the cortical rhythmic activities that mediate the animal's arousal state.

8.4. Medial Cortex

Medial cortex receives projections from a number of subcortical structures as well as each of the other three cortical areas. Its afferents show a complex organizational pattern, including a lamination of afferents such that many sets of afferents synapse upon a particular segment of the dendrites of layer 2 neurons. It is not possible to say much about the functional significance of these afferent systems except that medial cortex is in potential receipt of information from the main olfactory bulb via lateral cortex and of nonolfactory sensory information via dorsal cortex. Neurophysiological studies have confirmed such predictions made on the basis of the anatomical connections of medial cortex in that responses to electrical stimulation of the olfactory bulbs and activation of other sensory systems via either natural or electrical stimulation have been found to elicit neuronal activity in medial cortex of turtles (e.g., Pivavarov and Trepakov, 1972; Bass *et al.*, 1983) and lizards (see Belekhova, 1979). The latency of responses in medial cortex are always longer than those in dorsal cortex or lateral cortex.

Some of the efferent projections from medial cortex involve feedbacks to other cortical areas, but others leave the cortex and terminate in the hypothalamus, septum, or anterior thalamus. These presumably play some role in mediating behavior, but there is no real information on what this role might be.

8.5. Cortex as a Whole

There have been several studies examining the effect of cortical lesions on various aspects of behavior. The intention has usually been to investigate the role of a specific cortical area, but given the degree to which the cortical areas are interconnected and the fact that most systems afferent to one cortical area pass through one or more other areas en route to their eventual targets, it is unlikely that any of the experiments conducted to date can yield unambiguous conclusions about a given cortical area. The application of newer methods of inducing chemical lesions, which potentially circumvent the problems that attend fibers of passage, may help. In the meantime, work carried out so far may shed some light on the overall function of the cortex in reptiles on the assumption that any lesion in the cortex is likely to result in a global alteration of cortical activity.

One conclusion that comes from most of the lesion studies is that the cortex in reptiles does not play a simple role in discriminative function. Thus, lesions of dorsal cortex in lizards (Peterson, 1980) and turtles (Hayes and Hertzler, 1967; Hertzler and Hayes, 1967; Bass *et al.*, 1973; Morlock, 1972; Reiner and Powers, 1983) have generally failed to produce any noticeable deficit in the animal's ability to discriminate luminance levels or visual patterns, or perform a variety of visual locomotor tasks. On the other hand, cortical lesions do produce deficits in behaviors that require the animal to alter its pattern of behavior. Killackey *et al.*, (1972) found that lesions of dorsal cortex produce a deficit in the animal's ability to habituate to visual stimuli. Turtles were held so that they faced an instrument that released a Ping-Pong ball, causing it to fall toward the turtle's nose. Normal animals initially retract their neck and head when the ball is released, but then habituate to the stimulus. However, lesions of dorsal cortex eliminated the habituation response. Peterson (1980) studied the effects of lesions of dorsal cortex on spatial learning of desert iguanas (*Dipsosaurus dorsalis*) using mazes introduced by Lashley (1929). Normal animals were able to learn to locate a warm plate in a maze, but lesioned animals tended to wander into one cul-de-sac of the maze and stay there. Lesions of dorsal cortex also produced a deficit in reversal learning in these animals. Normal animals could be trained to go to one arm of a maze to find a warm plate. When they reached criterion on one arm, they could subsequently be trained to reverse themselves and go to the other arm of the maze. Lesioned animals could not learn this reversal task, and their errors tended to increase in number with subsequent reversals. Similarly, Cranny and Powers (1983) found that lesions of dorsal cortex in turtles produced a deficit in the animal's ability to perform visual reversal tasks or to shift the basis of the discrimination from a pattern to color or to reverse go/no-go discriminations. For example, turtles were first trained to respond to a horizontal stripe and then required to respond to a vertical stripe. Or, turtles might be trained to respond to a vertical stripe regardless of its color (red or green) and then to shift to a response made on the basis of color.

9. Reptilian and Mammalian Patterns of Cortical Organization

As noted in section 1, it has been clear for over 100 years that reptiles and mammals are the two groups of vertebrates with complex, laminated cerebral cortices. This apparent similarity prompted considerable and sustained interest in the cortex of reptiles because of the possibility that an understanding of the cortex of reptiles would provide clues to the evolution of the mammalian cortex. Until quite recently, reptiles were believed to lack the direct thalamocortical projections that seemed the hallmark of mammalian cortex, and the progressive establishment of such projections in early mammals was seen as a key step in the evolution of the mammalian brain. However, it now appears that reptiles do have direct thalamocortical projections and that cortex in both reptiles and mammals receives projections from a large list of subcortical structures. The cortices of reptiles and mammals may then be more alike than previously suspected. This last section briefly summarizes what now seem to be the most important similarities and differences between the cortices of reptiles and mammals.

The forebrains of both reptiles and mammals are similar in that the dorsal surface of their cerebral hemispheres is formed by a pallium that extends from the septum, medially, to the striatum laterally. In contrast to the situation in other groups of vertebrates whose forebrains develop by a process of evagination, the pallium in reptiles and mammals consists, at least partially, of a laminated cortex in which neurons have migrated away from the ventricle to form at least three layers of cell bodies. The pallium can be divided into three major segments in both groups (Fig. 37).

The first is an "olfactory cortex"—lateral cortex or pyriform cortex—which is situated on the lateral wall of the hemisphere and receives direct projections

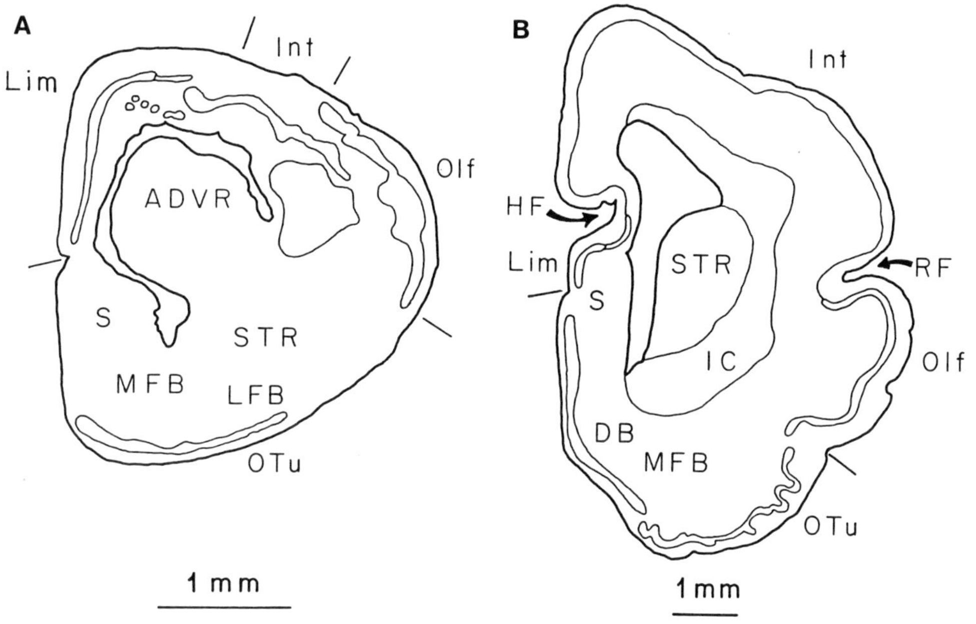

Figure 37. Comparison of pallial organization in reptiles and mammals. This figure shows transverse sections through the telencephalon of (A) a tegu lizard, *Tupinambis,* and (B) an opossum, *Didelphis.* Unlike placental mammals, marsupials (and monotremes) lack a corpus callosum, thereby simplifying comparisons with the lizard brain. Both telencephalons have a lateral ventricle and are divided into three major parts. The septum (S) makes up the ventromedial face of the telencephalon. The striatum (STR) makes up the ventrolateral wall of the telencephalon. Olfactory structures such as the diagonal band (DB) or olfactory tubercle (OTu) are present on the ventral surface. The forebrain bundles run rostrally into the telencephalon. They consist of the medial forebrain bundle (MFB) and the lateral forebrain bundle (LFB) or internal capsule (IC). The pallium or roof of the telencephalon has two components in reptiles: a cortex and the anterior dorsal ventricular ridge (ADVR). The pallium in mammals has only a single component: the cortex. In both groups, the pallium can be divided into three functional segments. The olfactory cortex (Olf) is situated laterally. The intermediate segment (Int) is present dorsally. It includes part of dorsal cortex and ADVR in reptiles and the cortex between the rhinal fissure (RF) and the hippocampal formation (HF) in mammals. The limbic cortex (Lim) occupies the medial wall of the pallium in both groups. It is involuted by the hippocampal fissure in mammals. Note that the two sections are drawn to different scales.

from the mitral cells of the main olfactory bulb. It is a trilaminar cortex in both groups, and there are several similarities in its neuronal organization (Ulinski and Rainey, 1980). Perhaps the major difference is that lateral cortex of reptiles receives direct, bilateral projections from the main olfactory bulbs, while in mammals there are only indirect connections between the two piriform cortices (Haberly and Price, 1978).

The second major pallial segment is a "limbic cortex" that lies dorsal to the septum and forms the dorsomedial wall of the hemisphere. This relationship is partly obscured in mammals by the formation of the hippocampal fissure, the ventral elaboration of limbic cortex that accompanies the formation of the temporal lobe in some species, and by the disruption of the medial wall of the hemisphere caused by the formation of the corpus callosum in placental mammals. Nevertheless, there are general similarities between limbic cortex in reptiles and that of mammals. These include the trilaminar organization of the medial and dorsomedial cortical areas in reptiles and the mammalian dentate gyrus and CA fields, the cytology of neurons in these areas (particularly that of the candelabra cells of medial cortex and the granule cells of the dentate gyrus), and the presence of a population of neurons that contain zinc in their axons. There is also a less obvious similarity in that limbic cortex in both groups received afferents from olfactory and nonolfactory cortical areas. These connections are direct in reptiles. Lateral cortex carries information from the main olfactory bulb to medial cortex, and dorsal cortex carries information from the dorsal lateral geniculate complex to medial cortex. The connections are indirect in mammals. Information from the main olfactory bulb reaches the dentate gyrus only after relay in both the piriform and entorhinal cortical areas. Information from thalamic sensory nuclei reaches the dentate gyrus only after passing through several cortical areas, including one or more "sensory" areas and the entorhinal cortex (Jones and Powell, 1970). However, there is a striking similarity in the organization of afferent systems within the limbic cortex of the two groups with inputs from the "olfactory" and "nonolfactory" sensory modalities terminating in a laminar fashion in the outer part of the first cortical layer and commissural and association systems terminating deeper in the most medially situated cortical areas of animals in the two groups. Finally, the connections of the limbic cortices in the two groups show a tendency toward a lamellar organization in which axon systems run roughly in the transverse plane at each level of the septotemporal axis. This is seen, for example, in the trajectory of the axons of candelabra cells that run from medial cortex to the other three cortical areas in reptiles and those that run from the granule cells of the dentate gyrus to field CA3 in mammals.

The major differences between limbic cortex in the two groups lie in the organization of their efferent systems. Both have projections to the septum, hypothalamus, thalamus and reciprocally back to those areas on the lateral surface of the hemisphere that project, directly or indirectly, to medial cortex. However, in reptiles medial cortex is the major source of efferents from limbic cortex, and dorsomedial cortex (which bears some resemblance to the CA fields of mammals in terms of its cytology) participates only in commissural and association projections, while in mammals the dentate gyrus projects almost exclusively to the CA fields which, together with the subicular complex, are the source of efferent projections from the limbic cortex.

The third segment of the pallium lies on the dorsal surface of the hemisphere, between limbic cortex and olfactory cortex. It is composed entirely of isocortex and adjacent areas in mammals, but in reptiles (and birds) it consists of at least part of dorsal cortex and the anterior dorsal ventricular ridge. This segment receives projections from the auditory, somatosensory, and visual sensory nuclei of the dorsal thalamus as well as a range of subpallial structures (e.g., locus coeruleus, raphe, basal telencephalon) in both reptiles and mammals. The organization of the anterior dorsal ventricular ridge appears to be very different from that of isocortex at first glance, but there are actually several similarities in organization between the two structures. These are discussed in some detail by Ulinski (1983), and will not be considered here. However, this intermediate segment of the pallium is laminated, contains modality-specific sensory areas, and gives rise to efferent projections to the striatum in both groups.

Some of the most important differences between the intermediate segment in reptiles and mammals pertain to motor systems. Isocortex of monotreme, marsupial, and placental mammals contains one or more cortical areas that elicit movements of restricted body parts following electrical stimulation, give rise to corticospinal and corticobulbar projections, and receive afferent projections from thalamic nuclei that carry information from the cerebellum and striatum to the cortex. There is no indication that such areas are present in reptiles. Thus, reptiles seem to have only corticostriate systems and lack direct projections from the pallium to the reticular formation and spinal cord. It is also the case that the striatum is relatively small in reptiles, while it is greatly expanded and bulges into the lateral ventricles in mammals. Ulinski (1986a) has speculated that these differences reflect the evolution of the ability to control complex movements of the jaws and limbs that occurred relatively late in the evolution of mammals.

10. References

Ariëns Kappers, C. U., 1921, *Vergleichende Anatomie des Nervensystems,* Bohn, Haarlem.

Ariëns Kappers, C. U., Huber, G. C., and Crosby, E. C., 1936, *The Comparative Anatomy of the Nervous System of Vertebrates, Including Man,* reprinted 1967, Hafner, New York.

Armstrong, J. A., Gamble, H. J., Goldby, F., 1953, Observations on the olfactory apparatus and telencephalon of *Anolis,* a microsmatic lizard, *J. Anat.* **87:**288–307.

Balaban, C. D., 1977, Olfactory projections in emydid turtles (*Pseudemys scripta elegans* and *Graptemys pseudogeographica*), *Am. Zool.* **17:**887.

Balaban, C. D., 1978, Structure of the pallial thickening in turtles (*Pseudemys scripta elegans*), Anat. Rec. **190:**330–331.

Balaban, C. D., 1979, Organization of thalamic projections to anterior dorsal ventricular ridge in two species of turtles (*Pseudemys scripta elegans* and *Chrysemys picta belli*), Ph.D. dissertation, University of Chicago.

Balaban, C. D., and Ulinski, P. S., 1981a, Organization of thalamic afferents to anterior dorsal ventricular ridge in turtles—I. Projections of thalamic nuclei, *J. Comp. Neurol.* **200:**95–130.

Balaban, C. D., and Ulinski, P. S., 1981b, Organization of thalamic afferents to anterior dorsal ventricular ridge in turtles—II. Properties of the rotundal–dorsal map, *J. Comp. Neurol.* **200:**131–150.

Bass, A. H., and Northcutt, R. G., 1981a, Retinal recipient nuclei in the painted turtle, *Chrysemys picta:* An autoradiographic and HRP study, *J. Comp. Neurol.* **199:**97–112.

Bass, A. H., and Northcutt, R. G., 1981b, Primary retinal targets in the Atlantic loggerhead sea turtle, *Caretta caretta, Cell Tissue Res.* **218:**253–264.

Bass, A. H., Pritz, M. B., and Northcutt, R. G., 1973, Effects of telencephalic and tectal ablations on visual behavior in the side-necked turtle, *Podocnemis unifils, Brain Res.* **55:**455–460.

Bass, A. H., Andry, M. L., and Northcutt, R. G., 1983, Visual activity in the telencephalon of the painted turtle, *Chrysemys picta, Brain Res.* **263:**201–210.

Baumgarten, H. G., and Braak, H., 1968, Catecholamine in Gehirn der Eidechse (*Lacerta viridis* and *Lacerta muralis*), Z. Zellforsch. **86:**574–608.

Baylor, D. A., and Fettiplace, R., 1976, Transmission of signals from photoreceptors to ganglion cells in the eye of the turtle, *Cold Spring Harbor Symp. Quant. Biol.* **40:**529–536.

Bear, M. F., and Ebner, F. F., 1983, Somatostatin-like immunoreactivity in the forebrain of *Pseudemys* turtles, *Neuroscience* **2:**297–307.

Beckers, H. J. A., Platel, R. and Nieuwenhuys, R., 1971/1972, Les aires corticales de quelques reptiles squamates *Lacerta viridis, Chamaeleo lateralis, Monopeltis guentheri, Acta Morphol. Neerl. Scand.* **9:**337–364.

Belekhova, M. G., 1979, Neurophysiology of the forebrain, in: *Biology of the Reptilia,* Volume 10 (C. G. Gans, R. G. Northcutt, and P. S. Ulinski, eds.), Academic Press, New York, pp. 287–359.

Belekhova, M. V., Kosareva, A. A., Vaselkin, N. P., and Ermakova, T. V., 1979, Telencephalic afferent connections in the turtle *Emys orbicularis:* A peroxidase study, *J. Evol. Biochem. Physiol.* **15:**97–103.

Bennett, A. F., and Dawson, W. R., 1976, Metabolism, in: *Biology of the Reptilia,* Volume 5 (C. Gans and W. R. Dawson, eds.), Academic Press, New York, pp. 127–223.

Benowitz, L., 1980, Functional organization of the avian telencephalon, in: *Comparative Neurology of the Telencephalon* (S. O. E. Ebbesson, ed.), Plenum Press, New York, pp. 389–421.

Berbel, P. J., Martinez-Guijarro, F. J., and Lopez-Garcia, C., 1987, Intrinsic organization of the medial cerebral cortex of the lizard *Lacerta pituysensis:* A Golgi study, *J. Morphol.* **194:**275–286.

Berquist, H., and Kallen, B., 1954, Notes on the early histogenesis and morphogenesis of the central nervous system in vertebrates, *J. Comp. Neurol.* **100:**627–659.

Berson, D. M., and Hartline, P. H., 1988, A tecto-rotundal-telencephalic pathway in the rattlesnake: Evidence for a forebrain representation of the infrared sense, *J. Neurosci.* **8:**1074–1088.

Blanton, M., and Kriegstein, A., 1987, Development of cortical inhibition: An in *vitro* HRP and GABA-immunocytochemical study in turtle, *Neurosci. Abstr.* **13:**953.

Blanton, M. G., Shen, J. M., and Kriegstein, A. R., 1986, Actions of acetylcholine in turtle dorsal cortex, *Neurosci. Abstr.* **12:**573.

Blanton, M. G., Shen, J. M., and Kriegstein, A. R., 1987, Evidence for the inhibitory neurotransmitter γ-aminobutyric acid in aspiny and sparsely spiny nonpyramidal neurons of the turtle dorsal cortex, *J. Comp. Neurol.* **259:**277–297.

Boiko, V. P., 1980, Responses to visual stimuli in thalamic neurons of the turtle *Emys orbicularis, Neurosci. Behav. Physiol.* **10:**183–188.

Braak, H., Baumgarten, H. G., and Falck, B., 1968, 5-Hydroxytryptamine in Gerhirn der Eidechse (*Lacerta viridis* and *Lacerta muralis*), *Z. Zellforsch.* **90:**161–185.

Braford, M. R., Jr., and Northcutt, R. G., 1974, Olfactory bulb projections in the bichir, *Polypterus, J. Comp. Neurol.* **156:**165–178.

Brauth, S. E., 1984, Enkephalin-like immunoreactivity within the telencephalon of the reptile *Caiman crocodilus, Neuroscience* **11:**345–358.

Brauth, S. E., 1988, Catecholamine neurons in the brainstem of the reptile *Caiman crocodilus, J. Comp. Neurol.* **270:**313–326.

Brauth, S. E., Reiner, A., Kitt, C. A., and Karten, H. J., 1983, The substance P-containing striatotegmental path in reptiles: An immunohistochemical study, *J. Comp. Neurol.* **219:**305–327.

Bruce, L. L., and Butler, A. B., 1984a, Telencephalic connections in lizards. I. Projections to cortex, *J. Comp. Neurol.* **229:**585–601.

Bruce, L. L., and Butler, A. B., 1984b, Telencephalic connections in lizards. II. Projections to anterior dorsal ventricular ridge, *J. Comp. Neurol.* **229:**602–616.

Butler, A. B., 1976, Telencephalon of the lizard *Gekko gekko* (Linnaeus): Some connections of the cortex and dorsal ventricular ridge, *Brain Behav. Evol.* **13:**396–417.

Butler, A. B., 1978, Forebrain connections in lizards and the evolution of sensory systems, in: *Behavior and Neurology of Lizards* (N. Greenberg and P. D. Maclean, eds.), NIMH, Washington, D.C., pp. 65–78.

Butler, A. B., 1980, Cytoarchitectonic and connectional organization of the lacertilian telencephalon with comments on vertebrate forebrain evolution, in: *Comparative Neurology of the Telencephalon* (S. O. E. Ebbesson, ed.), Plenum Press, New York, pp. 297–329.

Butler, A. B., and Northcutt, R. G., 1973, Architectonic studies of the diencephalon of *Iguana iguana* (Linnaeus), *J. Comp. Neurol.* **149:**439–462.

Cairney, J., 1926, A general survey of the forebrain of *Sphenodon punctatum, J. Comp. Neurol.* **42:**255–348.

Carey, J. H., 1967, The nuclear pattern of the telencephalon of the black snake, *Coluber constrictor*

constrictor, in: *Evolution of the Forebrain* (R. Hassler and H. Stephen, eds.), Plenum Press, New York, pp. 73–80.
Carroll, R. L., 1969, Origin of reptiles, in: *Biology of the Reptilia,* Volume 1 (C. Gans, A. Bellairs, and T. S. Parsons, eds.), Academic Press, New York, pp. 1–44.
Carroll, R. L., 1987, *Vertebrate Paleontology and Evolution,* Freeman, San Francisco.
Clark, J. M., and Ulinski, P. S., 1984, A Golgi study of anterior dorsal ventricular ridge in the alligator, *Alligator mississippiensis, J. Morphol.* **179:**153–174.
Connors, B. W., and Kriegstein, A. R., 1986, Cellular physiology of the turtle visual cortex: Distinctive properties of pyramidal and stellate neurons, *J. Neurosci.* **6:**164–177.
Connors, B. W., and Ransom, B. R., 1982, Electrophysiological studies on ependymal cells of turtle cortex, *Neurosci. Abstr.* **8:**237.
Conty, P., 1953, Richerche sulla gliorchitetonica dei reitili (cheloni, sauri e ofidi), *Arch. Ital. Anat. Embriol.* **58:**295–320.
Cosans, C., and Ulinski, P. S., 1990, Spatial organization of axons in turtle visual cortex: Intralamellar and interlamellar projections, *J. Comp. Neurol.* **295:** 548–558.
Cranny, J., and Powers, A. S., 1983, The effects of core nucleus and cortical lesions in turtles on reversal and dimensional shifting, *Physiol. Psychol.* **11:**103–111.
Crawford, A. C., and Fettiplace, R., 1980, The frequency selectivity of auditory nerve fibres and hair cells in the cochlea of the turtle, *J. Physiol. (London)* **306:**79–125.
Crawford, A. C., and Fettiplace, R., 1981a, An electrical tuning mechanism in turtle cochlear hair cells, *J. Physiol. (London)* **312:**377–412.
Crawford, A. C., and Fettiplace, R., 1981b, Non-linearities in the responses of turtle hair cells, *J. Physiol. (London)* **315:**317–338.
Crosby, E. C., 1917, The forebrain of *Alligator mississippiensis, J. Comp. Neurol.* **27:**325–403.
Crosby, E. C., and Showers, M. J., 1969, Comparative anatomy of the preoptic and hypothalamic areas, in: *The Hypothalamus* (W. Haymaker, E. Anderson, and W. J. H. Nauta, eds.), Thomas, Springfield, Ill., pp. 61–135.
Cruce, J. A. F., 1974, A cytoarchitectonic study of the diencephalon of the tegu lizard, *Tupinambis nigropunctatus, J. Comp. Neurol.* **153:**215–238.
Cruce, W. L. R., and Nieuwenhuys, R., 1974, The cell masses in the brain stem of the turtle *Testudo hermanni:* A topographical and topological analysis, *J. Comp. Neurol.* **156:**277–306.
Curwen, A. O., 1937, The telencephalon of *Tupinambis nigropunctatus.* I. Medial and cortical areas, *J. Comp. Neurol.* **66:**375–404.
Dacey, D. M., and Ulinski, P. S., 1983, Nucleus rotundus in snake, *Thamnophis sirtalis:* An analysis of a nonretinotopic projection, *J. Comp. Neurol.* **216:**175–191.
Dacey, D. M., and Ulinski, P. S., 1986a, Optic tectum in the eastern garter snake, *Thamnophis sirtalis:* I. Morphology of efferent cells, *J. Comp. Neurol.* **245:**1–28.
Dacey, D. M., and Ulinski, P. S., 1986b, Optic tectum of the eastern garter snake, *Thamnophis sirtalis:* V. Morphology of brainstem afferents and general discussion, *J. Comp Neurol.* **245:**423–453.
Davila, J. C., Guirado, S., De la Calle, A., and Marin-Giron, F., 1985, Electron microscopy of the medial cortex in the lizard *Psammodromus algirus, J. Morphol.* **185:**327–338.
Davila, J. C., Guirado, S., De la Calle, A., and Marin-Giron, F., 1986, Ultrastructure of the dorsal cortex of the lizard *Psammodromus algirus, J. Hirnforsch.* **27:**295–304.
Davydova, T. V., and Goncharova, N. V., 1979, Comparative characterization of the basic forebrain cortical zones in *Emys orbicularis* (Linnaeus) and *Testudo horsfieldi* (Gray), *J. Hirnforsch.* **20:**245–262.
Davydova, T. V., and Smirnov, G. D., 1969, Neuronal organization of the cortical plate in the tortoise cerebral hemispheres, *H. Hirnforsch.* **11:**333–346.
de Lange, S. J., 1911, Das Vorderhirn der Reptilien, *Folia Neurobiol.* **5:**548–597.
Desan, P. H., 1984, The organization of the cerebral cortex of the pond turtle, *Pseudemys scripta elegans,* Ph.D. dissertation, Harvard University.
Distel, H., and Ebbesson, S. O. E., 1981, Habenular projections in the monitor lizard (*Varanus benegalensus*), Exp. Brain Res. **43:**324–329.
du Lac, S., and Dacey, D. M., 1981, Relation of the retina and optic tectum to the lateral geniculate complex in garter snake (*Thamnophis sirtalis*), Neurosci. Abstr. **7:**460.
Dunser, K. R., Maxwell, J. H., and Granda, A. M., 1981, Visual properties of cells in anterior dorsal ventricular ridge of turtle, *Neurosci, Lett.* **25:**281–284.
Durward, A., 1930, The cell masses in the forebrain of *Sphenodon punctatum, J. Anat.* **65:**8–44.
Ebbesson, S. O. E., 1980, The parcellation theory and its relation to interspecific variability in brain organization, evolutionary and ontogenetic development and neuronal plasticity, *Cell Tissue Res.* **213:**179–212.

Ebbesson, S. O. E., and Voneida, T. J., 1969, The cytoarchitecture of the pallium in the tegu lizard, *Brain Behav. Evol.* **2:**431–466.

Ebner, F. F., and Colonnier, M., 1975, Synaptic patterns in the visual cortex of turtles: An electron microscopic study, *J. Comp. Neurol.* **160:**51–80.

Ebner, F. F., and Colonnier, M., 1978, A quantitative study of synaptic patterns in turtle visual cortex, *J. Comp. Neurol.* **179:**263–276.

Elprana, D., Wouterlood, F. G., and Alones, V. E., 1980, A corticotectal projection in the lizard *Agama agama, Neurosci, Lett.* **18:**251–256.

Flannigan, W. F., Jr., 1974a, Sleep and wakefulness in iguanid lizards, *Ctenosaura pectinata* and *Iguana iguana, Brain Behav. Evol.* **8:**401–436.

Flannigan, W. F., Jr., 1974b, Sleep and wakefulness in chelonian reptiles. II. The redfooted tortoise, *Geochelone carbonaria, Arch. Ital. Biol.* **112:**253–277.

Flannigan, W. F., Jr., Wilcox, R. H., and Rechtschaffen, A., 1973, The EEG and behavioral continuum of the crocodilian, *Caiman sclerops, Electroencephalogr. Clin. Neurophysiol.* **34:**521–538.

Flannigan, W. F., Jr., Knight, C. P., Hartse, K. M., and Rechtschaffen, A., 1974, Sleep and wakefulness in chelonian reptiles. I. The box turtle, *Terrapene carolina, Arch. Ital. Biol.* **112:**227–252.

Franzoni, M. F., and Fasolo, A., 1982, The hypothalamus of *Lacerta sicula.* A Golgi study on the caudal hypothalamus, *Cell Tissue Res.* **223:**61–71.

Gaidenko, G. V., 1978, Efferent connections of the dorsal cortex in tortoises, *J. Evol. Biochem. Physiol.* **13:** 268–270.

Gamble, H. J., 1956, An experimental study of the secondary olfactory connexions in *Testudo graeca, J. Anat.* **90:**15–29.

Gans, C., 1978, The characteristics and affinities of the Amphisbaenia, *Trans. Zool. Soc. London* **34:**347–416.

Garcia-Verdugo, J. M., Berbel, P. J., and Lopez-Garcia, C., 1981, Estudio con Golgi con microscopia electronica de los ependimocitos de la corteza cerebral del lagarto *Lacerta galloti, Trab. Inst. Cajal* **72:**269–278.

Garcia-Verdugo, J. M., Farinas, I., Molowny, A., and Lopez-Garcia, C., 1986, Ultrastructure of putative migrating cells in the cerebral cortex of *Lacerta galloti, J. Morphol.* **189:**189–198.

Goffinet, A. M., 1983, The embryonic development of the cortical plate in reptiles: A comparative study in *Emys orbicularis* and *Lacerta agilis, J. Comp. Neurol.* **215:**437–452.

Goffinet, A. M., Daumerie, C., Langerwerf, B., and Pieau, C., 1986, Neurogenesis in reptilian cortical structures: ^{3}H-thymidine autoradiographic analysis, *J. Comp. Neurol.* **243:**106–116.

Goldby, F., 1934, The cerebral hemispheres of *Lacerta viridis, J. Anat.* **68:**157–215.

Goldby, F., and Gamble, H. J., 1957, The reptilian cerebral hemispheres, *Biol. Rev.* **32:**383–420.

Goossens, N., Dierickx, K., and Vandesande, F., 1980, Immunocytochemical localization of somatostatin in the brain of the lizard, *Ctenosauria pectinata, Cell Tissue Res.* **208:**499–505.

Greenberg, N., 1982, A forebrain atlas and stereotaxic technique for the lizard, *Anolis carolinensis, J. Morphol.* **174:**217–236.

Guirado, S., De La Calle, A., Davila, J. C., and Marin-Giron, F., 1984, Light microscopy of the medial wall of the cerebral cortex of the lizard *Psammodromus algirus, J. Morphol.* **181:**319–331.

Guirado, S., Davila, J. C., De La Calle, A., and Marin-Giron, F., 1987, A Golgi study of the dorsal cortex in the lizard *Psammodromus algirus, J. Morphol.* **194:**265–274.

Gusel'nikov, V. I., and Pivavarov, A. S., 1978, Postsynaptic mechanism of habituation of turtle cortical neurons to moving stimuli, *Neurosci. Behav. Physiol.* **9:**1–7.

Haberly, L. B., and Price, J. L., 1978, Association and commissural fiber systems of the olfactory cortex of the rat. II. Systems originating in the olfactory peduncle, *J. Comp. Neurol.* **181:**781–808.

Hall, J. A., Foster, R. E., Ebner, F. F., and Hall, W. C., 1977, Visual cortex in a reptile, the turtle (*Pseudemys scripta* and *Chrysemys picta*), *Brain Res.* **130:**197–216.

Hall, W. C., and Ebner, F. F., 1970a, Parallels in the visual afferent projections of the thalamus in the hedgehog (*Paraechinas hypomelas*) and the turtle (*Pseudemys scripta*), *Brain Behav. Evol.* **3:**135–154.

Hall, W. C., and Ebner, F. F., 1970b, Thalamotelencephalic projections in the turtle *(Pseudemys scripta), J. Comp. Neurol.* **140:**101–122.

Halpern, M., 1974, An experimental demonstration of the fornix system in a snake, *Soc. Neurosci* 4th Annu. Meet.

Halpern, M., 1976, The efferent connections of the olfactory bulb and accessory olfactory bulb in the snakes, *Thamnophis sirtalis* and *Thamnophis radix, J. Morphol.* **150:**553–578.

Halpern, M., 1980, The telencephalon of snakes, in: *Comparative Neurology of the Telencephalon* (S. O. E., Ebbesson, ed.), Plenum Press, New York, pp. 257–295.

Halpern, M., and Frumin, N., 1973, Retinal projections in a snake (*Thamnophis sirtalis*), J. Morphol. **141**:359–382.

Hartse, K. M., and Rechtschaffen, A., 1974, Effect of atropine sulfate on the sleep-related EEG spike activity of the tortoise, *Geochelone carbonaria, Brain Behav. Evol.* **9**:81–94.

Hayes, W. N., and Hertzler, D. R., 1967, Role of the optic tectum and general cortex in reptilian vision, *Psychon. Sci.* **9**:521.

Heimer, L., 1969, Secondary olfactory connections, *Ann. N.Y. Acad. Sci.* **167**:129–146.

Heller, S. B., and Ulinski, P. S., 1987, Morphology of geniculocortical axons in turtles of the genera *Pseudemys* and *Chrysemys, Anat. Embryol.* **175**:505–515.

Harmen, H. M., Jouvet, M., and Klein, M., 1964, Etude polygraphique du sommeil chez la tortue, *C.R. Soc. Biol.* **258**:2175–2178.

Herrick, C. L., 1892, Embryological notes on the brain of the snake, *J. Comp. Neurol.* **2**:160–176.

Hertzler, D. R., and Hayes, W. N., 1967, Cortical and tectal function in visually guided behavior of turtles, *J. Comp. Physiol. Psychol.* **63**:444–447.

Hetzel, W. von, 1974, Die Ontogenese des Telencephalon bie *Lacerta sicula* (Rafinesque), mit Besonderer Berucksichtigung der Pallialler Entwicklung, *Zool. Beitr. N.S.* **20**:361–458.

Hines, M., 1923, The development of the telencephalon in *Sphenodon punctatum, J. Comp. Neurol.* **35**:483–537.

Hohmann, C. F., Carroll, P. T., and Ebner, F. F., 1983, Acetylcholine levels and choline acetyltransferase activity in turtle cortex, *Brain Res.* **258**:120–122.

Hoogland, P. V., ten Donkelaar, H. J., and Cruce, J. A. F., 1978, Efferent connections of the septal area in a lizard (*Tupinambis migropunctatus*), Neurosci. Lett. **7**:61–65.

Hotton, N., Maclean, P. D., Roth, J. J., and Roth, E. C., 1986, *The Ecology and Biology of Mammal-Like Reptiles,* Smithsonian Institution Press, Washington, D.C.

Inoue, Y., Inoue, T., Nishimura, Y., and Shimai, K., 1974, The glial cells in the reptilian brain and spinal cord—Golgi study, *Okajimas Folia Anat. Jpn.* **51**:161–188.

Johnston, J. B., 1915, The cell masses in the forebrain of the turtle *Cistudo carolina, J. Comp. Neurol.* **25**:393–468.

Johnston, J. B., 1916, The development of the dorsal ventricular ridge in turtles, *J. Comp. Neurol.* **26**:481–505.

Jones, E. G., and Powell, T. P. S., 1970, An anatomical study of connecting sensory pathways within the cerebral cortex of the monkey brain, *Brain* **93**:793–820.

Kallen, B., 1951, On the ontogeny of the reptilian forebrain. Nuclear structures and ventricular sulci, *J. Comp. Neurol.* **95**:307–347.

Kallen, B., 1956, Notes on the mode of formation of brain nuclei during ontogenesis, *C.R. Assoc. Anat. Reun.* **42P**:747–756.

Karmanova, I. G., and Churnosov, E. V., 1972, Electrophysiological Investigation of natural sleep and wakefulness of turtles and chickens, *Zh. Evol. Biochem. Fiziol.* **8**:59–65.

Khatchaturian, H., Dores, R. M., Watson, S. J., and Akil, H., 1984, β-Endorphin/ACTH immunocytochemistry in the CNS of a lizard, *Anolis carolinensis:* Evidence for a major mesencephalic cell group, *J. Comp. Neurol.* **229**:576–584.

Killackey, H., Pellmar, T., and Ebner, F., 1972, The effects of general cortex ablation on habituation in the turtle, *Fed. Proc.* **31**:819.

King, J. S., 1966, A comparative investigation of neuroglia in representative vertebrates: A silver carbonate study, *J. Morphol.* **119**:435–466.

Kirsche, W., 1972, Die Entwicklung des Telencephalons der Reptilien und deren Verziehung zur Hirn-Bauplanlehre, *Nova Acta Leopold.* **37**:1–78.

Knapp, H., and Kang, D. S., 1968a, The visual pathways of the snapping turtle (*Chelydra serpentina*), Brain Behav. Evol. **1**:19–42.

Knapp, H., and Kang, D. S., 1968b, The retinal projections of the sidenecked turtle (*Podocnemis unifilis*) with some notes on the possible origin of the pars dorsalis of the lateral geniculate body, *Brain Behav. Evol.* **1**:369–404.

Kriegstein, A. R., 1985, An intact vertebrate visual system *in vitro:* Visually evoked pyramidal cell responses in turtle cortex, *Neurosci. Lett.* **22**:S305.

Kriegstein, A. R., 1987, Synaptic responses of cortical pyramidal neurons to light stimulation in the isolated turtle visual system, *J. Neurosci.* **7**:2488–2492.

Kriegstein, A. R., and Connors, B. W., 1986, Cellular physiology of the turtle visual cortex: Synaptic properties and intrinsic circuitry, *J. Neurosci.* **7**:178–191.

Kriegstein, A. R., and Shen, J. M., 1987, Development of cortical inhibition: Appearance of synchronized neuronal discharges, *Neurosci. Abstr.* **13**:954.

Kriegstein, A. R., Shen, J. M., and Eshhar, N., 1986, Monoclonal antibodies to the turtle cortex reveal neuronal subsets, antigenic cross-reactivity with the mammalian neocortex, and forebrain structures sharing a pallial derivation, *J. Comp. Neurol.* **254:**330–340.

Kriegstein, A. R., Shen, J. M., and Blanton, M. G., 1988, Development of inhibitory function during corticogenesis in turtle, in: *Recent Advances in Understanding the Structure and Function of the Forebrain in Reptiles* (W. K. Schwerdtfeger and W. J. A. J. Smeets, eds.), Karger, Basel, pp. 131–141.

Kruger, L., and Berkowitz, E. C., 1960, The main afferent connections of the reptilian telencephalon as determined by degeneration and electrophysiological methods, *J. Comp. Neurol.* **115:**125–142.

Kruger, L., and Maxwell, D. S., 1967, Comparative fine structure of vertebrate neuroglia: Teleosts and reptiles, *J. Comp. Neurol.* **129:**115–141.

Kunzle, H., 1985, The cerebellar and vestibular nuclear complexes in the turtle. II. Projections to the prosencephalon, *J. Comp. Neurol.* **242:**122–133.

Kunzle, H., and Schnyder, H., 1984, Do retinal and spinal projections overlap within the turtle thalamus? *Neuroscience* **10:**161–168.

Lacey, D. J., 1978, The organization of the hippocampus of the fence lizard: A light microscopic study, *J. Comp. Neurol.* **182:**247–265.

Langston, W., Jr., 1973, The crocodilian skull in historical perspective, in: *Biology of the Reptilia,* Volume 4 (C. Gans and T. S. Parsons, eds.), Academic Press, New York, pp. 263–284.

Larson-Prior, L. J., and Slater, N. T., 1988, GABAergic inhibition and epileptiform discharges in the turtle hippocampus *in vitro, Brain Res.* **460:**369–375.

Larson-Prior, L., Ulinski, P. S., and Slater, N. T., 1989, Excitatory amino acid receptor mediated transmission in geniculocortical and intracortical pathways in visual cortex, *Soc. Neurosci. Abstr.* **15:** 947.

Lashley, K. S., 1929, *Brain Mechanisms and Intelligence,* Dover, New York.

Lissberger, S. G., Morris, E. J., and Tychsen, L., 1987, Visual motion processing and sensory motor integration for smooth pursuit eye movements, *Annu. Rev. Neurosci.* **10:**97–130.

Lohman, A. H. M., and Mentink, G. M., 1972, Some cortical connections of the tegu lizard *Tupinambis tequixin, Brain Res.* **45:**325–344.

Lohman, A. H. M., and van Woerden-Verkely, E., 1976, Further studies on the cortical connections of the tegu lizard, *Brain Res.* **103:**9–28.

Lohman, A. H. M., and van Woerden-Verkely, E., 1978, Ascending connections to the forebrain in the tegu lizard, *J. Comp. Neurol.* **182:**555–594.

Lopez-Garcia, C., Molowny, A., and Perez-Clausell, J., 1983a, Volumetric and densitometric study in the cerebral cortex and septum of a lizard (*Lacerta galloti*) using the Timm method, *Neurosci. Lett.* **40:**13–18.

Lopez-Garcia, C., Tineo, P. L., and Del Corral, J., 1984, Increase of the neuron number in some cerebral cortex areas of a lizard, *Podarcis hispanica* (Steind. 1870), during postnatal periods of life, *J. Hirn forsch.* **25:**255–260.

Lopez-Garcia, C., Martinez-Guijarro, F. J., Berbel, P., and Garcia-Verdugo, J. M., 1988, Long-spined polymorphic neurons of the medial cortex of lizards: A Golgi, Timm and electron-microscopic study, *J. Comp. Neurol.* **272:**409–423.

Martinez-Garcia, F., Amiguet, M., Olucha, F., and Lopez-Garcia, C., 1986, Connections of the lateral cortex in the lizard *Podarcis hispanica, Neurosci. Lett.* **63:**39–44.

Martinez-Guijarro, F. J., Berbel, P. J., Molowny, A., and Lopez-Garcia, C., 1984, Apical dendritic spines and axonic terminals in the bipyramidal neurons of the dorsomedial cortex of lizards (*Lacerta*), Anat. Embryol. **170:**321–326.

Martin-Perez, V., Garcia-Verdugo, J. M., and Llahi Sastre, S., 1981, Organization del septum die reptiles (*Podarcis pityusensis;* Lacertidae). Un estudio volumetrico y con Los metodos Die Golgi, *Trab. Inst. Cajal* **72:**215–225.

Mazurskaya, P. Z., 1973, Organization of neuronal receptive fields of the tortoise *Emys orbicularis* forebrain cortex, *J. Evol. Biochem. Physiol.* **8:**550–555.

Mazurskaya, P. Z., Davydova, T. V., and Smirnov, G. D., 1976, Functional organization of exteroceptive projections in the forebrain of the turtle, *Neurosci. Trans.* **1:**109–117.

Minelli, G., 1966, Architettura delle corteccie di alcuni rettili (*Lacerta muralis, Lacerta viridis, Testudo graecia, Crocodylus actuts*), Arch. Zool. Ital. **51:**543–573.

Moore, G. P., and Tschirigi, R. D., 1962, Non-specific responses of reptilian cortex to sensory stimuli, *Exp. Neurol.* **5:**196–209.

Mori, K., and Shepherd, G. M., 1979, Synaptic excitation and long-lasting inhibition of mitral cells in the *in vitro* turtle olfactory bulb, *Brain Res.* **172:** 155–159.

Mori, K., Nowycky, M. C., and Shepherd, G. M., 1981, Electrophysiological analysis of mitral cells in the isolated turtle olfactory bulb, *J. Physiol. (London)* **314:**281–294.

Morlock, H. C., 1972, Behavior following ablation of the dorsal cortex of turtles, *Brain Behav. Evol.* **5:**256–263.

Mufson, E. J., Desan, P. H., Mesulam, M. M., Wainer, B. H., and Levy, A. I., 1984, Choline acetyltransferase-like immunoreactivity in the forebrain of the red-eared pond turtle (*Pseudemys scripta elegans*), Brain Res. **323:**103–108.

Mulligan, K., and Ulinski, P. S., 1990, Organization of the geniculocortical projection in turtles: Isoatimuth lamellae in the visual cortex, *J. Comp. Neurol.* **296:** 531–547.

Naik, D. R., Sar, M., and Stumpf, W. E., 1981, Immunohistochemical localization of enkephalin in the central nervous system and pituitary of the lizard, *Anolis carolinensis, J. Comp. Neurol.* **198:**583–601.

Newman, D. B., and Cruce, W. L. R., 1982, The reptilian brainstem reticular formation: A comparative study using Nissl and Golgi techniques, *J. Morphol.* **173:**325–349.

Nieuwenhuys, R., 1982, An overview of the organization of the brain of actinopterygian fishes, *Am. Zool.* **22:**287–310.

Northcutt, R. G., 1967, Architectonic studies of the telencephalon of *Iguana iguana, J. Comp. Neurol.* **130:**109–148.

Northcutt, R. G., 1970, The telencephalon of the western painted turtle, *Chrysemys picta belli, Ill. Biol. Monogr.* No. 43.

Northcutt, R. G., 1978, Forebrain and midbrain organization in lizards and its evolutionary significance, in: *The Behavior and Neurology of Lizards* (N. Greenberg and P. D. MacLean, eds.), NIMH, Rockville, Md., pp. 11–64.

Northcutt, R. G., 1981, Evolution of the telencephalon in nonmammals, *Annu. Rev. Neurosci.* **4:**301–350.

Northcutt, R. G., 1986, Lungfish neural characters and their bearings on sarcopterygian phylogeny, *J. Morphol.* Suppl. **1:**277–298.

Northcutt, R. G., and Butler, A. B., 1974, Retinal projections in the northern water snake, *Natrix sipedon sipedon, J. Morphol.* **142:**117–136.

Northcutt, R. G., and Davis, R. E., 1983, Telencephalic organization in ray-finned fishes, in: *Fish Neurobiology,* Volume 2, (R. E. Davis and R. G. Northcutt, eds), University of Michigan Press, Ann Arbor, pp. 203–236.

Northcutt, R. G., and Kicliter, E., 1980, Organization of the amphibian telencephalon, in: *Comparative Neurology of the Telencephalon* (S. O. E., Ebbesson and H. Vanegas, eds.), Plenum Press, New York, pp. 203–256.

Olucha, F., Poch, L., Martinez-Garcia, F., Schwerdtfeger, W. K., and Lopez-Garcia, C., 1988, Projections from the medial cortex in the brain of lizards. 1. Correlation of anterograde and retrograde transport of horseradish peroxidase with Timm staining, *J. Comp. Neurol.* **276:**469–480.

Orrego, F., 1961, The reptilian forebrain. I. The olfactory pathways and cortical areas in the turtle, *Arch. Ital. Biol.* **99:**425–445.

Orrego, F., 1962, The reptilian forebrain. III. Cross connections between the olfactory bulbs and the cortical areas in the turtle, *Arch. Ital. Biol.* **100:**1–16.

Orrego, F., and Lissenby, D., 1962, The reptilian forebrain. IV. Electrical activity in the turtle cortex, *Arch. Ital. Biol.* **100:**17–30.

Ouimet, C. C., Patrick, R. L., and Ebner, F. F., 1981, An ultrastructural and biochemical analysis of norepinephrine-containing varicosities in the cerebral cortex of the turtle *Pseudemys, J. Comp. Neurol.* **195:**289–304.

Ouimet, D. C., Patrick, R. L., and Ebner, F. F., 1985, The projection of three extrathalamic cell groups to the cerebral cortex of the turtle *Pseudemys, J. Comp. Neurol.* **237:**77–84.

Papez, J. W., 1935, Thalamus of turtles and thalamic evolution, *J. Comp. Neurol.* **61:**433–475.

Parent, A., 1973. Distribution of monoamine-containing nerve terminals in the brain of the painted turtle, *Chrysemys picta, J. Comp. Neurol.* **148:**153–166.

Parent, A., 1979, Monoaminergic systems of the brain, in: *Biology of the Reptilia,* Volume 10 (C. Gans, R. G. Northcutt, and P. S. Ulinski, eds.), Academic Press, New York, pp 247–285.

Parent, A., 1986, *Comparative Neurobiology of the Basal Ganglia,* Wiley, New York.

Parent, A., and Poirer, L. J., 1971, Occurrence and distribution of monoamine-containing neurons in the brain of the painted turtle (*Chrysemys picta*), J. Anat. **110:**81–89.

Parent, A., and Poitras, D., 1974, The origin and distribution of catecholaminergic axon terminals in the cerebral cortex of the turtle (*Chrysemys picta*), Brain Res. **78:**345–358.

Parsons, T. S., 1970, The nose and Jacobson's organ, in: *Biology of the Reptilia,* Volume 2 (C. Gans and T. S. Parsons, eds.), Academic Press, New York, pp. 99–192.

Perez-Clausell, J., 1988, The organization of zinc-containing terminal fields in the brain of the lizard *Podarcis hispanica.* A histochemical study, *J. Comp. Neurol.* **267:**153–171.

Peters, S., Koh, J., and Choi, D. W., 1987, Zinc selectively blocks the action of N-methyl-D-aspartate on cortical neurons, *Science* **236:**589–593.

Peterson, E., 1980, Behavioral studies of telencephalic function in reptiles, in: *Comparative Neurology of the Telencephalon* (S. O. E. Ebbesson, ed.), Plenum Press, New York, pp. 343–388.

Pivavarov, A. S., and Trepakov, V. V., 1972, Intracellular analysis of unit responses to afferent stimulation in the general and hippocampal cortex of turtles, *Neurosci. Behav. Physiol.* **6:**144–150.

Platel, R., 1969, Etude cytoarchitectonique qualitative et quantitative des aires corticals d'un saurien *Scincus scinus* (L.) Scincides, *J. Hirnforsch.* **11:**31–66.

Platel, R., Beckers, H. J. A., and Nieuwenhuys, R., 1973, Les champs corticaux chez *Testudo hermanni* (reptile chélonien) et chez *Caiman crocodylus* (reptile crocodilien), *Acta Morphol. Neerl. Scand.* **11:**121–150.

Powell, T. P. S., and Kruger, L., 1960, The thalamic projection upon the telencephalon in *Lacerta viridis, J. Anat.* **94:**528–542.

Prasada Rao, P. D., and Subhedar, N., 1977, A cytoarchitectonic study of the hypothalamus of the lizard, *Calotes versicolor, Cell Tissue Res.* **180:**63–85.

Prasada Rao, P. D., Subhedar, N., and Davie Raju, P., 1981, Cytoarchitectonic patterns of the hypothalamus in the cobra, *Naja naja, Cell Tissue Res.* **217:**505–529.

Pritchard, P. C. H., 1967, *Living Turtles of the World,* T. F. H. Publications, Hong Kong.

Pritz, M. B., and Stritzel, M. E., 1987, Percentage of intrinsic and relay cells in a thalamic nucleus projecting to general cortex in reptiles, *Caiman crocodilus, Brain Res.* **409:**146–150.

Pyrethon, J., and Dusan-Pyrethon, D., 1969, Etude polygraphique du cycle vielle-sommeil chez trois genres de reptiles, *C.R. Soc. Biol.* **163:**181–186.

Quay, W. B., 1979, The parietal eye–pineal complex, in: *The Biology of the Reptilia,* Volume 9 (C. Gans, R. G. Northcutt, and P. S. Ulinski, eds.), Academic Press, New York, pp. 245–406.

Rainey, W. T., 1979, Organization of nucleus rotundus, a tectofugal thalamic nucleus in turtles. I. Nissl and Golgi analyses, *J. Morphol.* **160:**121–142.

Rainey, W. T., and Ulinski, P. S., 1982a, Organization of nucleus rotundus, a tectofugal thalamic nucleus in turtles. II. The tectorotundal projection, *J. Comp. Neurol.* **209:**187–207.

Rainey, W. T., and Ulinski, P. S., 1982b, Organization of nucleus rotundus, a tectofugal thalamic nucleus in turtles. III. Ultrastructural analyses, *J. Comp. Neurol.* **209:**208–223.

Rainey, W. T., and Ulinski, P. S., 1986, Morphology of neurons in the dorsal lateral geniculate complex in turtles of the genera *Pseudemys* and *Chrysemys, J. Comp. Neurol.* **253:**440–465.

Ramón, P., 1891, El encefalo de los reptiles, *Trab. Lab. Hist. Fac. Med. Zaragoza* pp. 1–30.

Ramón, 1896, Estructura del encefalo del camaleon, *Rev. Trim, Microgr.* **1:**146–182.

Reese, A. M., 1910, The development of the brain of the American alligator, *Smithson. Misc. Collect.* **54**(2)**:**1922.

Reiner, A., 1983, Comparative studies of opioid peptides: Enkephalin distribution in turtle central nervous system, *Neurosci. Abstr.* **9:**439.

Reiner, A., and Karten, H. J., 1985, Comparison of olfactory bulb projections in pigeons and turtles, *Brain Behav. Evol.* **27:**11–27.

Reiner, A., and Powers, A. S., 1983, The effects of lesions of telencephalic visual structures on visual discriminative performance in turtles (*Chrysemys picta picta*), J. Comp. Neurol. **218:**1–24.

Reiner, A., Brauth, S. E., and Karten, H. J., 1984, Evolution of the amniote basal ganglia, *Trends Neurosci.* **7:**320–325.

Reiner, A., Eldred, W. D., Beinfeld, M. C., and Krause, J. E., 1985, The co-occurrence of a substance P-like peptide and cholecystokinin-8 in a fiber system of turtle cortex, *J. Neurosci.* **5:**1527–1544.

Reperant, M. J., 1976, Afferences et efferences telencephaliques du cortex dorsal de la vipere (*Vipera aspis* L.) Donnees preliminaries, *C.R. Acad. Sci.* **283:**809–812.

Riss, W., Halpern, M., and Scalia, F., 1969, The quest for clues to forebrain evolution—The study of reptiles, *Brain Behav. Evol.* **2:**1–50.

Romer, A. S., 1956, *Osteology of the Reptiles,* University of Chicago Press, Chicago.

Rose, M., 1923, Histologische Lokalisation des Vorderhirns der Reptilien, *J. Psychol. Neurol.* **29:**219–272.

Scalia, F., Halpern, M., and Riss, W., 1969, Olfactory bulb projections in the South American caiman, *Brain Behav. Evol.***2:**238–262.

Schlegel, J. R., and Kriegstein, A. R., 1987, Quantitative autoradiography of muscarinic and benzodiazepine receptors in the forebrain of the turtle, *Pseudemys scripta, J. Comp. Neurol.* **265:**521–529.

Schmidt, K. P., and Inger, R. F., 1957, *Living Reptiles of the World,* Doubleday, New York.

Schwerdtfeger, W. K., and Lorente, M.-J., 1988a, GABA-immunoreactive neurons in the medial and dorsomedial cortices of the lizard. Some data on their structure, distribution and synaptic relations, in: *Recent Advances in Understanding the Structure and Function of the Forebrain in Reptiles* (W. K. Schwerdtfeger and W. J. Smeets, eds.), Karger, Basel, pp. 110–121.

Schwerdtfeger, W. K., and Lorente, M.-J., 1988b, Laminar distribution and morphology of gamma-aminobutryric acid (GABA)-immunoreactive neurons in the medial and dorsomedial areas of the cerebral cortex of the lizard *Podarcis hispanica, J. Comp. Neurol.* **278:**473–485.

Schwerdtfeger, W. K., and Smeets, W. J., (eds.), 1988, *Recent Advances in Understanding the Structure and Function of the Forebrain in Reptiles,* Karger, Basel.

Schwerdtfeger, W. K., Lopez-Garcia, C., Martinez-Guijarro, F. J., and Roberto, P. L. T., 1986, GABAergic neurons in the septum of the lizard, *Podarcis hispanica, Brain Res.* **384:**184–188.

Senn, D. G., 1979, Embryonic development of the central nervous system, in: *Biology of the Reptilia,* Volume 9 (C. Gans, R. G. Northcutt, and P. S. Ulinski, eds.), Academic Press, New York, pp. 173–244.

Senn, D. G., and Northcutt, R. G., 1973, The forebrain and midbrain of some squamates and their bearing on the origin of snakes, *J. Morphol.* **140:**135–152.

Shanklin, W. M., 1930, The central nervous system of *Chameleo vulgaris, Acta Zool. (Stockholm)* **11:**425–490.

Shen, J. M., and Kriegstein, A. R., 1987, Turtle hippocampal cortex contains distinct cell types, burst-firing neurons, and an epileptogenic subfield, *J. Neurophysiol.* **56:**1616–1649.

Shen, J. M., Huguenard, J. R., and Kriegstein, A. R., 1987, Development of cortical inhibition: Functional GABA A receptors at earliest stages of corticogenesis, *Neurosci. Abstr.* **13:**954.

Sjostrom, A. M., and Ulinski, P. S., 1985, Morphology of retinogeniculate terminals in the turtle, *Psuedemys scripta elegans, J. Comp. Neurol.* **238:**107–120.

Skeen, L. C., Pindzola, R. R., and Schofield, B. R., 1984, Tangential organization of olfactory, association, and commissural projections to olfactory cortex in a species of reptile (*Trionyx spiniferus*), bird (*Aix sponsa*), and mammal (*Tupaia glis*), Brain Behav. Evol. **25:**206–216.

Sligar, C. M., and Voneida, T. J., 1981, Efferent projections of the septum in the tegu lizard, *Tupinambis nigropunctatus, Anat. Rec.* **201:**197–202.

Smeets, W. J. A. J., 1988, The distribution of dopamine immunoreactivity in the forebrain and midbrain of the snake *Python regius:* A study with antibodies against dopamine, *J. Comp. Neurol.* **271:**115–129.

Smeets, W. J. A. J., and Steinbusch, H. W. M., 1988, Distribution of serotonin immunoreactivity in the forebrain and midbrain of the lizard *Gekko gecko, J. Comp. Neurol.* **271:**419–434.

Smeets, W. J. A. J., Nieuwenhuys, R., and Roberts, B. L., 1983, *The Central Nervous System of Cartilaginous Fishes,* Springer, Berlin.

Smeets, W. J. A. J., Hoogland, P. V., and Lohman, A. H. M., 1986, A forebrain atlas of the lizard *Gekko gekko, J. Comp. Neurol.* **254:**1–19.

Smeets, W. J. A. J., Jonker, A. J., and Hoogland, P. V., 1987, Distribution of dopamine in the forebrain and midbrain of the red-eared turtle, *Pseudemys scripta elegans,* reinvestigated using antibodies against dopamine, *Brain Behav. Evol.* **30:**121–142.

Smith, L. M., Ebner, F. F., and Colonnier, M., 1980, The thalamocortical projection in *Pseudemys* turtles: A quantitative electron microscope study, *J. Comp. Neurol.* **190:**445–462.

Sotelo, C., Llinas, R., and Baker, R., 1974, Structural study of inferior olivary nucleus of the cat: Morphological correlates of electronic coupling, *J. Neurophysiol.* **37:**541–559.

Starck, D., 1979, Cranio-cerebral relations in recent reptiles, in: *Biology of the Reptilia,* Volume 9 (C. Gans, R. G. Northcutt, and P. S. Ulinski, eds.), Academic Press, New York, pp. 1–38.

Stensaas, L. J., and Stensaas, S. S., 1968, Light microscopy of glial cells in turtles and birds, *Z. Zellforsch.* **91:**315–340.

Subhedar, N., and Rama Krishna, N. S., 1984, A Golgi-type study of the hypothalamus of the lizard, *Calotes versicolor, Cell Tissue Res.* **236:**399–411.

Tandler, J., and Kantor, H., 1907, Die Entwicklungsgeschicte des Geckogehirnes, *Anat. Hefte* **33:**553–665.

ten Donkelaar, H. J., and Nieuwenhuys, R., 1979, The brainstem, in: *Biology of the Reptilia,* Volume 10 (C. Gans, R. G. Northcutt, and P. S. Ulinski, eds.), Academic Press, New York, pp. 133–200.

ten Donkelaar, H. J., Bangma, G. C., Barbas-Henry, H. A., DeBoer-van Huizen, R., and Wolters, J. G., 1987, The brainstem in a lizard, *Varanus exanthematicus, Adv. Anat. Embryol. Cell Biol.* **107:**1–168.

Ulinski, P. S., 1974, Cytoarchitecture of cerebral cortex in snakes, *J. Comp. Neurol.* **158:**243–266.

Ulinski, P. S., 1975, Corticoseptal projections in the snakes *Natrix sipedon* and *Thamnophis sirtalis, J. Comp. Neurol.* **164:**375–388.

Ulinski, P. S., 1976, Intracortical connections in the snakes *Natrix sipedon* and *Thamnophis sirtalis, J. Morphol.* **150:**463–484.

Ulinski, P. S., 1977a, Intrinsic organization of snake medial cortex: An electron microscopic and Golgi study, *J. Morphol.* **152:**247–280.

Ulinski, P. S., 1977b, Tectal efferents in the banded water snake, *Natrix sipedon, J. Comp. Neurol.* **173:**251–274.

Ulinski, P. S., 1979, Intrinsic organization of snake dorsomedial cortex: An electronmicroscopic and Golgi study, *J. Morphol.* **161:**185–210.

Ulinski, P. S., 1981, Thick caliber projections from brainstem to cerebral cortex in the snakes *Thamnophis sirtalis* and *Natrix sipedon, Neuroscience* **6:**1725–1743.

Ulinski, P. S., 1983, *Dorsal Ventricular Ridge: A Treatise on Forebrain Organization in Reptiles and Birds,* Wiley–Interscience, New York.

Ulinski, P. S., 1986a, Neurobiology of the therapsid–mammal transition, in: *The Ecology and Biology of Mammal-Like Reptiles* (J. J. Roth, E. C. Roth, P. D. MacLean, and N. Hotton III, eds.), Smithsonian Institution, Washington, D.C., pp. 149–172.

Ulinski, P. S., 1986b, Organization of corticogeniculate projections in the turtle, *Pseudemys scripta, J. Comp. Neurol.* **254:**529–542.

Ulinski, P. S., 1987a, Ultrastructure of the dorsal lateral geniculate complex in turtles of the genera *Pseudemys* and *Chrysemys, Brain Behav. Evol.* **29:**117–142.

Ulinski, P. S., 1987b, Laminar organization of efferents from turtle dorsal lateral geniculate complex, *Am. Zool.* **27:**166A.

Ulinski, P. S., 1987c, Distribution of GABA accumulating neurons in the diencephalon of the turtle, *Pseudemys scripta, Anat. Rec.* **218:**141A.

Ulinski, P. S., 1988, Functional organization of turtle visual cortex, in: *Recent Advances in Understanding the Structure and Function of the Forebrain in Reptiles* (W. K. Schwerdtfeger and W. J. Smeets, eds.), Karger, Basel, pp. 151–161.

Ulinski, P. S., and Kanarek, D. A., 1973, Cytoarchitecture of nucleus sphericus in the common boa, *Constrictor constrictor, J. Comp. Neurol.* **151:**159–174.

Ulinski, P. S., and Mulligan, K. A., 1987, Representation of visual space in the visual cortex of turtles, *Neurosci. Abstr.* **13:**1048.

Ulinski, P. S., and Nautiyal, J., 1988, Organization of retinogeniculate projections in turtles of the genera *Pseudemys* and *Chrysemys, J. Comp. Neurol.* **276:**92–112.

Ulinski, P. S., and Peterson, E. H., 1981, Patterns of olfactory projections in the desert iguana, *Dipsosaurus dorsalis, J. Morph.* **168:**189–228.

Ulinski, P. S., and Rainey, W. T., 1980, Intrinsic organization of snake lateral cortex, *J. Morphol.* **165:**85–116.

Ulinski, P. S., du Lac, S., and Dacey, D. M., 1983, Descending projections of the geniculate complex, *Invest. Ophthalmol. Vis. Sci.* **24:**64.

Umeda, S., Takeuchi, Y., and Sano, Y., 1983, Immuno-histochemical demonstration of serotonin neurons in the central nervous system of the turtle (*Clemmys japonica*), Anat. Embryol. **168:**1–19.

Unger, L., 1906, Untersuchungen über die Morphologie und Faserung des Reptiliengehirns. I. Das Vorderhirn des Geckos, *Anat. Hefte* **31:**273–348.

Vasilescu, E., 1970, Sleep and wakefulness in the tortoise (*Emys orbicularis*), Rev. Roum. Biol. Ser. Zool. **15:**177–179.

Voneida, T. J., and Ebbesson, S. O. E., 1969, On the origin and distribution of axons in the pallial commissures in the tegu lizard (*Tupinambis nigropunctatus*), Brain Behav. Evol. **2:**467–481.

Walker, J. M., and Berger, R. J., 1973, A polygraphic study of the tortoise (*Testudo denticulata*). Absence of electrophysiological signs of sleep, *Brain Behav. Evol.* **8:**453–467.

Ware, C. B., 1974, Projections of dorsal cortex in the side necked turtle (*Podocnemis unifilis*, Soc. Neurosci., 4th Annu. Meet., p. 466.

Warner, F. J., 1931, The cell masses in the telencephalon and diencephalon of the rattlesnake, *Crotalus atrox, Proc. Acad. Sci. (Amsterdam)* **34:**1156–1163.

Warner, F. J., 1946, The development of the forebrain of the American watersnake (*Natrix sipedon*), J. Comp. Neurol. **84:**385–418.

Weiss, J. C., and Ulinski, P. S., 1985, Synaptic organization of dorsal area of turtle, *Pseudemys scripta elegans, J. Morphol.* **184:**135–154.

Westbrook, G. L., and Mayer, M. L., 1987, Micromolar concentrations of Zn^{2+} antagonize NMDA and GABA responses of hippocampal neurons, *Nature* **328:**640–643.

Wolters, J. G., ten Donkelaar, H. J., and Verhofstad, A. A. J., 1984, Distribution of catecholamines in the brain stem and spinal cord of the lizard *Varanus exanthematicus:* An immunohistochemical study based on the use of antibodies to tyrosine hydroxylase, *Neuroscience* **13:**469–497.

Wolters, J. G., ten Donkelaar, H. J., Steinbusch, H. W. M., and Verhofstan, A. A. J., 1985, Distribution of serotonin in the brain stem and spinal cord of the lizard *Varanus exanthematicus:* An immunohistochemical study, *Neuroscience* **14:**169–193.

Wouterlood, F. G., Alones, V. E., and Lohman, A. H. M., 1981, An electron microscopic study of the mediodorsal cerebral cortex in the lizard *Agama agama, Anat. Embryol.* **163:**185–200.

Wouterlood, F. G., Alones, V. E., Elprana, D., and Lohman, A. H. M., 1982, Terminal degeneration in the mediodorsal cerebral cortex of the lizard *Agama agama:* Light and electron microscopy, *J. Morphol.* **172:**45–58.

Yamamoto, K. M., and Shimizu, N., 1977, Comparative anatomy of the topography of catecholamine containing neuron system in the brain stem from birds to teleosts, *J. Hirnforsch.* **18:**229–240.

Zangerl, R., 1969, The turtle shell, in: *Biology of the Reptilia,* Volume 1 (C. Gans, A. d'A. Bellairs, and T. S. Parson, eds.), Academic Press, New York, pp. 311–340.

6

Neurobiology of the Reptile–Bird Transition

PHILIP S. ULINSKI
and DANIEL MARGOLIASH

1. Introduction

Birds evolved from archosaurian reptiles during the Mesozoic era, between 230 and 65 million years ago. The archosaurs, or "ruling" reptiles, dominated the terrestrial fauna throughout the Mesozoic. They include several evolutionary lineages of which the dinosaurs, the pterosaurs or flying reptiles, and the various lineages of crocodilians are the most important. Which particular group of archosaurs gave rise to birds has been a matter of controversy. Heilmann (1926) suggested that birds evolved from the protosuchians, a group of archosaurs related to crocodilians. Ostrom (1976) reexamined the evidence related to the origin of birds and suggested they evolved from small, bipedal dinosaurs. This view now is widely accepted by paleontologists (see Feduccia, 1980; Hecht *et al.*, 1985). The conclusion that birds are derived from some group of archosaurian reptiles comes from detailed comparisons of the skeletal anatomy of the earliest fossil bird, *Archaeopteryx lithographica,* with members of the various groups of archosaurs. *Archaeopteryx* was a pigeon-sized bird that lived in a barren or desert environment, as judged by the presence of cacti in the sediments near which some *Archaeopteryx* specimens were found (Viohl, 1985, 1990). *Archaeopteryx* is found in the upper Jurassic (the second of the three periods in the Mesozoic era), in sediments about 130 million years old.

PHILIP S. ULINSKI • Department of Organismal Biology and Anatomy, and Committee on Neurobiology, University of Chicago, Chicago, Illinois 60637. DANIEL MARGOLIASH • Department of Organismal Biology and Anatomy, and Committees on Neurobiology and Biopsychology, University of Chicago, Chicago, Illinois 60637.

Analysis of the six specimens (Wellnhofer, 1974, 1988, 1990) of *Archaeopteryx* suggests it had some morphological adaptations associated with flight (e.g., Ostrom, 1974; Padian, 1986). The structure of its hindlimb shows it was a bipedal animal with a reversed hallux that allowed it to grasp branches. The forelimbs served as wings, and some specimens show imprints of feathers that are essentially identical to the primary feathers of Recent birds. The shafts of the feathers are displaced toward one edge, suggesting the feathers could withstand the distribution of stresses created during powered flight (Feduccia and Tordoff, 1979). Some bony elements of the shoulder girdle were modified from the typical reptilian pattern to form a furcula or "wishbone." It was heavier than the furcula of Recent birds and could have played a role in stabilizing the shoulder joint, but not as an elastic element that assists the respiratory system during flight as it may in Recent birds (Jenkins *et al.*, 1988). The ventral surface of the pectoral girdle probably served as the origin for the pectoralis muscles that depress the wing during flight (Dial *et al.*, 1988; Goslow *et al.*, 1989), but the dorsal aspect of the girdle indicates *Archaeopteryx* lacked the supracoracoideus muscle that functions as an elevator of the wing in Recent birds. The presence of feathers suggests *Archaeopteryx* was homeothermic (Regal, 1975) with the energetic capability for sustained flight.

Although the fossil record of birds later in the Mesozoic era is limited, it is sufficient to suggest that the evolution of birds continued through the remainder of the era. A new fossil bird from the lower Cretaceous of Las Hoyas, Spain, retained some of the primitive features (such as unfused bones in its feet) found in *Archaeopteryx,* but had a strutlike coracoid bone in its shoulder similar to that seen in Recent birds (Sanz *et al.*, 1988). *Archaeopteryx* had a long tail containing many vertebral elements, but the Las Hoyas bird shows a reduction of caudal vertebrae to form a pygostyle, also like that seen in Recent birds. Most of the fossil birds from the late Cretaceous are diving birds (e.g., Martin, 1983) that were secondarily flightless (Cracraft, 1986, 1988). They are very specialized forms, comparable to Recent loons, that must represent only that fraction of the total Cretaceous avifauna that happened to fossilize. Representatives of most of the modern orders of birds are present in the fossil record by the beginning of the Cenozoic era, about 65 million years ago. Owls date from the Eocene period (Walker, 1973), and members of the order Passeriformes, the large order that includes the songbirds, are found in the Miocene period, about 30 million years ago (Simpson, 1980; Kavanau, 1987). In very general terms, it appears that birds were present by the upper Jurasic, underwent some morphological modifications during the lower Cretaceous, and gave rise to a major adaptive radiation that extended throughout the upper Cretaceous period and early Cenozoic era.

The fossil record gives a general idea of the morphological steps involved in the evolution of flight and the adaptive radiation of birds, but does not indicate if *Archaeopteryx* had anything like the social organization and communication systems that are important features of avian biology. There are, however, hints in the fossil record that the evolution of such behaviors is a major theme in the archosaurs. Crocodilians show some simple forms of parental care such as defense of their nests and transport of their young (Kushlan, 1973; Hunt, 1975; Pooley and Gans, 1976). Analyses of fossil dinosaur footprints suggest that some species of dinosaurs traveled in herds structured with adults on the outside of the group and young individuals on the inside (Bakker, 1968). The presence of horns and other head ornaments in a variety of dinosaur species suggests that they participated in agonistic behaviors during courtship and the establishment of dominance hierarchies (Hopson, 1975; Alexander, 1989). However, there is no indication that any of the groups of living or extinct archosaurs had the capabilities for the complex behaviors widespread in birds, including some that can be reasonably categorized as cognitive (Pepperberg, 1983, 1987, 1988).

The changes in morphology, physiology, and behavior that occurred during the reptile–bird transition must have involved changes in virtually all components of the nervous system. The goal of this essay is to provide a brief overview of what is known about the forebrains of birds by considering the nervous system in relation to the biology of the reptile–bird transition. One of the specimens of *Archaeopteryx* provides a glimpse of a natural endocast of the brain and reveals that its external morphology was intermediate between that of reptiles and Recent birds (Edinger, 1926; Dechaseaux, 1968; Jerison, 1968; Wellnhofer, 1974), but there is otherwise no clear evidence relevant to the neural changes involved in the reptile–bird transition in the fossil record. However, the methods of phylogenetic systematics or cladistics (e.g., Eldredge and Cracraft, 1980; Wiley, 1981) provide an indirect line of analysis based on the distribution of anatomical and behavioral features among the various lineages, or clades, of living reptiles and birds. Features shared by both Recent birds and several clades of reptiles can reasonably be held to be primitive characters, while those shared among various lineages of birds but not consistently seen in reptiles can be viewed as shared–derived characters acquired during the reptile–bird transition. Features of neural evolution within the birds themselves can be analyzed by considering the distribution of particular characters within the various clades of birds, and characters widely distributed in Recent reptiles but absent or reduced in birds can be presumed to have been lost or reduced during the reptile–bird transition. This latter point is particularly relevant to this volume in that the cerebral cortex of birds is generally not as well developed as that of reptiles, and there appears to have been some modification of the cortex during the evolution of birds.

The essay begins by summarizing the basic organization of the forebrains of birds, placing emphasis on features common to reptiles and birds that presumably reflect the archosaurian ancestry of the birds. Second, it outlines the embryological processes that set apart the organization of the brains of birds from those of reptiles. Third, it deals with the features of brain organization that appear characteristic of birds generally and are probably shared–derived characters associated with the evolution of birds. The fourth section considers the modification of the cerebral cortex. The final section deals with two sets of neural characters associated with particular groups of birds, rather than with birds generally. The analysis indicates that the brains of birds evolved in three phases. The first consisted of changes that were occurring widely within the archosaurs. The second involved the evolution of a characteristic avian pattern of forebrain organization through a modification of developmental events. The third involved modifications of the avian pattern within different clades of birds demonstrating specific behavioral specializations.

No attempt is made to provide a complete account of the neurobiology of birds. Several recent reviews (e.g., Pearson, 1972; Goodman and Schein, 1974; Cohen and Karten, 1974; Granda and Maxwell, 1979; Benowitz, 1980; Nottebohm, 1980; Ulinski, 1983; Konishi, 1985; Parent, 1986) are available and should be consulted for detailed treatments of specific topics. Bakker (1986) and Alexander (1989) provide up-to-date and readable accounts of archosaur evolution, including the origin of birds.

2. Reptile–Bird Pattern of Forebrain Organization

The telencephalon of reptiles and birds consists of a pair of cerebral hemispheres formed by a bilateral evagination of the rostral end of the neural tube (Fig. 1). Each hemisphere contains a lateral ventricle that communicates with its opposite member through an interventricular foramen. The walls of each hemisphere can be divided into three components. The septum forms the ventromedial wall of the hemisphere. The striatum forms the ventrolateral wall of the hemisphere and is traversed by major fiber systems that interconnect the brain stem and telencephalon. The largest of these is the fasciculus prosencephali lateralis, or lateral forebrain bundle, which includes fibers coursing from the dorsal thalamus to the telencephalon. The roof of each hemisphere is formed by the pallium, which contains two distinct parts in both reptiles and birds. The medial part is a cortex, consisting of several alternating layers of cells and fiber systems. The lateral part is the dorsal ventricular ridge (DVR), which protrudes into the lateral ventricle from the ventrolateral wall of the hemi-

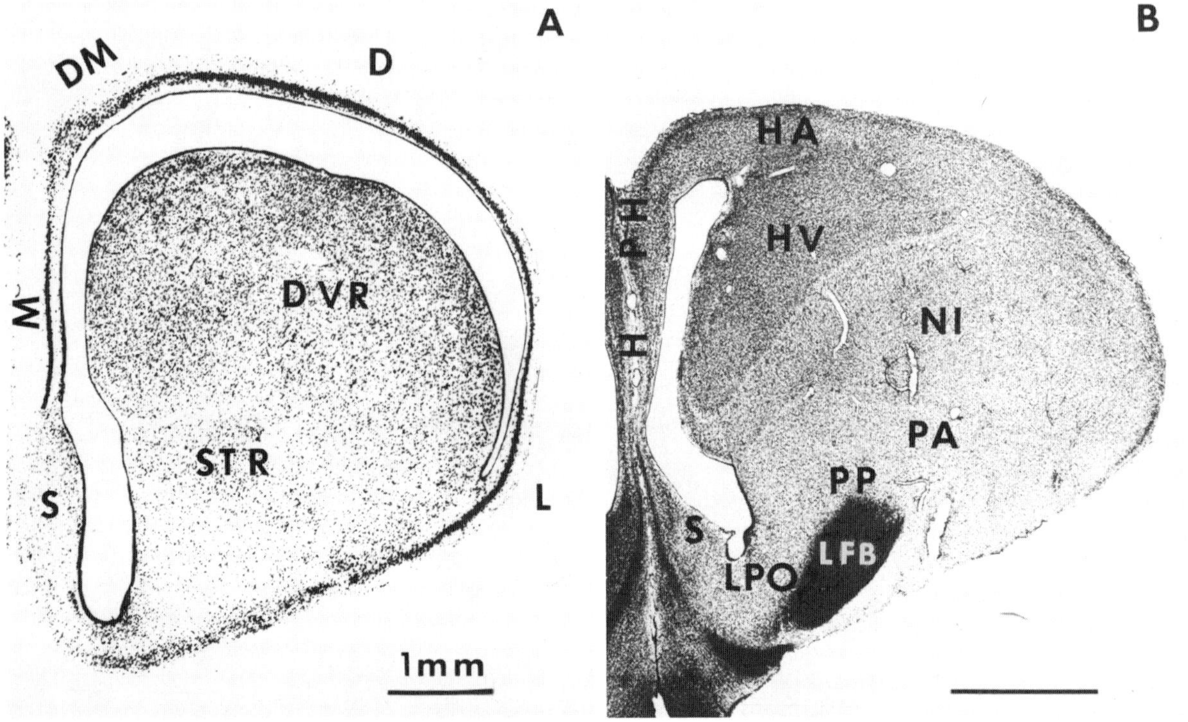

Figure 1. Cross sections of cerebral hemispheres of (A) an alligator and (B) a pigeon. Transverse sections through the cerebral hemispheres of an archosaurian reptile and a bird are shown to illustrate the general organization of the telencephalons in reptiles and birds. The hemisphere consists of a septum (S), striatum (STR), and pallium in both species. The pallium consists of a cortex and the dorsal ventricular ridge (DVR) in *Alligator.* In pigeons, the dorsal ventricular ridge is expanded and contains a number of subdivisions, while the cortex is reduced in size. D, dorsal cortex; DM, dorsomedial cortex; H, hippocampal area of cortex; HA, hyperstriatum accessorium; HV, hyperstriatum ventral; L, lateral cortex; LFB, lateral forebrain bundle; LPO, parolfactory lobe; M, medial cortex; NI, intermediate neostriatum; PA, paleostriatum augmentatum; PP, paleostriatum primitivum. The two hemispheres are shown at different magnifications, so that the scale bar represents 1 mm in both cases.

sphere. It contains neurons with oval or round somata that tend to form clusters of cells with apposed somata (e.g., Stingelin, 1958; Rehkamper et al., 1984, 1985). These neurons have stellate dendritic fields formed by several dendrites (Saini and Leppelsack, 1977, 1981; Watanabe *et al.,* 1983; Tombol *et al.,* 1988a,c; Nixdorf *et al.,* 1989). Many cells have very spiny dendrites. Some have sparsely spiny or smooth dendrites and stain for GABA-related markers (Csillag *et al.,* 1989).

The nomenclature applied to nuclear groups within the cerebral hemispheres of birds is confusing because it is based upon a fundamental misunderstanding of the boundaries of the pallium and striatum. Ariëns Kappers (e.g., Ariëns Kappers, 1921; Ariëns Kappers *et al.,* 1936) surveyed the morphology of the telencephalon in a variety of vertebrates and proposed a linear model of telecephalic evolution (see Northcutt, 1981) in which animals "low on the phylogenetic tree" have relatively simple, and old, components. Components were added to the forebrain as evolution proceeded "up" the scale. This was consistent with the classic notion that much of behavior is mediated by the roof of the midbrain in "primitive" vertebrates and that the control of behavior is transferred to the telencephalon in more "advanced" vertebrates through a process called encephalization. Ariëns Kappers suggested the oldest, and simplest, vertebrates had only a single pallial component and a single striatal component, which he called the archicortex and archistriatum, respectively. During the course of evolution, a paleocortex and neocortex were added sequentially to the pallium and a paleostriatum and neostriatum were added sequentially to the striatum.

DVR was the source of confusion because it protrudes into the lateral ventricle and bears a superficial resemblance to the caudate and putamen (viewed as components of the neostriatum) of mammals. Ariëns Kappers misidentified the ventrolateral wall of the hemisphere in reptiles and birds as the paleostriatum and considered DVR of reptiles to represent an incipient neostriatum. DVR is distinctly larger in birds than in reptiles, and Ariëns Kappers postulated an independent hypertrophy of the striatum in birds. He used the term *hyperstriatum* to designate this additional tissue in birds. Consequently, much of the forebrain of birds is divided into nuclear groups, often with indistinct boundaries, that are designated as components of the striatum (Fig. 2). Hyperstriatum is traditionally subdivided into several components [e.g., hyperstriatum ventrale (HV)]. Neostriatum is divided into caudal (NC), intermediate (NI), and frontal parts (NF). The paleostriatum is subdivided into a paleostriatum primitivum (i.e., derived earliest in evolution) which was subsequently completed by the paleostriatum augmentatum (PA). A few additional parts, such as nucleus basalis (BAS) and the ectostriatal (E) complex, are also recognized. Maxmillian Rose (1914) recognized the difficulties involved in drawing precise boundaries within the telencephalons of birds and defined, instead, a series of "fields" designated by the letters A through S. Recent interest in the organization of the brains of songbirds has led to the introduction of several idiosyncratic terms (e.g., "nucleus X," NIf, Uva) into the literature. The nomenclature in current use is an amalgamation of the Ariëns Kappers and Rose systems, plus idiosyncratic terms in some species. Thus, investigators will speak of field L of Rose (a component of the neostriatum) as projecting to HVc (probably a component of the neostriatum), which projects in turn to nucleus X. Table 1 in the atlas of the pigeon brain prepared by Karten and Hodos (1967) provides a comparison of the Ariëns Kappers and Rose nomenclatures.

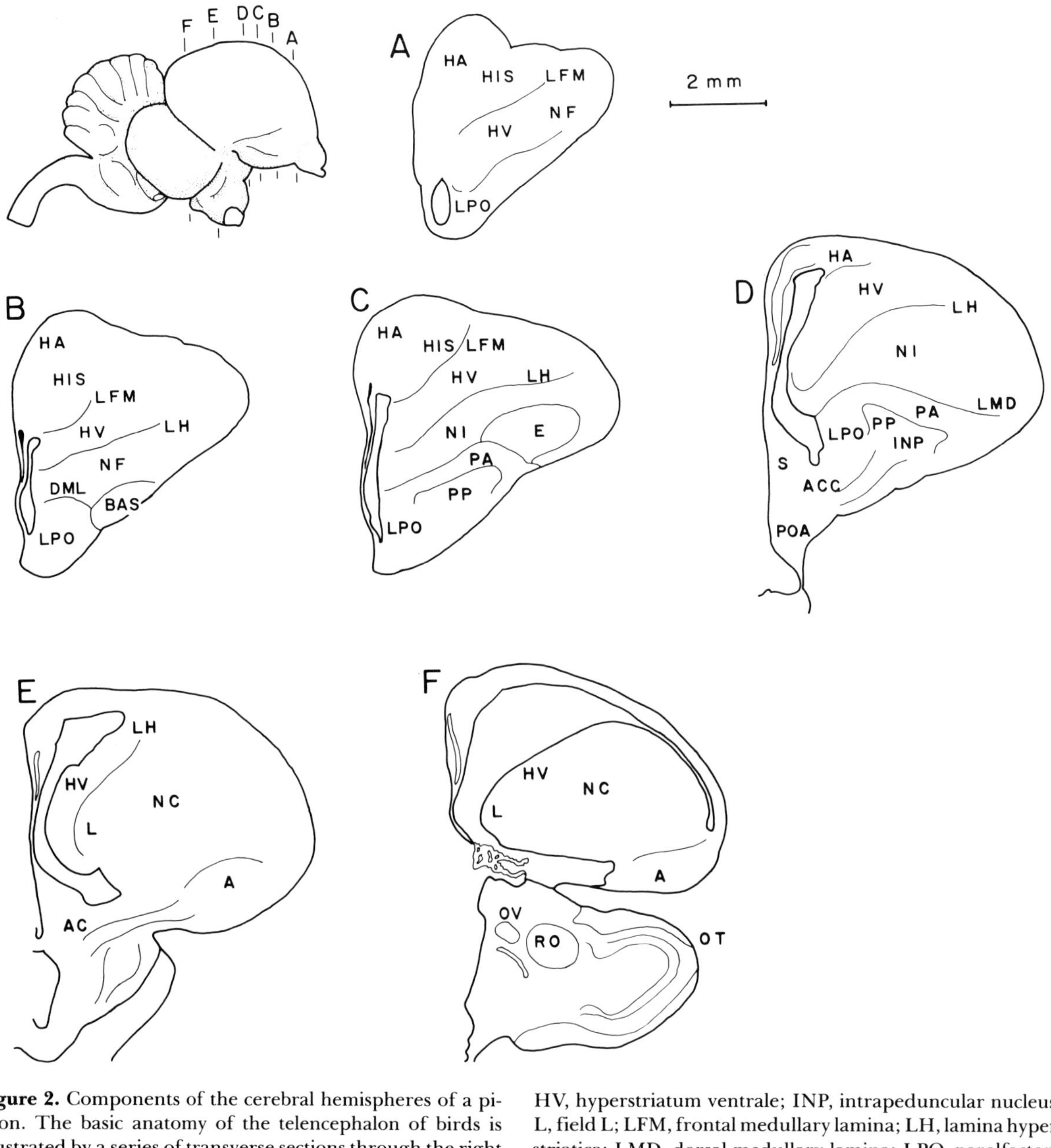

Figure 2. Components of the cerebral hemispheres of a pigeon. The basic anatomy of the telencephalon of birds is illustrated by a series of transverse sections through the right cerebral hemisphere. The planes of the sections are illustrated on the inset (upper left), which shows a lateral view of a pigeon brain. A, archistriatum; AC, anterior commissure; ACC, nucleus accumbens; BAS, nucleus basalis; DML, dorsomedial lateral nuclei; E, ectostriatum; HA, hyperstriatum accessorium; HIS, hyperstriatum intercalatus superioris; HV, hyperstriatum ventrale; INP, intrapeduncular nucleus; L, field L; LFM, frontal medullary lamina; LH, lamina hyperstriatica; LMD, dorsal medullary lamina; LPO, parolfactory lobe; NC, caudal neostriatum; NF, frontal neostriatum; NI, intermediate neostriatum; OT, optic tectum; OV, nucleus ovoidalis; PA, paleostriatum augmentatum; POA, preoptic area of hypothalamus; PP, paleostriatum primitivum; RO, nucleus rotundus; S, septum. Reproduced from Ulinski (1983).

Evidence based on the pattern of thalamotelencephalic projections, embryology, and the distribution of transmitter-related markers in the cerebral hemispheres of reptiles and birds (see Ulinski, 1983; Parent, 1986) has now made it certain that Ariëns Kappers and many subsequent workers misidentified the boundary between the striatum and pallium in reptiles and birds. All modern workers agree that DVR is a component of the pallium and that the striatum of both reptiles and birds is relatively smaller than that of mammals. Thus, some of the divisions of the striatum in the Ariëns Kappers nomenclature (e.g., hyperstriatum, neostriatum, ectostriatum) are actually parts of the pallium, while others (paleostriatum, archistriatum) are parts of the striatum.

This recent clarification of the composition of the telencephalon is of considerable value because it has led to an overall concept of the basic pattern, and its variations, present in the brains of vertebrates. Although our knowledge of forebrain organization in some groups of fishes is still fragmentary, it appears that septal, striatal, and pallial components of the forebrain can be identified in all vertebrates. In all cases in which modern experimental studies have been conducted, the pallium has been found to receive inputs from the olfactory bulb complex on its lateral surface and from auditory, visual, and somatosensory nuclei in the dorsal thalamus in its more dorsal components. Each major group of vertebrates has a distinct configuration of the three telencephalic components, the key feature of the forebrain of reptiles and birds being the division of the pallium into cortical and DVR subcomponents. The cortex is relatively large in reptiles, but is a smaller constituent of the forebrain of birds that occupies principally the caudal and medial surfaces of the hemispheres. Although all reptiles and birds have a DVR, there is a considerable variation in the morphology of this structure between various taxonomic groups. There is a general resemblance between the organization of DVR in crocodilians (see Clark and Ulinski, 1984) and birds that presumably reflects their common archosaurian ancestries. However, DVR in birds has a distinct configuration that can be understood in terms of the embryology of the telencephalon.

3. Embryological Transformations in Avian Brains

Several important features of the brains of birds result from transformations in the timing and spatial pattern of neuron generation during development.

3.1. Avian Pattern of DVR Organization

The early embryology of the telencephalon is virtually identical in reptiles and birds (Fig. 3). The cerebral hemispheres form from the rostral neural tube by a process of evagination, and there is a gradual thickening of the telencephalic walls as neuroblasts migrate radially from the neuroupithelium lining the ventricular surface of the hemispheres. In reptiles, the timing of neurogenesis and migration in the pallium results in a relative thickening of the lateral pallial wall and an eventual rotation of DVR into the ventricle. DVR protrudes vertically into the lateral ventricle in all reptiles, separated ventrally from the striatum by a cell-poor zone. There is an overall expansion of the cortical component of the pallium that occurs in concert with the rotation of DVR, so cortex comes to span the dorsal hemispheric surface from the septum to the ventrolateral surface of the hemisphere. By contrast, the migration of neuroblasts occurs more evenly in the striatum and pallium of embryonic birds. DVR consequently becomes very thick, and can actually occlude parts of the lateral ventricle, but never undergoes the rotation seen in reptiles. The lateral ventricle is located primarily along the medial surface of DVR. Although a cortical component of the pallium does form, it does not undergo an extensive lateral expansion and is confined principally to the medial surface of the hemisphere. Sections through the hemisphere show a "corticoid" formation consisting of a layer of densely packed somata situated on the lateral surface of the hemisphere. However, this is not a true cortex in the sense of a laminated structure situated over a ventricle.

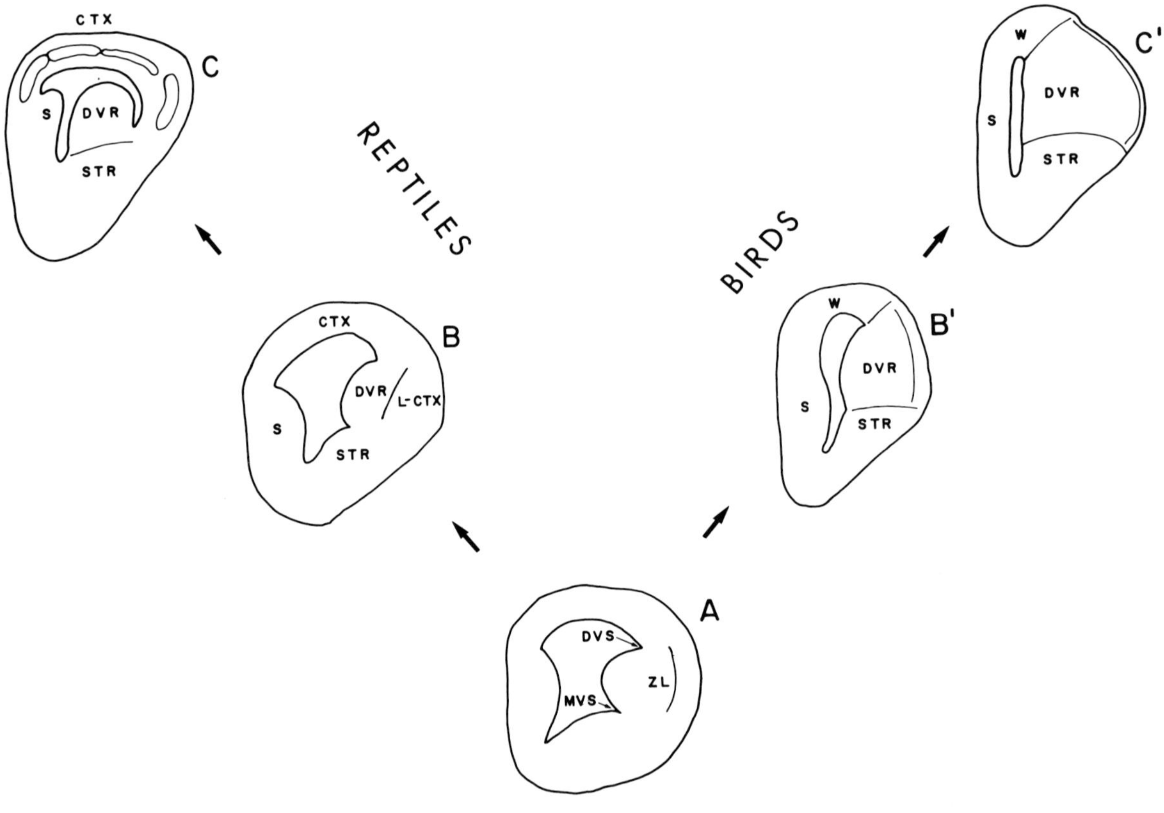

Figure 3. Embryology of telencephalon. This figure illustrates the different developmental pathways that occur in the cerebral hemispheres of reptiles and birds. The early embryonic configuration (A) of the hemisphere is similar in reptiles and birds. The dorsal ventricular ridge (DVR) is marked dorsally and ventrally, respectively, by the dorsal ventricular sulcus (DVS) and the middle ventricular sulcus (MVS). The lateral edge of DVR is marked by a cell-poor area known as the zona limitans (ZL). In reptiles, DVR undergoes a rotation that gives it a vertical orientation in adult reptiles (C). In birds, this rotation does not take place and DVR expands laterally so that adult birds (C′) retain the embryonic configuration of the ventricle. CTX, cortex; L-CTX, lateral cortex; S, septum; STR, striatum; W, Wulst. Reproduced from Ulinski (1983).

3.2. Embryology of the Wulst

A secondary consequence of the retention into adulthood of the embryonic configuration of DVR is the formation of the Wulst. This is a German descriptive term meaning a "bump" that is applied to a protuberance or swelling present on the dorsal surface of the forebrain of many birds. There is considerable variation in the size and position of the Wulst. It is a relatively gently swelling in birds such as pigeons, but is a very prominent protuberance in other birds. In owls, the Wulst is situated rostrally in the hemisphere. In parrots, it is situated caudally. The morphology of the Wulst has been surveyed by Stingelin (1958) and reviewed by Pearson (1972). The Wulst is distinctly laminated in many species, and its dorsal position in the hemisphere might lead the casual observer to conclude that it represents a cortical component of the pallium. However, it is actually a region of DVR that incorporates several subdivisions of the hyperstriatum, such as the hyperstriatum accessorium. It is visible on the dorsal surface of the brain, in a position occupied by cortex in reptiles, because DVR does not change its position during development and because the cortical component of the pallium does not undergo a lateral expansion.

Tritiated thymidine studies in chickens (Tsai *et al.*, 1981a,b) demonstrate the embryonic transformation involved in the formation of the Wulst. Neurons eventually situated in the adult Wulst are initially located in the lateral part of DVR external to the ectostriatum. They undergo a progressive dorsomedial shift during the later stages of embryogenesis and gradually gain their adult position. The spatial shift reflects a reorganization of telencephalic visual areas. A principal source of afferents to the Wulst is the dorsal thalamic nucleus that receives direct retinal projections. It is one of two thalamic nuclei that convey visual information to the telencephalon in both reptiles and birds. The second nucleus is called nucleus rotundus, and receives visual information indirectly through the optic tectum. In reptiles, the retinorecipient and tectorecipient thalamic nuclei project to two adjacent targets in the telencephalon. Nucleus rotundus projects to a region in the lateral part of DVR; the retinorecipient nucleus projects to the pallial thickening," which is situated immediately external to the first region. In birds, precisely the same relationships exist *in embryo*. Neurons destined to lie in the ectostriatum and the hyperstriatum accessorium lie adjacent to each other in the lateral part of DVR. The ectostriatum will receive projections from nucleus rotundus; the hyperstriatum accessorium will receive projections from the principal optic complex. It is only later in embryogenesis that the hyperstriatum accessorium and adjacent structures shift in position and come to be positioned on the dorsal surface of the hemisphere. Aspects of the neural organization of the Wulst are considered in Section 4.3.

3.3. Development of Cerebellum and Precerebellar Nuclei

The cerebellum develops from the rhombic lip in reptiles and birds. This is a subventricular zone that lines the margins of the fourth ventricle. Germinative cells in the rhombic lip continue to divide for a relatively long time into embryogenesis, and the neuroblasts produced form the cerebellar plate, which is situated above the fourth ventricle. In addition, some neuroblasts from the rhombic lip migrate ventrally over the surface of the brain stem, proceeding either rostrally or caudally, and form precerebellar nuclei such as the inferior olivary and pontine nuclei. Neurogenesis is relatively short-lived in the rhombic lips of most reptiles, the cerebellum is a simple platelike structure, and the olivary and pontine nuclei are represented by a few scattered cells. A prominent cerebellum is an archosaurian feature, with the cerebellum of crocodilian reptiles containing two mound-shaped lobes. The cerebellum in birds comprises both an extensive vermis and small, paired hemispheres. It has a lobular structure much like that seen in mammals. Birds have distinct inferior olivary and pontine nuclei that project to the cerebellum as climbing and mossy fibers, respectively. Experimental embryological studies in chicks demonstrate that these nuclei are derived from the rhombic lip (Harkmark, 1954), so the presence of the cerebellum and precerebellar nuclei in birds can be viewed as a relative prolongation of neurogenesis within this part of the brain stem. The functional implications of these nuclei are considered in Section 4.4.1b.

4. Shared–Derived Features of Avian Brains

The preceding sections outline some ways in which the basic reptilian brain pattern is modified in birds and point out, in particular, that changes in the timing of developmental events seem to play a key role in these modifications. This section discusses several of the features of avian brains that appear to be uniquely associated with birds.

4.1. Brain Size

Brain size (as measured by brain weight) varies generally as a power function of body size (as measured by body weight) across a wide range of vertebrates. This relationship can be conveniently demonstrated by plotting log (brain weight) as a function of log (body weight). For any given taxonomic group of animals, such plots typically show a scatter of points around a regression line. The points can be included within a "minimum convex polygon" (Jerison, 1973). Thus, there is a polygon of reptile points, a polygon of bird points, one of mammal points, and so forth. The slopes of the regression lines and polygons tend to be about the same for each taxonomic group, but the *y* intercepts vary so that some animals have larger brains relative to their body sizes than do others. In particular, birds have greater brain weights per unit body weights than do reptiles. Estimates of brain and body weights for *Archaeopteryx* indicate that its brain weight/body weight ratio falls within the avian minimum convex polygon, overlapping the points for theropod dinosaurs (Hopson, 1977). Further analysis indicates that relative enlargement of the telencephalon accounts for a significant part of the overall increase of brain weight per unit body weight in birds over that seen in reptiles (Portman and Stingelin, 1961; Pearson, 1972). Birds, such as owls, crows, and parrots, have particularly high brain weight/body weight ratios. The precise significance of high ratios is not clear, although there is some indication that brain size reflects metabolic demands (Armstrong and Bergeron, 1985). In any case, it appears that an increase in relative brain size was a feature of the reptile–bird transition that occurred prior to *Archaeopteryx*.

4.2. Optic Tectum

The optic tectum is relatively large in birds compared to reptiles. It is expanded laterally to form two prominent lobes (often called the optic lobes) that cover the lateral surface of the midbrain. Some expansion of the optic tectum relative to the configuration seen in crocodilians can be seen in endocasts of *Archaeopteryx* and of pterosaurs (Hopson, 1979). The optic tectum receives visual information directly from the retinal ganglion cells, and contains a topographically organized representation of the contralateral retinal surface (Cowan *et al.*, 1961; Reperant, 1973). There is considerable interspecific variation in retinal organization in birds, but many species have two areas of retinal specialization. One is a "pecking field" in the dorsal temporal retina that faces forwards into frontal, binocular space. The second is a field in the central retina that faces laterally. Behavioral studies demonstrate that birds use the two retinal areas differently, depending on the specific task at hand (Bloch and Mantinoya, 1982; Bloch *et al.*, 1984; Mantinoya *et al.*, 1983; Maldonado *et al.*, 1988). The temporal field is used to fixate on stationary or slowly moving objects located relatively close to the bird. The central field is used to follow moving objects some distance from the bird. The presence of both fields is reflected in the organization of the optic tectum of pigeons (Hamdi and Whiteridge, 1954) as an increase in the magnification factors within the tectal representation of the two fields.

The tectum is divided into a series of alternating layers of fibers and cell bodies. Its organization is generally similar in reptiles (see Northcutt, 1984; Ulinski et al., 1990) and birds (see Hunt and Brecha, 1984), although the number and clarity of layers in birds is generally greater than that seen in most reptiles. The principal efferent projections of the tectum are to the brain-stem reticular formation and to nucleus rotundus in the dorsal thalamus (Hunt and Kunzle, 1976). The reticular formation is the source of premotor pathways that coordinate and regulate the activity of motor neurons in the spinal cord and brainstem. Nucleus rotundus receives tectal input and projects in turn to the lateral part of DVR, which is the ectostriatum in birds (see Ulinski, 1983, for references). The organization of these projections appears to be very similar in birds and reptiles. Birds, therefore, show a quantitative expansion of the optic tectum, without major reorganization of its circuitry.

4.3. Visual Wulst

The Wulst is a distinctly avian structure that reflects an embryological reorganization of DVR from the reptilian configuration. It is laminated (Watanabe *et al.*, 1983) and comprises four cytoarchitectonic components. The most dorsal is the hyperstriatum accessorium, which contains medium-sized cells; ventral to hyperstriatum accessorium proper is the intercalated nucleus of the hyperstriatum accessorium, which consists of smaller cells; ventral to this is the hyperstriatum intercalatus superior, which contains scattered cells; most ventral is the hyperstriatum dorsale, which contains large cells. The relative development and cytoarchitectonic clarity of these layers varies between species, the Wulst being relatively small and indistinct in pigeons and chickens, for example, and very large and distinct in owls. All of the neurons in the Wulst have a stellate configuration in Golgi preparations and have several spinous dendrites. There is a distinct tendency for somata to clump together to form columns oriented perpendicular to the surface of the brain, and bundles of axons course between the columns through the various layers of the Wulst (Pettigrew, 1979).

Axonal tracing experiments demonstrate that the Wulst receives projections from the principal optic nucleus of the thalamus (Karten, 1979; Bagnoli and Burkhalter, 1983; Hunt and Webster, 1972; Karten *et al.*, 1973; Meier *et al.*, 1974; Miceli *et al.*, 1975; Reperant *et al.*, 1974; Nixdorf and Bischof, 1982). This is the avian equivalent of the reptilian dorsal lateral geniculate complex, and—like the reptilian structure—receives direct projections from retinal ganglion cells (e.g., Cowan *et al.*, 1961; Reperant, 1973) and projects to the telencephalon. The retinal projection to the principal optic nucleus is retinotopically organized, and cells in the nucleus have relatively small visual receptive fields (De Britto *et al.*, 1975; Jassik-Gerschenfeld *et al.*, 1976; Maxwell and Granda, 1979; Pettigrew, 1979).

An important difference from the geniculate complex of reptiles concerns the laterality of the projections. Retinogeniculate projections in reptiles are typically bilateral. The contralateral projection is the most extensive, but there is nonetheless a convergence of information from the two retinas at the level of dorsal thalamus. The subsequent projection from the geniculate to the telencephalon is entirely ipsilateral. In birds, by contrast, the retinal projection to the principal optic nucleus is exclusively contralateral. Thus, the principal nucleus contains an organized representation of the entire contralateral retina (Pettigrew, 1979), and individual neurons in the nucleus respond exclusively to stimuli presented to the contralateral eye. Binocular visual units are found in the Wulst and photic stimulation of one eye causes bilateral activity in the hyperstriatum in 2-deoxyglucose studies (Streit *et al.*, 1980). The necessary convergence of information from the two eyes occurs via a second decussation within the projection from the principal optic nucleus to the Wulst (Fig. 4). Anterograde tracing experiments indicate that the thalamic projection to the Wulst is bilateral (Meier *et al.*, 1974; Miceli *et al.*, 1975). There is a dense projection from the principal nucleus to the ipsilateral hyperstriatum accessorium intercalatus, and a somewhat sparser projection to the same layer of the contralateral Wulst that crosses the midline via the dorsal supraoptic decussation. Retrograde tracing studies (Reperant *et al.*, 1974) confirm the various cytoarchitectonic subdivisions of the principal nucleus projection to the Wulst, with some giving rise to ipsilateral projections, some to contralateral projections, and some to bilateral projections. Double labeling studies with fluorescent dyes reveal that some of the bilateral projections arise from the two branches of the same neurons (Miceli and Reperant, 1982).

Projections from the principal nucleus to the Wulst are topographically organized, and electrophysiological studies indicate an organized representation of the contralateral visual field in the Wulst. This has been demonstrated most clearly in owls, in which the nasal–temporal axis of the visual field is mapped along the lateral–medial axis of the Wulst, and the superior–inferior axis is mapped along the caudal–rostral axis of the Wulst (Pettigrew, 1979). Details of the map are less certain in other species in which the Wulst is smaller (Denton, 1981; Perisic *et al.*, 1971; Revzin, 1969; Wilson, 1980a). Electrode penetrations perpendicular to the Wulst encounter units with receptive fields in the same region of visual space, so there is a form of columnar organization in which the same point in visual space is represented in a line extending through the various layers of the Wulst (Pettigrew, 1979). Neurons in the Wulst tend to have circular or oval receptive fields (Pettigrew, 1979; Wilson, 1980b), which are relatively small in owls, but larger in pigeons and chickens. In owls, many of the units resemble the simple cells described by Hubel and Wiesel in the visual cortex of cats in responding preferentially to bars with particular orientations (Pettigrew and Konishi, 1976). Electrode penetrations passing tangentially through the Wulst provide some evidence for a weak form of orientation columns in owls (Pettigrew, 1979).

Most birds have some degree of binocular overlap, and those regions of the Wulst that represent binocular visual space receive inputs from the contralateral principal optic nucleus as well as from the ipsilateral nucleus. There is considerable interspecific variation in the extent of the binocular representation, and

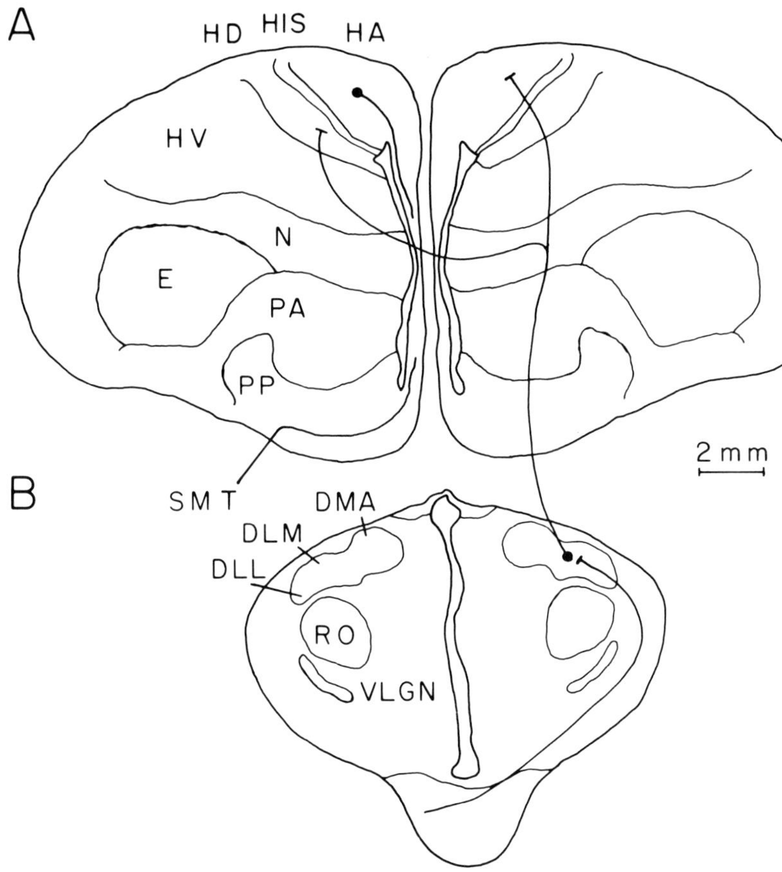

Figure 4. Organization of the Wulst. This figure illustrates the basic organization of the visual pathway to the Wulst in pigeons. (A) Transverse section through the rostral telencephalon. The Wulst consists of the hyperstriatum accessorium (HA), hyperstriatum intercalatus superioris (HIS), and hyperstriatum dorsale (HD). It receives bilateral projections from the principal optic complex of the dorsal thalamus, and projects through the septomesencephalic tract (SMT) to the brain stem. (B) Transverse section through the diencephalon. The principal optic complex consists of the dorsolateral lateral (DLL), dorsolateral medial (DLM), and dorsomedial anterior (DMA) nuclei of the dorsal thalamus. It receives visual input from the contralateral retina. E, ectostriatum; HV, hyperstriatum ventrale; N, neostriatum; PA, paleostriatum augmentatum; PP, paleostriatum primitivum; RO, nucleus rotundus; VLGN, ventral lateral geniculate nucleus.

those species with a large binocular representation tend to have many units that respond to stimuli presented to both eyes (Pettigrew and Konishi, 1976; Pettigrew, 1979). Such units show preference for one eye, and tangential electrode penetrations indicate some alternation in ocular preference along the electrode tract, suggestive of ocular dominance columns. Some units show a form of disparity tuning that indicates the Wulst plays a role in stereopsis. Pettigrew and colleagues (Pettigrew, 1986) have performed a survey of binocular response properties in a sample of birds. They find such properties in the Wulst of owls in the owl families Strigidae and Tytonidae. There is some heterogeneity in the order Caprimulgiformes. The oilbird, *Steatornis caripensis,* for example, is a large neotropical bird that lives in totally dark caves, which it leaves at night to hunt for fruit. It has a well-developed visual system, but does not seem to have binocular vision. Its Wulst has a large number of orientation-selective cells that can be driven only by the contralateral eye. By contrast, birds in the families Podargidae (frogmouths) and Aegothelidae (owlet frogmouths) have highly developed visual systems used for predation, mostly on invertebrates, at night.

It is intuitively reasonable that an increase in the size and complexity of the visual forebrain occurred in parallel with the evolution of flight, but it is actually not at all clear what biological role is played by the Wulst. There is some indication that the principal optic nucleus–Wulst pathway receives input from particular sets of retinal ganglion cells. In owls, ganglion cells in the temporal retina project to the principal optic nucleus (Bravo and Pettigrew, 1981), indicating the Wulst receives information from the pecking field in the dorsal temporal retina. This is consistent with the idea that the Wulst is involved in binocular vision and stereopsis, and units in the Wulst with binocular response properties seem to be abundant principally in those species—independent of their taxonomic affinities—with well-developed binocular visual fields. A reasonable hypothesis is that the principal optic nucleus–Wulst pathway may be particularly related to foraging from the substrate.

However, lesions of nucleus rotundus or its telencephalic target, the ectostriatum, lead to deficits in the ability of pigeons to perform on spatial discrimination tasks (see Hodos, 1976; Hodos and Bonbright, 1974; Hodos and Karten, 1970; Benowitz, 1980; Weiss and Hodos, 1986), while lesions of the Wulst seem to have little or no effect on traditional pattern discrimination tasks (Hodos *et al,* 1973). More sophisticated psychophysical methods show that Wulst lesions do produce some deficits on intensity and pattern discrimination tasks, but the degree of deficit would probably not lead to severe difficulties in most natural circumstances. For example, Wulst lesions do not impair the abilities of homing pigeons to return to their lofts (Bingman *et al.,* 1985). On the other hand, Wulst lesions do produce deficits in the bird's ability to do reversal learning tasks such as learning to respond to triangles in one set of experiments and to squares in another set (e.g., Hodos, 1976). Interestingly, such lesions produce greater deficits in species in which the Wulst is well developed (Zeier, 1971). Thus, the Wulst probably plays a role in binocular vision in some species of birds, but is also likely involved in more sophisticated aspects of visual processing in all birds.

4.4. Motor System Organization

The reptile–bird transition involved a number of distinct modifications in motor control of the limbs and head. This section summarizes characteristic features of the neural circuitry related to motor control, and then discusses their roles in more specific behaviors.

4.4.1. Neural Circuitry

4.4.1a. Striatum. The major flow of information from the telencephalon to the premotor centers of the brain stem of reptiles and birds involves connections through the striatum (review: Reiner *et al.*, 1984b). The striatum receives projections from sensory regions of the pallium (e.g., Brauth *et al.*, 1978; Bonke *et al.*, 1979), and gives rise to two pathways that link the striatum with the optic tectum (Fig. 5). The first involves striatal projections (Karten and Dubbeldam, 1973; Kitt and Brauth, 1981) to a dopaminergic cell group in the midbrain tegmentum called the substantia nigra in reptiles (Parent, 1986; Brauth, 1988) and the pedunculopontine nucleus in birds (Dube and Parent, 1981; Kitt and Brauth, 1986). This nucleus is one of a set of midbrain dopaminergic groups (including also the ventral tegmental area) that give rise to reciprocal projections to the striatum (Brauth *et al.*, 1978; Kitt and Brauth, 1986). In addition, the substantia nigra and pedunculopontine nucleus project to the deep layers of the optic tectum. The second link between the striatum and the optic tectum involves a pretectal nucleus called the nucleus of the posterior commissure in reptiles and the lateral spiriform nucleus in birds (Reiner *et al.*, 1982a). This projection also terminates in the deep layers of the tectum and is known to be enkephalinergic (Reiner *et al.*, 1982b). The tectum projects to the brain-stem reticular formation (Reiner and Karten, 1982), which projects in turn to the spinal cord (Cabot *et al.*, 1982; Webster and Steeves, 1988).

The striatum of both reptiles and birds contains two major populations of neurons. The most numerous have medium-sized somata and bear several spiny dendrites (Tombol *et al.*, 1988b). Other neurons have relatively large, fusiform somata and bear large, aspiny dendrites that form extensive dendritic trees (Palacios, 1976). In reptiles, both types of neurons are found in all parts of the striatum, but the large aspiny neurons show a tendency to occur most frequently in the caudal striatum, embedded within the fibers of the lateral forebrain bundle (Ulinski, 1989; Vyas and Ulinski, 1990). In birds, there is a strict segregation with the medium-sized neurons occurring in a band of neurons lying immediately ventral to DVR that includes the paleostriatum augmentatum, the parolfactory lobe, and nucleus accumbens. The large aspiny neurons are restricted to the paleostriatum primitivum and, in a slightly smaller form, in the intrapeduncular nucleus. Both the paleostriatum primitivum and the intrapeduncular nucleus are embedded entirely within the lateral forebrain bundle. The two populations of neurons participate in parallel connections from the striatum to the brain stem (Fig. 5). The medium-sized neurons use substance P (Reiner *et al.*, 1983) or enkephalin (Reiner *et al.*, 1984a) as transmitter substances and project to the pedunculopontine nucleus (TPC) and ventral tegmental area. The large aspiny neurons show immunoreactivity to the hexapeptide lysine-asparagine-neurotensin or LANT6 (Reiner and Carraway, 1985; Brauth *et al.*, 1986) and also contain GABA (Csillag *et al.*, 1987). They project to the lateral spiriform nucleus (SPL; Reiner *et al.*, 1982b). The segregation of large aspiny neurons from medium-sized spiny neurons suggests that the two populations of neurons have somewhat different input–output relationships in birds than in reptiles.

This is a relatively minor difference, but a major difference in striatal organization between reptiles and birds is the presence of striatal projections to dorsal thalamus, and from dorsal thalamus to DVR in birds. These projections involve a set of nuclei situated at the medial edge of the dorsal thalamus that project topographically to a strip of DVR, the intermediate neostriatum, that extends from the lateral ventricle medially to the "periectostriatal belt" that surrounds the dorsal surface of the ectostriatum at the lateral edge of DVR (Fig. 6). The dorsal medial posterior nucleus of the thalamus (DMP) receives a sparse projection from the medial parts of the striatum (Kitt and Brauth, 1982) and projects to the medial region of intermediate neostriatum (NI). The dorsal intermediate posterior nucleus (DIP) receives a dense projection from the paleostriatum primitivum and intrapeduncular nucleus (Karten and Dubbeldam, 1973; Brauth *et al.*, 1978; Kitt and Brauth, 1981), and projects to the central part of the intermediate neo-

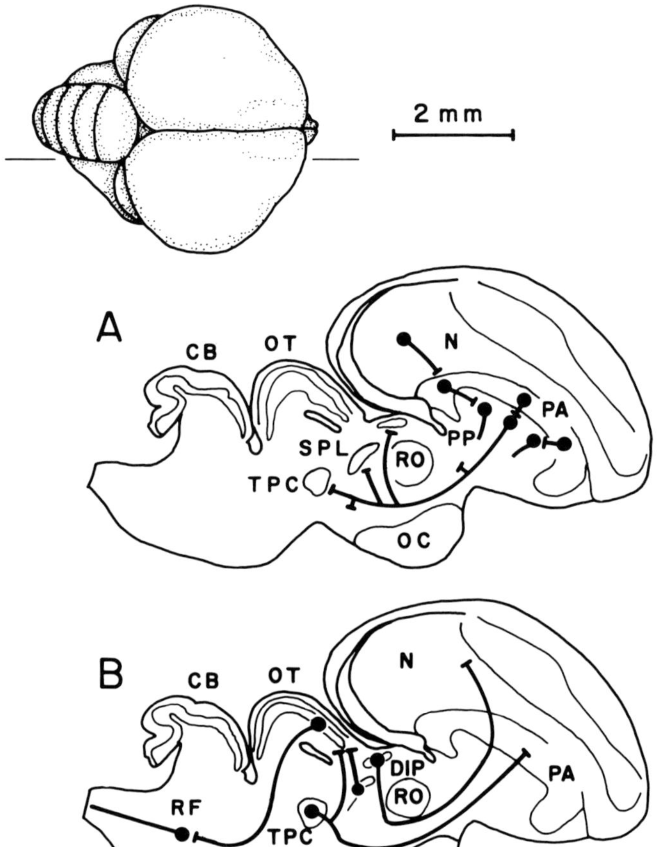

Figure 5. Striatal–tectal pathways. This figure illustrates the basic organization of the pathways that carry information from the striatum to the optic tectum in pigeons. The projections are shown on two views of the same parasagittal section. The plane of the section is demonstrated in the inset, which is a dorsal view of the brain. (A) The striatum (PA and PP) receives inputs from the neostriatum (N) in the dorsal ventricular ridge. The striatum has major projections to the lateral spiriform nucleus (SPL) and the pars compacta of the pedunculopontine nucleus (TPC). There is also a projection to the dorsal intermediate posterior nucleus of the dorsal thalamus. (B) The lateral spiriform nucleus and pedunculopontine nucleus project to the optic tectum (OT), which gives rise to the tectoreticular pathways. The dorsal intermediate posterior nucleus (DIP) projects to the neostriatum. CB, cerebellum; OC, optic chiasm; RF, reticular formation; RO, nucleus rotundus. Reproduced from Ulinski (1983).

striatum. The dorsal lateral posterior nucleus (DLP) is the most lateral of the three nuclei in the posterior complex. It is divided into a caudal, large-celled part and a rostral, small-celled part. The caudal DLP receives afferent projections from the central layers of the optic tectum, and projects to the lateral part of the intermediate neostriatum and adjacent periectostriatal belt (Kitt and Brauth, 1982; Gamlin and Cohen, 1986; Brauth *et al.,* 1987). Units in this nucleus respond to visual stimuli. The rostral DLP receives afferents from the spinal cord and dorsal column nuclei, and projects to the lateral part of the intermediate neostriatum, rostral to the terminal zone of DLP. It contains units that respond to somatosensory stimuli (Delius and Bennetto, 1972). Finally, the ventromedial part of nucleus ovoidalis (Ov) of the dorsal thalamus receives ascending auditory projections and projects, in turn, to the intermediate neostriatum (Brauth *et al.,* 1987). Thus, the intermediate neostriatum receives projections from the cerebellum, striatum, optic tectum, and auditory and somatosensory systems, arrayed in a topographic fashion, from medial to lateral. Like the Wulst, it is a distinctly avian character.

4.4.1b. Cerebellar Circuitry. The cerebellum in birds is generally larger than that of reptiles, and was slightly larger in *Archaeopteryx* and pterosaurs than in reptiles. The vermis was divided by Larsell into a series of lobules numbered I–X, in a manner corresponding to Larsell's nomenclature for the cerebellum of mammals. Detailed accounts of the morphology of the cerebellum in a range of birds can be found in Larsell (1948, 1967), Larsell and Whitlock (1952), and Pearson (1972). Afferents to the cerebellum include those from the spinal cord, vestibular nuclei, and so forth (Clarke, 1977; Whitlock, 1952). Similar projec-

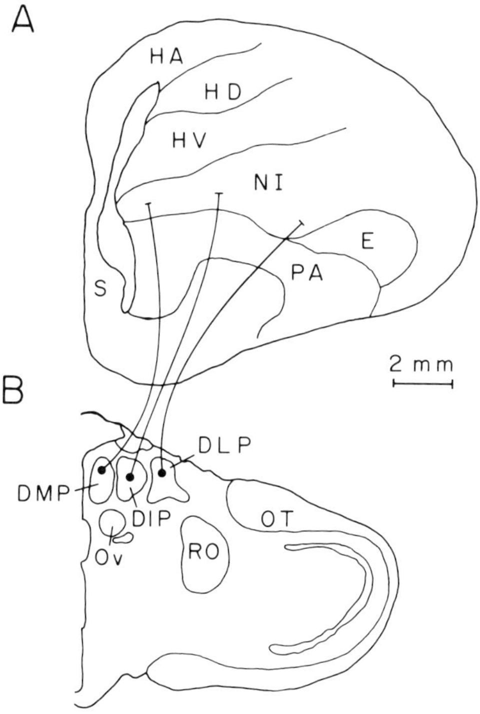

Figure 6. Projections of the dorsal posterior complex. This figure illustrates the basic organization of the projections of the dorsal posterior nuclear complex in pigeons. (A) Transverse section through the cerebral hemisphere. The intermediate neostriatum (NI) receives topographically organized projections from the dorsal posterior complex of the thalamus. (B) Transverse sections through the diencephalon. The dorsal posterior nuclear complex projects topographically to the intermediate neostriatum. The dorsal medial posterior nucleus (DMP) receives projections from the medial striatum and projects to the medial intermediate neostriatum. The dorsal intermediate posterior nucleus (DIP) receives a dense projection from the striatum and projects to the central part of the intermediate neostriatum; the dorsal lateral posterior nucleus (DLP) has distinct caudal and rostral parts. The caudal part receives projections from the central layers of the optic tectum and projects to the lateral part of the intermediate neostriatum. The rostral part receives projections from the spinal cord and dorsal column nuclei and projects rostrally in the neostriatum. E, ectostriatum; HA, hyperstriatum accessorium; HD, hyperstriatum dorsale; HV, hyperstriatum ventrale; OT, optic tectum; Ov, nucleus ovoidalis; PA, paleostriatum augmentatum; RO, nucleus rotundus; S, septum.

tions are seen in reptiles (reviews: ten Donkelaar *et al.*, 1987; ten Donkelaar, 1990). However, the brain stem of reptiles does not include distinct inferior olivary and pontine nuclei, both of which are present in birds. The pontine nuclei of birds consist of scattered cells distributed in medial and lateral groups on the ventral surface of the rostral metencephalon. The inferior olivary nuclei consist of a U-shaped group of cells in the medulla, comprising dorsal and ventral lamellae (Kooy, 1917; Vogt-Nilsen, 1954). Retrograde tracing studies demonstrate that the inferior olivary complex (Clarke, 1977; Armstrong and Clarke, 1979; Furber, 1983) and pontine nuclei (Brodal *et al.*, 1950) project to the cerebellum in an organized way. The inferior olivary complex is a source of climbing fibers (Freedman *et al.*, 1977); the pontine nuclei are a source of mossy fibers.

Both the olivary complex and pontine nuclei convey information from visual nuclei in the telencephalon and brain stem to the cerebellum. The basal optic nucleus of the midbrain tegmentum receives inputs from large, displaced retinal ganglion cells (Karten *et al.*, 1977; Reiner *et al.*, 1979). It projects directly to folium IX and the paraflocculus of the cerebellum (Brauth and Karten, 1977) via mossy fibers, and indirectly to folia IX and X via climbing fibers originating in the olive (Armstrong and Clarke, 1979; Freedman *et al.*, 1977; Furber, 1983). The nucleus lentiformis mesencephali of the pretectum receives retinal input and projects to folia VI, VII, and VIII, as well as the vestibulocerebellum (Clarke, 1977; Gamlin and Cohen, 1988). Its influence is direct through mossy fiber projections to the cerebellum, or indirect via mossy fibers from the lateral pontine nucleus or climbing fibers from the anterior olive. The optic tectum also receives direct retinal input and projects to the medial and lateral pontine nuclei (Hunt and Kunzle, 1976). Neurons in the lateral pontine nuclei have very large receptive fields and are directionally selective (Clarke, 1977). The visual Wulst exerts an influence on the cerebellum via projections to the pontine nuclei, nucleus lentiformis mesencephali, and the basal optic nucleus. Thus, the precerebellar nuclei in birds convey visual information to the cerebellum directly from retinorecipient nuclei of the brain stem, as well as from a visual center in the telencephalon. All of these nuclei contain units that respond to stimuli moving over relatively large regions of the visual world, but several of the nuclei contain cells that show a distinct preference for stimuli moving along a particular axis of visual space. McKenna and Wallman (1985) hypothesize that they are involved in a coordinate transformation from the spatial coordinates of visual space to the nonorthogonal coordinate system formed by the extraocular muscles. The system as a whole would stabilize the head and eyes as the visual world moves past the two retinas, and could play a particular role in visual guidance during flight.

4.4.1c. Descending Telencephalic Pathways. Several laboratories have searched for the presence of descending telencephalic pathways in reptiles using modern anatomical tracing techniques. The only telencephalic efferent systems that have been well documented involve projections from the visual cortex (Ulinski, 1986) or DVR (Hoogland, 1977; Ulinski and Nautiyal, 1987) to the visual thalamus and optic tectum (Elprana *et al.*, 1980). There is no evidence for projections from the pallium to the brain stem in reptiles. By contrast, there is clear evidence that the pallium of birds projects to the brain stem.

The best documented pallial efferent system involves projections that originate in the visual Wulst and project caudally through the septomesencephalic tract (Adamo, 1967; Bagnoli *et al.*, 1980; Karten, 1971; Karten *et al.*, 1973; Miceli *et al.*, 1987). This tract exits the telencephalon and reaches a variety of brain stem structures including the ventral lateral geniculate nucleus, optic tectum, and basal optic nucleus (Rio *et al.*, 1983). Retrograde transport experiments indicate that projections from the Wulst to the optic tectum originate from the hyperstriatum accessorium (Bagnoli *et al.*, 1980; Reiner and Karten, 1983). The projections are topographically organized (Karten *et al.*, 1973), and have complex physiological effects on tectal neurons (Bagnoli *et al.*, 1979). They may play a role in modulating the activity of tectal efferents to premotor structures in the brain stem reticular formation.

In addition to these projections from the caudal or visual Wulst, there are descending projections from an anterior portion of the Wulst in burrowing owls, *Speotyto cunnicularia.* Electrophysiological studies indicate that the anterior Wulst in these birds contains a topographically organized representation of the surface of the talons (Karten *et al.*, 1978). Somatosensory information reaches the anterior Wulst through the cuneate–gracile nuclei and a relay in the dorsal thalamus. Anterograde degeneration experiments show that the anterior Wulst projects to the medial spiriform nucleus of the thalamus, the red nucleus, pontine nuclei, cuneate–gracile nuclei, and the contralateral spinal cord (Karten, 1971). This is apparently a system related specifically to motor control of the talons. Unfortunately, these findings have not been extended to other species of birds, so it is not known if this is a general avian feature, or something found only in some groups of birds of prey.

Finally, birds have distinct projections from the archistriatum to the brain stem. The archistriatum is a region of the ventrocaudal telencephalon that corresponds to the basal DVR of reptiles (see Ulinski, 1983) and the amygdala of mammals. It contains several distinct cytoarchitectonic components (Zeier and Karten, 1971). The posterior and medial parts of the archistriatum have connections, such as efferents to the hypothalamus, that suggest it is part of the limbic system. Comparable projections are seen in reptiles. The anterior and intermediate components of the archistriatum receive afferent projections (Fig. 7) from several regions of DVR (Brauth and McHale, 1988; Ritchie, 1979), and give rise to a major descending projection to the brain stem via the occipitomesencephalic tract. This tract contains thick axons that course through the brain stem, terminating in the thalamus, optic tectum, brain-stem tegmentum, lateral reticular formation, lateral pontine nuclei, sensory nuclei of the brain stem and rostral spinal cord. These projections from the archistriatum provide a major route by which information from DVR exits the telencephalon in that several areas of DVR project to the archistriatum (e.g., Brauth and McHale, 1988; Ritchie, 1979). Thus, the intermediate neostriatum, which receive inputs from the nuclei of the posterior dorsal thalamus, projects to the lateral neostriatum, which projects in turn to the archistriatum. Although some minor projections from the pallium and striatum to the brain stem are present in reptiles, an increase in their size and complexity is characteristic of birds.

4.4.2. Control of Actions

4.4.2a. Bipedal Locomotion and Flight. Reptiles are typically quadrupedal animals that move with their limbs extended in a sprawled posture. Locomotion

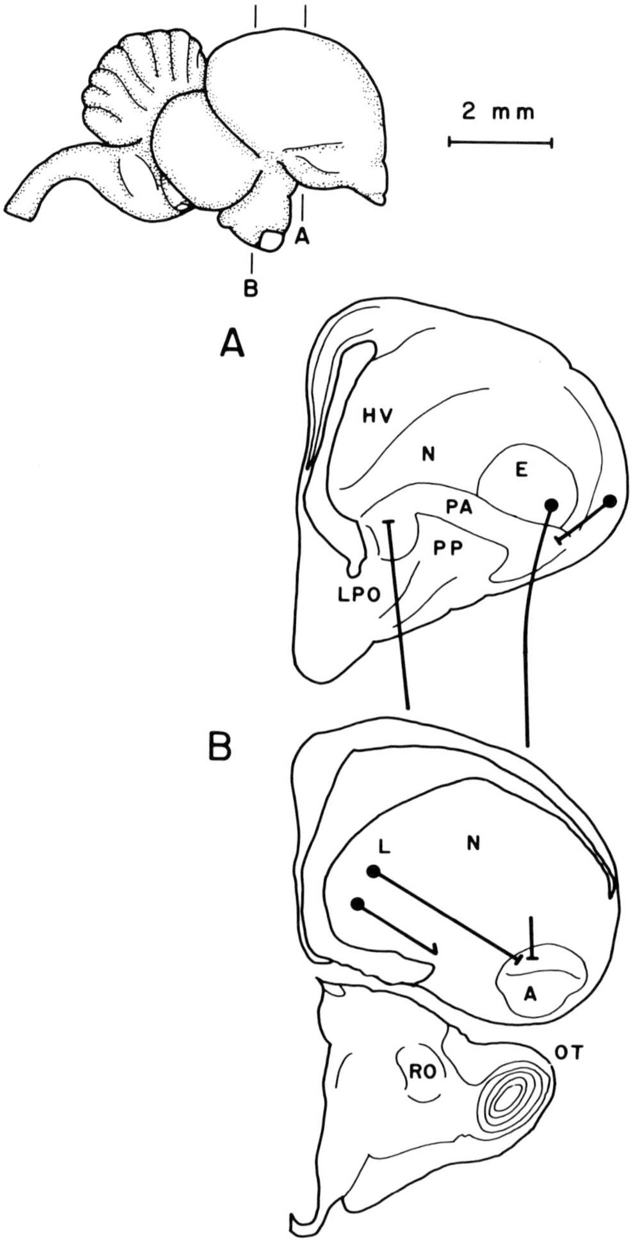

Figure 7. Afferent connections of the dorsal intermediate archistriatum. This figure illustrates the basic organization of afferent connections to the dorsal intermediate archistriatum on two sections through the cerebral hemisphere of a pigeon. The inset shows the plane of the sections plotted on a lateral view of a pigeon brain. The archistriatum (A) receives afferents from the auditory area in DVR (field L), the visual area in DVR (ectostriatum, E) and from the temporoccipitoparietal area that forms the external surface of DVR. The archistriatum gives rise to descending pathways, which are described in the text. Reproduced from Ulinski (1983).

involves lateral undulations of the trunk combined with alternate limb movements such that the two forelimbs move out of phase with each other, as do the two hindlimbs. There is a great deal of variability in locomotor patterns within the reptiles with snakes and many lizards showing an elongation of the body, loss of limbs, and reliance entirely upon lateral undulation for forward progression. Some lizards exhibit a degree of bipedalism for short spurts when running. However, there is a general tendency for modification of the basic reptilian posture within the archosaurs. In many cases, this involves the four legs being positioned vertically beneath the body, rather than extending laterally. There is some indication that early crocodilians had an erect posture and that the sprawled posture of Recent crocodilians is a secondary adaptation to an aquatic environment (Parrish, 1987). An erect posture is particularly noticeable in the large dinosaurs, ostensibly to help support the animal's considerable weight, but this posture also permits development of a true bipedal posture in which the hindlimbs increase in size and the forelimbs are freed to move in front of the animal. This transformation occurs repeatedly within the dinosaurs, and is a particular feature of the small, theropod dinosaurs that were probably close to the ancestors of birds. Bipedality is a good way (but not necessarily the only way) for an animal to make the forelimb modifications needed to form wings, as well as a way for it to gain the ground speed needed for powered flight. Paleontologists are divided on the issue, but it is conceivable that *Archaeopteryx* could have attained flight speed by running along the ground.

In addition to these skeletal and muscular modifications associated with flight, the reptile–bird transition involved a major change in the neural circuitry associated with limb movements. There is now good evidence that rhythmic movements of the limbs of reptiles are controlled by central pattern generators in the spinal cord (Stein and Grossman, 1980). The activity of these generators is modulated by descending projections from the brain stem to the spinal cord (Lennard and Stein, 1977; Kazennikov *et al.*, 1980). Thus, stimulation of cervical spinal cord in turtles produces rhythmic, swimming movements of the four limbs in which the two forelimbs move out of phase with each other, as do the two hindlimbs. This gait pattern is the typical tetrapod pattern and is general within reptiles. By contrast, birds show two modifications from the reptilian gait pattern. The movements of the wings are entirely decoupled from those of the hindlimbs, so the wings are folded next to the body during normal walking movements and the hindlimbs are folded up next to the body during flight. The hindlimbs continue to move out of phase with each other, but the wings typically move synchronously. There is a transition associated with becoming airborne in which the bird first runs, using alternating hindlimb movements, then begins to flap its wings while still running, and finally folds its hindlimbs under its body as free flight is achieved.

Experiments in which the spinal cord is stimulated in spinal birds demonstrate that alternating movements of the hindlimbs are generated by central pattern generators in the caudal spinal cord (ten Cate, 1960). This general finding was confirmed by Jacobson and Hollyday (1982a,b) who made detailed electromyographic studies of the activity of the leg muscles in walking chicks. They then showed that electrical stimulation of the white matter of the rostral spinal cord produced normal walking movements of the legs in decerebrate chicks. In some instances, the walking–flight transition could be induced by increasing the strength of the stimulating current. At low intensities, the chick demonstrated walking movements. As the current strength increased, the wings began to move synchronously and the hindlimbs were then folded under the body. There was some indication that movements of the hindlimbs were generated by stimulation in one region of the lateral funiculus of the spinal cord, while wing movements were generated by other stimulation points. Steeves *et al.*, (1987) reached similar conclusions in geese, and suggest that distinct reticulospinal pathways are involved in control of the wings and hindlimbs. Wing movements seem to be produced selectively by stimulation of the lateral, small-celled reticular formation in the subtrigeminal region. Hindlimb movements seem to result from stimulation of the large-celled reticular formation. Normal phase relations between walking and breathing can be produced in decerebrate geese by stimulation of the brain stem (Funk *et al.*, 1989). Thus, the telencephalon is not necessary for the generation of locomotion in birds, but the reptile–bird transition may have involved some modifications in the pattern of reticulospinal connections so that the wings became decoupled from the legs.

4.4.2b. Orientation to and the Manipulation of Prey. A secondary consequence of bipedalism in birds is that it permits the use of the hindlimbs to manipulate objects in the bird's environment. Such movements require the ability to fractionate or individuate the activation of individual muscle groups out of the rhythmic sequences produced by spinal cord central pattern generators. From a connectional viewpoint, animals that can manipulate objects with their limbs tend to have direct telencephalic projections to the vicinity of motor neuron pools in the spinal cord or brain stem. The corticospinal and corticobulbar paths play this role in mammals. The projection from the anterior Wulst to the spinal cord may play this role in certain birds.

Although manipulation of objects by the hindlimbs is limited to only certain groups of birds, most birds use their beaks to manipulate objects. Most reptiles do relatively little prey manipulation with their jaws, engaging instead in various forms of inertial feeding in which prey items are seized by the jaws and then swallowed by a rapid movement. Other reptiles, such as snakes, are capable of very complex and precise movements of the jaws during prey manipulation prior to ingestion. Pecking movements involved in routine prey acquisition fall well within the range of jaw movements seen in reptiles, but delicate manipulations of objects used in a tool-like fashion, such as the use of thorns to manipulate prey items by Galápagos finches or egg-rolling behaviors seen in herring gulls, probably do not occur in reptiles. The process of orienting to a prey item, pecking at it, and then manipulating the item with the beak (mandibulation) is a very complex behavior involving a variety of sensory and motor systems (Zeigler and Karten, 1973; Zeigler, 1974, 1976; Zeigler *et al.*, 1980; Deich *et al.*, 1985).

The visual system circuitry involved in orienting to and acquiring a prey item is apparently common to both reptiles and birds. It involves projections from the visual areas of the telencephalon to the striatum, and from the striatum to optic tectum via linking nuclei in the brain stem. Brauth (1987), for example, has shown that chemical lesions in the striatum of alligators impair their ability to follow and snap at goldfish. Bugbee (Bugbee, 1979; Bugbee and Hodos, 1979) has shown that the pathway from the striatum to the tectum through the lateral spiriform nucleus plays particular roles in orienting behaviors in pigeons. Birds can normally peck at either stationary or moving food items, such as grains of corn. However, lesions of the lateral spiriform nucleus leave the bird's ability to peck at stationary food items intact while imparing its ability to peck successfully at moving food items. Analysis of videotapes of pigeons pecking at stationary objects shows that the visually guided component of the peck consists of two fixations (Goodale, 1983). The first occurs about 140 msec before the beak reaches the target and involves the target object falling on any part of the retina. The second fixation occurs 40 msec later, and involves the target object falling upon the dorsal temporal retina (the "pecking" field). It ends by the pigeon opening its beak enough to grasp the object as it makes contact.

The circuitry involved in mandibulation, although perhaps present in reptiles, is highly developed in many birds (e.g., Dubbeldam, 1984; Wild *et al.*, 1984). Information from sensory receptors of the head is conveyed to trigeminal nuclei in the brain stem in both reptiles and birds. There are projections from the trigeminal complex to the cerebellum in reptiles (ten Donkelaar *et al.*, 1987), but these projections are more extensively developed in birds, where the trigeminal sensory complex (Dubbeldam, 1980; Arends and Dubbeldam, 1984; Arends *et al.*, 1984) plays important roles in the control of movements of the head and beak during mandibulation. Subnucleus oralis of the descending trigeminal nucleus is the source of mossy fiber projections to lobules VIII and IX of the cerebellum (Arends and Zeigler, 1989). Subnucleus interpolaris projects to the inferior olivary complex, which is the source of a climbing fiber projection to the cerebellar cortex that conveys trigeminal information. There may also be projections from the trigeminal sensory complex to the adjacent lateral reticular formation, which Steeves *et al.* (1987) implicate as a source of descending pathways to the spinal cord, and is also a likely source of projections to jaw premotor neurons. Lobules VII and IX of the cerebellar cortex project to the caudal portion of the medial cerebellar nucleus, which projects in turn to the brain stem reticular formation.

In addition to these circuits linking the trigeminal complex to the cerebellum and premotor regions of the brain stem reticular formation (Wild *et al.*, 1984), birds have a large and distinct pathway (Wild *et al.*, 1985) from the principal nucleus of the trigeminal sensory complex directly to DVR by way of the quintofrontal tract. This projection terminates in a region of the rostral, lateral DVR called nucleus basalis (review: Veenman and Gottschaldt, 1986). There is some indication that a trigeminotelencephalic projection may be present in reptiles (ten Donkelaar and de Boer-van Huizen, 1981), and the DVR of alligators contains an indistinct field of cells in a region comparable to nucleus basalis (Clark and Ulinski, 1984). This region can be seen generally in birds (Witkovsky *et al.*, 1973), but is particularly well developed in water birds that use their beaks to search for food in the sediments of lakes and rivers. In ducks, for example, nucleus basalis is hypertrophied and contains a detailed, somatotopic representation of the sensory surface of the beak (Dubbeldam and Wijsman, 1975; Dubbeldam *et al.*, 1981; Berkhoudt *et al.*, 1981). Nucleus basalis projects to the frontal neostriatum, which is situated directly above it (Wild *et al.*, 1985). This region of DVR projects to the anterior intermediate component of the archistriatum, which, as noted above, is the source of a major descending telencephalic projection to the brain-stem reticular formation. The sequence of projections from the trigeminal complex to nucleus basalis, and ultimately to the brain-stem reticular formation, thereby forms a second route by which the activity of premotor neurons can be modulated by sensory inputs from the beak and oral surfaces (Klein *et al.*, 1985).

In summary, the reptile–bird transition involved a decoupling of hindlimb and forelimb central pattern generators, a subsequent coupling of wing pattern generators (both associated with the acquisition of flight), some increase in cerebellar circuitry, and the establishment of major descending pathways from the pallium and striatum to the brain stem and spinal cord.

5. Cerebral Cortex

Reptiles have a well-differentiated cerebral cortex, which is discussed in detail in Chapter 5 of this volume. Four cytoarchitectonic areas can be distinguished in all Recent reptiles. They have the form of elongate strips that extend rostrocaudally over the surface of the hemisphere, and are designated medial, dorsomedial, dorsal, and lateral cortex according to their positions.

Medial and dorsomedial cortex are generally limbic in character and seem to be present in birds, in which they are usually called the hippocampal and parahippocampal areas, respectively. The hippocampal area is the most medial and in many species is a trilaminar cortex in which the second, or intermediate, layer contains many, densely packed neurons. The parahippocampal area is thicker and contains five layers (e.g., Ramón y Cajal, 1955). Outermost is a plexiform layer containing primarily fibers. Second is a layer of small stellate cells. Third is a layer of stellate cells and pyramidal cells. Fourth is a deep layer of stellate cells. Innermost is an ependymal layer. The second, third, and fourth layers merge with each other, so that the cortex appears to be trilaminar at first glance. The cytoarchitecture of the hippocampal and parahippocampal areas has been described in a variety of clades of birds by Craigie (1928, 1932, 1935, 1936, 1939, 1940a,b). Several minor subdivisions can be distinguished within the major areas, and these vary slightly in character between species. However, Craigie was unable to discern any overall pattern of variation in the cytoarchitecture of the medial part of cortex within the species he studied.

The few hodological studies carried out to date indicate that the hippocampal and parahippocampal areas connect with components of the limbic system. Both areas project to the septum in a roughly topographic way, so that their rostral parts project to the precommissural septum while their caudal parts project to the postcommissural septum (Krayniak and Siegel, 1978a). It is not certain whether or not the hippocampal or parahippocampal areas project to the diencephalon (Bons *et al.*, 1976; Krayniak and Siegel, 1978b). The septum projects, in turn, to the hypothalamus and back to the hippocampal area (Krayniak and Siegel, 1978b). The parahippocampal area receives afferents from the raphe nuclei (Benowitz and Karten, 1976) and from the nuclei of the diagonal band (Krayniak and Siegel, 1978b; Benowitz and Karten, 1976). A major gap in our knowledge concerns the manner in which the hippocampal and parahippocampal areas are interconnected with other components of the pallium. There are in reptiles major interconnections between the medial and dorsomedial cortex and the other areas of cortex (see Chapter 5, this volume), but these connections appear not to have been studied (or reported) in birds. Lesions of the hippocampal and parahippocampal areas in homing pigeons produce a deficit in the bird's ability to find its home loft. The particular deficit involves the final phase of the homing flight in which the bird uses visual clues to locate its loft (Bingman *et al.*, 1984, 1985). This implies that information from telencephalic visual areas must reach the cortex and contribute to the formation of a "cognitive map." However, the routes used by this information are not known.

The lateral cortex of reptiles is an olfactory cortex. It receives projections from the main olfactory bulb and projects to central components of the olfactory system, as well as contributing direct projections to medial cortex. The cytoarchitecture and organization of lateral cortex varies considerably between species of reptiles, being quite extensive in some and reduced to a thin, noncellular pallial membrane in some lizards of the genus *Anolis*. The pyriform cortex, or pyriform area, of birds is located on the ventrolateral surface of the cerebral hemisphere. It receives projections from the olfactory bulb (Skeen *et al.*, 1984; Reiner and Karten, 1985), and generally resembles the lateral cortex of reptiles in its internal organization (Skeen *et al.*, 1984). There is a considerable variation in the relative size of the pyriform cortex between birds (Cobb, 1960a,b), but a general trend is for birds that live in aquatic environments to have relatively larger olfactory bulbs and pyriform cortices than those that live in terrestrial environments. This has been called "Cobb's wet rule."

The fourth cortical area in reptiles is dorsal cortex, which is clearly divided into two or three cytoarchitectonic components that vary considerably in character between species. The medial part seems to be generally limbic in character and has its principal connections with the medial and dorsomedial areas. The lateral component is a visual cortex. It is most extensive in turtles in which it receives a large projection from the dorsal lateral geniculate complex. The organization of this projection is, as far as is known, unique to turtles and is nontopographic in nature (Ulinski, 1988). The region of the telencephalon that receives input from the dorsal lateral geniculate complex is relatively small in lizards (Bruce and Butler, 1984), and is located between the major component of dorsal cortex and lateral cortex. There is no clear equivalent to dorsal cortex in birds. This region is, instead, occupied by the visual Wulst which—in contrast to the situation in reptiles—receives bilateral, retinotopically organized projections from the principal optic complex.

In summary, the medial, dorsomedial, and lateral components of the cortex of reptiles are continued in a relatively reduced form in birds. Dorsal cortex is replaced by those components of the hyperstriatum present in the Wulst.

6. Shared–Derived Features Found within Particular Clades of Birds

The preceding sections considered major modifications in the nervous system that appear to be general features of the reptile–bird transition. However, there are clear indications that additional modifications occured within the nervous systems of particular groups of birds. This section considers two examples of such changes that have attracted a great deal of attention by neurobiologists during the past several years.

6.1. Sound Localization in Owls

The common barn owl (*Tyto alba*) exhibits a remarkable acuity for sound localization, orienting to novel acoustic stimuli by executing stereotyped, rapid head saccades. Barn owls localize frontal targets to within approximately 5° in both azimuth and elevation, the accuracy of localization decreasing with increasing eccentricity of target (Knudsen *et al.*, 1979). They can orient accurately to sound sources that terminate before the start of a head movement. Once initiated, a head saccade typically goes to completion even if a second, conflicting, sound is later presented. This behavioral capability results from modifications of both the peripheral anatomy and central auditory pathways of barn owls.

Cues for sound localization generally include differences in the phase and time of arrival of sounds resulting from the different path lengths to the two ears, and intensity differences resulting from the frequency-dependent acoustic shadow cast by the head. In most birds, the interaural distance and head size are less than one wavelength for the frequencies used in sound localization, and there is little intensity attenuation or time difference between the two ears. Many birds appear to rely on a pressure-gradient system that uses the air-filled interaural pathway to communicate sound between the middle ears (Schwarzkopff, 1952). In the Japanese quail, differences between phase interactions in sound reaching the tympanic membranes of the contralateral and ipsilateral ears result in relatively large interaural intensity differences (Hill *et al.*, 1980). This results in frequency- and intensity-dependent directional tuning of the cochlear microphonic, which may be the basis of sound localization (Coles *et al.*, 1980).

Where time and intensity cues interact in the computation of the position of a sound source, changes in either can affect the perceived azimuthal and elevational location. Accurate localization both in azimuth and in elevation is facilitated by processing separate cues for the two dimensions independently. Barn owls, which hunt rodents from overhead perches or on the wing by listening for prey movements, do not rely on a pressure-gradient system for sound localization (cf. Coles and Guppy, 1988). In the barn owl, the middle ears are acoustically uncoupled because the middle ear canal exhibits large attenuation for the frequencies used in sound localization (Moiseff and Konishi, 1981b). A series of associated peripheral specializations accentuate the separation of time cues mapped onto azimuthal position and intensity cues mapped onto elevational position. The facial ruff acts as a sound collector that comprises acoustically dense feathers along the circumference and midline of the face, and acoustically transparent feathers within the center. Associated with the ruff is a vertical asymmetry in the bony openings of the external ear canals, so that the left ear opening is higher than the right (Payne, 1971). The preaural flaps that partially occlude the ear openings are also placed asymmetrically. These adaptations result in interaural intensity difference (IID) changing primarily with elevation at high frequencies, and changing primarily with azimuth at frequencies less than 4 kHz. Interaural time differences (ITD) resulting from different path lengths to the two ears vary systematically as the position of a sound source moves in the azimuthal plane (Knudsen, 1980, Moiseff, 1988; Payne, 1971). Since the sounds localized by owls often have significant power above 4 kHz (Konishi, 1973), IID is the major determinant of elevational position and ITD is the major determinant of azimuthal position (Knudsen and Konishi, 1979; Moiseff, 1988; Moiseff and Konishi, 1981a).

The independence of time and intensity cues generated in the periphery is maintained throughout much of the brain-stem auditory system of barn owls (Moiseff and Konishi, 1983). The basic pattern of auditory nuclei in barn owls reflects the general avian plan, including parallel brain-stem pathways (Takahashi and Konishi, 1988a, 1988b), but the auditory nuclei are large and well differentiated. Eighth nerve fibers bifurcate prior to reaching the cochlear nucleus complex, which consists of the nuclei angularis and magnocellularis. Although these two nuclei receive input from the same auditory fibers, they exhibit dramatically different properties. Neurons in nucleus magnocellularis respond in synchrony with the phase of the signal. For a given neuron, the magnitude of the response is independent of amplitude over a range of approximately 20–30 dB. Phase locking extends to very high frequencies (9 kHz; Sullivan and Konishi, 1984), possibly because of cochlear specializations including a dense fibrous thickening along part of the basilar membrane (Smith *et al.,* 1985). In contrast, nucleus angularis neurons are sensitive to sound intensity, exhibit monotonic increases in firing rate with increasing intensity over a range of 30–40 dB, and do not exhibit phase locking at high frequencies (Sullivan and Konishi, 1984).

Neurons in nucleus magnocellularis project bilaterally to nucleus laminaris, which is the first site of binaural interaction in the brain stem. Nucleus laminaris is a single sheet of cells in chickens, but is a hypertrophied ovoid with large, relatively sparse cells and extensive neuropil in barn owls. Ipsilateral input enters laminaris dorsally and interacts with corresponding contralateral input, entering ventrally. This forms a map of stimulus phase superimposed on the tonotopic map as ascertained through a "neurophonic" potential recording, although single unit isolation has been difficult to achieve (Sullivan and Konishi, 1986). The position of an axon's projections to laminaris is systematically related to time delay (Carr and Konishi, 1988), so that laminaris neurons may act as coincidence detectors that fire maximally when the time difference between the two ears is equal and opposite to the difference in time delays between axonal inputs. Position in laminaris may thus encode ITD as in the Jeffress model (Jeffress, 1948).

Both nucleus laminaris and nucleus angularis project to the nuclei of the lateral lemniscus. Nucleus laminaris projects to the anterior subdivision of the ventral nucleus of the lateral lemniscus, and neurons in this structure exhibit sensitivity to ITD (Moiseff and Konishi, 1983). Nucleus angularis projects contralaterally to the posterior subdivision of the ventral nucleus of the lateralis lemniscus. Neurons in this structure are excited by contralateral stimulation and inhibited by ipsilateral stimulation (Moiseff and Konishi, 1983). The degree of ipsilateral inhibition systematically decreases within an isofrequency lamina from dorsal to ventral to form a map of intensity difference (Manley *et al.,* 1988). The different subdivisions of ventral nucleus of the lateral lemniscus, thus, represent components of the separate time and intensity pathways. These nuclei, along with nucleus angularis and nucleus laminaris, project to the nucleus mesencephalicus lateralis dorsalis (MLD) of the midbrain.

MLD is situated in the torus semicircularis of the midbrain and contains a central nucleus and an external nucleus.* The central nucleus is further sub-

*The auditory component of the midbrain is called the torus semicircularis in nonmammalian vertebrates. The MLD is a component of the torus that is characteristic of birds generally. Recently, Knudsen (1983a) introduced a nomenclature (i.e., central and external nuclei of the inferior colliculus) for components of MLD in barn owls which implies homologies with components of the inferior colliculus of mammals. Although this nomenclature is now widely used, we prefer to retain the term MLD because it is not clear if central and external components, as used by Knudsen, are generally present in birds and reptiles. If they are not, the relationship of components of MLD in barn owls to components of the inferior colliculus of mammals is not clear.

divided into a core and surrounding shell. Nucleus angularis projects to the contralateral shell (Takahashi and Konishi, 1988a) whereas nucleus laminaris projects to the ipsilateral core (Takahashi and Konishi, 1988b). The core projects to a restricted portion of the contralateral shell, which in turn projects to the external nucleus (Takahashi *et al.,* 1989). Thus, neurons in the external nucleus have access to input exclusively from contralateral structures. The various subdivisions of MLD appear to have different functional properties.

One role of MLD is the resolution of phase ambiguity. At a given frequency there can be several azimuthal locations that share the same ITD. This represents ambiguous information, and barn owls localizing high-frequency tone bursts occasionally make large errors. Such errors do not occur with noise stimuli, which are rich in phase information. The resolution of phase ambiguity occurs in the projections of the central nucleus onto the external nucleus. The central nucleus is tonotopically organized. Its neurons are sharply tuned to frequency (Knudsen and Konishi, 1978c) and form a series of isofrequency lamellae. Neurons in the central nucleus are sensitive to ITDs, but their ITD tuning curves exhibit side peaks. Thus, individual central neurons do not encode for a characteristic delay and cannot resolve the phase ambiguity associated with tonal stimuli. All of the neurons within a given electrode penetration perpendicular to the isofrequency lamellae have one peak at the same ITD, but side peaks at varying frequencies. Characteristic delay is thus represented within central MLD by an array of neurons (Wagner *et al.,* 1987).

In contrast to the central nucleus, neurons in the external nucleus are arranged in a topographic map systematically representing contralateral auditory space, and show little or no tonotopic organization (Knudsen and Konishi, 1987b). They have spatial receptive fields whose borders are largely independent of absolute intensity and are presumable sharpened by lateral interaction with inhibitory surrounds of neighboring cells (Knudsen and Konishi, 1978a). Space-specific neurons are tuned to specific combinations of ITD and IID. The sharpness of tuning of individual neurons determines the azimuthal and elevational extent, respectively, of the spatial receptive fields (Moiseff and Konishi, 1981a). The ITD and IID tuning of space-specific neurons can be manipulated independently by injection of local anesthetic into nucleus magnocellularis or nucleus angularis (Konishi *et al.,* 1985). The properties of these neurons result from projections from the shell of the central subdivision. The projection is organized such that neurons in different insofrequency lamellae of the shell project to a single point in the external nucleus (Knudsen, 1983a). A focal injection of HRP into a physiologically identified site in the external nucleus primarily labels a band of cells in the central nucleus. The best ITD of the site in the external nucleus is the same as the characteristic delay of the band of cells in the central nucleus (Wagner *et al.,* 1987). As with neurons of the central nucleus, the ITD tuning curves of space-specific neurons exhibit multiple peaks when measured with tone bursts (Takahashi and Konishi, 1986). These peaks include a central peak (best ITD) which defines the characteristic delay for that neuron and is independent of frequency, and side peaks whose positions are related to the frequency of the tone burst. In contrast to neurons of the central nucleus, however, noise stimuli or even two tones presented simultaneously to space-specific neurons are sufficient to suppress the side peaks (Takahashi and Konishi, 1986).

The receptive fields of space-specific neurons are quantitatively related to behavior. The linear relationship between the azimuthal component of the head-turning vectors and the ITD of the noise stimuli that elicit head turns does not differ from the linear relationship between the azimuthal centers of neuronal receptive field centers and the ITD of the noise stimuli (Moiseff and Konishi, 1981a). When owls are prepared with a monaural occlusion, they make systematic errors in localizing sound sources (Knudsen and Konishi, 1979). Correspondingly, the receptive fields of space-specific neurons shift, with a magnitude and direction consistent with the errors in localization (Knudsen and Konishi, 1980).

The external nucleus projects in a topographically organized fashion to the optic tectum. Since the optic tectum is the source of tectoreticular pathways that control head movements, the projection from the external nucleus of MLD to the tectum plays a crucial role in orientation of sounds. Many neurons in the optic tectum respond bimodally to sound and visual stimuli. The auditory and visual spatial receptive fields of these neurons are aligned, and the neurons are arranged in a map of space. However, the auditory and visual maps are not in register where peripheral regions of space are represented (Knudsen, 1982). The registry of auditory and visual maps present in much of the tectum is a product of developmental processes sensitive to environmental cues. There is a sensitive period ending at about 8 weeks of age during which young barn owls can adjust to abnormal binaural cues and recover accurate sound localization after monaural occlusion (Knudsen *et al.,* 1984a). After restoration of normal auditory cues by removal of earplugs, barn owls can also regain accurate localization, but only if the earplugs are removed during a critical period ending at 38–42 weeks of age (Knudsen *et al.,* 1984b). Barn owls raised with earplugs have abnormal optic tectum maps, with neurons exhibiting auditory spatial fields shifted relative to visual spatial fields (Knudsen, 1983b). The auditory map in the tectum is dependent on normal vision and tectal neurons in barn owls raised with closed eyelids but normal hearing exhibit misalignment of auditory and visual receptive fields and abnormal map topography (Knudsen, 1988).

The overall sequence of projections in the auditory pathways of barn owls conforms to the basic avian plan. It is the details of the response properties in neurons in the external nucleus of MLD that result in the barn owl's ability to localize sounds. These properties have been studied in several species of owls chosen to represent distinct clades within the order Strigiformes (Volman and Konishi, 1988). In the great horned (*Bubo virginianus*) and burrowing owls (*Athene cunicularia*), the ears are symmetrical, and the space map in the external nucleus is based entirely on ITDs and encodes only the azimuthal location. Curiously, the neurons require binaural input although all neurons prefer nearly balanced binaural intensity, representing a sound source directly in front of the animal. These neurons respond to relatively low frequencies (2–5 kHz). In contrast, the long-eared owl (*Asio otus*) has asymmetrical ears. Neurons in the external nucleus in this species are also exclusively binaural, but exhibit weak elevation tuning and are tuned to a range of ITDs. They respond to higher frequencies (2.5–7 kHz). Thus, sensitivity to binaural time differences may be a general feature of owls. However, the use of intensity cues for elevational localization is apparently correlated with higher frequency of hearing and ear asymmetry. These features have evolved at least twice within the owls, once in the barn owls within the family Tytonidae and separately within the family Strigidae, which includes all other owls.

6.2. Vocal Learning and the Song System

6.2.1. Vocal Behavior in Birds

In contrast to most reptiles, birds are highly vocal animals that produce sound with a syrinx situated near the juncture of the primary bronchi and trachea. The syrinx contains intrinsic muscles that change its conformation and extrinsic muscles that change the position of the trachea. The position of the syrinx and number of muscle pairs vary among species so that different syrinxes act on the basis of different physical principles (Gaunt and Gaunt, 1985). There is a rough correlation between number of muscle groups and vocal complexity as assessed by inspection of sonograms. In oscine birds ("true" passerines), the sound-producing medial tympaniform membranes exist bilaterally. Birds can produce two voices simultaneously when the membranes act independently (Greenewalt, 1968; Suthers, 1989), although the membranes may also be acoustically coupled during some vocalizations (Nowicki and Marler, 1988; Suthers, 1989).

Most birds have a distinct series of stereotyped calls that resemble in complexity the vocalizations produced by reptiles such as alligators. These are often produced in association with specific behaviors, are of short duration and simple morphology, and are primarily influenced by innate factors. For example, the calls of chickens and ringdoves are not affected by deafening at hatching (Konishi, 1963; Nottebohm and Nottebohm, 1971). However, some birds also produce songs, which can be complex, learned vocalizations. Song typically depends on ontogenetic factors. In oscine passerines, for example, song development in deaf birds is highly abnormal (Konishi, 1964, 1965a, 1985; Marler and Sherman, 1982). Vocal learning is also known for parrots in the order Psittaciformes, and in hummingbirds in the order Apodiformes (Baptista and Schuchmann, 1989; Snow, 1968). In species that have song, calls can also be modified in response to ontogenetic influences (Dooling et al., 1987; Mundinger, 1970; Nowicki, 1989; Simpson and Vicario, 1990).

6.2.2. Song System

A system of nuclei associated with vocal production is present in oscine passerines and parakeets (Nottebohm, 1980; Paton *et al.,* 1981). Telencephalic components of the song system include the nucleus hyperstriatum ventrale, pars caudale (HVc) in the hyperstriatum, nucleus interfacialis (NIf) and nucleus magnocellularis neostriatalis anterioris (MAN) in the neostriatum, area X in the medial striatum, and nucleus robustus archistriatalis (RA) in the archistriatum. All of these structures are characterized by rather distinct cytoarchitectonic borders.

The nuclei are linked together in a series of descending pathways. The major pathway is entirely ipsilateral and unidirectional, with NIf projecting to HVc which in turn projects to RA (Nottebohm *et al.,* 1976, 1982). RA projects ultimately to the intercollicularis component of the midbrain (Nottebohm *et al.,* 1976), especially to its dorsomedial subdivision (DM) (Gurney, 1981), and to the posterior or tracheosyringeal division of the hypoglossal motor nucleus (nXIIts). nXIIts gives rise to axons that unilaterally innervate the syrinx in songbirds. Lesions of most of the nuclei in this pathway disrupt singing (Nottebohm *et al.,*

1976), and neurons in at least some of the nuclei exhibit time-locked activity synchronous with singing (McCasland, 1987). A second major pathway involves projections from HVc to area X, which projects to the dorsal posterior nuclear complex (Okuhata and Saito, 1987). This nucleus gives rise to input to the lateral part of MAN, which projects to RA (Bottjer *et al.*, 1989). Lesions of nuclei in this pathway do not affect adult song (Nottebohm et al., 1976), but young birds with MAN lesions do not crystallize normal songs (Bottjer et al., 1984). Area X lesions also affect development of normal song (Schaarff and Nottebohm, 1989; Sohrabji *et al.*, 1989). A third pathway involves the prethalamic nucleus uvaeformis (Nottebohm *et al.*, 1982), whose inputs are not known. Projections in these pathways are arranged topographically, except that no simple organization for HVc's inputs or outputs has been determined (Gurney, 1980; Nottebohm *et al.*, 1982). HVc projects to both the dorsal and ventral parts of RA. Axons entering the dorsal part bifurcate extensively while axons entering the ventral part bifurcate infrequently (Vicario and Simpson, 1988). Dorsal RA projects to DM, while ventral RA projects to nXIIts in a topographic fashion. The muscles of the syrinx are represented as overlapping, but distinct, domains in nXIIts (Vicario and Nottebohm, 1988).

In many oscines, unilateral sectioning of the tracheosyringeal nerve demonstrates a laterality effect: the left nerve cut produces dramatic degradation of many or most song elements while the right nerve cut has little effect (Nottebohm, 1971). This effect is also observed with unilateral lesions of HVc and RA (Nottebohm *et al.*, 1976), which has been interpreted as evidence for left hemispheric dominance or control of singing (Nottebohm, 1980). Physiological data, however, demonstrate that both sides are active during singing (McCasland, 1987). Recent measurement of the airflow in each bronchus and the near-field acoustic signal from each medial tympaniform membrane suggest that some birds have contributions to most sounds from both sides, although birds with strong left hypoglossal dominance have yet to be tested (Suthers, 1989).

Singing requires coordination of several motor systems. Changes in the configuration of the vocal tract, tongue, and beak all effect the produced sound (Nowicki and Marler, 1988). Singing is also associated with stereotyped posture. Canaries with bilateral lesions of HVc or RA can participate in "silent song," assuming the singing posture but producing no throat movements or sound (Nottebohm *et al.*, 1976). Song is coordinated with respiration, and a conspicuous respiratory rhythm is superimposed on the spontaneous activity in the nerves that innervate the syrinx. This respiratory rhythm "gates" the efficacy of electrical stimulation of HVc or RA, so that stimulation during expiration elicits greater activity in the nerve than does stimulation during inspiration (Manogue and Paton, 1982). Sound is produced only during expiration, and birds sustain long bouts of continuous singing by taking "minibreaths" at the end of each syllable (Hartley and Suthers, 1989). Mechanisms for call and song production must also be closely linked because HVc neurons are recruited during both activities (McCasland, 1983). The intercollicular nucleus is associated with call production, and stimulation in a restricted, medial part of the intercollicular nucleus elicits call production in species such as Japanese quail (Potash, 1970) that do not learn vocalizations, as well as in songbirds such as red-winged blackbirds (Newman, 1970). In the Java sparrow, intercollicular stimulation is said to produce only poorly modulated noises (Seller, 1980). How the two motor systems interact, and whether one is subordinate in songbirds, has yet to be explored.

6.2.3. Auditory Input to the Song System

Auditory input reaches the song system via projections from nucleus ovoidalis in the dorsal thalamus to field L in the caudal neostriatum (Karten, 1968; Nottebohm *et al.*, 1976; Kelley and Nottebohm, 1979). The projection is topographic in both parakeets (Brauth *et al.*, 1987) and zebra finches (Bell *et al.*, 1989). The majority of auditory neurons within field L and the surrounding caudal neostriatum respond to tone bursts, and are organized into a tonotopic map in field L that extends over the entire caudal neostriatum and adjacent hyperstriatum (Bonke *et al.*, 1979; Heil and Scheich, 1985; Margoliash, 1986; Müller and Leppelsack, 1985; Rübsamen and Dörrscheidt, 1986; Theurick *et al.*, 1984). Field L and the surrounding caudal neostriatum is subdivided into three functional zones: an intermediate zone with relatively simple tone-burst responses, and superficial and deep layers with more complex and selective response properties (Bonke *et al.*, 1979; Müller and Leppelsack, 1985; Müller and Scheich, 1985; Scheich *et al.*, 1979a). Many neurons in the caudal neostriatum respond to tone bursts in starlings, but about 17% require more complex stimuli that include amplitude modulation or limited bandwidth noise bursts (Müller and Leppelsack, 1985). Similarly, the majority of neurons in guinea fowl respond to tone bursts, while 16% are selective for spectral or temporal parameters found in a guinea fowl vocalization, the lambus call (Scheich *et al.*, 1979b). Within field L, neurons with complex response properties tend to segregate to the superficial and deep layers. In starlings, a small number of field L neurons are highly selective for elements of song (Leppelsack, 1983; Leppelsack and Vogt, 1976). Some field L neurons apparently are stimulated only by stimuli that combine two separate frequencies in guinea fowl (Scheich *et al.*, 1979b) and oscine mynah birds (Langner *et al.*, 1981).

Field L projects to two nuclei in the oscine song system. There is first a projection from central field L to a cell-sparse "shelf" medial and contiguous with the ventral border of HVc (Kelley and Nottebohm, 1979). Interactions between ventral HVc neurons and field L axons may be a source of auditory input into HVc (L. Katz, see Margoliash, 1987). Second, a lateral component of field L projects to RA, although direct interaction between field L axons and RA neurons has not been demonstrated. Which functional classes of field L neurons give rise to the projections to song system nuclei has yet to be determined, although the anatomical evidence indicates that the ascending projection neurons that surround the intermediate lamina are involved (Saini and Leppelsack, 1977).

Auditory responses are widespread within the song system (Katz, 1982; Katz and Gurney, 1981; Margoliash, 1983, 1986; McCasland and Konishi, 1981; Paton and Nottebohm, 1984), and an auditory-evoked response can even be recorded in the hypoglossal nerve to the syrinx (Williams and Nottebohm, 1985). In contrast to field L, the response properties of auditory neurons are shaped by the song-learning process. A small subset of HVc neurons in white-crowned sparrows do not respond to tone bursts, but require a temporal sequence of two notes for excitation. These temporal combination-sensitive neurons are optimally stimulated by sequential notes from the individual bird's own autogenous song, and therefore are "song-specific" (Margoliash, 1983). Song-specific neurons are highly selective, responding only to autogenous song or a limited subset of conspecific songs from the same dialect. Some respond to a sequence of two whistles or a sequence of simple tone bursts. Their optimal frequencies correspond to the center frequencies for the corresponding whistles, so that the response for some units is elicited by the second of two identical tone bursts. Harmonic combination-sensitive neurons selective for autogenous song have been demonstrated in the zebra finch HVc, suggesting that combination sensitivity may be common in the HVc of songbirds (Fortune and Margoliash, 1989). As in field L, the majority of HVc neurons have simpler properties including responses to tone bursts. In contrast to field L, however, most HVc neurons are selective for autogenous song (Margoliash, 1986; Margoliash and Konishi, 1985; Margoliash and Fortune, in preparation). Field L does not seem to contain neurons selective for autogenous song (Margoliash, 1986). The individual-specific or idiosyncratic feature of the response properties of HVc neurons suggests that auditory–motor interactions in HVc may play an important role in the song crystallization process (Margoliash, 1987; Volman and Konishi, 1987).

6.2.4. Plasticity in Song System Nuclei

The ability to sing varies considerably between birds, depending on their sex and hormonal status as well as species. The male bird typically sings, whereas the female sings less frequently or not at all. These behavioral differences are under hormonal control and are associated with distinct patterns of sexual dimorphism in song system nuclei. There is a correlation between the amount of female singing and the volume of song system nuclei (Brenowitz and Arnold, 1986). Song system nuclei are either rudimentary or not apparent in species such as the zebra finch where the female does not sing (Nottebohm and Arnold, 1976). Canary females sing infrequently, and their song system nuclei are present but reduced compared to the male (Nottebohm and Arnold, 1976). Bay wren females are competent singers, and their song system nuclei are comparable to males (DeVoogd *et al.,* 1988). The female song system may be competent to respond to testosterone, as in the canaries where exogenous hormone elicits a burst of singing from the female. The activational effects of testosterone result in increases in the level of protein synthesis (Konishi and Akutagawa, 1981), and increases in the soma size and the size of dendritic arbors of neurons in RA (DeVoogd and Nottebohm, 1981, DeVoogd *et al.,* 1985). The increased dendritic length of RA neurons does not change the type or distribution of synapses on these neurons, however (Canady *et al.,* 1988). The effects of hormone are enhanced by auditory stimulation, so that a modest increase in nuclear volume results from increased neuronal spacing in deaf female canaries treated with testosterone, whereas hormone stimulates a substantial increase in nuclear volume, which results principally from increased neuronal number, in hearing-intact females (Bottjer *et al.,* 1986; Bottjer and Dignan, 1988).

Effects on nuclear volume are also seen in males. Male birds have high hormone titers and large song system nuclei during the breeding season, but their hormone titer is low and their song system nuclei are smaller during the winter months. This effect is seen in red-winged blackbirds and canaries (Kirn *et al.,* 1989; Nottebohm, 1981), but has not been observed in the HVc of white-crowned sparrows (Baker *et al.,* 1984). These data are consistent with the hypothesis that seasonal changes in nuclear volume are related to changes in song repertoire (Nottebohm, 1981). In rufous-sided towhees, however, seasonal changes in nuclear volume are observed although these birds maintain constant repertoires between seasons (Brenowitz *et al.,* 1989). Thus, hormone alone or its absence may stimulate growth and regression of nuclear volumes. In the canary, a weak correlation has been observed between the size of an individual's song repertoire and the size of the brain nuclei (Nottebohm *et al.,* 1981). This correlation is much stronger in geographically separated populations of marsh wrens (Canady *et al.,* 1984). In contrast, however, no correlation between repertoire size and nuclear volume was observed in red-winged blackbirds (Kirn *et al.,* 1989).

Many song system nuclei, and the syrinx, have steroid hormone-concentrating cells (Arnold *et al.,* 1976; Lieberburg and Nottebohm, 1979). The gender differences in brain structure arise from a developmental sequence that unfolds in the presence of a changing hormonal environment, with two peaks in the level of testosterone occurring during the periods of song acquisition and song development (Marler *et al.,* 1987). For example, both male and female zebra finches have large numbers of neurons and large nuclei in RA up to about 30 days of age. At about this time, the female neurons undergo a round of atrophy and cell death whereas the male neurons continue to grow (Konishi and Akutagawa, 1985; Bottjer *et al.,* 1985). The female song system can be rescued by administration of estrogen early in life. The effect of hormone can be organizational, as in the case of the projection of MAN onto RA, present in males and early estrogen-treated females but absent in untreated females (Gurney, 1981). Early estrogen treatment of female zebra finches results in song system neurons and nuclei that are intermediate in size to the male. In adulthood, neurons in such birds respond to exogenous testosterone by increasing in size, and the bird can sing (Gurney, 1982). The action of hormone may be transient, indirect, and species specific. Thus, there are fewer hormone-concentrating cells in HVc of zebra finches than in canaries (Nordeen *et al.,* 1987). In adult zebra finches, none of the neurons in the song system have estrogen receptors, whereas these are present in adult canary HVc neurons (Gahr *et al.,* 1987). Neurons with estrogen receptors are also present in the HVc of juvenile zebra finches, but decline in number at about 40 days of age (Gahr and Konishi, 1988).

The plastic changes in the song system nuclei also reflect, in part, a process of ongoing neurogenesis in the forebrains of adult songbirds. This was initially demonstrated in canaries using [^{3}H]thymidine labeling (Goldman and Nottebohm, 1983). Labeled neurons are present if birds are permitted long survival times after thymidine injection, but only the precursor cells in the overlying ventricular zone are labeled if birds are sacrificed shortly after thymidine labeling. Thymidine-labeled cells in HVc receive morphologically normal synapses (Burd and Nottebohm, 1985), exhibit normal spiking activity, and respond to auditory stimuli in some cases (Paton and Nottebohm, 1984). Many new HVc neurons in adult canaries are intrinsic (Paton *et al.,* 1985), but even newly formed projection neurons can survive in adult brains. Different classes of HVc projection neurons have different patterns of neurogenesis. Area X-projecting cells are almost exclusively born *in ovo,* while RA-projecting cells continue to be incorporated past 240 days (Alvarez-Buylla *et al.,* 1988; Nordeen and Nordeen, 1988). The stability of the X-projecting neurons in adult canaries was confirmed by labeling cells with latex beads and comparing the number of surviving cells 1 and 2 years after the injections (Gahr, 1990). Some new neurons migrating ventrally from the ventricular zone remain in HVc, whereas others pass into other forebrain structures.

Since the brain does not increase in size, new neurons must replace old ones.

The number of HVc neurons in the fall is lower than in spring. Neurons apparently accumulate throughout the year, and the population is reduced through one round of cell death in the fall (Nottebohm *et al.*, 1986). This suggests that neurogenesis might be regulated by testosterone levels, which also decline in the fall. Accordingly, the rate at which new neurons are born in intact females is not increased by giving testosterone (Goldman and Nottebohm, 1983), but testosterone replacement increases neurogenesis if females are ovariectomized (Nottebohm, 1989). Thus, hormonal regulation can affect the rate of neurogenesis, either through direct stimulation of neuronal production or an indirect effect. Since testosterone is known to increase dendritic growth and synaptogenesis in the song system (Clower *et al.*, 1989; DeVoogd *et al.*, 1985), perhaps an effect of the hormone is to make postsynaptic space available to new neurons. If so, this would suggest that new neurons do not have a predetermined fate, instead being "captured" by a particular circuit that has space to accommodate extra elements. Nuclei such as HVc that are physically closer to the germinal zone would be better situated to recruit new neurons.

Although the fact of postnatal neurogenesis is now well established, its functional role remains uncertain. Neurogenesis suggests a clear capacity to reorganize neuronal networks. In canaries, the size of the vocal repertoire, as measured by number of different syllable types, increases in the adult years (Nottebohm and Nottebohm, 1978). This prompted hypotheses that new neurons were incorporated into motor circuits to establish new vocal patterns. However, female songbirds and species lacking adult vocal plasticity also exhibit adult neurogenesis, and neurogenesis is not limited to the song system (Nottebohm, 1985). Thus, it appears that many species of birds have access to new neurons postnatally and in adulthood, and presumably can take advantage of this in many ways. A unitary explanation for the functional significance of neurogenesis is unlikely to obtain.

6.2.5. Evolution of the Song System

Vocal learning is thought to have evolved independently in songbirds, parrots, and hummingbirds (Nottebohm, 1972). Although the vocal control system has not been completely characterized in each of these clades of birds, it appears that the song system evolved as the result of relatively minor modifications of existing forebrain pathways. Thus, the basic pattern of descending pathways in the song system resembles the general pattern of DVR projections to the archistriatum, and the subsequent projections of the archistriatum to the brain stem. Also, the projections of field L onto the motor system conform to the pattern of field L projections in birds generally. The most striking feature of the nuclei of the song system in oscine songbirds is their cytoarchitectural clarity. However, there is some reason to believe that the basic elements of the song system are present in suboscine passerine birds.

A song system is not cytoarchitectonically apparent in suboscine passerines (Nottebohm, 1980). However, as with many oscine females, female suboscine flycatchers sing in response to exogenous testosterone (Kroodsma, 1984). This results, presumably, from the action of testosterone on a homologous but cytoarchitectonically indistinct, suboscine song system. Also, oscine birds exhibit seasonal fluctuations in the volumes of song system nuclei. The fluidity with which birds can orchestrate dramatic changes in morphology of existing brain structures during both ontogeny and adulthood suggests that modifications of preexisting sensory-motor pathways in the brain during evolution resulted in a song system with cytoarchitectonically distinct nuclei. Recent evidence that the song of juvenile-deafened suboscine eastern phoebes (*Sayoruis phoebe*) develops normally suggests that differentiation of the song system and song learning co-evolved (Kroodsma and Konishi, unpublished results).

7. Conclusions

The neural changes involved in the reptile–bird transition can be grouped into three conceptual stages. The first is represented by *Archaeopteryx* and is characterized by a slight increase in the size of the cerebellum and optic tectum and some increase in the breadth of the cerebral hemispheres. If, as the fossil record indicates, *Archaeopteryx* was bipedal, it must have had a decoupling of the central pattern generators controlling the movements of the hindlimbs and the forelimbs. If, as seems likely, it was capable of powered flight, the forelimbs must have moved synchronously. The expansion of the cerebellum and optic tectum suggests some increase in interaction between the cerebellum and accessory optic system related to flight. The relatively small size of the cerebral hemispheres provides no indication of a lateral expansion of DVR into an avian configuration. All of these changes probably occurred in parallel several times within the archosaurs: the adoption of a bipedal posture, with the attendant neural changes, occurred several times within the various groups of dinosaurs, and the evolution of flight occurred at least twice (in birds and in pterosaurs). The endocasts of pterosaurs suggest a gradual broadening of the cerebral hemispheres, increase in size of the optic tectum and cerebellum, and decrease in size of the olfactory bulbs during the course of pterosaur evolution. The endocast of the advanced pterosaur, *Pteranodon,* is actually somewhat more "avian" than that of *Archaeopteryx.* The first stage of avian neural evolution was, thus, essentially an archosaurian one.

The second stage involves further enlargement of the cerebellum and optic tectum and the development of a typically avian configuration in the cerebral hemispheres. There are, unfortunately, no endocasts of avian brains available from this critical stage of evolution. However, the fact that the cerebral hemispheres of all clades of Recent birds have the same basic configuration indicates that the avian forebrain pattern had envolved prior to the separation of the major groups of Recent birds. The clearly avian skeletal morphology in the Cretaceous birds *Ichthyornis* and *Hesperornis* makes it likely that these birds also achieved an avian forebrain configuration. This change is characterized by a lateral expansion of DVR, particularly of the intermediate neostriatum, that occurs in concert with the development of the dorsal posterior nuclear complex in the thalamus. These structures are involved in an increase in the complexity of the striatal and cerebellar circuitry over that seen in reptiles. A second key change is the replacement of the dorsal cortex with the Wulst that occurs in relation to the development of bilateral visual projections from the dorsal thalamus to the lateral telencephalon. Neither of these two cerebral features occurs in reptiles; both are as typically avian as feathers.

The third stage involves particular modifications associated with behavioral adaptations within the various clades of birds. Examples of these include the development of a spatial representation of auditory space and descending, telencephalic control of the talons within the owls; an expansion of nucleus basalis of DVR associated with the use of the beak as a tactile organ within the wading birds; and elaboration of the song system for the generation of species-specific songs within the oscine perching birds, parrots, and hummingbirds. There is presently no way of placing dates on these modifications, but the presence of a diversity of birds within the Cretaceous period and early periods of the Cenozoic era indicate that the major groups of birds radiated rapidly after the establishment of the basic avian configuration.

ACKNOWLEDGEMENT. This work was supported in part by grants to DM from the NIH (NS25677-01), ONR (N00014-89-J-1509), and the Searle Scholars Program/Chicago Community Trust.

8. References

Adamo, N. J., 1967, Connections of efferent fibers from hyperstriatal areas in chicken, raven and African lovebird, *J. Comp. Neurol.* **131:**337–356.

Alexander, R. M., 1989, *Dynamics of Dinosaurs and Other Extinct Giants,* Columbia University Press, New York.

Alvarez-Buylla, A., Theelan, M., and Nottebohm, F., 1988, Birth of projection neurons in higher vocal centers of the canary forebrain before, during, and after song learning, *Proc. Natl. Acad. Sci. USA* **85:**8722–8726.

Arends, J. J. A., and Dubbeldam, J. L., 1984, The subnuclei and primary afferents of the descending trigeminal system in the mallard (*Anas platyrhynchos*), *Neuroscience* **13:**781–795.

Arends, J. J. A., and Zeigler, H. P., 1989, Cerebellar connections of the trigeminal system in the pigeon (*Columba livia*), Brain Res. **487:**69–78.

Arends, J. J. A., Woelders-Blok, A., and Dubbeldam, J., 1984, The efferent connections of the nuclei of the descending trigeminal tract in the mallard (*Anas platyrhynchos*), *Neuroscience* **13:**797–817.

Ariëns Kappers, C. U., 1921, *Vergleichende Anatomie des Nervensystems,* Bohn, Haarlem.

Ariëns Kappers, C. U., Huber, G. C., and Crosby, E. C., 1936, *The Comparative Anatomy of the Nervous System of Vertebrates, Including Man,* MacMillan, New York.

Armstrong, E., and Bergeron, R., 1985, Relative brain size and metabolism in birds, *Brain Behav. Evol.* **26:**141–153.

Armstrong, R. C., and Clarke, P. G. H., 1979, Neuronal death and the development of the pontine nuclei and inferior olive in the chick, *Neuroscience* **4:**1635–1647.

Arnold, A., Nottebohm, F., and Pfaff, D. W., 1976, Hormone concentrating cells in vocal control and other areas of the brain of the zebra finch (*Poephila guttata*), *J. Comp. Neurol.* **165:**487–512.

Bagnoli, P., and Burkhalter, A., 1983, Organization of the afferent projections to the Wulst in the pigeon, *J. Comp. Neurol.* **214:**103–113.

Bagnoli, P., Francesconi, W., and Magni, F., 1979, Interaction of optic tract and visual Wulst impulses on single units of the pigeon's optic tectum, *Brain Behav. Evol.* **16:**19–37.

Bagnoli, P., Grassi, S., and Magni, F., 1980, A direct connection between visual Wulst and tectum opticum in the pigeon (*Columba livia*) demonstrated by horseradish peroxidase, *Arch. Ital. Biol.* **118:**72–88.

Baker, M. C., Bottjer, S. W., and Arnold, A. P., 1984, Sexual dimorphism and lack of seasonal changes in vocal control regions of the white-crowned sparrow brain, *Brain Res.* **295:**85–89.

Bakker, R. T., 1968, The superiority of dinosaurs, *Discovery, New Haven* **3:**11–22.

Bakker, R. T., 1986, *The Dinosaur Heresies,* Morrow, New York.

Baptista, L. F., and Schuchmann, K. -L., 1989, Song learning in the Anna hummingbird (*Calypte anna*), *Ethology* **84:**15–26.

Bell, P. M., Margoliash, D., and Ulinski, P. S., 1989, Cytoarchitecture and differential projections of nucleus ovoidalis in the zebra finch, *Poephila guttata, Soc. Neurosci. Abstr.* **15:**617.

Benowitz, L., 1980, Functional organization of the avian telencephalon, in: *Comparative Neurology of the Telencephalon* (S. O. E. Ebbesson, ed.), Plenum Press, New York, pp. 389–421.

Benowitz, L. I., and Karten, H. J., 1976, The tractus infundibuli and other afferents to the parahippocampal region of the pigeon, *Brain Res.* **102:**174–180.

Berkhoudt, H., Dubbeldam, J. L., and Zeilstra, S., 1981, Studies on the somatotopy of the trigeminal system in the mallard, *Anas platyrhynchos* L. IV. Tactile representation in the nucleus basalis, *J. Comp. Neurol.* **196:**407–420.

Bingman, V. P., Bagnoli, P., Ioloa, P., and Gasini, G., 1984, Homing behavior of pigeons after telencephalic ablations, *Brain Behav. Evol.* **24:**94–108.

Bingman, V. P., Ioale, P., Casini, G., and Bagnoli, P., 1985, Dorsomedial forebrain ablations and home loft association behaviors in homing pigeons, *Brain Behav. Evol.* **26:**1–9.

Bloch, S., and Mantinoya, C., 1982, Comparing frontal and lateral viewing in the pigeon. I. Tachistoscopic visual acuity as a function of distance, *Behav. Brain Res.* **5:**221–244.

Bloch, S., Rey, J., and Mantinoya, C., 1984, Visual acuity as a function of distance for frontal and lateral viewing in the pigeon. III. Different patterns of eye movements for binocular and monocular fixation, *Behav. Brain Res.* **13:**56–61.

Bonke, B. A., Bonke, D., and Scheich, H., 1979, Connectivity of the auditory forebrain nuclei in the guinea fowl (*Numida meleagris*), Cell Tissue Res. **200:**101–121.

Bonke, D., Scheich, H., and Langner, G., 1979, Responsiveness of units in the auditory neostriatum of the guinea fowl (*Numida meleagris*) to species-specific calls and synthetic stimuli. I. Tonotopy and functional zones, *J. Comp. Physiol.* **132:**243–255.

Bons, N., Bouille, C., Bayle, J. D., and Assenmacher, I., 1976, Light and electron microscopic evidence of hypothalamic afferences originating from the hippocampus in the pigeon, *Experientia* **32:**1443–14451.

Bottjer, S. W., and Dignan, T. P., 1988, Joint hormonal and sensory stimulation modulate neuronal number in adult canary brains, *J. Neurobiol.* **19:**624–635.

Bottjer, S. W., Miesner, E. A., and Arnold, A. P., 1984, Forebrain lesions disrupt development but not maintenance of song in passerine birds, *Science* **224:**910–903.

Bottjer, S. W., Glaessuer, S. L., and Arnold, A. P., 1985, Ontogeny of brain nuclei controlling song learning and behavior in zebra finches, *J. Neurosci.* **6:**1556–1562.

Bottjer, S. W., Schoonmaker, J. N., and Arnold, A. P., 1986, auditory and hormonal stimulation interact to produce neural growth in adult canary, *J. Neurobiol.* **17:**605–612.

Bottjer, S. W., Halseman, K. A., Brown, S. A., and Miesner, E. A., 1989, Axonal connections of a forebrain nucleus involved with vocal learning in zebra finches, *J. Comp. Neurol.* **279:** 312–326.

Brauth, S. E., 1987, Motor functions of the caiman basal ganglia, *Soc. Neurosci. Abstr.* **13:**978.

Brauth, S. E., 1988, Catecholamine neurons in the brainstem of the reptile *Caiman crocodilus, J. Comp. Neurol.* **270:**313–326.

Brauth, S. E., and Karten, H. J., 1977, Direct accessory optic projections to the vestibulo-cerebellum: A possible channel for oculomotor control systems, *Exp. Brain Res.* **28:**73–84.

Brauth, S. E., and McHale, C. M., 1988, Auditory pathways in budgerigar. II. Intratelencephalic pathways, *Brain Behav. Evol.* **32:**193–207.

Brauth, S. E., Ferguson, J. L., and Kitt, C. A., 1978, Prosencephalic pathways related to the paleostriatum of the pigeon (*Columba livia*), Brain Res. **147:**205–221.

Brauth, S. E., Kitt, C. A., Reiner, A., and Quiron, R., 1986, Neurotensin binding sites in the forebrain and midbrain of the pigeon, *J. Comp. Neurol.* **253:**358–373.

Brauth, S. E., McHale, C. M., Brasher, C. A., and Dooling, R. J., 1987, Auditory pathways in the budgerigar. I. Thalamo-telencephalic projections, *Brain Behav. Evol.* **30:**174–199.

Bravo, H., and Pettigrew, J. D., 1981, The distribution of neurons projecting from retina and visual cortex to the thalamus and tectum opticum of the barn owl, *Tyto alba,* and the burrowing owl, *Speotyto cunnicularia, J. Comp. Neurol.* **199:**418–441.

Brenowitz, E. A., and Arnold, A. P., 1986, Interspecific comparisons of the size of neural song control regions and song complexity in duetting birds: Evolutionary implication, *J. Neurosci.* **6:**2875–2879.

Brenowitz, E. A., Kroodsma, D., Nalls, B., and Wingfield, J., 1989, Seasonal changes in avain song control nuclei without seasonal changes in song repertoires, *Soc. Neurosci. Abstr.* **15:**619.

Brodal, A., Kristiansen, K., and Jansen, J., 1950, Experimental demonstration of a pontine homologue in birds, *J. Comp. Neurol.* **92:**23–69.

Bruce, L. L., and Butler, A. B., 1984, Telencephalic connections in lizards. I. Projections to cortex, *J. Comp. Neurol.* **229:**585–601.

Bugbee, N., 1979, The effect of bilateral lesions in the nucleus spiriformis lateralis on visually guided behavior in the pigeon, Ph.D. thesis, University of Maryland.

Bugbee, N. M., and Hodos, W., 1979, The basal ganglia–tectal pathway: Its role in visually guided behavior in the pigeon, *Soc. Neurosci. Abstr.* **5:**68.

Burd, G. D., and Nottebohm, F., 1985, Ultrastructural characterization of synaptic terminals formed on newly-generated neurons in a song control nucleus of the adult canary forebrain, *J. Comp. Neurol.* **240:**143–152.

Cabot, J. B., Reiner, A., and Bogan, N., 1982, Avian bulbospinal pathways: Anterograde and retrograde studies of cells of origin, funicular trajectories and laminar terminations, *Prog. Brain Res.* **57:**79–108.

Canady, R. A., Kroodsma, D. E., and Nottebohm, F., 1984, Population differences in complexity of a learned skill are correlated with the brain space involved, *Proc. Natl. Acad. Sci. USA* **81:**6232–6234.

Canady, R. A., Burd, G. D., DeVoogd, T. J., and Nottebohm, F., 1988, Effect of testosterone on input received by an identified neuron type of the canary song system: A Golgi/electron microscopy/degeneration study, *J. Neurosci.* **8:**3770–3784.

Carr, C. E., and Konishi, M., 1988, Axonal delay lines for time measurement in the owl's brainstem, *Proc. Natl. Acad. Sci. USA* **85:**8311–8315.

Clark, J. M., and Ulinski, P. S., 1984, A Golgi study of anterior dorsal ventricular ridge in the alligator, *Alligator mississippiensis, J. Morphol.* **179:**153–174.

Clarke, P. G. H., 1977, Some visual and other connections to the cerebellum of the pigeon, *J. Comp. Neurol.* **174:**535–552.

Clower, R. P., DeVoogd, T. J., and Nixdorf, B., 1989, Synaptic plasticity in the hypoglossal nucleus of female canaries: Structural correlates of season, hemisphere, and testosterone treatment, *Behav. Neural Biol.* **52:**63–77.

Cobb, S., 1960a, A note on the size of the avian olfactory bulb, *Epilepsia* **1:**394–402.

Cobb, S., 1960b, Observations on the comparative anatomy of the avian brain, *Perspec. Biol. Med.* **3:**383–408.

Cohen, D., and Karten, H. J., 1974, The structural organization of avian brain: An overview, in *Birds, Brain and Behavior* (I. J. Goodman and M. Schein, eds.), Academic Press, New York, pp. 29–73.

Coles, R. B., and Guppy, A., 1988, Directional hearing in the barn owl (*Tyto alba*), *J. Comp. Physiol.* **163:**117–133.

Coles, R. B., Lewis, D. B., Hill, K. G., Hutchings, M. E., and Gower, D. M., 1980, Directional hearing in the Japanese quail (*Coturnix coturnix japonica*). II. Cochlear physiology, *J. Exp. Biol.* **86:**153–170.

Cowan, W. M., Adamson, L., and Powell, T. P. S., 1961, An experimental study of the avian visual system, *J. Anat.* **95:**545–563.

Cracraft, J., 1986, The origin and early diversification of birds, *Paleobiology* **12:**383–399.

Cracraft, J., 1988, Early evolution of birds, *Nature* **331:**389–390.

Craigie, E. H., 1928, Observations on the brain of the humming bird (*Chrysolampis mosquitus* Linn. and *Chlorostilbon caribaeus* Lawr.), *J. Comp. Neurol.* **45:**377–481.

Craigie, E. H., 1932, The cell structure of the cerebral hemisphere in the humming bird, *J. Comp. Neurol.* **56:**135–168.

Craigie, E. H., 1935, The cerebral hemispheres of the kiwi and emu (*Apteryx* and *Dromiceius*), J. Anat. **69:**380–393.

Craigie, E. H., 1936, The cerebral cortex of the ostrich (*Struthio*), J. Comp. Neurol. **64:**389–415.

Craigie, E. H., 1939, The cerebral cortex of *Rhea americana, J. Comp. Neurol.* **70:**331–353.

Craigie, E. H., 1940a, The cerebral cortex in some tinamidae, *J. Comp. Neurol.* **72:**299–328.

Craigie, E. H., 1940b, The cerebral cortex in paleognathine and neognathine birds, *J. Comp. Neurol.* **73:**179–234.

Csillag, A., Stewart, M. G., and Curtis, E. M., 1987, GABAergic structures in the chick telencephalon: GABA immunohistochemistry combined with light and electron microscope autoradiography, and Golgi impregnation, *Brain Res.* **437:**283–297.

Csillag, A., Bourne, R. C., Patel, N., Stewart, M. G., and Tombol, T., 1989, Localization of GABA-like immunoreactivity in the ectostriatum of domestic chicks: GABA immunocytochemistry combined with Golgi impregnation, *J. Neurocytol.* **18:**369–379.

De Britto, L. R. G., Brunelli, M., Francesconi, W., and Magni, F., 1975, Visual response pattern of thalamic neurons in the pigeon, *Brain Res.* **97:**337–343.

Dechaseaux, C., 1968, Le cerveau d'Archaeopteryx est-il de "type reptilien"? *C.R. Acad. Sci. Ser. D* **267:**2108–2110.

Deich, J. R., Klein, B. G., and Zeigler, H. P., 1985, Grasping in the pigeon: Mechanisms of motor control, *Brain Res.* **337:**362–367.

Delius, J. D., and Bennetto, K., 1972, Cutaneous sensory projections to the avian forebrain, *Brain Res.* **37:**205–221.

Denton, C. J., 1981, Topography of the hyperstriatal visual projection area in the young domestic chicken, *Exp. Neurol.* **74:**482–498.

DeVoogd, T. J., and Nottebohm, F., 1981, Sex differences in dendritic morphology of a song control nucleus in the canary: A quantitative Golgi study, *J. Comp. Neurol.* **196:**309–316.

DeVoogd, T. J., Nixdorf, B., and Nottebohm, F., 1985, Synaptogenesis and changes in synaptic morphology related to acquisition of a new behavior, *Brain Res.* **329:**304–308.

DeVoogd, T. J., Brenowitz, E. A., and Arnold, A. P., 1988, Small sex differences in song control dendrites are associated with minimal differences in song capacity, *J. Neurobiol.* **19:**199–209.

Dial, K. P., Kaplan, S. R., Goslow, G. E., Jr., and Jenkins, F. A., Jr., 1988, A functional analysis of the primary upstroke and downstroke muscles in the domestic pigeon (*Columba livia*) during flight, *J. Exp. Biol.* **134:**1–16.

Dooling, R. J., Gephart, B. F., Price, P. H., McHale, C., and Brauth, S. E., 1987, Effects of deafening on the contact call of the budgerigar, *Melopsittacus undulatus, Anim. Behav.* **35:**1264–1266.

Dubbeldam, J. L., 1980, Studies on the somatotopy of the trigeminal system in the mallard, *Anas platyrhynchos* L. II. Morphology of the principal sensory nucleus, *J. Comp. Neurol.* **191:**557–571.

Dubbeldam, J. L., 1984, Brainstem mechanisms for feeding in birds: Interaction or plasticity, *Brain Behav. Evol.* **25:**85–98.

Dubbeldam, J. L., and Wijsman, J. P. M., 1975, The ascending projections of the principal sensory trigeminal nucleus in the mallard (*Anas platyrhynchos*), Acta Morphol. Neerl. Scand. **13:**230–231.

Dubbeldam, J. L., Brauch, C. S. M., and Don, A., 1981, Studies on the somatotopy of the trigeminal system in the mallard, *Anas platyrhynchos* L. III. Afferents and organization of the nucleus basalis, *J. Comp. Neurol.* **196:**391–405.

Dube, L., and Parent, A., 1981, The monoamine-containing neurons in avian brain: I. A study of the brainstem of the chicken (*Gallus domesticus*) by means of fluorescence and acetylcholinesterase histochemistry, *J. Comp. Neurol.* **196:**695–708.

Edinger, T., 1926, The brain of *Archaeopteryx, Ann. Mag. Nat. Hist.* **18:**151–156.

Eldredge, N., and Cracraft, J., 1980, *Phylogenetic Patterns and the Evolutionary Process,* Columbia University Press, New York.

Elprana, D., Wouterlood, F. G., and Alones, V. E., 1980, A corticotectal projection in the lizard *Agama agama, Neurosci. Lett.* **18:**251–256.

Feduccia, A., 1980, *The Age of Birds,* Harvard University Press, Cambridge, Mass.

Feduccia, A., and Tordoff, H. B., 1979, Feathers of *Archaeopteryx:* Asymmetric vanes indicate aerodynamic function, *Science* **203:**1021–1022.

Fortune, E. S., and Margoliash, D., 1989, Combination-sensitive neurons in the zebra finch's HVc, *Soc. Neurosci. Abstr.* **15:**617.

Freedman, S. L., Voogd, J., and Vielvoyie, G. J., 1977, Experimental evidence for climbing fibers in the avian cerebellum, *J. Comp. Neurol.* **175:**243–252.

Funk, G. D., Milsom, W. K., Sholomenko, G. N., and Steeves, J. D., 1989, Role of the telencephalon in the synchronization of locomotor and respiratory frequencies during walking in Canada geese, *J. Exp. Biol.* in press.

Furber, S. E., 1983, The organization of the olivocerebellar projection in the chicken, *Brain Behav. Evol.* **22:**198–211.

Gahr, M., 1990, The delineation of a bird nucleus: Comparisons of cytochemical, hodological, and cytoarchitectural views of the song control nucleus HVc of the adult canary, *J. Comp. Neurol.* **294:**30–36.

Gahr, M., and Konishi, M., 1988, Developmental changes in estrogen-sensitive neurons in the forebrain of the zebra finch, *Proc. Natl. Acad. Sci. USA* **85:**7380–7383.

Gahr, M., Flügge, F., and Güttinger, H.-R, 1987, Immunocytochemical localization of estrogen-binding neurons in the songbird brain, *Brain Res.* **402:**173–177.

Gamlin, P. D. R., and Cohen, D. H., 1986, A second ascending visual pathway from the optic tectum to the telencephalon in the pigeon (*Columba livia*), J. Comp. Neurol. **250:**296–310.

Gamlin, P. D. R., and Cohen, D. H., 1988, Projections of the retinorecipient pretectal nuclei in the pigeon (*Columba livia*), J. Comp. Neurol. **269:**18–46.

Gaunt, A. S., and Gaunt, S. L. L., 1985, Syringeal structure and avian phonation, in: *Current Ornithology,* Volume 2 (R. F. Johnston, ed.), Plenum Press, New York, pp. 213–245.

Goldman, S. A., and Nottebohm, F., 1983, Neuronal production, migration and differentiation

in a vocal control nucleus of the adult female canary brain, *Proc. Natl. Acad. Sci. USA* **80:** 2390–2394.
Goodale, M. A., 1983, Visually guided pecking in the pigeon *(Columba livia), Brain Behav. Evol.* **22:**22–41.
Goodman, I. J., and Schein, M. W., 1974, *Birds: Brain and Behavior,* Academic Press, New York.
Goslow, G. E., Jr., Dial, K. P., and Jenkins, F. A., Jr., 1989, The avian shoulder: An experimental approach, *Am. Zool.* **29:**287–301.
Granada, A. M., and Maxwell, J. H., 1979, *Neural Mechanisms of Behavior in the Pigeon,* Plenum Press, New York.
Greenewalt, C. H., 1968, *Bird Song: Acoustics and Physiology,* Smithsonian Institute Press, Washington, D.C.
Gurney, M. E., 1980, Sexual Differentiation of Brain and Behavior in the Zebra Finch (*Poephila guttata*): A Cellular Analysis, Ph.D. thesis, California Institute of Technology.
Gurney, M. E., 1981, Hormonal control of cell form and number in the zebra finch song system, *J. Neurosci.* **1:**658–673.
Gurney, M. E., 1982, Behavioral correlates of sexual differentiation in the zebra finch song system, *Brain Res.* **231:**153–172.
Hamdi, F. A., and Whiteridge, D., 1954, The representation of the retina on the optic tectum of the pigeon, *Q. J. Exp. Physiol.* **39:**111–119.
Harkmark, W., 1954, Cell migrations from the rhombic lip to the inferior olive, the nucleus raphe and the pons. A morphological and experimental investigation in chick embryos, *J. Comp. Neurol.* **100:**115–209.
Hartley, R. S., and Suthers, R. A., 1989, Airflow and pressure during canary song: Direct evidence of mini-breaths, *J. Comp. Physiol.* **165:**15–26.
Hecht, M. K., Ostrom, J. H., Viohl, G., and Wellnhofer, P., 1985, *The Beginnings of Birds,* Freunde Des Jura-Museums Eichstatt, Eichstatt.
Heil, P., and Scheich, H., 1985, Quantitative analysis and two-dimensional reconstruction of the tonotopic organization of the auditory field L in the chick from 2-deoxyglucose data, *Exp. Brain Res.* **58:**532–543.
Heilmann, G., 1926, *The Origin of Birds,* Appleton, New York.
Hill, K. G., Lewis, D. B., Hutchings, M. E., and Coles, R. B., 1980, Directional hearing in the Japanese quail (*Coturnix coturnix japonica*). I. Acoustic properties of the auditory system, *J. Exp. Biol.* **86:**135–151.
Hodos, W., 1976, Vision and the visual system: A bird's eye view, in: *Progress in Psychobiology and Physiological Psychology* (J. M. Sprague and A. N. Epstein, eds.), Academic Press, New York, pp. 29–62.
Hodos, W., and Bonbright, J. C., Jr., 1974, Intensity difference thresholds in pigeons after lesions of the tectofugal and thalamofugal visual pathway, *J. Comp. Physiol. Psychol.* **87:**1013–1031.
Hodos, W., and Karten, H. J., 1970, Visual intensity and pattern discrimination deficits after lesions of ectostriatum in pigeons, *J. Comp. Neurol.* **140:**53–68.
Hodos, W., Karten, H. J., and Bonbright, J. C., Jr., 1973, Visual intensity and pattern discrimination after lesions of the thalamofugal visual pathway in pigeons, *J. Comp. Neurol.* **148:**447–468.
Hoogland, P. V., 1977, Efferent connections of the striatum in *Tupinambis nigropunctatus, J. Morphol.* **152:**229–246.
Hopson, J. A., 1975, The evolution of cranial display structures in hadrosaurian dinosaurs, *Paleobiology* **1:**21–43.
Hopson, J. A., 1977, Relative brain size and behavior in archosaurian reptiles, *Annu. Rev. Ecol. Syst.* **8:**429–448.
Hopson, J. A., 1979, Paleoneurology, in: *Biology of the Reptilia,* Volume 9 (C. C. Gans, R. G. Northcutt and P. S. Ulinski, eds.), Academic Press, New York, pp. 39–146.
Hunt, R. H., 1975, Maternal behavior in the Morelet's crocodile, *Crocodylus moreleti, Copeia* **1975:**763–764.
Hunt, S. P., and Brecha, N., 1984, The avian optic tectum: A synthesis of morphology and biochemistry, in: *Comparative Neurology of the Optic Tectum* (H. Vanegas, ed.), Plenum Press, New York, pp. 619–648.
Hunt, S. P., and Kunzle, H., 1976, Observations on the projections and intrinsic organization of the pigeon optic tectum: An autoradiographic study based on anterograde and retrograde axonal and dendritic flow, *J. Comp. Neurol.* **170:**153–172.
Hunt, S. P., and Webster, K. E., 1972, Thalamo-hyperstriate interrelations in the pigeon, *Brain Res.* **44:**647–651.

Jacobson, R. D., and Hollyday, M., 1982a, A behavioral and electromyographic study of walking in the chick, *J. Neurophysiol.* **48:**238–256.

Jacobson, R. D., and Hollyday, M., 1982b, Electrically evoked walking and fictive locomotion in the chick, *J. Neurophysiol.* **48:**257–270.

Jassik-Gerschenfeld, D., Teulon, J., and Ropert, N., 1976, Visual receptive field types in the nucleus dorsolateralis anterior of the pigeon's thalamus, *Brain Res.* **108:**295–306.

Jeffress, L. A., 1948, A place theory of sound localization, *J. Comp. Physiol. Psychol.* **41:**35–39.

Jenkins, F. A., Jr., Dial, K. P., and Goslow, G. E., Jr., 1988, A cineradiographic analysis of bird flight: The wishbone is a spring, *Science* **241:**1495–1498.

Jerison, H. J., 1968, Brain evolution and *Archaeopteryx, Nature* **219:**1381–1382.

Jerison, H. J., 1973, *Evolution of the Brain and Intelligence,* Academic Press, New York.

Karten, H., 1968, The ascending auditory pathway in the pigeon (*Columba livia*). II. Telencephalic projections of the nucleus ovoidalis thalami, *Brain Res.* **11:**134–153.

Karten, H. J., 1971, Efferent projections of the Wulst of the owl, *Anat. Rec.* **169:**353.

Karten, H. J., 1979, Visual lemniscal pathways in birds, in: *Neural Mechanisms of Behavior in the Pigeon* (A. M. Granda and J. H. Maxwell, eds.), Plenum Press, New York, pp. 409–430.

Karten, H. J., and Dubbeldam, J. L., 1973, The organization and projections of the paleostriatal complex in the pigeon (*Columba livia*), J. Comp. Neurol. **148:**61–89.

Karten, H. J., and Hodos, W., 1967, *A Stereotaxic Atlas of the Brain of the Pigeon (Columba livia),* Johns Hopkins University Press, Baltimore.

Karten, H. J., Hodos, W., Nauta, W. J. H., and Revzin, A. M., 1973, Neural connections of the "visual wulst" of the avian telencephalon. Experimental studies in the pigeon (*Columba livia*) and owl (*Speotyto cunnicularia*), J. Comp. Neurol. **150:**253–278.

Karten, H. J., Fite, K. V., and Brecha, N., 1977, Specific projection of displaced retinal ganglion cells upon the accessory optic system in the pigeon (*Columba livia*), Proc. Natl. Acad. Sci. USA **74:**1753–1756.

Karten, H. J., Konishi, M., and Pettigrew, J., 1978, Somatosensory representation in the anterior wulst of the owl (*Speotyto cunnicularia*), Soc. Neurosci, Abstr. **4:**554.

Katz, L. C., 1982, The avian motor system for song has multiple sites and types of auditory input, *Soc. Neurosci. Abstr.* **8:**1021.

Katz, L. C., and Gurney, M. E., 1981, Auditory responses in the zebra finch's motor system for song, *Brain Res.* **211:**192–197.

Kavanau, J. L., 1987, *Lovebirds, Cockatiels, Budgerigars: Behavior and Evolution,* Science Software Systems, Los Angeles.

Kazennikov, O. V., Selinov, V. A., Shik, M. L., and Yakokleva, G. V., 1980, The rhombencephalic "locomotor region" in turtles, *Neurophysiology* **12:**251–257.

Kelley, D. B., and Nottebohm, F., 1979, Projections of a telencephalic auditory nucleus—field L—in the canary, *J. Comp. Neurol.* **183:**455–470.

Kirn, J. R., Clower, R. P., Kroodsma, D. E., and DeVoogd, T. J., 1989, Song related brain regions in the red-winged blackbird are affected by sex and season but not repertoire size, *J. Neurobiol.* **20:**139–163.

Kitt, C. A., and Brauth, S. E., 1981, Projections of the paleostriatum upon the midbrain tegmentum in the pigeon, *Neuroscience* **6:**1551–1566.

Kitt, C. A., and Brauth, S. E., 1982, A paleostriatal–thalamic–telencephalic path in pigeons, *Neuroscience* **7:**2735–2731.

Kitt, C. A., and Brauth, S. E., 1986, Telencephalic projections from midbrain and isthmal cell groups. II. The nigral complex, *J. Comp. Neurol.* **247:**92–110.

Klein, B. G., Deich, J. D., and Zeigler, H. P., 1985, Grasping in the pigeon (*Columba livia*): Final common path mechanisms, *J. Comp. Neurol.* **241:**180–190.

Knudsen, E. I., 1980, Sound localization in birds, in: *Comparative Studies of Hearing in Vertebrates* (A. N. Popper and R. R. Fay, eds.), Springer-Verlag, Berlin, pp. 289–322.

Knudsen, E. I., 1982, Auditory and visual maps of space in the optic tectum of the owl, *J. Neurosci.* **2:**1177–1194.

Knudsen, E. I., 1983a, Subdivisions of the inferior colliculus in the barn owl (*Tyto alba*), J. Comp. Neurol. **218:**174–186.

Knudsen, E. I., 1983b, Early auditory experience aligns the auditory map of space in the optic tectum of the barn owl, *Science* **222:**939–942.

Knudsen, E. I., 1988, Early blindness results in a degraded auditory map of space in the optic tectum of the barn own. *Proc. Natl. Acad. Sci. USA* **85:**6211–6214.

Knudsen, E. I., and Konishi, M., 1978a, Center-surround organization of auditory receptive fields in the owl, *Science* **202:**778–780.

Knudsen, E. I., and Konishi, M., 1978b, A neural map of auditory space in the barn owl, *Science* **200:**795–797.

Knudsen, E. I., and Konishi, M., 1978c, Space and frequency are represented separately in auditory midbrain of the owl, *J. Neurophysiol.* **41:**870–884.

Knudsen, E. I., and Konishi, M., 1979, Mechanisms of sound localization in the barn owl (*Tyto alba*), J. Comp. Physiol. **133:**13–21.

Knudsen, E. I., and Konishi, M., 1980, Monaural occlusion shifts receptive-field locations of auditory midbrain units in the owl, *J. Neurophysiol.* **44:**687–695.

Knudsen, E. I., Blasdel, G. G., and Konishi, M., 1979, Sound localization by the barn owl (*Tyto alba*) measured with the search coil technique, *J. Comp. Physiol.* **133:**1–11.

Knudsen, E. I., Esterly, S. D., and Knudsen, P. F., 1984a, Monaural occlusion alters sound localization during a sensitive period in the barn owl, *J. Neurosci.* **4:**1001–1011.

Knudsen, E. I., Knudsen, P. F., and Esterly, S. D., 1984b, A critical period for the recovery of sound localization accuracy following monaural occlusion in the barn owl, *J. Neurosci.* **4:**1012–1020.

Konishi, M., 1963, The role of auditory feedback in the vocal behavior of the domestic fowl, *Z. Tierpsychol.* **20:**349–367.

Konishi, M., 1964, Effects of deafening on song development in two species of juncos, *Condor* **66:**85–102.

Konishi, M., 1965a, Effects of deafening on song development in American robins and blackheaded grosbeaks, *Z. Tierpsychol.* **22:**584–599.

Konishi, M., 1965b, The role of auditory feedback in the control of vocalization in the white-crowned sparrow, *Z. Tierpsychol.* **22:**770–783.

Konishi, M., 1973, Locatable and nonlocatable acoustic signals for barn owls, *Am. Nat.* **107:**775–785.

Konishi, M., 1985, Birdsong: From behavior to neuron, *Annu. Rev. Neurosci.* **8:**125–170.

Konishi, M., and Akutagawa, E., 1981, Androgen increases protein synthesis within the avian brain vocal control system, *Brain Res.* **222:**442–446.

Konishi, M., and Akutagawa, E., 1985, Neuronal growth, atrophy, and death in a sexually dimorphic song nucleus in the zebra finch brain, *Nature* **315:**145–147.

Konishi, M., Sullivan, W. E., and Takahaski, T., 1985, The owl's cochlear nuclei process different sound localization cues, *J. Acoust. Soc. Am.* **78:**360–364.

Kooy, F. H., 1917, The inferior olive in vertebrates, *Folia Neuro-Biol.* **10:**205–369.

Krayniak, P. F., and Siegel, A., 1978a, Efferent connections of the hippocampus and adjacent regions in the pigeon, *Brain Behav. Evol.* **15:**372–388.

Krayniak, P. F., and Siegel, A., 1978b, Efferent connections of the septal area in the pigeon, *Brain Behav. Evol.* **15:**389–404.

Kroodsma, D. E., 1984, Songs of the alder flycatcher (*Empidonax alnorum*) and willow flycatcher (*Empidonax traillii*) are innate, *Auk* **101:**13–24.

Kushlan, J. A., 1973, Observations on maternal behavior in the American alligator, *Alligator mississippiensis, Herpetologica* **29:**256–257.

Langner, G., Bonke, D., and Scheich, H., 1981, Neuronal discrimination of natural and synthetic vowels in field L of trained mynah birds, *Exp. Brain Res.* **43:**11–24.

Larsell, O., 1948, The development and subdivisions of the cerebellum of birds, *J. Comp. Neurol.* **89:**123–190.

Larsell, O., 1967, *The Comparative Anatomy and Histology of the Cerebellum from Myxinoids through Birds*, University of Minnesota, Minneapolis.

Larsell, O., and Whitlock, D. G., 1952, Further observations on the cerebellum of birds, *J. Comp. Neurol.* **97:**545–566.

Lennard, P. R., and Stein, P. S. G., 1977, Swimming movements elicited by electrical stimulation of turtle spinal cord. I. Low spinal and intact preparations, *J. Neurophysiol.* **40:**768–778.

Leppelsack, H.-J., 1983, Analysis of song in the auditory pathway of song birds, in: *Advances in Vertebrate Neuroethology* (J. P. Ewert, R. R. Capranica, and D. J. Ingle, eds.), Plenum Press, New York, pp. 783–799.

Leppelsack, H.-J., and Vogt, M., 1976, Responses of auditory neurons in the forebrain of song bird to stimulation with species-specific sounds, *J. Comp. Physiol.* **107:**263–274.

Lieberburg, I., and Nottebohm, F., 1979, High-affinity androgen binding proteins in syringeal tissues of songbirds, *Gen. Comp. Endocrinol.* **37:**286–293.

McCasland, J. S., 1983, Neuronal control of bird song production, Ph.D. thesis, California Institute of Technology.

McCasland, J. S., 1987, Neuronal control of bird song production, *J. Neurosci.* **7:**23–39.

McCasland, J. S., and Konishi, M., 1981, Interaction between auditory and motor activities in an avian song control nucleus, *Proc. Natl. Acad. Sci. USA* **78:**1815–1819.

McKenna, O. C., and Wallman, J., 1985, Accessory optic system and pretectum of birds: Comparisons with those of other vertebrates, *Brain Behav. Evol.* **26:**91–116.

Maldonado, P. E., Maturana, H., and Varela, F. J., 1988, Frontal and lateral visual system in birds, *Brain Behav. Evol.* **32:**57–62.

Manley, G. A., Koppl, C., and Konishi, M., 1988, A neural map of interaural intensity differences in the brain stem of the barn owl, *J. Neurosci.* **8:**2665–2676.

Manogue, K. R., and Paton, J. A., 1982, Respiratory gating of activity in the avian vocal control system, *Brain Res.* **247:**383–387.

Mantinoya, C., Rivaud, S., and Bloch, S., 1983, Comparing frontal and lateral viewing in the pigeon. II. Velocity thresholds for movement discrimination, *Behav. Brain Res.* **8:**375–358.

Margoliash, D., 1983, Acoustic parameters underlying the responses of song-specific neurons in the white-crowned sparrow, *J. Neurosci.* **3:**1039–1057.

Margoliash, D., 1986, Preference for autogenous song by auditory neurons in a song system nucleus of the white-crowned sparrow, *J. Neurosci.* **6:**1643–1661.

Margoliash, D., 1987, Neural plasticity in birdsong learning, in: *Imprinting and Cortical Plasticity* (J. P. Rauschecker and P. Marler, eds.), Wiley, New York, pp. 23–54.

Margoliash, D., and Konishi, M., 1985, Auditory representation of autogenous song in the song-system of white-crowned sparrows, *Proc. Natl. Acad. Sci. USA* **82:**5997–6000.

Marler, P., and Sherman, V., 1982, Structure in sparrow song without auditory feedback: An emendation of the auditory template hypothesis, *J. Neurosci.* **3:**517–531.

Marler, P., Peters, S., and Wingfield, J., 1987, Correlations between song acquisition, song production, and plasma levels of testosterone and estradiol in sparrows, *J. Neurobiol.* **18:**531–548.

Martin, L. D., 1983, The origin and early radiation of birds, in: *Perspectives in Ornithology* (A. H. Brash and G. A. Clark, Jr., eds.), Cambridge University Press, London, pp. 291–338.

Maxwell, J. H., and Granda, A. M., 1979, Receptive fields of movement-sensitive cells in the pigeon thalamus, in: *Neural Mechanisms of Behavior in the Pigeon* (A. M. Granda and J. H. Maxwell, eds.), Plenum Press, New York, pp. 177–197.

Meier, R. E., Mihailovic, J., and Cuenod, M., 1974, Thalamic organization of the retino-thalamo-hyperstriatal pathway in the pigeon (*Columba livia*), Exp. Brain Res. **19:**351–364.

Miceli, D., and Reperant, J., 1982, Thalamo-hyperstriatal projections in the pigeon (*Columba livia*) as demonstrated by retrograde double-labeling with fluorescent tracers, *Brain Res.* **245:**365–371.

Miceli, D., Peyrichoux, J., and Reperant, J., 1975, The retino-thalamo-hyperstriatal pathway in the pigeon, *Brain Res.* **100:**125–131.

Miceli, D., Reperant, J., Villalobos, J., and Dionne, L., 1987, Extratelencephalic projections of the avian visual wulst. A quantitative autoradiographic study in the pigeon (*Columba livia*), J. Hirnforsch. **28:**45–57.

Moiseff, A., 1988, Binaural disparity cues available to the barn owl for sound localization, *J. Comp. Physiol.* **164:**629–636.

Moiseff, A., and Konishi, M., 1981a, Neuronal and behavioral sensitivity to binaural time differences in the owl, *J. Neurosci.* **1:**40–48.

Moiseff, A., and Konishi, M., 1981b, The owl's interaural pathway is not involved in sound localization, *J. Comp. Physiol.* **144:**299–304.

Moiseff, A., and Konishi, M., 1983, Binaural characteristics of units in the owl's brainstem auditory pathway: Precursors of restricted spatial fields, *J. Neurosci.* **3:**2553–2562.

Müller, C. M., and Leppelsack, H.-J., 1985, Feature extraction and tonotopic organization in the avian auditory forebrain, *Exp. Brain Res.* **59:**587–599.

Müller, S. C., and Scheich, H., 1985, Functional organization of the avian auditory field L, *J. Comp. Physiol.* **156:**1–12.

Mundinger, P. C., 1970, Vocal imitation and individual recognition of finch calls, *Science* **168:**480–482.

Newman, J. D., 1970, Midbrain regions relevant to auditory communication in songbirds, *Brain Res.* **22:**259–261.

Nixdorf, B. E., and Bischof, H. J., 1982, Afferent connections of the ectostriatum and visual wulst in the zebra finch (*Taeniopygia guttata castanotis* Gould)—An HRP study, Brain Res. **248:**9–17.

Nixdorf, B. E., Davis, S. S., and DeVoogd, T. J., 1989, Morphology of Golgi-impregnated neurons in hyperstriatum ventralis, pars caudalis in adult male and female canaries, *J. Comp. Neurol.* **284:**337–349.

Nordeen, K. W., and Nordeen, E. J., 1988, Projection neurons within a vocal motor pathway are born during song learning in zebra finches, *Nature* **334:**149–151.

Nordeen, K. W., Nordeen, E. J., and Arnold, A. P., 1987, Estrogen accumulation in zebra finch song control nuclei: Implications for sexual differentiation and adult activation of song behavior, *J. Neurobiol.* **18:**569–582.

Northcutt, R. G., 1981, Evolution of the telencephalon in non mammals, *Annu. Rev. Neurosci.* **4:**301–350.

Northcutt, R. G., 1984, Anatomical organization of the optic tectum in reptiles, in: *Comparative Neurology of the Optic Tectum* (H. Vanegas, ed.), Plenum Press, New York, pp. 547–600.

Nottebohm, F., 1971, Neural lateralization of vocal control in a passerine bird. I. Song, *J. Exp. Zool.* **177:**229–261.

Nottebohm, F., 1972, The origins of vocal learning, *Am. Nat.* **106:**116–140.

Nottebohm, F., 1980, Brain pathways for vocal learning in birds: A review of the first 10 years, *Prog. Psychobiol. Physiol. Psychol.* **9:**85–124.

Nottebohm, F., 1981, A brain for all seasons: Cyclical anatomical changes in song control nuclei in the canary brain, *Science* **194:**211–213.

Nottebohm, F., 1985, Neuronal replacement in adulthood. *Hope for a new neurology, Ann. N.Y. Acad. Sci.* **457:**143–161.

Nottebohm, F., 1989, Hormonal regulation of synapses and cell number in the adult canary brain and its relevance to theories of long-term memory storage, in: *Neural Control of Reproductive Function,* Liss, New York, pp. 583–601.

Nottebohm, F., and Arnold, A. P., 1976, Sexual dimorphism in vocal control areas of the songbird brain, *Science* **194:**211–213.

Nottebohm, F., and Nottebohm, M., 1971, Vocalizations and breeding behaviour of surgically deafened ring doves *Streptopelia risoria, Anim. Behav.* **19:**313–328.

Nottebohm, F., and Nottebohm, M. E., 1978, Relationship between song repertoire and age in the canary, *Serinus canarius, Z. Tierpsychol.* **46:**298–395.

Nottebohm, F., Stokes, T. M., and Leonard, C. M., 1976, Central control of song in the canary, *Serinus canarius, J. Comp. Neurol.* **165:**457–486.

Nottebohm, F., Kasparian, S., and Pandazis, C, 1981, Brain space for a learned task, *Brain Res.* **213:**99–109.

Nottebohm, F., Kelley, D. B., and Paton, J. A., 1982, Connections of vocal control nuclei in the canary telencephalon, *J. Comp. Neurol.* **207:**344–357.

Nottebohm, F., Nottebohm, M. E., and Crane, L., 1986, Developmental and seasonal changes in canary song and their relation to changes in the anatomy of song-control nuclei, *Behav. Neural Biol.* **46:**455–471.

Nowicki, S., 1989, Vocal plasticity in captive black-capped chickadees: The acoustic basis and rate of call convergence, *Anim. Behav.* **37:**64–73.

Nowicki, S., and Marler, P., 1988, How do birds sing? *Music Percep.* **5:**391–426.

Okuhata, S., and Saito, N., 1987, Synaptic connections of thalamo-cerebral vocal nuclei of the canary, *Brain Res. Bull.* **18:**35–44.

Ostrom, J. H., 1974, *Archaeopteryx* and the origin of flight, *Q. Rev. Biol.* **49:**27–47.

Ostrom, J. H., 1976, *Archaeopteryx* and the origin of birds, *Biol. J. Linn. Soc.* **8:**91–182.

Padian, K., 1986, *The Origin of Birds and the Evolution of Flight,* California Academy of Sciences, San Francisco.

Palacios, E., 1976, Golgi studies of the telencephalon of the domestic chicken (*Gallus domesticus*), Anat. Rec. **184:**495.

Parent, A., 1986, *Comparative Neurobiology of the Basal Ganglia,* Wiley, New York.

Parrish, J. M., 1987, The origin of crocodilian locomotion, *Paleobiology* **13:**369–414.

Paton, J. A., and Nottebohm, F., 1984, Neurons born in adult brain are recruited into functional circuits, *Science* **225:**1046–1048.

Paton, J. A., Manogue, K. R., and Nottebohm, F., 1981, Bilateral organization of the vocal control pathway in the budgerigar, *Melopsittacus undulatus, J. Neurosci.* **1:**1279–1288.

Paton, J. A., O'Loughlin, B. E., and Nottebohm, F., 1985, Cells born in adult canary forebrain are local interneurons, *J. Neurosci.* **5:**3088–3093.

Payne, R. S., 1971, Acoustic location of prey by barn owls (*Tyto alba*), J. Exp. Biol. **54:**535–573.

Pearson, R., 1972, *The Avian Brain,* Academic Press, New York.

Pepperberg, I. M., 1983, Cognition in the African grey parrot: Preliminary evidence for auditory/vocal comprehension of the class concept, *Anim. Learn. Behav.* **11:**179–185.

Pepperberg, I. M., 1987, Evidence for conceptual quantitative abilities in the African grey parrot: Labeling of cardinal sets, *Ethology* **75:**37–61.

Pepperberg, I. M., 1988, The importance of social interaction and observation in the acquisition of communicative competence: Possible parallels between avian and human learning, in: *Social*

Learning: Psychological and Biological Perspectives (T. R. Zentall and B. G. Galef, Jr., eds), Erlbaum, Hillsdale, N. J., pp. 279–299.

Perisic, M., Mihailovic, J., and Cuenod, M., 1971, Electrophysiology of contralateral and ipsilateral visual projections to the Wulst in pigeon (*Columba livia*), Int. J. Neurosci. **2:**7–14.

Pettigrew, J. D., 1979, Binocular visual processing in the owl's telencephalon, *Proc. R. Soc. London Ser. B* **204:**435–454.

Pettigrew, J. D., 1986, The evolution of binocular vision, in: *Visual Neuroscience* (J. D. Pettigrew, K. J. Sanderson, and W. R. Levick, eds.), Cambridge University Press, London, pp. 208–222.

Pettigrew, J. D., and Konishi, M., 1976, Neurons selective for orientation and binocular disparity in the visual Wulst of the barn owl (*Tyto alba*), Science **193:**675–678.

Pooley, A. C., and Gans, C., 1976, The Nile crocodile, *Sci. Am.* **234:**114–124.

Portman, A., and Stingelin, W., 1961, The central nervous system, in: *Biology and Comparative Physiology of Birds,* Volume 2 (A. J. Marshall, ed.), Academic Press, New York, pp. 1–36.

Potash, L. M., 1970, Vocalizations elicited by brain stimulation in *Coturnix coturnix japonica, Behaviour* **36:**149–167.

Ramón y Cajal, S., 1955, Histologie du systéme nerveux de l'homme et des vertébrès. Consejo superior de investigaciones cientificas, Instituto Ramón y Cajal, Madrid.

Regal, P. J., 1975, The evolutionary origin of feathers, Quart, Rev. Biol. **50:**35–66.

Rehkamper, G., Zilles, K., and Schleicher, A., 1984, A quantitative approach to cytoarchitectonics. IV. The areal pattern of the hyperstriatum ventrale in the domestic pigeon, *Columba livia, Anat. Embryol.* **169:**319–327.

Rehkamper, G., Zilles, K., and Schleicher, A., 1985, A quantitative approach to cytoarchitectonics. X. The areal pattern of the neostriatum in the domestic pigeon, *Columba livia.* A cyto- and myeloarchitectonical study, *Anat. Embryol.* **171:**345–355.

Reiner, A., and Karten, H. J., 1982, Cells of origin of descending tectofugal pathways in the pigeon (*Columba livia*), J. Comp. Neurol. **204:**165–187.

Reiner, A., and Karten, H. J., 1983, The laminar source of efferent projections from the avian wulst, *Brain Res.* **275:** 349–354.

Reiner, A., and Karten, H. J., 1985, Comparison of olfactory bulb projections in pigeons and turtles, *Brain Behav. Evol.* **27:**11–27.

Reiner, A., Brecha, N., and Karten, H. J., 1979, A specific projection of retinal displaced ganglion cells to the nucleus of the basal optic root in the chicken, *Neuroscience* **4:**1679–1688.

Reiner, A., Brecha, N. C., and Karten, H. J., 1982a, Basal ganglia pathways to the tectum: The afferent and efferent connections of the lateral spiriform nucleus of pigeons, *J. Comp. Neurol.* **208:**16–36.

Reiner, A., Karten, H. J., and Brecha, N. C., 1982b, Enkephalin-mediated basal ganglia influences over the optic tectum: Immunohistochemistry of the tectum and the lateral spiriform nucleus in pigeons, *J. Comp. Neurol.* **208:**37–53.

Reiner, A., Karten, H. J., and Solina, A. R., 1983, Substance P: Localization within paleostriatal–tegmental pathways in the pigeon, *Neuroscience* **9:**61–85.

Reiner, A., Davis, B. M., Brecha, N. C., and Karten, H. J., 1984a, The distribution of enkephalin-like immunoreactivity in the telencephalon of the adult and developing domestic chicken, *J. Comp. Neurol.* **228:**245–262.

Reiner, A., Brauth, S. E., and Karten, H. J., 1984b, Evolution of the amniote basal ganglia, *Trends Neurosci.* **7:**320–325.

Reperant, J., 1973, Nouvelles donnees sur les projections visuelles chez le pigeon (*Columba livia*), J. Hirnforsch. **14:**152–287.

Reperant, J., Raffin, J.-P., and Miceli, D., 1974, La voie retino-thalamo-hyper-striatale chez le poussin (*Gallus domesticus* L), *C.R. Acad. Sci.* **279:**279–282.

Revzin, A. M., 1969, A specific visual projection area in the hyperstriatum of the pigeon, *Brain Res.* **15:**246–249.

Rio, J. P., Villalobos, J., Miceli, D., and Reperant, J., 1983, Efferent projections of the visual wulst upon the nucleus of the basal optic root in the pigeon, *Brain Res.* **271:**145–151.

Ritchie, T. C., 1979, Intratelencephalic visual connections and their relationship to the archistriatum in the pigeon (*Columba livia*), Ph.D. thesis, University of Virginia, Charlottesville.

Rose, M., 1914, Uber die cytoarchitektonische Gliederung des Vorderhirns der Vogel, *J. Psychol. Neurol.* **21:**278–352.

Rübsamen, R., and Dörrscheidt, G. J., 1986, Tonotopic organization of the auditory forebrain in a songbird, the European starling, *J. Comp. Physiol.* **158:**639–646.

Saini, K. D., and Leppelsack, H.-J., 1977, Neuronal arrangement in the auditory field L of the neostriatum of the starling, *Cell Tissue Res.* **176:**309–316.

Saini, K. D., and Leppelsack, H.-J., 1981, Cell types of the auditory caudomedial neostriatum of the starling (*Sturnus vulgaris*), J. Comp. Neurol. **198:**209–230.
Sanz, J. L., Bonaparte, J. F., and Lacasa, A., 1988, Unusual early Cretaceous birds from Spain, *Nature* **331:**433–435.
Scharaff, C., and Nottebohm, F., 1989, Lesions in area X affect song in juvenile but not adult male zebra finches, *Soc. Neurosci. Abstr.* **15:**618.
Scheich, H., Bonke, B. A., and Langner, G., 1979a, Functional organization of some auditory nuclei in the guinea fowl demonstrated by the 2-deoxyglucose technique, *Cell Tissue Res.* **204:**17–27.
Scheich, H., Langner, G., and Bonke, D., 1979b, Responsiveness of units in the auditory neostriatum of the guinea fowl (*Numida meleagris*) to species-specific calls and synthetic stimuli. II. Discrimination of iambus-like calls, *J. Comp. Physiol.* **132:**257–276.
Schwarzkopff, J., 1952, Untersuchungen uber die Arbeitsweise des Mittelohres und das Richtungshoren der Singvogel unter Verwendung von Choclea-Potentialen, *Z. Vgl. Physiol.* **34:**46–68.
Seller, T. J., 1980, Midbrain regions involved in call production in Java sparrows, *Behav. Brain Res.* **1:**257–265.
Simpson, G. G., 1980, Fossil birds in evolution, in: *Papers in Avian Paleontology, Contrib. Sci. Nat. Hist. Mus. Los Angeles County* **330:**3–8.
Simpson, H. B., and Vicario, D. S., 1990, Brain pathways for learned and unlearned vocalizations differ in zebra finches, *J. Neurosci* **10:**1541–1556.
Skeen, L. C., Pindzola, R. R., and Schofield, B. R., 1984, Tangential organization of olfactory, association, and commissural projections to olfactory cortex in a species of reptile (*Trionyx spiniferus*), bird (*Aix sponsa*), and mammal (*Tupaia glis*), Brain Behav. Evol. **25:**206–216.
Smith, C. A., Konishi, M., and Schuff, N., 1985, Structure of the barn owl's (*Tyto alba*) inner ear, *Hearing Res.* **17:**237–247.
Snow, D. W., 1968, The singing assemblies of Little Hermits, *Living Bird* **3:**47–55.
Sohrabji, R., Nordeen, W. J., and Nordeen, K. W., 1989, Selective impairment of song development in zebra finches following lesions of the song control nucleus, area X, *Soc. Neurosci. Abstr.* **15:**681.
Steeves, J. D., Sholomenko, G. N., and Webster, D. M. S., 1987, Stimulation of the pontomedullary reticular formation initiates locomotion in decerebrate birds, *Brain Res.* **401:**205–212.
Stein, P. S. G., and Grossman, M. L., 1980, Central program for scratch reflex in turtle, *J. Comp. Physiol.* **140:**287–294.
Stingelin, W., 1958, Vergleichend morphologische Untersuchungen am Vorderhin der Vogel auf cytologischer und cytoarchitektonischer Grundlage, Helbing & Lichtenhahn, Basal.
Streit, P., Burkhalter, A., Stella, M., and Cuenod, M., 1980, Patterns of activity in pigeon brain's visual relays as revealed by the [^{14}C]-2-deoxyglucose method, *Neuroscience* **5:**1053–1066.
Sullivan, W. E., and Konishi, M., 1984, Segregation of stimulus phase and intensity coding in the cochlear nucleus of the barn owl, *J. Neurosci.* **47:**1787–1799.
Sullivan, W. E., and Konishi, M., 1986, Neural map of interaural phase difference in the owl's brainstem, *Proc. Natl. Acad. Sci. USA* **83:**8400–8404.
Suthers, R. A., 1989, Respiratory dynamics and vocal asymmetry in birdsong, in: *2nd International Congress of Neuroethology* (J. Erber, R. Menzal, H.-J. Pflüger, and D. Todt, eds.), Thieme, Stuggart.
Takahashi, T., and Konishi, M., 1986, Selectivity of interaural time difference in the owl's midbrain, *J. Neurosci.* **6:**3413–3422.
Takahashi, T. T., and Konishi, M., 1988a, Projections of the nucleus angularis and nucleus laminaris to the lateral lemniscal nuclear complex of the barn owl, *J. Comp. Neurol.* **274:**212–238.
Takahashi, T. T., and Konishi, M., 1988b, Projections of the cochlear nuclei and nucleus laminaris to the inferior colliculus of the barn own, *J. Comp. Neurol.* **274:**190–211.
Takahashi, T. T., Wagner, H., and Konishi, M., 1989, Role of commissural projections in the representation of bilateral auditory space in the barn owl's inferior colliculus, *J. Comp. Neurol.* **281:**545–554.
ten Cate, J., 1960, Locomotor movements in the spinal pigeon, *J. Exp. Biol.* **37:**609–613.
ten Donkelaar, H. J., 1990, The cerebellum, in: *Biology of the Reptilia* (C. G. Gans and P. S. Ulinski, eds.), University of Chicago Press, Chicago, in press.
ten Donkelaar, H. J., and de Boer-van Huizen, R., 1981, Ascending projections of the brain stem reticular formation in a non mammalian vertebrate (the lizard *Varanus exanthematicus*), with notes on the afferent connections of the forebrain, *J. Comp. Neurol.* **200:**501–528.
ten Donkelaar, H. J., Bangma, G. C., Barbas-Henry, H. A., de Boer-van Huizen, R., and Wolters, J. G., 1987, The brain stem in a lizard, *Varanus exanthematicus, Adv. Anat. Embryol. Cell Biol.* **107:**1–168.
Theurick, M., Muller, C. M., and Scheich, H., 1984, 2-Deoxyglucose accumulation parallels extracellularly recorded spike activity in the avian auditory neostriatum, *Brain Res.* **322:**157–161.

Tombol, T., Csillag, S., and Stewart, M. G., 1988a, Cell types of the hyperstriatum ventrale of the domestic chicken (*Gallus domesticus*): A Golgi study, *J. Hirnforsch.* **29:**319–334.

Tombol, T., Csillag, A., and Stewart, M. G., 1988b, Cell types of the paleostriatal complex of the domestic chicken (*Gallus domesticus*): A Golgi study, *J. Hirnforsch.* **29:**493–507.

Tombol, T., Magloczky, Z., and Csillag, A., 1988c, The structure of the chicken ectostriatum. I. Golgi study, *J. Hirnforsch.* **29:**525–546.

Tsai, H. M., Garber, B. B., and Larramendi, L. M. H., 1981a, Thymidine autoradiographic analysis of telencephalic histogenesis in the chick embryo: I. Neuronal birthdates of telencephalic compartments *in situ, J. Comp. Neurol.* **198:**275–292.

Tsai, H. M., Garber, B. B., and Larramendi, L. M. H., 1981b, Thymidine autoradiographic analysis of telencephalic histogenesis in the chick embryo: II. Dynamics of neuronal migration, displacement, and aggregation, *J. Comp. Neurol.* **198:**293–306.

Ulinski, P. S., 1983, *Dorsal Ventricular Ridge,* Wiley, New York.

Ulinski, P. S., 1986, Organization of corticogeniculate projections in the turtle, *Pseudemys scripta, J. Comp. Neurol.* **254:**529–542.

Ulinski, P. S., 1988, Functional architecture of turtle visual cortex, in: *The Forebrain of Reptiles* (W. A. K. Schwerdtfeger and W. J. A. J. Smeets, eds.), Karger, Basel, pp. 151–161.

Ulinski, P. S., 1989, Neuronal organization of striatum in the alligator, *Alligator mississippiensis,* in: *Comparative Aspects of the Structure and Development of the Forebrain in Lower Vertebrates* (W. K. Schwerdtfeger and W. J. A. J. Smeets, eds.), Springer, Berlin.

Ulinski, P. S., and Nautiyal, J., 1988, Organization of retinogeniculate projections in turtles of the genera *Pseudemys* and *Chrysemys, J. Comp. Neurol.* **276:**92–112.

Ulinski, P. S., Dacey, D. M., and Sereno, M. I., 1990, Optic tectum, in: *Biology of the Reptilia,* Volume 17 (C. G. Gans and P. S. Ulinski, eds.), University of Chicago Press, Chicago, in press.

Veenman, C. L., and Gottschaldt, K. M., 1986, The nucleus basalis–neostriatum complex in the goose (*Anser anser* L.), *Adv. Anat. Embryol. Cell Biol.* **96:**1–85.

Vicario, D. S., and Nottebohm, F., 1988, Organization of the zebra finch song control system: I. Representation of syringeal muscles in the hypoglossal nucleus, *J. Comp. Neurol.* **271:**346–354.

Vicario, D. S., and Simpson, H. B., 1988, Control of syringeal muscles in nucleus RA of zebra finches, *Soc. Neurosci. Abstr.* **14:**89.

Viohl, G., 1985, Geology of the Sohnofen lithographic limestone and the habitat of *Archaeopteryx,* in: *The Beginnings of Birds* (M. K. Hecht, J. H. Ostrom, G. Viohl, and P. Wellnhoffer, eds.), Freunde des Jura-Museums Eichstatt, Eichstatt, pp. 31–44.

Vogt-Nilsen, L., 1954, The inferior olive in birds. A comparative morphological study, *J. Comp. Neurol.* **101:**447–481.

Volman, S. F., and Konishi, M., 1987, Auditory selectivity in the song-control nucleus HVc appears with the onset of plastic song, *Soc. Neurosci. Abstr.* **13:**870.

Volman, S. F., and Konishi, M., 1988, Comparative physiology of auditory localization in owls, *Soc. Neurosci. Abstr.* **14:**322.

Wagner, H., Takahashi, T., and Konishi, M., 1987, Representation of interaural time difference in the central nucleus of the barn owl's inferior colliculus, *J. Neurosci.* **7:**3105–3116.

Walker, C. A., 1973, The origins of owls, in: *Owls of the World* (J. A. Burton, ed.), Dutton, New York, pp. 27–33.

Watanabe, M., Ito, H., and Masai, H., 1983, Cytoarchitecture and visual receptive neurons in the wulst of the Japanese quail (*Coturnix coturnix japonica*), J. Comp. Neurol. **213:**188–198.

Webster, D. M. S., and Steeves, J. D., 1988, Origins of brainstem-spinal projections in the duck and goose, *J. Comp. Neurol.* **273:**573–583.

Weiss, S. R. B., and Hodos, W., 1986, Discrimination of mirror-image stimuli after lesions of the visual system in pigeons, *Brain Behav. Evol.* **29:**207–222.

Wellnhofer, P., 1974, Das funfte Skelettexemplar von *Archaeopteryx, Beitr. Naturgesch. Vorzeit Abt. A* **147:**169–216.

Wellnhoffer, P., 1988, A new specimen of *Archaeopteryx, Science* **240:**1790–1792.

Wellnhoffer, P., 1990, *Archaeópteryx, Sci. Amer.* **262:**70–77.

Whitlock, D. G., 1952, A neurohistological and neurophysiological study of afferent fiber tracts and receptive areas of the avian cerebellum, *J. Comp. Neurol.* **97:**567–623.

Wild, J. M., Arends, J. J. A., and Zeigler, H. P., 1984, A trigeminal sensorimotor circuit for pecking, grasping and feeding in the pigeon (*Columba livia*), Brain Res. **300:**146–151.

Wild, J. M., Arends, J. A. A., and Zeigler, H. P., 1985, Telencephalic connections of the trigeminal system in the pigeon (*Columba livia*): A trigeminal sensorimotor circuit, *J. Comp. Neurol.* **234:**441–464.

Wiley, E. O., 1981, *Phylogenetics: The Theory and Practice of Phylogenetic Systematics,* Wiley, New York.

Williams, H., and Nottebohm, F., 1985, Auditory responses in avian vocal motor neurons: A motor theory for song perception in birds, *Science* **229:**279–282.

Wilson, P., 1980a, The organization of the visual hyperstriatum in the domestic chick. I. Topology and topography of the visual projection, *Brain Res.* **188:**319–332.

Wilson, P., 1980b, The organization of the visual hyperstriatum in the domestic chick. II. Receptive field properties of single units, *Brain Res.* **188:**333–345.

Witkovsky, P., Zeigler, H. P., and Silve, R., 1973, The nucleus basalis of the pigeon: A single-unit analysis, *J. Comp. Neurol.* **147:**119–128.

Zeier, H., and Karten, H. J., 1971, The archistriatum of the pigeon: Organization of afferent and efferent connections, *Brain Res.* **31:**313–326.

Zeigler, H. P., 1974, Feeding behavior in the pigeon: A neurobehavioral analysis, in: *Birds: Brain and Behavior* (I. J. Goodman and M. W. Schein, eds.), Academic Press, New York, pp. 101–132.

Zeigler, H. P., 1976, Feeding behavior of the pigeon (*Columba livia*), Adv. Study Behav. **7:**285–389.

Zeigler, H. P., and Karten, H. J., 1973, Brain mechanism and feeding behavior in the pigeon (*Columba livia*). II. Analysis of feeding behavior deficits after lesions of the quintofrontal structures, *J. Comp. Neurol.* **152:**83–102.

Zeigler, H. P., Levitt, P. W., and Levine, R. R., 1980, Eating in the pigeon (*Columba livia*): Movement patterns, stereotypy, and stimulus control, *J. Comp. Physiol. Psychol.* **94:**783–794.

Zeier, H., 1971, Behavioral adaptation on operant schedules after forebrain lesions in the pigeon, in: *Birds: Brain and Behavior* (I. J. Goodman and M. W. Schein, eds.), Academic Press, New York, pp. 153–164.

II

Introduction to Mammalian Cortical Evolution

7

Evolution of Neocortex

JOHN ALLMAN

1. Introduction

In this chapter I shall attempt to derive the outlines of a theory of neocortical evolution from a series of observations based mainly on the anatomy and physiology of cerebral cortex in living animals. Such an exercise in evolutionary inference is by its nature a speculative enterprise. Hopefully, it will serve to guide future comparative, developmental and biophysical studies that might shed some additional light on this intriguing but inaccessible topic. To illustrate organizational features of cortex, I have drawn examples mainly from visual cortex.

2. General Features of Cortical Organization

2.1. Horizontal Expansion

Compared across all mammals, the surface area of the neocortical sheet varies by nearly five orders of magnitude* while the thickness of neocortex varies by less than one order of magnitude.† Therefore, evolutionary changes in

**Sorex minutus* has a cortical surface area of approximately 30 mm^2 (Brauer and Schober, 1970). Elias and Schwartz (1969) provide data for brain volume and cortical surface area in six cetacean species. On the bases of these data it is possible to estimate the cortical surface area of the largest mammal, the blue whale (*Balaenoptera musculus*) for which there is only a direct measurement of brain weight (6800 g) available (Crile and Quiring, 1940). Extrapolating from the data for the smaller whales, the blue whale would be expected to have a cortical surface area of about 1,350,000 mm^2.

†The vertical thickness of neocortex varies from a minimum of about 0.6 mm in the long-fingered bat (*Miniopterus schreivers'i*) as measured from a published photomicrograph with brain shrinkage due to celloidin embedding taken into acount (Igarashi and Kamiya, 1972) to a maximum of about 5 mm.

JOHN ALLMAN • Division of Biology, California Institute of Technology, Pasadena, California 91125.

neocortex have occurred mainly parallel to the cortical surface (horizontally) and have been sharply constrained in the vertical dimension. The neocortical sheet stands in marked contrast to other telencephalic structures, which have a radial or nuclear organization.

2.2. Constraints on Vertical Expansion

The common feature of vertical cortical architecture is the apical dendrite of the pyramidal neuron, which receives stratified inputs from different sources. Possible active processes involving dendritic spines may enable the apical dendrite to perform a number of basic logic operations that form a crucial part of cortical circuitry (Shepherd and Brayton, 1987). The apical dendrite may be subject to biophysical constraints that limit its length, which in turn may limit the maximum vertical thickness of neocortex.

2.3. Architectural Uniformity

The cellular architecture of neocortex is remarkably uniform throughout its horizontal extent* (Creutzfeldt, 1978). This basic neocortical uniformity suggests that all of neocortex arose from a single precursor structure, most probably in the dorsolateral telencephalon of an ancestral reptile.†

2.4. Developmental Uniformity

The entire neocortical sheet appears to develop as a unit in each hemisphere in the embryo. Synaptogenesis proceeds along the same time course throughout the cortex (Rakic *et al.*, 1986). The cortex does not appear to develop on an area-by-area basis. The one apparent exception is that the myelination of primary sensory areas occurs much earlier than in higher areas (Flechsig, 1904).

*It sometimes has been suggested that the number of neurons under a particular sized patch of cortical surface (striate cortex excepted) is constant for all mammals (Bok and Taalman Kip, 1939; Rockel *et al.*, 1980). However, Haug's (1987) measurements in 29 species indicate that the actual density of cortical neurons under a square millimeter of cortical surface varies by about a factor of 10 with the lowest being in the elephant and the highest in the woolly monkey.

†The line of ancient reptiles that were ancestral to mammals separated from basal groups of reptiles very soon after their appearance in the fossil record approximately 300 million years ago (Carroll, 1988). The dorsolateral telencephalon in these early reptiles may have resembled the comparable structure in living amphibians, which contains pyramidal cells located near the ventricle with their apical dendrites radiating out toward the external surface of the brain (Kemali and Braitenberg, 1969). This basic pattern is retained in mammalian neocortex with the proliferation of additional cellular laminae. The dorsal ventricular ridge found in living reptiles and derivative structures in birds may also have emerged from portions of the dorsolateral telencephalon in the ancestral reptiles and represent a parallel development to portions of neocortex in these animals. These structures lack pyramidal neurons with their characteristic apical dendrites and instead have their own distinctive neural architecture (Ulinski, 1983, 1986). Thus, the dorsal ventricular ridge may constitute a more derived and specialized condition than neocortex.

2.5. Functional Diversity

In contrast to the horizontal uniformity of the neocortex in architecture and development, some of the functions of different cortical areas are remarkably diverse. Much of the neocortical surface is devoted to topographically organized representations of the sensory domains of vision, touch, and hearing and to the muscles, although other portions appear to be nontopographic. The functional differences probably are largely the result of differences in connections with other brain structures.

2.6. Expansion of White Matter

One consequence of the horizontal extension of cortex in evolution is that a large portion of the telencephalon must be devoted to fiber pathways in the white matter that connect different parts of the neocortical sheet.* This is strikingly evident when one compares the brain of a bird such as an owl with a mammal with a similarly sized brain (see Fig. 1). The Wulst, the structure in the owl with analogous connections and neural response properties to visual cortex, is a massive cellular aggregation that *lacks* pyramidal cells with apical dendrites characteristic of neocortex (Pettigrew, 1979). The avian Wulst contains a single map of the visual field unlike the multiple areas present in mammalian visual cortex (Pettigrew, 1979).

3. Comparisons with Reptiles

In searching for the origin of neocortex it is useful to compare cortical organization in reptiles and mammals, although no living reptile is truly representative of the reptilian ancestors of mammals.

3.1. Laminar Connections

Reptilian cortex receives its input in layer 1, where incoming fibers terminate at different levels on apical dendrites. For example, in reptilian visual cortex, the lateral geniculate nucleus input terminates on the distal third of apical dendrites while intercortical connections terminate on the proximal two-thirds

*When the volumes of different components of the brains of insectivores and primates are plotted relative to body size, the white matter associated with neocortex has the steepest slope of all the components (Frahm *et al.*, 1982). For this group of mammals, neocortical volume expands with an exponent of 0.67 relative to body size while the neocortical white matter has an exponent of 0.86, which indicates that proportionally more white matter is required to serve an increasingly large cortex. For example, when the neocortex and associated white matter are summed together, the white matter constitutes 6.7% of the total in *Sorex minutus* and 41.9% in *Homo sapiens*. About one-third of the entire human brain is devoted to white matter associated with neocortex.

of the apical dendrites (Ulinski, 1988). Similar connections are present in layer 1 in mammalian primary visual neocortex, but the main ascending input terminates in layer 4 (Tigges *et al.*, 1977; Diamond *et al.*, 1985). Reptilian visual cortex (area D2) contains two cellular layers both of which contain cells that project back to the lateral geniculate nucleus (Heller and Ulinski, 1987). Apparently intermixed with these cells are neurons that project to other cortical regions and to other subcortical structures (Desan, 1984). Mammalian neocortex contains many cellular layers with specialized inputs and outputs for each layer. Neurons in layers 2 and 3 of primary visual neocortex project to higher cortical areas while neurons in layer 6 project back to the lateral geniculate nucleus (Lund *et al.*, 1975). This laminar pattern of ascending and descending connections is characteristic of many cortical areas (Rockland and Pandya, 1979; Maunsell and Van Essen, 1983; Van Essen, 1985; Pandya *et al.*, 1988). Neurons with ascending outputs and feedback neurons are present in both reptiles and mammals, but in mammals they are organized in distinct laminae.

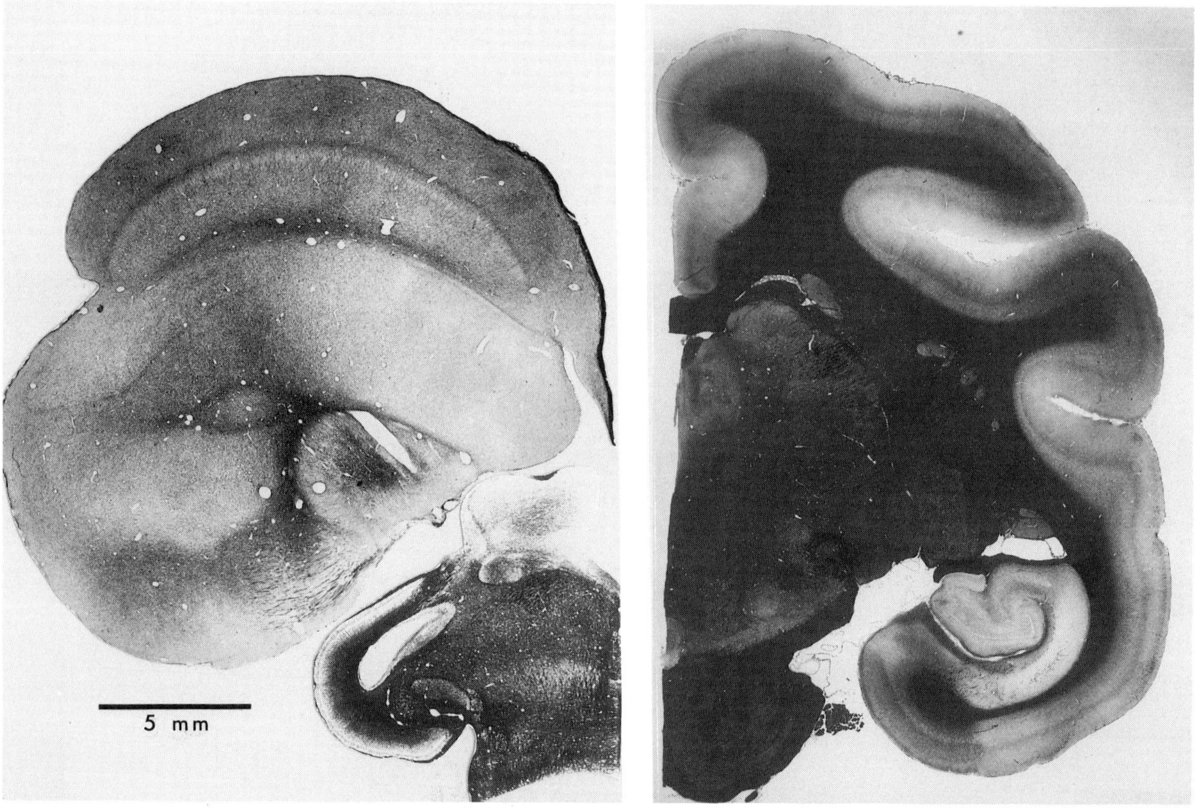

Figure 1. Fiber-stained coronal sections through the forebrain of an owl (left) and an owl monkey (right). Note the much more extensive portion of the brain devoted to fiber pathways in the white matter in the monkey as compared to the owl. The owl section was stained with the Gallyas silver technique; the owl monkey section was stained with the Heidenhain hematoxylin technique.

3.2. The Derivation of Neocortical Laminae from Ancestral Reptilian Cortex

The laminar organization of the mammalian cortex could have been derived from ancestral reptilian cortex by mutations that regulated neural proliferation and the migration of cells in the cortical plate during development. For example, the laminar segregation of neurons projecting back to the lateral geniculate nucleus could have been achieved by modulating cell adhesion processes during neural migration (Pinto-Lord *et al.*, 1982; Edelman, 1987; McConnell, 1988).

3.3. Organization for Identification and Storage of Environmental Constancies

In reptiles, there is a crude topographic order in the projection from the lateral geniculate nucleus to visual cortex (Ulinski, 1988), while in mammals there is a high degree of topographic order in the geniculocortical projection (Polyak, 1932). Similarly, the receptive field organization of visual cortex in turtles is crudely topographic with very large receptive fields (Mazurskaya, 1971, 1972), while the receptive field organization of primary visual cortex in mammals is highly topographic (Kaas *et al.*, 1970; Allman and Kaas, 1971b; Sousa *et al.*, 1978; Van Essen, 1979). The original cortical organization probably was either crudely topographic or nontopographic with global receptive fields. Cortex was probably devoted to the identification and storage of behaviorally relevant environmental constancies. For example, in the olfactory domain, individual conspecifics might be identified on the basis of different volatile organic compounds emitted by each animal. Since these volatile compounds tend to occur in mixtures that are characteristic of that individual and since they are received over a broad range of concentrations and in a complex chemical environment, the crucial task of the olfactory system is to identify the *pattern* characteristic on an individual independent of concentration and against the confusing background of other odors. Thus, the basic functions of olfactory cortex are to identify constancies within the environment and to store these constancies for future behavioral contingencies (Haberly, 1985).

4. Neocortex Retains Ancient Features of Global Organization

The ancient nontopographic organization has been retained in neocortex in spite of the high degree of topographic order present in many neocortical areas. Even in primary visual cortex, the true receptive field for many neurons is very much larger than the receptive field mapped in conventional studies (Allman *et al.*; in press). The extensive nonclassical surround is revealed by presenting stimuli simultaneously within and outside the classical receptive field and determining the influence of the outside stimuli. The stimuli presented beyond the classical receptive field often have profound and highly selective effects on the responses within the classical receptive field (Allman *et al.*, 1985, in press). The large nonclassical receptive fields probably subserve integrative functions such as perceptual constancies and the separation of figure from background. For example, neurons in the fourth visual area (V4) appear to extract color constancy using their nonclassical receptive fields (Zeki, 1983; Desimone *et al.*, 1985) in a manner similar to Land's (1986) retinex theory.* Thus, just as olfactory cortex would extract constancies concerning the edibility of a particular fruit on the basis of a pattern of volatile compounds emitted at a certain stage of ripeness over a broad range of chemical concentrations, similarly V4 would extract parallel constancies concerning edibility of a particular fruit on the basis for the spectral pattern of reflectances at a particular stage of ripeness over a broad range of lighting conditions. In each case the cortex would determine that the sweet-smelling red fruit is good to eat in the wet early morning as well as in the dry afternoon. The development of the capacity of cerebral cortex to extract behaviorally significant constancies over a very broad range of physical conditions fits into the general context of vertebrate evolution. Kluver (1942) proposed that just as mechanisms have evolved to maintain a constant internal physiological environment in higher organisms, which create for them an independence from external environmental conditions, similarly the perceptual constancy functions of visual cortex "guarantee varying degrees of freedom from the visual milieu." These metabolic and perceptual constancy mechanisms are of enormous adaptive value but are energetically and computationally very expensive and highly interdependent (see Section 8).

5. Evolution of New Areas

It has sometimes been suggested that the ancient neocortex was nontopographic and that the topographically organized areas developed more recently and the primary sensory areas the most recently (Sanides, 1970; Morgane *et al.*, 1986; Pandya *et al.*, 1988). According to the evidence presented previously, it is likely that in the reptilian precursors of mammals the cortex was nontopographic or only crudely topographic and that aspects of this ancient organization are preserved in neocortex. However, the presence of highly organized sensory maps in neocortex of monotremes (Bohringer and Rowe, 1977), marsupials (Sousa *et al.*, 1978), and primitive eutherian mammals (Kaas *et al.*, 1970) suggests that neocortical maps are old features going back to the common origin of mammals early in the Mesozoic era more than 200 million years ago. Similarly, in the hedgehog, a mammal that has retained many primitive features of early eutherian mammals, most of the neocortex is devoted to a small number of topographically organized areas (Kaas *et al.*, 1970). These data suggest that the neocortex present in Mesozoic mammals contained a small number of areas, some of which contained well-defined topographic maps while others may have lacked topography. In these early mammals, the topographic organization was superimposed on an ancient pattern of global cortical organization.

The comparative data suggest that many new areas have developed in the

*The achievement of color constancy does not require a cerebral cortex; it is present in goldfish (Ingle, 1985).

course of neocortical evolution in different lines of descent. These areas have been more easily delineated within highly organized sensory representations, but they may also exist in the less topographically organized neocortex. About 50 million years ago, in the early stages of primate evolution, there was a great expansion in the size of the occipital and temporal lobes indicative of an expansion of visual cortex (Radinsky, 1967; Allman, 1977). The skulls of these early primates possessed a number of other features characteristic of modern primates such as the presence of orbits indicative of large, frontally directed eyes (Allman, 1977). The visual cortical expansion in the early primates may have been due both to an enlargement of existing areas and to the development of new areas (Allman and McGuinness, 1988). The enlargement of existing areas was probably linked to the expansion of the representation of the central retina associated with frontally directed eyes and higher acuity in the central visual field. In addition some of the array of higher cortical visual areas present in both prosimian and simian primates may have emerged at this early stage in primate evolution (Allman and McGuinness, 1988; Sereno and Allman, 1990). The great similarity in the number and organization of the cortical visual areas in old and new world monkeys suggests that the basic pattern for simian visual cortex was established before the divergence of these lines about 40 million years ago (Sereno and Allman, 1990).

While primary visual area can be easily homologized in different orders of mammals and thus appears to be relatively ancient (Kaas *et al.*, 1970), it is much more difficult to establish interorder homologies for higher cortical areas (Baker *et al.*, 1981; Tusa *et al.*, 1981). The greater variability in the patterns of organization in the higher cortical visual areas in different mammalian orders suggests that these areas are more recent phylogenetic developments that arose independently in different mammalian orders (Sereno and Allman, 1990).

6. Mechanisms for the Emergence of New Areas

A common mode of evolution appears to have been the duplication of existing structures and the subsequent divergence of the duplicates (Gregory, 1935; Allman and Kaas, 1971a; Allman, 1987). Presumably, an initial mutation produced the duplication and mutations in subsequent generations differentially influenced the development of the duplicates so that they diverged in structure and function. Duplication also appears to be the principal means whereby new genes have emerged in evolution (Lewis, 1951; Ohno, 1970; Li, 1983).

A possible model for the differentiation of existing cortical areas and the acquisition of new areas is provided by the mechanisms for segmental differentiation in *Drosophila* embryogenesis. Early in the development of *Drosophila*, the actions of genes establish a polar gradient and a set of distinct stripes corresponding to half segments, which in turn are differentiated from a common replicated pattern by the action of the homeotic genes (see Fig. 2). Scott and Carroll (1987) have proposed an evolutionary scenario for insect development. "The development of the largely identical segments of an annelid-like ancestor may have depended on the function of a gene similar to *ftz* that was expressed in stripes and directed the formation of repeating pattern elements. The homeotic genes could have arisen by duplication and divergence of the *ftz*-like gene—an extension of the proposal by E. Lewis (1951) that homeotic genes arose by duplication and divergence" (Scott and Carroll, 1987, p. 695). Sets of homeotic genes controlling the development of the head and body are located in the *Antennapedia* and *Bithorax* chromosomal complexes, respectively. The order of the homeotic genes in the *Bithorax* complex corresponds approximately to the anatomical arrangement of the segments controlled by these genes (Lewis, 1978). The regulatory actions of the homeotic genes bring about the differentiation of the initially identical segments (Lewis, 1978).

Each of the homeotic genes contains a 180-bp sequence of DNA, called the "homeobox," which shows a high degree of correspondence. Multiple homeobox-containing genes are also found in vertebrates, including frogs, mice, and humans (McGinnis *et al.*, 1984; Gehring, 1985; McGinnis, 1985). In mice, where they have

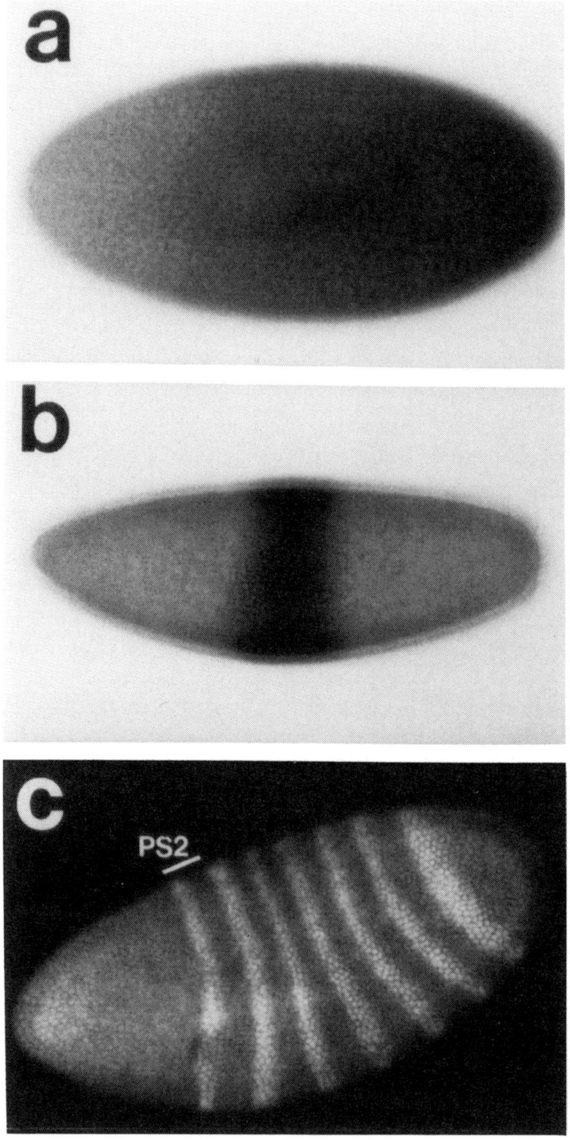

Figure 2. Expression patterns of gene products in *Drosophila* development from Scott and Carroll (1987). Whole mount embryos were permeabilized and stained with antibodies against gene products. Anterior is to the left and ventral is down in each panel. The embryos are about 0.5 mm long. The embryos are at the blastoderm stage (2.5–3.5 hr of development) when the approximately 6000 cells form a monolayer covering the yoke. (a) Expression of the *caudal* gene product (MacDonald and Struhl, 1986), a nuclear protein that is distributed in a gradient in the early blastoderm embryo. Staining is with peroxidase, so the protein signal appears black. (b) Expression of the *Kruppel* gap gene nuclear protein (Gaul *et al.*, 1987) detected with peroxidase stain. The broad band corresponds approximately to parasegments 4–6. (c) Expression of the *fushi tarazu* (*ftz*) pair-rule gene product detected by immunofluorescence so the nuclear protein signal appears white. The anterior edges of the stripes mark the anterior edges of the parasegments; the extent of parasegment 2 is indicated.

been most extensively studied, there are two families of homeobox-containing genes, each containing six members, located on chromosomes 6 and 11 (Martin *et al.*, 1987). There are also isolated homeobox-containing genes on other chromosomes. The homeobox encodes for a 60-amino-acid-long "homeo domain" that is rich in the basic amino acids lysine and arginine, which would enable the homeo domain to bind to DNA (Laughton and Scott, 1984). The homeo domain apparently has been strongly conserved in evolution, possibly to perform the function of binding the regulatory gene product to regulation sites in the DNA. The *in situ* hybridization of the cloned DNA of the homeotic genes is located primarily in the developing nervous system in *Drosophila,* where the genes may regulate the differentiation of neural circuitry for each segment (Scott, 1984). Homeobox-containing genes also are expressed in mammals in specific regions of the CNS (Awgulewitsch *et al.*, 1986; Davis *et al.*, 1988). Recently, they have been found in sets of segmentlike structures (rhombomeres) that innervate particular brachial arches in the developing hindbrain of the mouse (Lewis, 1989; Wilkinson *et al.*, 1989). The protein encoded by the homeobox-containing gene, *Hox 1.3,* is present in neocortical pyramidal neurons (Odenwald *et al.*, 1987).

An analogous, or even possibly homologous, hierarchical set of genes could regulate the development of cortical areas in mammals by establishing rostral–caudal and dorsal—ventral polarity in the cortex, by segmenting the cortical sheet into areas, and by differentiating the areas according to their connections. Gene duplications and subsequent divergence within the set of controlling genes could produce new areas, new connectivities, and new functions.

Turing (1952) suggested that the spatial patterns in embryogenesis could be established if two substances (morphogens) with different diffusion rates reacted with each other. The morphogens might be functionally related to the proteins encoded by the homeobox-containing genes, and interactions like those postulated by Turning may be responsible for pattern formation in *Drosophila* (Meinhardt, 1982; Akam, 1987). Major changes in the pattern, such as changes in the number of segments, could be produced by mutations altering the diffusion rates of morphogens or their interactions. Similar mechanisms might be responsible for the specification of the number of cortical areas in mammalian development.* Thus, mutations altering the physical or chemical properties of morphogens might have produced the major expansion in the number of cortical visual areas that occurred early in primate evolution.

7. The Role of Experience in Cortical Organization

The experiments of Merzenich *et al.*, (1988) in somatosensory cortex indicate that *use* influences topographic mapping in cortex, which presumably reflects the capacity to perform neural computations within different parts of the map. Thus, cortical organization is linked to behavior and has some capacity to change throughout the life span of a mammal. The neocortex may participate in epigenetic evolution. Changes in behavior that resulted in changes in cortical

*In *Drosphila* these patterns are established during a period in which the embryo is a syncytium (Akam, 1987; Ingham, 1988), which is not the case in the developing cortex (Rakic, 1988). However, it is possible that morphogens communicate among cells at this stage in cortical development.

organization could be transmitted by observational learning from one generation to the next. For example, a foraging technique might require a particular type of difficult discrimination which would result in use-dependent changes in cortical organization. This behavior and its attendant cortical manifestations could be propagated through the generations through observational learning.

8. Mammalian Energetics and Neocortex as Linked Homeostatic Mechanisms

A mammal requires 5 to 10 times more energy expenditure than does an equivalent-sized reptile (Bennett and Ruben, 1979). Most of this increased energy expenditure is devoted to homeostatic mechanisms of thermoregulation (Muller, 1985). Thus, mammals must eat a great deal more than reptiles and do so on a regular basis to avoid starvation.* My hypothesis is that neocortex evolved in tandem with the higher energy metabolism characteristic of mammals so as to assure a continuous localization of food resources. Neocortex is a topographically organized network that stores representations of the animal's interactions with its environment. The neocortical network requires an initial period of training, which I believe is manifested by infant play behavior. During this period of network training the developing infant's energy requirements are provided by its mother's milk. Thus, in this formulation, mammalian lactation and play behavior are necessary adaptations to support the linked homeostatic mechanisms of mammalian energetics and neocortex.†

9. The Expansion of Neocortex and Longevity

The relative amount of neocortex to body size tends to be greatest in mammals with the longest life span.‡ This probably relates to the high energy re-

*There are mammals such as the fat-tailed dwarf lemur (*Cheirogaleus medius*) that can go for a very long time without eating during which period their body temperature drops substantially. However it is likely that these mammals have very specialized biochemical regulatory mechanisms that enable their enzyme systems to operate over a broad range of body temperatures.

†I believe that somewhat analogous developments occurred in the line leading to birds in the Mesozoic. These developments probably included the elaboration of the Wulst and dorsal ventricular ridge, thermoregulation at higher body temperature, and parenting behavior.

‡The tendency for neocortical enlargement to be correlated with longevity is evident when the neocortical progression indices of Stephan and Andy (1970) are compared with maximum life spans for insectivores and primates. The neocortical progression index is intended to show how much neocortex a mammal has relative to a comparably sized "basal insectivore" thought to represent the primitive mammalian condition. Basal insectivores have an average neocortical progression index of 1 and a relatively short life span. For example, the basal insectivores *Hemicentates* and *Tenrec* have maximum life spans of about 2 years (Eisenberg, 1981). *Cebus* and *Pan* have a neocortical index of 59. The longevity record for *Cebus* is 54 years and there exist several very elderly chimpanzees still alive in their late 50s (M. Groves, personal communication). For humans the neocortical index is 156 and the maximum documented life span is about 115 years (C. Finch, personal communication). It is difficult to establish the maximum life span for long-lived nonhuman primates since it has only been in the past few decades that they have been kept in captivity under relatively optimal conditions of nutrition and housing.

quirements of mammals and to different strategies for coping with occasional food shortages. Short-lived mammals tend to have a high reproductive potential and relatively less neocortex. In these mammals, food shortages will produce catastrophic population declines that quickly recover with improved conditions which has been termed r-selection (MacArthur and Wilson, 1967; Wilson, 1975). Long-lived mammals have smaller litters with much greater parental investment in individual offspring (K-selection). The expanded neocortex enables metabolically active, nonhibernating, mammals to store the necessary experiences, foraging strategies, and cognitive processes to cope with periods of severe food shortage that will inevitably happen during a long life span. Viewed in this way the expansion of neocortex can be regarded as the ultimate K-selection strategy enabling organisms to survive the vicissitudes of the environment encountered during a long life span.

10. Summary

Neocortex probably arose as a specialization of the dorsolateral telencephalon in mammal-like reptiles early in the Mesozoic. My hypothesis is that the neocortex evolved in tandem with the higher energy metabolism characteristic of mammals as linked homeostatic mechanisms. These mechanisms, while computationally and energetically expensive, provided a powerful selective advantage for the early mammals. The comparative data from living primitive mammals suggest that the early neocortex contained a relatively small number of areas, most of which were topographically organized representations of the sensory modalities of vision, touch, and hearing. The topographic organization was superimposed on a preexisting global organization. Thus, neocortex possesses the capacity for very specific local representations combined with more global representations that subserve perceptual and temporal constancies. The phylogenetic expansion of the cortical sheet has been mainly horizontal with the vertical thickness limited by the information processing capabilities of the apical dendrites of pyramidal neurons. The organization of cortex is governed by a set of genes that control cell proliferation and migration. By analogy with the homeotic genes that control the differentiation of parasegments in *Drosophila* embryogenesis, the genes controlling the parcellation and expansion of cortical areas may be a family of replicated genes derived from a common primitive gene. It has been conventional to assert that different cortical areas within a modality such as vision have different functions; however, these areas share common circuitry and may to a significant degree share the capacity to perform common functions. Cortical organization is the product of the interaction and genetic and epigenetic factors. Behavioral specializations in use influence cortical organization and these behaviors and perhaps some aspects of cortical organization are transmitted from generation to generation through observational learning. Neocortex is most expanded relative to body size in species with the greatest life span. The expanded neocortex enables metabolically active mammals to store the necessary experiences, foraging strategies, and cognitive processes to cope with periods of food shortage that will inevitably occur during a long life span. Viewed in this manner, the expansion of neocortex can be regarded as the ultimate K-selection strategy.

ACKNOWLEDGMENTS. I thank Ed Lewis for stimulating discussions of the evolution of the genetic control of development. The members of the Helmholtz Club, the Montreal Neurological Institute, the Neural Systems and Behavior Course at the Marine Biology Laboratory, and the Woods Hole Symposium on Computational Neuroscience honed the ideas presented here through vigorous discussion. The work was supported by a Markey Foundation Grant in developmental biology and by the Sloan Foundation.

11. References

Akam, M., 1987, The molecular basis for metameric pattern in the *Drosophila* embryo, *Development* **101:**1–22.

Allman, J., 1977, Evolution of the visual system in the early primates, *Prog. Psychobiol. Physiol. Psychol.* **7:**1–53.

Allman, J. M., 1987, Maps in context: Some analogies between visual cortical and genetic maps, in: *Matters of Intelligence* (L. Vaina, ed.), Reidel, Dordrecht, pp. 369–393.

Allman, J. M., and Kaas, J. H., 1971a, A representation of the visual field in the caudal third of the middle temporal gyrus of the owl monkey (*Aotus trivirgatus*), Brain Res. **31:**84–105.

Allman, J. M., and Kaas, J. H., 1971b, Representation of the visual field in striate and adjoining cortex of the owl monkey (*Aotus trivirgatus*), Brain Res. **35:**89–106.

Allman, J., and McGuinness, E., 1988, Visual cortex in primates, in: *Comparative Primate Biology,* Volume 4 (H. Steklis and J. Erwin, eds.), Alan Liss, New York, pp. 279–326.

Allman, J., Miezin, F., and McGuinness, E., 1985, Stimulus specific responses from beyond the classical receptive field: Neurophysiological mechanisms for local–global comparisons in visual neurons, *Annu. Rev. Neurosci.* **8:**407–430.

Allman, J., Miezin, F., and McGuinness, E., 1988, The effects of background motion on the responses of neurons in the first and second cortical visual areas, in: *Signal and Sense,* Neuroscience Research Program, New York, in press.

Awgulewitsch, A., Utset, M., Hart, C., McGinnis, W., and Ruddle, F., 1986, Spatial restriction in expression of a mouse homeo box locus within the central nervous system, *Nature* **320:**328–335.

Baker, J., Petersen, S., Newsome, W., and Allman, J., 1981, Visual response properties of neurons in four extrastriate visual areas of the owl monkey (*Aotus trivirgatus*): A quantitative comparison of medial, dorsomedial, dorsolateral and middle temporal areas, *J. Neurophysiol.* **45:**397–416.

Bennett, A. F., and Ruben, J. A., 1979, Endothermy and activity in vertebrates, *Science* **206:**649–654.

Bohringer, R. C., and Rowe, M. J., 1977, The organization of the sensory and motor areas of cerebral cortex in the platypus (*Ornithorhynchus anatinus*), *J. Comp. Neurol.* **174:**1–14.

Bok, S., and Taalman Kip, M., 1939, The size of the body and the size and number of nerve cells in the cerebral cortex, *Acta Neerl. Morphol.* **3:**1–22.

Brauer, K., and Schober, W., 1970, *Catalogue of Mammalian Brains,* Fischer, Jena.

Carroll, R., 1988, *Vertebrate Paleontology and Evolution,* Freeman, New York.

Creutzfeldt, O., 1978, The neocortical link: Thoughts on the generality of structure and function in the neocortex, in: *Architectonics of the Cerebral Cortex* (M. Brazier and H. Petsche, eds.), Raven Press, New York, pp. 367–384.

Crile, G., and Quiring, D., 1940, A record of the body weight and certain organ and gland weights of 3690 animals, *Ohio J. Sci.* **40:**219–259.

Davis, C., Noble-Topham, S., Rossant, J., and Joyner, A., 1988, Expression of the homeo box-containing gene *En-2* delineates a specific region of the developing mouse brain, *Genes Dev.* **2:**361–371.

Desimone, R., Schein, S., Moran, J., and Ungerleider, L., 1985, Contour, colour, and shape analysis beyond the striate cortex, *Vision Res.* **25:**441–452.

Diamond, I. T., Conley, M., Itoh, K., and Fitzpatrick, D., 1985, Laminar organization of geniculocortical projections in *Galago senegalensis* and *Aotus trivirgatus, J. Comp. Neurol.* **242:**610.

Edelman, G. M., 1987, *Neural Darwinism,* Basic Books, New York.

Eisenberg, J., 1981, *The Mammalian Radiations,* University of Chicago Press, Chicago.
Elias, H., and Schwartz, D., 1969, Surface area of the cerebral cortex of mammals determined by stereological methods, *Science* **166:**111–113.
Flechsig, P., 1904, Einige bemerkungen uber die untersuchungsmethoden der grosshirnrinde insbesondere des menschen, *Ber. Verh. Saechs. Ges. Wiss. Leipzig Math. Phys. K1.* **56:**5–104, 177–248.
Frahm, H., Stephan, H., and Stephan, M., 1982, Comparison of brain structure volumes in Insectivora and primates. I. Neocortex, *J. Hirnforsch.* **23:**375–389.
Gaul, U., Seifert, E., Struhl, R., and Jackle, H., 1987, Analysis of *Kruppel* protein distribution during early Drosophila development reveals posttranscriptional regulation, *Cell* **50:**639–647.
Gehring, W., 1985, The molecular basis of development, *Sci. Am.* 153–162.
Gregory, W., 1935, Reduplication in evolution, *Rev. Biol.* **10:**272–290.
Haberly, L. B., 1985, Neuronal circuitry in olfactory cortex: Anatomy and functional implications, *Chem. Senses* **10:**219–238.
Haug, H., 1987, Brain sizes, surfaces, and neuronal sizes in the cortex cerebri: A stereological investigation of man and his variability and a comparison with some mammals (primates, whales, marsupials, insectivores, and one elephant), Am. J. Anat. **180:**126–142.
Heller, S. B., and Ulinski, P. S., 1987, Morphology of geniculocortical axons in turtles of the genera *Pseudemys* and *Chrysemys, Anat. Embryol* **175:**505–515.
Igarashi, S., and Kamiya, T., 1972, *Atlas of the Vertebrate Brain,* University Park Press, Baltimore.
Ingham, P., 1988, The molecular genetics of embryonic pattern formation in *Drosophila, Nature* **335:**28–34.
Ingle, D., 1985, The goldfish as a retinex animal, *Science* **227:**651–654.
Kaas, J. H., Hall, W. C., and Diamond, I. T., 1970, Cortical visual areas I and II in the hedgehog: Relation between evoked potential maps and architectonic subdivisions, *J. Neurophysiol.* **33:**595–615.
Kemali, M., and Braitenberg, V., 1969, *Atlas of the Frog's Brain,* Springer, Berlin.
Kluver, H., 1942, Functional significance of the geniculostriate system, *Biol. Symp.* **7:**253–300.
Land, E. H., 1986, An alternative technique for the computation of the designator in the retinex theory of color vision, *Proc. Natl. Acad. Sci. USA* **83:**3078–3080.
Laughton, A., and Scott, M., 1984, Sequence of a *Drosophila* segmentation gene: Protein structure homology with DNA-binding proteins, *Nature* **310:**25–31.
Lewis, E., 1951, Pseudoallelism and gene evolution, *Cold Spring Harbor Symp. Quant. Biol.* **16:**159–174.
Lewis, E., 1978, A gene complex controlling segmentation in *Drosophila, Nature* **276:**565–570.
Lewis, J., 1989, Genes and segmentation, *Nature* **341:**382–383.
Li, W. H., 1983, Evolution of duplicate genes and pseudogenes, in: *Evolution of Genes and Proteins* (M. Nei and R. Koehn, eds.), Sinauer, Sunderland, p. 14.
Lund, J., Lund, R., Hendrickson, A., Bunt, A., and Fuchs, A., 1975, The origin of efferent pathways from the primary visual cortex, area 17, of the macaque monkey as shown by retrograde transport of horseradish peroxidase, *J. Comp. Neurol.* **164:**287–305.
MacArthur, R., and Wilson, E., 1967, *The Theory of Island Biogeography,* Princeton University Press, Princeton, N.J.
MacDonald, P. M., and Struhl, G., 1986, A molecular gradient in early Drosophila embryos and its role in specifying the body pattern, *Nature* **324:**537–545.
Martin, G., Boncinelli, E., Duboule, D., Gruss, P., Jackson, I., Krumlauf, R., Lonai, P., McGuinness, W., Ruddle, F., and Wolgemuth, D., 1987, Nomenclature for homeobox-containing genes, *Nature* **325:**31–32.
Maunsell, J., and Van Essen, D., 1983, The connections of the middle temporal visual area (MT) and their relationship to a cortical hierarchy in the macaque monkey, *J. Neurosci.* **3:**2563–2586.
Mazurskaya, P. Z., 1971, Study of the projection of the retina to the forebrain of the tortoise *Emys orbicularis, J. Evol. Biochem. Physiol.* **7:**532–536. (English translation).
Mazurskaya, P. Z., 1972, Organization of the neuronal receptive fields of the tortoise *Emys orbicularis* forebrain cortex, *J. Evol. Biochem. Physiol.* **8:**550–555.
McConnell, S. K., 1988, Development and decision-making in the mammalian cerebral cortex, *Brain Res. Rev.* **13:**1–23.
McGinnis, W., 1985, Homeo box sequences of the antennapedia class are conserved only in higher animal genomes, *Cold Spring Harbor Symp. Quant. Biol.* **50:**263–270.
McGinnis, W., Garber, H., Wirz, H., Kuroiwa, A., and Gehring, W., 1984, A homologous protein-

coding sequence in *Drosophila* homeotic genes and its conservation in other metazoans, *Cell* **37**:403–408.

Meinhardt, H., 1982, *Models of Biological Pattern Formation,* Academic Press, New York.

Merzenich, M. M., Recanzone, G., Jenkins, W. M., Allard, T. T., and Nudo, R. J., 1988, Cortical representational plasticity, in: *Neurobiology of Neocortex* (P. Rakic and W. Singer, eds.), Wiley, New York, pp. 41–68.

Morgane, P., Jacobs, M., and Galaburda, A., 1986, Evolutionary morphology of the dolphin brain, in: *Dolphin Cognition and Behavior: A Comparative Approach* (R. Schusterman, J. Thomas, and F. Wood, eds.), Erlbaum, Hillsdale, N.J., pp. 5–30.

Muller, E., 1985, Basal metabolic rates in primates—The possible role of phylogenetic and ecological factors, *Comp. Biochem. Physiol.* **81A**:707–711.

Odenwald, W., Taylor, C., Palmer-Hill, F., Freidrich, V., Tani, M., and Lazzarini, R., 1987, Expression of a homeo domain protein in noncontact-inhibited cultured cells and postmitotic neurons, *Genes Dev.* **1**:482–496.

Ohno, S., 1970, *Evolution by Gene Duplication,* Springer, Berlin.

Pandya, D., Seltzer, B., and Barbas, H., 1988, Input–output relations in primate cerebral cortex, in: *Comparative Primate Biology: Neurosciences* (H. Steklis and J. Erwin, eds.), Liss, New York, pp. 39–80.

Pettigrew, J., 1979, Binocular visual processing of the owl's telencephalon, *Proc. R. Soc. London Ser. B* **204**:435–454.

Pinto-Lord, M., Evrard, P., and Caviness, V., 1982, Obstructed neuronal migration along radial glial fibers in the neocortex of the reeler mouse: A Golgi–EM analysis, *Dev. Brain Res.* **4**:379–393.

Polyak, S., 1932, *The Main Afferent Fiber Systems of the Cerebral Cortex in Primates,* University of California Press, Berkeley.

Radinsky, L., 1967, The oldest primate endocast. *Am. J. Phys. Anthropol.* **27**:385–388.

Rakic, P., 1988, Specification of cerebral cortical areas, *Science* **241**:170–176.

Rakic, P., Bourgeois, J.-P., Eckenhoff, M., Zecevic, N., and Goldman-Rakic, P., 1986, Concurrent overproduction of synapses in diverse regions of the primate cerebral cortex, *Science* **232**:232–235.

Rockel, A., Hiornes, R., and Powell, T., 1980, The basic uniformity in structure of the neocortex, *Brain* **103**:221–244.

Rockland, K., and Pandya, D., 1979, Laminar origins and terminations of cortical connections of the occipital lobe in the rhesus monkey, *Brain Res.* **179**:3–20.

Sanides, F., 1970, Functional architecture of motor and sensory cortices in primates in light of a new concept of neocortical evolution, in: *The Primate Brain* (C. R. Noback and W. Montagna, eds.), Appleton–Century–Crofts, New York, pp. 137–208.

Scott, M., 1984, Homeotic gene transcripts in the neural tissue of insects, *Trends Neurosci.* **7**:221–223.

Scott, M., and Carroll, S., 1987, The segmentation and homeotic gene network in early drosophila development, *Cell* **51**:689–698.

Sereno, M., and Allman, J., 1990, Cortical and visual areas in mammals, in: *Neural Basis of Visual Function* (A. Levinthal, ed.), MacMillan, London, in press.

Shepherd, G. M., and Brayton, R. K., 1987, Logic operations are properties of computer-simulated interactions between excitable dendritic spines, *Neuroscience* **21**:151–165.

Sousa, A., Gattass, R., and Oswaldo-Cruz, E., 1978, The projection of the opossum's visual field on the cerebral cortex, *J. Comp. Neurol.* **177**:569–588.

Stephan, H., and Andy, O. J., 1970, The allocortex in primates, in: *The Primate Brain* (C. R. Noback and W. Montagna, eds.), Appleton–Century–Crofts, New York, pp. 109–135.

Tigges, J., Tigges, M., and Perachio, A., 1977, Complementary laminar terminations of afferents to area 17 originating in area 18 and the lateral geniculate nucleus in squirrel monkey, *J. Comp. Neurol.* **176**:87–100.

Turing, A., 1952, The chemical basis of morphogenesis, *Trans R. Soc. London Ser. B* **237**:37–72.

Tusa, R., Palmer, L., and Rosenquist, A., 1981, Multiple cortical visual areas: visual field topography in the cat, in: *Multiple Visual Areas* (C. Woolsey, ed.), Humana Press, Clifton, N.J., pp. 1–32.

Ulinski, P., 1983, *Dorsal ventricular Ridge,* Wiley, New York.

Ulinski, P. S., 1986, Neurobiology of the therapsid–mammal transition, in: *The Ecology and Biology of Mammal-like Reptiles* (N. Hotton, P. MacLean, J. Roth, and E. Roth, eds.), Smithsonian Institution, Washington, D.C., pp. 149–171.

Ulinski, P. S., 1988, Functional architecture of turtle visual cortex, in: *Forebrain in Reptiles* (W. Schwerdtfeger and W. Smeets, eds.), Karger, Basel, pp. 151–161.

Van Essen, D., 1979, Visual cortical areas, *Annu. Rev. Neurosci.* **2**:227–263.

Van Essen, D., 1985, Functional organization of primate visual cortex, in: *Cerebral Cortex,* Volume 3 (E. Jones, ed.), Plenum Press, New York, pp. 259–329.
Wilson, E., 1975, *Sociobiology,* Harvard University Press, Cambridge, Mass.
Wilkinson, D., Bhatt, S., Cook, M., Boncinelli, E., and Krumlauf, R., 1989, Segmental expression of Hox-2 homeobox-containing genes in the developing mouse hindbrain, *Nature* **341**:405–409.
Zeki, S., 1983, Colour coding in the cerebral cortex: The reaction of cells in monkey visual cortex to wavelength and colours, *Neuroscience* **9**:741–765.

8

Fossil Evidence on the Evolution of the Neocortex

HARRY J. JERISON

1. Introduction

Yes, there is fossil evidence. That is the most surprising thing. The story the fossils tell is less surprising, because it is consistent with what one learns from living species by accepting the consensus about which are primitive and which are advanced (Sarnat and Netsky, 1974). There are a few other surprises, however, which may make this chapter worth reading.

There is fossil evidence because the cranial cavity of a mammal's skull can mold an endocast on which neocortex is easily distinguished from other structures in the brain. This works because an endocast is usually an excellent copy of a brain; neocortex is identified as dorsal to the rhinal fissure and posterior to the olfactory tubercle. The errors made on fossil endocasts are about the same as those that would be made on living brains when these superficial markings rather than histological data are the basis of the identification.

Endocasts can be made from any intact skull. The most interesting, of course, are from fossils because they provide direct evidence on the evolution of the brain. A natural endocast from *Bathygenys reevesi,* a small fossil ungulate from the Big Bend region of Texas, is shown in Fig. 1. Although this endocast is a piece of rock, it is unmistakably a picture of the brain of *Bathygenys* as it was in life, a 35 million-year-old fossil brain.

Comparisons between brains and endocasts in living mammals indicate that

HARRY J. JERISON • Department of Psychiatry and Biobehavioral Sciences, University of California, School of Medicine, Los Angeles, California 90024. This chapter is dedicated to the memory of Professor Leonard Radinsky (July 16, 1937–August 30, 1985), premier scholar of the endocranial anatomy of fossil mammals.

only minor errors occur in treating endocasts as brains. Hundreds of "fossil brains" have been collected throughout the world, either as natural endocasts or as latex or plaster casts made from fossil crania (Edinger, 1975; Jerison, 1973; Radinsky, 1979). The quality of an endocast as a model of the brain differs in different taxa. In fish, amphibians, and reptiles, the model is usually poor and may be useful only for gross estimates of total brain size. In mammals and birds, on the other hand, endocasts usually provide accurate and detailed pictures of the external surface of the brain. Olfactory bulbs, forebrain, and hindbrain are readily identifiable, as are most superficial cortical gyri and sulci of the mammalian brain. It is natural to use the same names for these landmarks in endocasts as in brains.

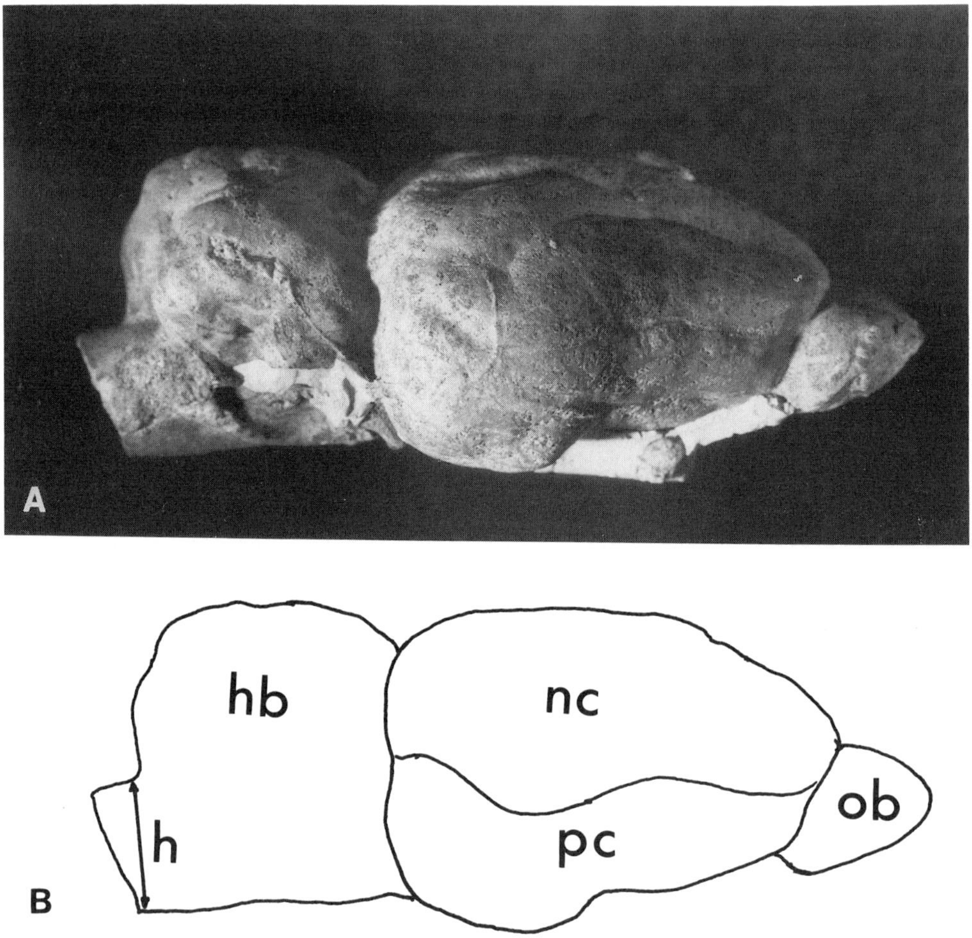

Figure 1. Natural endocast of *Bathygenys reevesi,* a Lower Oligocene oreodon from the Big Bend region of Texas (Wilson, 1971). Fossil bone appears as white; endocast is extremely hard rock, originally compressed and petrified sand that had been packed into the skull by natural forces before fossilization. (A) Lateral view. (B) Lateral view divided by areas to show portions allotted to neocortex (nc), paleocortex (pc), hindbrain (hb), olfactory bulbs (ob), and the height of the foramen magnum (h). (C) Dorsal view. (D) Ventral view. (Specimen UT 40209-431; courtesy of J. A. Wilson and the Department of Paleontology, University of Texas.)

In large-brained living mammals, such as cetaceans, elephants, the great apes, and man, the brain's convolutions are usually smoothed over on the endocasts, and a rhinal fissure may not be identifiable. The quantitative analysis presented later is based on a sample of fossil and living species of carnivores and ungulates in which this fissure is unmistakable. Both qualitatively and quantitatively, one can recognize differences among species in the relative amount of brain that can be identified as neocortex, and it is usually assumed that there have been progressive increases in neocortex during the evolution of the mammals. Neocorticalization is highly correlated with encephalization, i.e., with changes in total brain size beyond changes correlated with body size, and the fossil record of the brain provides convincing evidence for the progressive evolution of encephalization (Jerison, 1985a,b). This chapter emphasizes neocor-

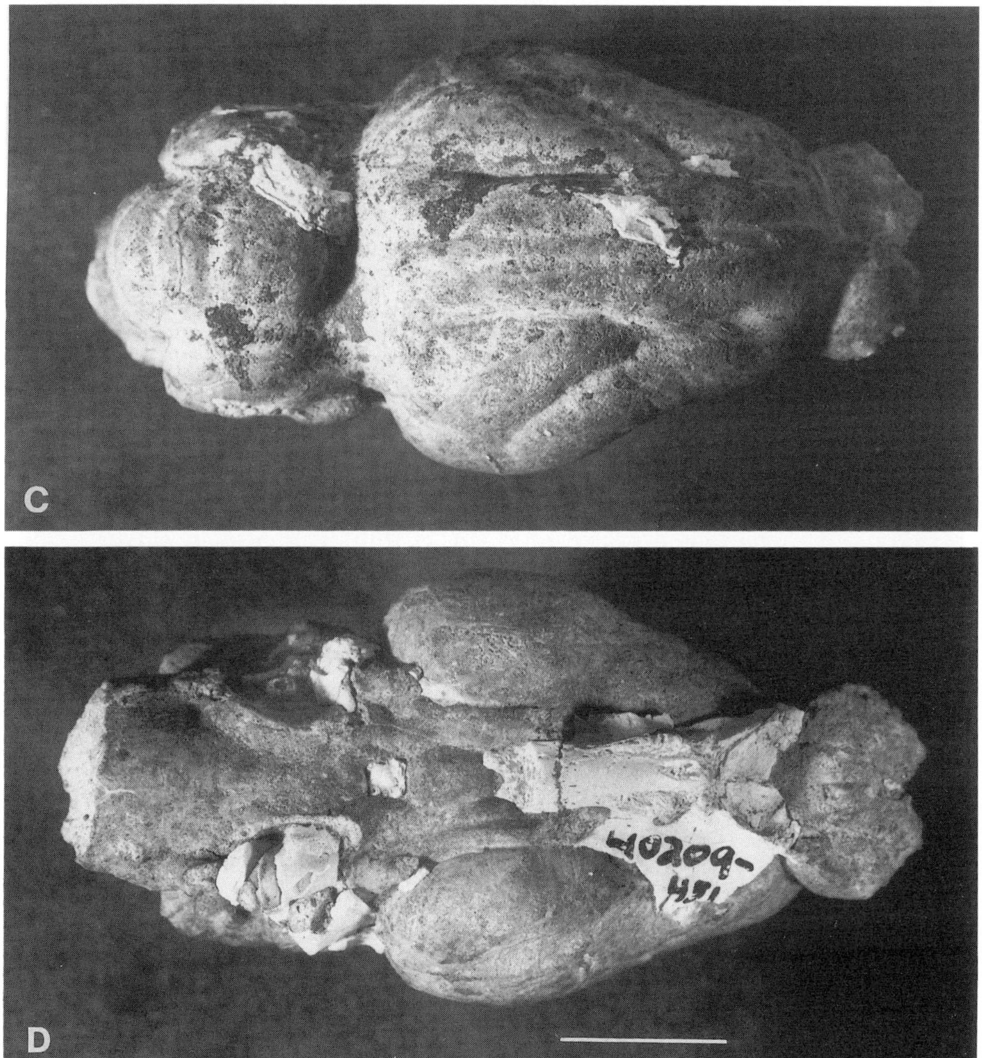

Figure 1. *(Continued)*

ticalization, and the quantitative exercise on neocorticalization in mammals is undertaken in the same spirit as the more often seen analyses of encephalization.

2. Primate Fossil Brains

As members of the order Primates, we are naturally especially interested in neocortical evolution in our group, and it is appropriate to begin with a review of the fossil evidence on primate neocortex.

The earliest primates for which endocasts are available are Eocene lemurlike species. Their endocasts have been interpreted as indicating expansion of occipital and temporal neocortex before such expansion is identifiable in other mammals of the period. Frontal neocortex expanded later (Gurche, 1982; Radinsky, 1979). The brain as a whole in these early primates of 37–54 million years ago was enlarged compared with that of other mammals of similar body size. In other words, these early primates were more encephalized than their contemporaries among other orders of mammals. From other evidence it is reasonable to conclude that this encephalization was correlated with neocorticalization.

Primate endocasts are generally less revealing about the morphology of sulci and gyri than are those of other mammals of similar size; even the Sylvian fissure can be incorrectly represented in the endocast relative to the brain (Connolly, 1950). Human endocasts are often almost completely smooth and resemble fresh brains with dura intact. The only unequivocal information from these endocasts is about gross size, although there are often useful hints about the size and position (i.e., organization) of parts of the neocortex.

The evidence on organization is best in some of the endocasts of smaller-brained hominid species in our lineage (Tobias, 1971). *Australopithecus* of about 3 million years ago, and *Homo habilis* of about 2 million years ago, do have visible gyri and sulci, and there have been interesting controversies about their interpretation, in particular concerning the evolution of human capacities for language. These are worth summarizing for a flavor of the problems that are probably unavoidable in working with these materials.

2.1. Fossil Brains and the Evolution of Language

A lunate sulcus (*Affenspalte*) has been described in an australopithecine (Holloway, 1976, 1981b, 1984), and a third frontal convolution (Broca's area) has been described in *Homo habilis* (Falk, 1983a). The first controversy is about whether the brain systems for human language that are anterior to the lunate (e.g., Wernicke's area) could have begun their enlargement in the australopithecines.

Holloway believes that the lunate in *Australopithecus africanus* was displaced posteriorly relative to its position in chimpanzees. (The brains in these species were similar in size, about 500 and 400 ml, respectively.) Language systems homologous to those of living humans could then have appeared in *Australopithecus* anteriorly to the sulcus in the vicinity of the angular gyrus. Falk (1980, 1983b) interprets the "lunate" differently, and would have the hominid and chimpanzee *Affenspalte* in similar positions. I have argued still another position, based not on detailed morphology but rather on the morphometric consequences of packing primary visual cortex into a primate brain with a posteriorly displaced lunate (Jerison, 1976). The anterior border of area 17 is near the lunate, which limits the extent of visual cortex on the lateral surface of the brain. If Holloway's interpretation is correct, I argued, and if hominids and chimpanzees required similar amounts of primary visual cortex to process visual information, the representation of vision in *medial* neocortex would have been much greater in early hominids than in living chimpanzees. Packing this additional tissue medially would have required major changes in the medial profile of the brain, especially around the calcarine fissure and the splenium of the corpus callosum, which are among the more uniform landmarks in mammalian brains.

My argument was uniformitarian (Simpson, 1970): Since the medial callosal region of the brain in placental mammals is similar in appearance in different species, this region in *Australopithecus* should have looked like that of "good" placental mammals, and visual cortex should have been packed normally. Falk's analysis is consistent with uniformitarianism; Holloway's is not. But the medial surface of the brain is not visible in endocasts, and this controversy cannot be resolved by the fossil evidence alone.

A different kind of problem arises in interpretations of a third frontal convolution as a possibly functional Broca's area. Falk (1983a) has been able to recognize the convolution in *Homo habilis,* but is it Broca's area? This correlation between structure and function could only be established if a third frontal convolution always was Broca's area in human brains. However, only half of the third frontal convolutions in human brains are functionally Broca's areas, usually but not always those in the left hemispheres. It would be possible to identify these in endocasts only if there were morphological asymmetries in this part of the brain that were correlated with the lateralization of speech and language. Connolly (1950) and Falzi *et al.,* (1982) have found no such difference between the hemispheres in superficially visible neocortex, which is all that could be seen on an endocast.* It is, therefore, impossible to reason from increased fissurization in a "Broca's area" that a language area had evolved. Increased fissurization is one of the effects of increased brain size (Jerison, 1982), and the appearance of a third frontal convolution is probably an inevitable consequence of the enlargement of the frontal lobes in hominids. Calling the convolution a Broca's area is, in effect, accepting the (perhaps correct) hypothesis that the increase in size in hominid brains was due to the evolution of the capacity for language.

There is uncertainty but no controversy on a third matter in the fossil evidence on the evolution of neocortical language systems, a possible asymmetry of the Sylvian fissure that has been noted in the Chapelle aux Saints specimen of Neandertal man (*Homo sapiens*). This would be evidence that a Wernicke's area had appeared in Neandertals (Geschwind and Galaburda, 1984; LeMay and Culebras, 1972). It would, of course, be surprising if language areas had not evolved by that stage of human history, about 40,000 years ago (Trinkaus and

*Falzi *et al.,* (1982) reported asymmetry in total neocortical surface in Broca's area, i.e., tissue hidden in the sulci as well as the superficially visible tissue. There appears to be an error in their statistical analysis, however; the hemispheric difference that they found in their small sample of 12 brains fell short of significance by conventional statistical criteria.

Howells, 1979). The Neandertals' brains tended to be a bit larger and somewhat differently shaped from brains typical of living people, but they were probably within the range of variation of living brains (Kochetkova, 1978; Holloway, 1981a). Brain enlargement alone, without the uncertain evidence on its shape in one specimen, would be enough to suggest that the major neurological adaptation involved in the construction of morphological language areas had been completed by that time. This does not mean that language as we know it was necessarily supported by the brain system. The language system of the brain, like the visual system and other major brain systems, undoubtedly depends on appropriate experience (i.e., appropriate external stimulation) to mold its structure into a functionally adequate neocortical system.

These issues in hominid brain evolution concern specializations of major and extensive neocortical systems in species that were at an advanced modern grade of encephalization, well beyond that achieved by "average" living mammals and somewhat above that of living great apes. Although controversial on functional specializations associated with that grade of encephalization, there is no disagreement that an extraordinarily high grade of neocortical evolution had been achieved by the time that hominid species became differentiated from the great apes, i.e., by, perhaps, 5 million years ago. Hominid and human specializations are wrinkles in the evolutionary web.

2.2. Encephalization and Neocorticalization

Because the rhinal fissure is not clearly identifiable in primate endocasts, some of our understanding of primate neocortical evolution depends on the evidence that the brain as a whole became larger, or encephalized, relative to expectations from changing body size. There is a straightforward relationship between the volume of neocortex and total brain size among living species (Fig. 2), and the uniformitarian expectation is that there was also such a relationship in the past. It is evident from Fig. 2 that the exact relationship may be different in different groups. Primates are neocortical specialists, as it were; insectivores are not. If this pattern occurred in the past, we have even stronger reasons for recognizing gross brain enlargement in primates as evidence for enlargement of neocortex, more or less as suggested by Fig. 2.

3. Neocorticalization in Mammals

The history of the mammals is part of the history of the reptiles–the fate of a group of "reptiles" that had evolved into "mammals." In their early history the mammals were small insectivorous creatures, known mainly from their teeth. The fossil record of neocortex is better represented in some orders of mammals than in others. The first topic in this section is the oldest evidence of the mammalian brain. I then digress slightly to place that evidence within the framework of vertebrate evolution. I conclude with a review of the last 60 million years of the brain during the Tertiary period, the age of mammals. In the section that follows, I present a detailed quantitative analysis of neocortical evolution during the Tertiary.

The oldest interpretable mammalian endocast is of the Upper Jurassic *Triconodon mordax,* which lived about 150 million years ago. This endocast displays only the dorsal surface of the brain, but it is complete enough to enable one to estimate gross brain volume (Jerison, 1973; Kielan-Jaworowska, 1986). *Triconodon* was comparable to living small-brained insectivores such as hedgehogs (*Erinaceus europaeus*) in relative brain size, and comparable to smaller living shrews and moles in absolute brain and body size and in general appearance. There is too little of the lateral surface of the "brain" visible in this specimen to enable one to judge whether a rhinal fissure was present, but from the similarity of the *Triconodon* endocast in size and appearance to that of living species in which at least some significant neocortex is present (the insectivore sample in Fig. 2), it seems likely that mammalian neocortex had appeared as an adaptation in *Triconodon.*

Quiroga (1980) has described a rhinal fissure in the endocast of an even earlier animal, a Middle Triassic mammallike reptile, although Kielan-Jaworowska (1986) believes that this is probably the impression of a blood vessel. The evidence of relative brain size is equivocal for Quiroga's specimen; it was probably at a reptilian rather than early mammalian grade of encephalization. Yet it cannot be ruled out that neocortex had evolved at that point in the reptile–mammal transition, before the first true mammals had appeared.

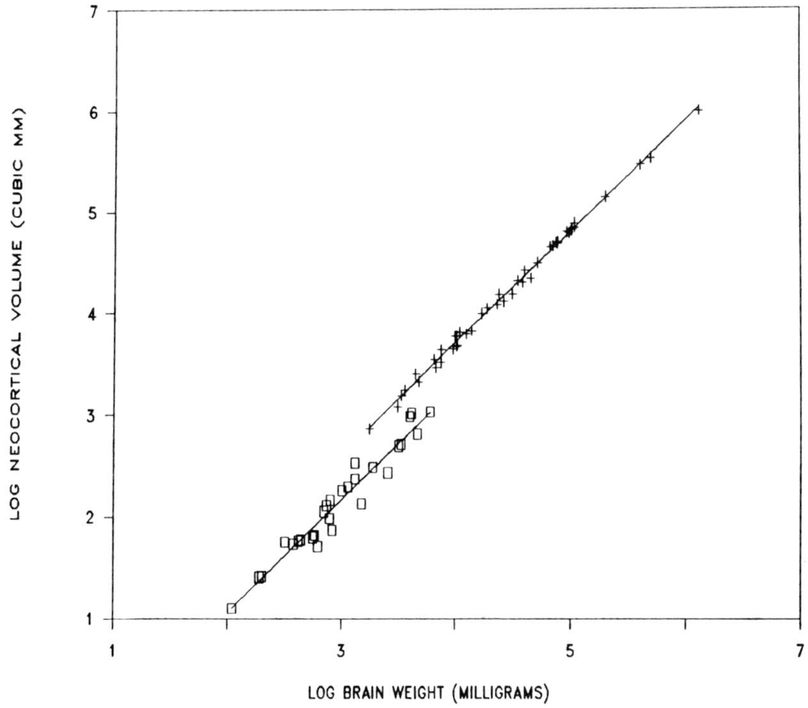

Figure 2. Neocortical volume as a function of brain weight in 28 living insectivore species (squares) and 45 living primate species (plus signs). Data from Stephan et al. (1981). Regression equations fitted to the data sets are, insectivores: $\log Y = 1.1 \log X - 1.15$; primates: $\log Y = 1.1 \log X - 0.68$.

The first reasonable, though not definite, evidence of a rhinal fissure on an endocast is in a unique assemblage of late Cretaceous mammals collected and described by Kielan-Jaworowska (1983, 1984, 1986). The group includes the earliest placentals, but I will illustrate her collection with a more unusual species for modern readers, a multituberculate, *Chulsanbaatar vulgaris,* which best typifies the Mesozoic mammals. The multituberculates were a rodentlike order of mammals, known from late Jurassic to early Tertiary deposits—a time range from 150 to 30 million years ago (Savage and Long, 1986). *Chulsanbaatar* weighed only 15 g or so, a small mammal even for the Mesozoic, smaller than most living species of mice. Its endocast is the best preserved of all Mesozoic evidence of the brain; Fig. 3 presents a lateral view accompanied by an outline sketch to illustrate possible positions of the rhinal fissure. Additional views of the endocast are in Kielan-Jaworowska (1983) and Kielan-Jaworowska *et al.* (1986). It is compared in Fig. 3 with the brain and endocast of the Madagascar insectivore, *Tenrec ecaudatus* (Bauchot and Stephan, 1967). At a body weight of 750 g, the tenrec is, of course, a much larger animal, but its brain is considered a living representative of the primitive condition. The difference between brain and endocast in this species is instructive for judgments about the position of the rhinal fissure.

Kielan-Jaworowska suggested that a rhinal fissure might have been present in *Chulsanbaatar* at about the position indicated by the dotted line on my sketch of the endocast (Fig. 3). When I saw the original, I thought that it was as likely to be at the position indicated by the dashed line that I sketched. (The more marked line on the photograph above that is the impression of the longitudinal fissure; the endocast is rotated slightly about its long axis, making part of that

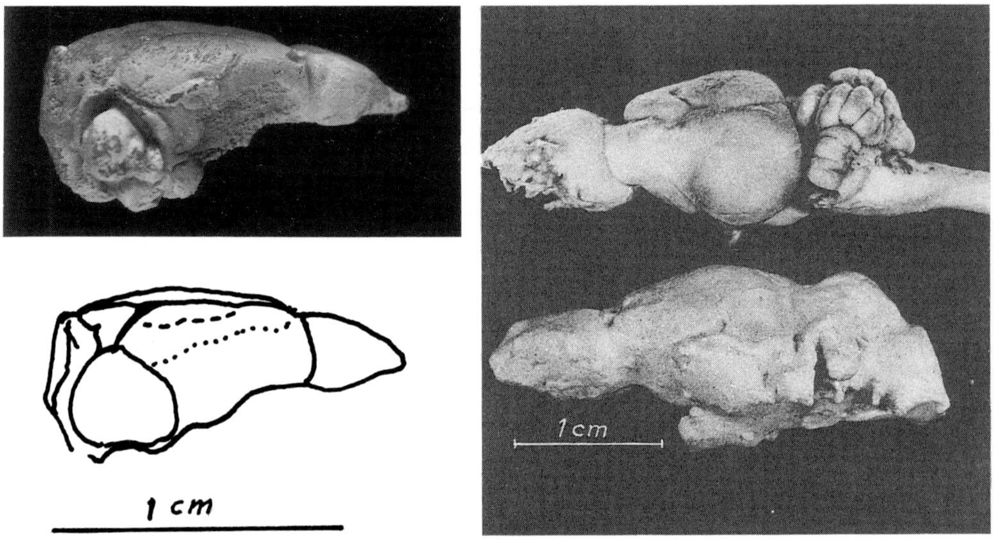

Figure 3. (Left) Natural endocast of *Chulsanbaatar vulgaris,* with sketch and scale (1 cm) below. (Right) Brain and endocast of *Tenrec ecaudatus,* from Bauchot and Stephan (1967). Note the difference between position of rhinal fissure in brain of tenrec relative to clearest marking on endocast, which is of a lateral sinus (see Kielan-Jaworowska *et al.*, 1986, concerning nomenclature). Sketch of endocast of *Chulsanbaatar* places rhinal fissure as suggested by Kielan-Jaworowska (dotted line), and alternative (dashed line) from hints on endocast. Note that alternative implies displacement as in tenrec. (Photograph of *Chulsanbaatar* endocast courtesy of Dr. Z. Kielan-Jaworowska; photographs of tenrec from Bauchot and Stephan, 1967, reprinted by permission.)

fissure visible.) The tenrec's brain and endocast suggest that questions like these cannot really be resolved. The rhinal fissure in tenrec would be wrongly identified and placed much more ventrally than its true position. Endocasts in which fissurization is not pronounced may be misleading in this respect. In those cases, it may be better to reason from gross brain size and encephalization that neocorticalization had occurred, relying on the logic of the data of Fig. 2 rather than on the morphology of the endocasts.

3.2. Some Vertebrate Paleontology

The story of fossil brains would be incomplete without a few paleontological facts and figures. The immediate ancestors of the mammals were therapsids, or mammallike reptiles, which were the dominant land animals at the end of the Paleozoic and beginning of the Mesozoic era. As a group, the mammallike reptiles became extinct during the Jurassic period and appear to have been replaced in their niches by the archosaurians ("ruling reptiles"), including the dinosaurs. In a sense, the first mammals should be thought of as therapsids that escaped extinction by evolving unique adaptations (probably as nocturnal species) that enabled them to survive in niches that were inaccessible to reptiles of the period (Jerison, 1973). Table I provides a summary framework in which to place these events in vertebrate evolution.

Mammals and dinosaurs were contemporaries throughout most of the Mesozoic era. The earliest mammalian fossils are known from late Triassic strata that are about 220 million years old. During the remaining 150 million years of the Mesozoic, mammals became diversified, evolving some of the major lineages that are present today. These include the earliest marsupials and placentals, but all the species were small; their giants were no larger than house cats, and most

Table I. Synopsis of Vertebrate Evolution

Era	Period and epoch	Age (years $\times 10^6$)	Fauna (first appearance)
Cenozoic	Quaternary		
	Recent	0.01–present	
	Pleistocene	1.9–0.01	*Homo erectus, H. sapiens*
	Tertiary		
	Pliocene	5.1–1.9	Hominids, *Homo habilis*
	Miocene	24–5.1	Hominoids
	Oligocene	37–24	Anthropoids
	Eocene	54–37	Progressive ungulates
	Paleocene	65–54	Primates[a] and carnivores
Mesozoic	Cretaceous	145–65	First definite rhinal fissure
	Jurassic	215–145	Earliest mammalian endocast
	Triassic	250–215	Earliest mammals
Paleozoic	Permian	285–250	
	Carboniferous	360–285	Reptiles
	Devonian	410–360	Bony fish and amphibians
	Silurian	440–410	Jawless fish
	Ordovician	505–440	First vertebrates
	Cambrian	590	Large-shelled invertebrates

[a]Primate teeth have been reported in late Cretaceous deposits.

were shrewlike or mouselike in size and habitus. Savage and Long (1986) have written a well-illustrated popular account of the entire history of the mammals, including a review of their therapsid ancestors.

Massive extinctions at the Cretaceous–Tertiary (K–T) time boundary included most large land vertebrates, such as dinosaurs. Environmental niches that were emptied were evidently invaded and filled by newly evolved early Tertiary mammalian species that were much more diverse than their Cretaceous predecessors in size and habitus. Early Tertiary mammals included species that reached weights of several hundred kilograms. These species were heavily built, and most appeared to be large herbivorous hooved animals—ungulate species that were "archaic" in that their taxonomic orders were replaced by presently living ("progressive") orders of ungulates. Most of the living orders of mammals had appeared by the end of the Paleocene or in the Lower Eocene, such as the odd-toed and even-toed ungulates (Perissodactyla and Artiodactyla, e.g., zebra and deer). Two specialized carnivorous orders evolved: first the order Creodonta, which is known from the late Cretaceous and is archaic in that its last known species became extinct during the Pliocene. The "true" carnivores, order Carnivora, appeared somewhat later than the creodonts, during the Upper Paleocene; species from the two orders coexisted during much of the Tertiary.

3.3. Mammalian Brains of the Tertiary Period

Endocasts of Paleocene and Lower Eocene species are often smooth, but rhinal fissures are usually visible. The position of the rhinal fissure in the more primitive of these fossils suggests a grade of neocortical development comparable to the basal insectivores of Bauchot and Stephan (1966, 1967). There are problems with this comparison, however, because most living insectivores are small animals (less than 1 kg) whereas the archaic mammals with known endocasts were usually large (10 kg or more), and because of allometry larger species tend to have larger brains. This raises difficulties in using small smooth insectivore brains as models for larger fossil brains, which, unlike the endocasts, may have been convoluted.

Smooth endocasts do not necessarily imply absence of fissurization in the brain (see Fig. 3). However, since even large-brained mammals of the early Tertiary seem to have had relatively little neocortex, the fissurization of neocortex in their brains may have been reduced, as indeed it is in living beavers (*Castor*).

A possible case in point is the Eocene *Uintatherium anceps,* an archaic ungulate that weighed about 1.4 metric tons. Its 300-ml endocast is less than half as large as brains of living mammals of comparable body size, but the absolute size of the uintathere brain was within the range of living chimpanzees. Brains of that size in living mammals are always convoluted, and this could have been true for *Uintatherium,* even though its endocast is quite smooth. Yet the carefully prepared uintathere endocast tells us little about the brain beyond its gross size—essentially nothing about the configuration of neocortex. *Uintatherium* could not be used in the quantitative analysis presented later in this chapter, because a rhinal fissure is not clearly marked. Reasonably clear rhinal fissures were present in a few other archaic ungulates; these are in our fossil sample and they contribute significant information on neocortical evolution.

Neocortex is a prominent feature in the endocasts of Eocene and later mammalian species in which there were specialized adaptations for carnivorous habits. Among these, there was enlarged neocortex in the archaic order Creodonta, the last surviving species of which became extinct perhaps 5 million years ago, and in early species of the living order Carnivora. Radinsky's (1977) illustration of endocasts in a group of such species is reproduced in Fig. 4. He properly included the archaic ungulate *Mesonyx obtusidens* among these early carnivores because its skeleton and dentition indicate carnivorous habits.

The complexity of trends in neocortical evolution is evident in Fig. 4. In the sequence of creodonts (*Tritemnodon, Cynohyaenodon, Hyaenodon, Apterodon,* and *Megistotherium*) there is progressive neocorticalization, but the trend is not uni-

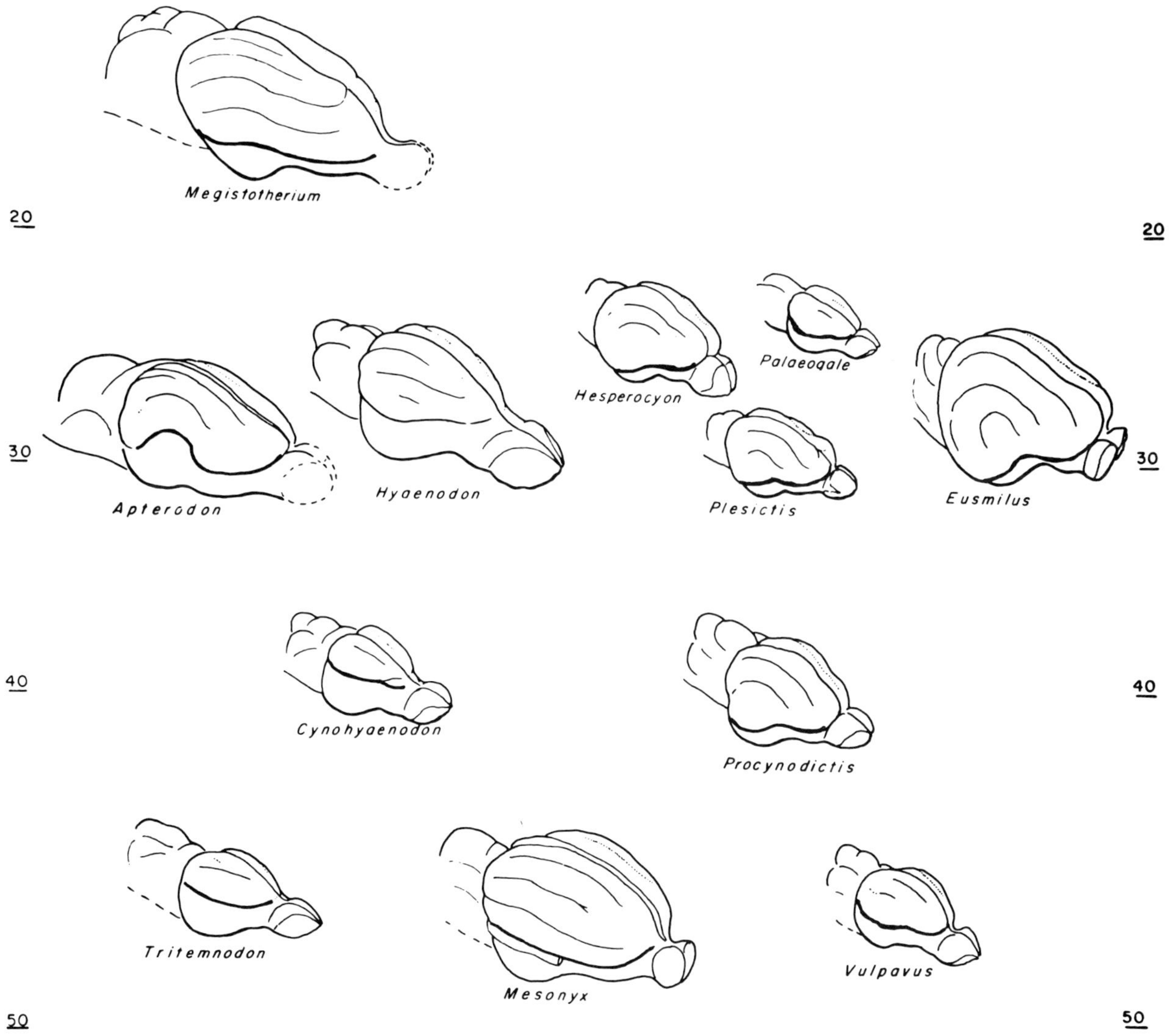

Figure 4. Carnivore brain evolution as revealed in fossil endocasts. This illustration was prepared under the direction of L. Radinsky; emphasis added by the author (darker line on each endocast) to indicate rhinal fissure. Vertical scales are millions of years ago; earliest fossils at bottom and latest at top. (From Radinsky, 1977, reprinted by permission.)

versal. *Cynohyaenodon* appears to be less corticalized than the older *Tritemnodon,* and *Apterodon* may be more neocorticalized than the somewhat later *Hyaenodon.* Progressive trends are less evident in the pictures of the endocasts of true carnivores: the miacid, *Vulpavus,* the canids, *Procynodictis, Plesictis, Hesperocyn, Palaeogale,* and the felid, *Eusmilus,* an early saber-tooth. The endocast of *Eusmilus* is strikingly similar to that of living cats in the configuration of the supersylvian and ectosylvian gyri. "Progress" in neocorticalization is evident in another way in Fig. 4. The less neocorticalized creodonts were eventually replaced by true carnivores, which is progress "above the species level" (Rensch, 1959).

Patterns like those in Fig. 4 occurred in other groups. They occurred in primates (Radinsky, 1979) and in ungulates (Jerison, 1973). As in the carnivorous species of Fig. 4, however, these trends were complex in all groups. Some Eocene prosimians (*Adapis, Smilodectes*) were notably less neocorticalized than living species, but the Eocene and Oligocene tarsioid species (*Tetonius, Necrolemur, Rooneyia*) were more modern in this respect. Among the ungulates the trends are even more regular, suggesting something comparable to the history of the creodonts in Fig. 4. It would be useful to quantify the story, to see the extent to which orderly change occurred.

4. Neocorticalization: A Quantitative Exercise

Neocorticalization proceeded in fits and starts, progressing steadily in some groups and, perhaps, not at all in others. These evolutionary trends can be quantified by measuring the size of neocortex and other parts of the endocasts, seeing how these were related to one another, and determining the changes that occurred with the passage of time.

There are some technical difficulties. First, many cortical structures connected with, or part of, neocortical systems are either invisible or partly obscured in endocasts. These include the hippocampus and piriform lobes, and the large fraction of neocortex that is folded into and hidden in the depths of the convolutions. Second, most visible structures on endocasts are surfaces on an irregular solid, which are difficult to measure directly. Computerized measurements are being developed, based on tomography of the endocasts and the reconstruction of three-dimensional images from the tomograph (R. B. Livingston, personal communication). They were not available in time for this report. A two-dimensional analysis is presented here to indicate the kinds of questions that can be put quantitatively and to provide provisional answers.

4.1. Method and Materials

For this exercise I used lateral projections of endocasts (see Fig. 1) to measure the height of the foramen magnum ("medulla" of endocasts) and four "brain" areas: neocortex, paleocortex, hindbrain, and olfactory bulbs. These were to be related to geographical age in millions of years. The complete data are in the Appendix.

Except for one measurement of the foramen magnum on a deer skull (*Odocoileus*), all the analysis was on specimens represented by drawings or photographs that have been published; 55 of these are from Radinsky (1973, 1975,

1976, 1977, 1981). Radinsky's sample included 38 Carnivora, 7 Creodonta, 4 Condylarthra, 5 species from extinct South American Ungulate orders, and 1 Eocene perissodactyl (*Hyrachyus,* an ancestral rhinoceros). To these I added four other progressive ungulates (artiodactyls): *Bathygenys* of Fig. 1, and two fossil and one living species from Jerison (1973). The complete sample consisted of 35 fossil species and 24 living species, a total of 59 species.

The variables were geological age, the four brain areas, and the height of the foramen magnum. The most straightforward variable is geological age, about which there is broad agreement (Radinsky, 1978; Savage and Long, 1986). The area of neocortex is an area as shown in Fig. 1. It is not the true area because it is a two-dimensional projection of a solid and because in convoluted brains a significant amount of neocortex is buried in the sulci. The amount of hidden neocortex appears to be correlated with the superficial area in living species (Jerison, 1982), however, and the measure here of neocortex is approximately linearly related to the true area. The area of the olfactory bulbs on the endocast is, similarly, a reasonably good estimator of the volume of the bulbs on the brain, and the height of the foramen magnum estimates the height of the medulla on the brain. In the latter cases the measures on the endocast are substantially larger than on the brain (compare brain and endocast of *Tenrec* in Fig. 3), but if one required the information (it is not needed here), the estimation of these brain data from endocast data by linear regression analysis would raise no problems.

There are much more serious problems in the interpretation of data on paleocortex and hindbrain from lateral projections of endocasts. In neocorticalized species paleocortex always appears reduced on lateral projections, because as neocortex expands paleocortical tissue becomes hidden in the rhinal fissure, and the piriform lobe and other "old cortex" structures are rotated ventrally. A lateral view of the brain loses data about the ventral surface, and paleocortex will always be underestimated. The visible area of hindbrain, which consists of "medulla" (foramen magnum), pons, and cerebellum, will also be underestimated as species become neocorticalized, because neocortex expands posteriorly to cover some of the cerebellum.

Another set of problems is related to the evolution of diversity in body size. Species differ in body size for many reasons, most of which are unrelated to brain functions. Nevertheless, body size must be taken into account in the analysis of the evolution of the size of brain structures, because of the allometric relationships among the sizes of organs. Other things being equal, larger species have larger brains and more neocortex than smaller species. We must, therefore, analyze the amount of neocortex relative to the amount expected in species of a given body size and determine how this *relative* amount changed over time.

Although it is possible to estimate body size from postcranial skeletons in most of the species in this sample, for this exercise the foramen magnum was used to estimate body size (Radinsky, 1967). In the data of Stephan *et al.* (1981), the volume of the medulla is the brain measure most highly correlated with body size (log units: $r = 0.983$); a scatter diagram of that correlation differs from the one in Fig. 2 in that insectivores and primates are not significantly different from one another in either intercept or slope and are adequately described by a single regression line. One equation that estimated body size from the size of the medulla worked for both insectivores and primates, despite the substantial difference between these groups in encephalization. Harman (1957) reported a similar result for a mixed sample of primates, ungulates, and carnivores (see

Jerison, 1973). This encourages one to use the foramen magnum on an endocast, which represents the effect of the volume of the medulla, as the statistical control for body size in the carnivores and ungulates in our sample.

The control was accomplished by using residuals from a linear regression analysis: the effect of foramen magnum height was the independent variable, and the measures (log units) of areas in the brain were dependent variables. Because the regression analysis is on logarithmic data, the residuals are ratios: the areas of the brain relative to the "expected," or "average," areas as estimated by the regression analysis. In the graphs below (Figs. 5 and 6) I present the residuals as quotients. A quotient of 1.0 means that the part of the brain is exactly as large as expected; 0.5 means that it is half as large; 2.0 means twice as large; and so on. If there was an evolutionary increase or decrease in the relative amount of a particular part of the brain in mammals, this would appear as a positive or negative relationship between the quotient for that part of the brain and geological recency.

4.2. Results: Correlated Changes in the Endocast Sample

A summary of the relationships among the measurements in the living and fossil species in the endocast sample is presented in Table II. This is a matrix of correlation coefficients from which one can read the extent to which measurements are related to one another. The pattern may be interpreted as follows: The only brain measurement that became greater (had a positive correlation coefficient with age) during the Tertiary was neocortex. The measured amount of paleocortex decreased with the passage of time, and the other measures were unchanged. The "reduction" in paleocortex may be an artifact of enlarged neocortex that partly covered and forced a ventral rotation of the paleocortical surface. The portion of paleocortex visible in lateral projection would then be reduced, but a reduction in the fraction of paleocortex shown as a lateral projection does not necessarily mean that there was a reduction in total paleocortex. There could be more at the ventral surface, and there could be some hidden under overlying neocortex in the region of the rhinal fissure.

The 59 species in this analysis differed substantially in body size. To examine the brain's evolution independently of its known correlation with body size, we may analyze the correlations of the residuals for each brain measure when the foramen magnum height is "partialled out," i.e., controlled statistically. The results of that analysis are summarized in Table III. In addition to the matrix of

Table II. Correlation Coefficients among Endocast Measures and Geological Age[a]

	Age	Neocortex	Paleocortex	Hindbrain	Olfactory bulbs	Foramen
Age	1.000					
nc	0.481	1.000				
pc	−0.358	0.332	1.000			
hb	−0.245	0.582	0.822	1.000		
ob	−0.084	0.599	0.747	0.820	1.000	
h	−0.166	0.627	0.794	0.908	0.837	1.000

[a]Data on brain parts in log units. At $N = 59$, $r > |0.26|$ is significantly different from 0 ($p < 0.05$). Abbreviations as in Fig. 1. Raw data in the Appendix.

correlation coefficients, Table III lists the values of the regression coefficients (slopes) for the relationship between each brain measure and geological age.

There is an interesting contrast between Tables II and III. Regardless of whether the effect of body size is or is not controlled, the patterns of correlation with age are similar. On the other hand, the correlations of the sizes of the parts of the brain with one another are quite high in Table II and low and generally insignificant in Table III. The difference probably indicates the importance of size as a general factor in the brain measures. A large endocast, whether ancient or modern, has a large neocortex, hindbrain, paleocortex, olfactory bulbs, and foramen magnum. Removing the effect of size by "controlling" the foramen magnum measure removes this size factor from all the brain measures, reducing their shared variance, and this is reflected in reduced correlation coefficients.

Removing the size factor with this statistical trick has an interesting effect on the strongest correlate of geological age, the relative amount of neocortex, and provides another lesson for interpreting a quantitative analysis. I interpret it as correcting a flaw in our sample. It happened that our sample of living species consisted mainly of 23 species of viverrids (Radinsky, 1975). Mongooses, civets, and their relatives happen to be smaller animals than most of our fossils, and they bias our sample to confound body size with geological age. A correlated confounding of age with brain size and, therefore, with the absolute (not relative) sizes of the parts of the brain, including neocortex, is inevitable. When we analyzed the absolute measurements (Table II), the confounding reduced the correlation coefficient. When we controlled body size by controlling the effect of the foramen magnum (Table III), we obtained a truer picture of neocorticalization, because we could estimate the correlation between *relative* amount of neocortex and age.

From data on living species (see next section) we know that the sizes of neocortex, paleocortex, and hindbrain are positively related to one another, independently of the correlation due to a general size factor. The correlation matrix in Table III shows neocorticalization, but it also shows a reduction in relative amount of paleocortex and hindbrain, a reduction that can be under-

Table III. Correlation Coefficients of Residuals and Regression Coefficients of Endocast Measures with Geological Age[a]

	Age	NC1	PC1	HB1	OB1
Age	1.000				
NC1	0.752	1.000			
PC1	−0.372	−0.35	1.000		
HB1	−0.224	0.039	0.394	1.000	
OB1	0.100	0.174	0.247	0.262	1.000
Beta[b]	—	0.027	−0.010	−0.004	0.002

[a]Correlations are among the residuals for neocortex (NC1), paleocortex (PC1), hindbrain (HB1), and olfactory bulbs (OB1) when height of foramen magnum is "controlled," and between these residuals and geological age. At $N = 59$, $r > |0.26|$ is significantly different from 0 ($p < 0.05$). The significant negative correlations are probably artifacts of the masking of parts of the brain by expanding neocortex (see text).

[b]The regression coefficient, beta, is the slope of a "best fitting" line in a graph relating each measure to geological age. Such lines are drawn in Figs. 5 and 6 for the regressions of NC1 and OB1 ("surface quotients") with respect to age and provide a sense of the appearance of the slopes. They can be reproduced by using the slopes, above, for a line through the mean age (−26 million years) and a quotient of 1.0.

stood as occurring for the fraction of hindbrain and paleocortex visible in lateral projections of the brain. Neocortex was increasing in relative size, and one must assume that structures correlated with neocortex were also increasing. Evidently the increase in the size of neocortex took the form of a lateral expansion that either covered the other brain structures as in the case of hindbrain (cerebellum) or forced their ventral rotation (paleocortex).

4.3. Graphic Analysis of the Endocast Sample

Neocorticalization in this sample is graphed in Fig. 5. Code letters there represent the groups to which the species belong. Although better data will become available from tomographic analysis of endocasts, the trends shown in Fig. 5 suggest several strong effects that are likely to be supported by future research.

There is, first, the fact of neocorticalization. Later species tended to have significantly more neocortex than earlier species. This is what the significantly positive slope of the regression line means. Second, species from archaic orders tended to have less neocortex than species from progressive orders. Thus, three of the four archaic ungulate species, six of the seven archaic carnivore (creodont)

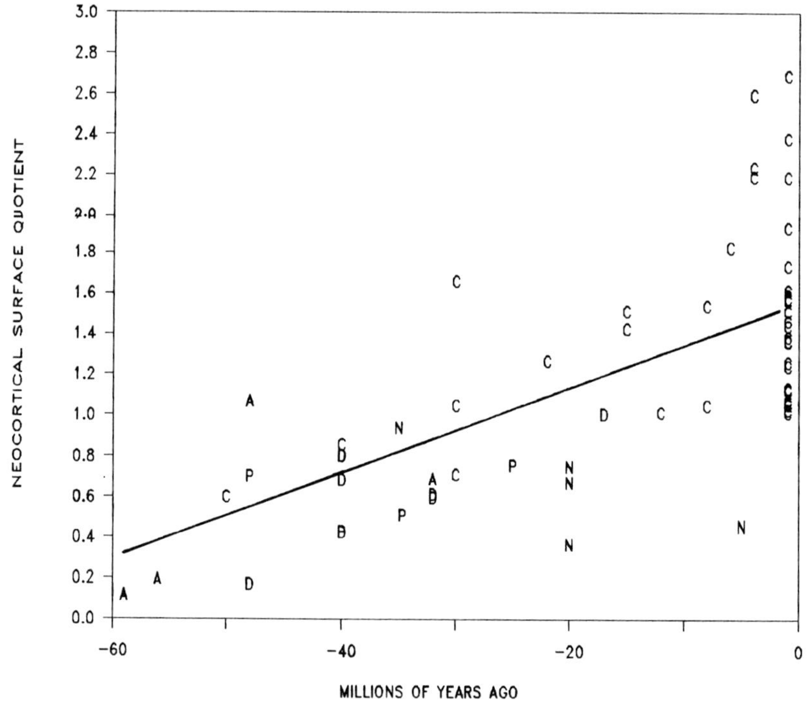

Figure 5. Change in relative neocortical surface area on a lateral projection of the endocast (see Fig. 1) as a function of geological age. Neocortical quotient is residual from regression of neocortex area on foramen magnum height and is interpreted as a ratio of measured neocortical area relative to area to be expected on the basis of body size in this sample. "Progressive" change noted here (positive slope of regression line) indicates increased neocorticalization over time. Each point is a species. A, archaic ungulates; C, progressive carnivores; D, archaic carnivores; N, neotropical (archaic) ungulates; P, progressive ungulates. For more details see text, Table III, and the Appendix.

species, and four of the five neotropical ungulate species (also archaic in that their orders are extinct) are below the regression line. Thirteen of our sixteen "archaic" species had less neocortex than would be predicted *for their geological age* by an unbiased regression analysis. For those who enjoy playing with statistics, a simple analysis contrasts this with an expected even split; $\chi^2 = 6.25$, $df = 1$. The probability that the observed split was an accident of sampling is 0.012, and amateur statisticians call that significant. They would support the hypothesis that archaic species were less neocorticalized than progressive species.

Fossil species from the progressive order Carnivora are overrepresented above the regression line, and it is only when one reaches the sample of living species that there is the expected even division of species above and below the line. (The five progressive ungulates are too small a sample for statistical decisions. Their 2 : 3 division above and below the regression line is consistent with their having been "average" species throughout their history.) Evolutionists traditionally avoid value-laden words like "archaic" and "progressive" that can be misinterpreted as teleological. Nevertheless, defined objectively by the fossil evidence of extinction to distinguish archaic from progressive species, these words also distinguished species with respect to neocorticalization. Neocorticalization is, thus, correlated with the survival of species. It is a progressive trait.

Not all brain systems that could be assessed validly from the size (areas) of their lateral projection were affected by natural selection. This is an important point, because it shows that our approach is fine enough to discriminate between the presence and absence of change. Figure 6 presents the graphic analysis of

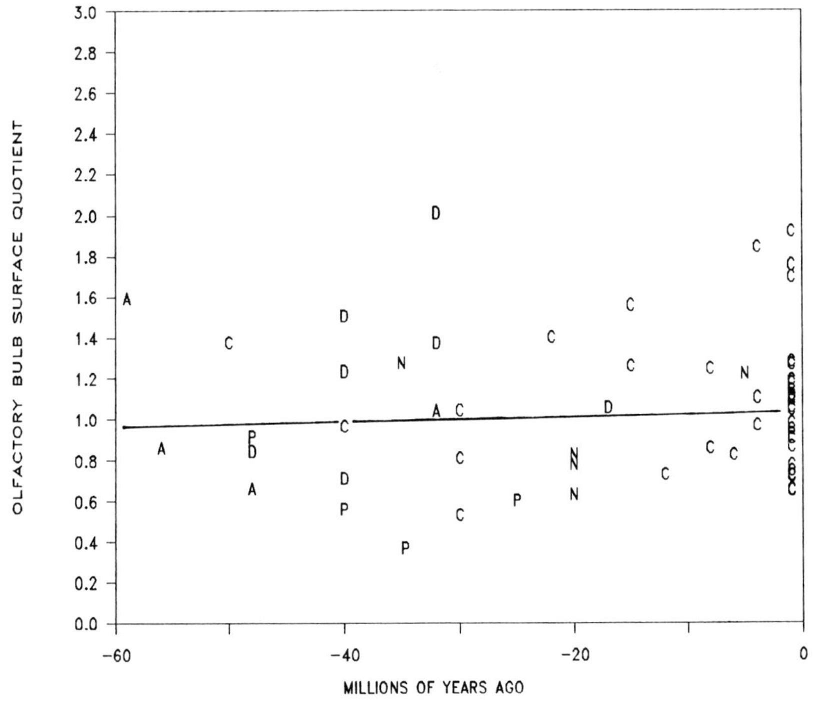

Figure 6. Absence of change in relative area surface quotient of olfactory bulbs on a lateral projection of the endocast (see Fig. 1) as a function of geological age. Labels as in Fig. 5. For more details see text, Table III, and the Appendix.

the evolution of the olfactory bulbs. As in the analysis of neocortex, body size was controlled by working with residual olfactory bulb size, from the regression of olfactory bulbs on foramen magnum. Unlike neocortex, the size of the olfactory bulbs relative to expected size was unchanged during the Tertiary period.

The absence of change in the relative size of the olfactory bulbs during 60 million years of evolution is dramatic and is worth special discussion. Although careful students (Baron *et al.*, 1983; Sarnat and Netsky, 1974) do not make the error, neurobiologists often assume that having large olfactory bulbs is a primitive mammalian trait and that the olfactory bulbs became relatively smaller as the mammals evolved. Figure 6 corrects this error by showing that olfactory bulbs have been a stable feature of the brain in Tertiary mammals. The misconception is a bit of primate chauvinism, as it were. Primates (at least the anthropoids) are neocortical specialists, but they are deficient mammals in olfactory development. A reduced role for olfaction is part of the adaptive mosaic of the adaptive zone of "higher" primates and is not a broad feature of mammalian evolution. Neocorticalization, on the other hand, appeared as a general trend in many mammalian groups and its relative absence in, e.g., the insectivores and many marsupials is correctly recognized as a primitive feature in these groups.

4.4. Results: The Living Brain

More detailed analysis of living brains adds dimensions of understanding to the fossil evidence. The living species in Figs. 5 and 6 that are represented by endocasts are viverrid carnivores and a single deer. No data on the parts of their brains are available that would clarify the relationship between superficially visible structures and parts of the brain that are hidden in the fissures or covered by other parts. It is technically feasible to get such data and then to compare brains and endocasts, but it would be a major anatomical research project with standard anatomical methods. I know of no data on large carnivores and ungulates. There is, however, a remarkable and extensive data set on insectivores and primates that has been collected over the years by Stephan and his colleagues and tabulated in a major quantitative publication (Stephan *et al.*, 1981). Some of these have already been used for Fig. 2, and I will summarize some suggestions from their data for interpreting the fossil evidence.

Stephan *et al.* present careful measurements on the volumes of many brain structures from reconstructions of serial sections prepared for histology. The sample consists of 76 species: 28 insectivores, 3 tree shrews, and 45 primates. Although they prefer finer divisions among the species, e.g., into basal and other insectivores, it is inappropriate for the present purpose to introduce these. I used their tabulated data on the volumes of cerebellum and medulla for the hindbrain measure and also to demonstrate the utility of medulla as an estimator of body size. I used their measure of the volume of neocortex, which includes underlying white matter, and the sum of the volumes of three structures: piriform lobe, hippocampus, and schizocortex, which I call "old cortex." In addition, I used the body weights that they provided, which are from the biological literature. Their data are the best and most complete that are available for my purpose, but there are unavoidable metrical problems in controlling for shrinkage, which they review at some length.

It would be possible to obtain accurate morphometric data rapidly on more species, specifically for relating data from endocasts to gross measures on brains, by using modern brain scanning techniques. The scans could be taken on fresh brains *in situ* at necropsy, with high-gain tomography or an equivalent technique, and the scans could be analyzed directly for data on both endocast (internal surface of the cranial cavity) and the brain. These would be adequate for partitioning brains into units as analyzed in Table III, for example, with complete measurements of volume and of both hidden and superficial surfaces of those units.

The requirement is simple. One needs correlations between complete major structures in living brains (e.g., paleocortex) and the corresponding superficial features of the brain that can be studied on endocasts. Additional and finer correlations among brain structures, though surely desirable, are not directly applicable to the analysis of the fossil evidence. In view of the bias of fossil samples toward large species, especially carnivores and ungulates, data on living specimens from these groups would be especially useful. R. B. Livingston's group, as mentioned earlier, may be able to collect such data as part of a larger brain-mapping project.

Although limited to insectivores and primates, the data of Stephan *et al.* show several general relationships that can be analyzed in evolutionary terms. Two of these were shown in Fig. 2. Here the relationship between neocortex and brain size in a large number of insectivore and primates species was summarized by the parameters of just two regression lines. The identical slopes of the lines in these two very different orders of mammals suggest that this feature is a characteristic of mammals. Harman (1957; see Jerison, 1973, p. 65) had reported a similar result on a sample of primates, ungulates, and carnivores, but using a different measure of cortical volume. He found that a single line fit the data of all three groups. In addition, he found that opossums were below the "advanced" mammal line, perhaps like the insectivores in Fig. 2. The differentiation of primates from insectivores was by the different intercepts.

Without introducing the cladists' jargon of plesiomorphies and the like, we can follow their style of thinking (Cracraft and Eldredge, 1979) by considering the slopes of Fig. 2 as "primitive" traits, reflecting some fundamental genetic control shared by all mammals. The intercepts must then be recognized as specializations that characterize particular groups, suggesting something about how the groups branched away from one another or from a common ancestor. The concern in presenting Fig. 2 was with the question of whether an enlarged brain implied an enlarged neocortex. This can be recognized as an evolutionary question as well as a functional one. It is essentially the question of whether the body plan of mammals is of a particular kind, which is, of course, explained by the evolutionary idea of common origins.

Another use of the data from Stephan's group is to provide additional dimensions that complement analyses like those shown in the two-dimensional graphs of Figs. 5 and 6. One can imagine an analysis like that in Fig. 2 as representing a plane at right angles to the plane of Fig. 5, with the intersection of the planes occurring at AGE = 0. The graph would have brain size (quotients) as a z dimension, as it were, and we would fit a plane to the cloud of points, graphing neocorticalization and encephalization at the same time. We have in a sense done this sort of thing in Tables II and III by presenting a matrix of correlation coefficients. The matrix can be the basis of any of several multivariate analyses, which could present an integrated picture of the interactions among the many measures of the brain and their changes over time.

The style of thought of the previous paragraph underlies the biometric method of "factor analysis." Although the results are not presented in detail, the brain data of Stephan *et al.* were factor-analyzed (the method of "principal components," Harman, 1976) to determine the number of factors that account for the variability in the data and for hints on the meaning of the factors. Just two factors accounted for almost all the variance. One of these was a general size factor, accounting for about 80% of the variance, and the second was a "special" (unique) olfactory bulb factor, accounting for most of the remaining 20%. (Insectivores and primates were treated as a single population for this purpose; a more complex and detailed picture can be developed if the groups are separated.) The high correlation between the volume of the medulla and body size mentioned earlier is an example of the size effect. Only the olfactory bulbs had a correlation (r = 0.420) less than 0.93 with body size, and this probably was an effect of combining primates and insectivores: Within the primates the correlation between olfactory bulbs and body size was 0.38, but within the insectivores the correlation was 0.93. I mention these details to emphasize that the complete picture available from the work on living brains is more complex than the portion useful for our understanding of the fossil evidence.

As indicated in Tables II and III, there was an inverse relationship of neocortex with paleocortex and possibly also with hindbrain in the data on endocasts. This was considered an artifact primarily because the data of Stephan *et al.* on brains showed the opposite effect. In reanalyzing the data of Stephan *et al.* to show this, their data had to be treated in a way comparable to the analysis of endocasts in Table III. When insectivores and primates were treated as a single sample and the brain measures were residuals after partialling out body weight, neocortex and paleocortex were positively correlated, and the other correlations were in the same direction but much higher than those in Table III. The highest correlation (r = 0.91) was between neocortex and hindbrain; second was that between "old cortex" (= paleocortex on an endocast) and hindbrain (r = 0.60), and lowest but still significant was that between neocortex and "old cortex" (r = 0.45). There were some significant negative correlations between the olfactory bulbs and other parts of the brain, and there were no significant positive correlations, reflecting perhaps the microsmatic adaptations of primates that made up so much of the sample.

Like the factor analyses, the correlational statements show the olfactory bulbs as developing under control of factors that seem to be at least partly independent of the single factor that is evidently so important for the size of other parts of the brain. This helps explain the contrast between the evolution of neocorticalization and of the olfactory bulbs pictured in Figs. 5 and 6.

5. Conclusions

This chapter has summarized some of the evidence reported in the large literature on fossil "brains." The quantitative analysis was an exercise to show how the qualitative evidence can be made more precise. Mammalian brain evolution was neocortical evolution as far as the fossil evidence goes. When the likelihood of artifacts in measuring other parts of the brain on endocasts is taken into account, we see that other parts must also have expanded along with the expanding neocortex, at least to the extent that one can judge this from the evidence on living species. The fossil evidence, therefore, confirms conventional wisdom in neurobiology. It would be consistent with that evidence to look for neocortical functions and assume that the evolutionary advantage conferred by these functions was the engine driving *progressive* brain evolution in mammals. But it would be a mistake to make much of the idea of progress. It is true that neocorticalized species are more prevalent now than in the distant past. It must also be true that some fitness is associated with this aspect of the brain's evolution. But there are many successful living species that are at a very ancient grade of mammalian neocorticalization. The tenrec of Fig. 3 is not the best example. Hedgehogs in Europe and opossums in America are better because they are so fit in the evolutionary sense. They may litter our highways because of their "stupidity" in refusing to yield the right of way to cars and trucks, but the litter is part of the evidence of their reproductive success. And they manage this at a grade of neocorticalization and encephalization that some mammalian species reached 150 million years ago.

If our conclusions on neocortex are expected, those on the olfactory bulbs are not. These structures are surprisingly constant features in mammalian brains. There is no reason to have predicted that they would not evolve to large or small size relative to other parts of the brain, depending on the extent of olfactory specializations in particular species. But according to present evidence from fossil brains, the olfactory bulbs have been constant and relatively unchanging features that make a brain a mammal's brain. They are not unusually enlarged in any species; most mammals are olfactory specialists. Progress in the evolution of the olfactory bulbs occurred mainly in a negative way, by reduction. The reduced state of olfactory bulbs in humans and other primates (and their complete absence in some cetaceans) merely reflects the extremes of diversity that are possible as the brain evolved to control the activities of mammals in the variety of niches in which they function.

Finally, although I have been concerned entirely with the facts of neocorticalization, like all neurobiologists I am interested in these facts because of their direct bearing on the evolution of mammalian behavioral adaptations. It is appropriate to conclude with a few words about the nature of the adaptations associated with neocorticalization. The most general statement of neocortical structure–function relations is, perhaps, in the work of Jones and Powell (1970), who showed that much if not all mammalian neocortex can be accounted for by the representation of broadly defined sensory–motor projection systems. One may go beyond this, to redefine projection (sensation and motor) systems versus association systems structurally, with respect to their localization in the brain (Diamond, 1979) and also functionally. The classical view of the function of projection systems is as relatively low–level information processing systems, data from which are then analyzed further by association systems. That view really applies the language of the "mental chemistry" of 19th century psychology to a problem that can be handled in more sophisticated ways. I interpret Jones and Powell's results as suggesting that neocortical expansion was related to processing information that is handled by sensory and motor systems. Information is the key word.

Elaboration of neocortex during mammalian evolution should be related to the elaboration of the capacity to handle information, but we should be careful not to restrict ourselves by prematurely characterizing and categorizing the information, e.g., as sensory or associative. Although it is correct to describe that information as sensory and motor, it may be misleading to emphasize the continuity with spinal processing of such information. Processing at a neocortical level should be thought of as higher in a hierarchy, an elaboration that may involve mechanisms not present in simpler systems. The vocabulary to describe the processing is likely to involve words like "perception," "cognition," "memory," and so forth, and in handling its neural information the brain may be translating it into "knowledge." The words are important because they imply a level of analysis for which the normal vocabulary of physiology may be inadequate. In speculating on this theme (Jerison, 1973, 1985a, 1986) I have emphasized that it is appropriate to think of the brain's work as building a reality and that the external and internal worlds of immediate experience are truly constructions of the brain. The basis of this view is that the information actually handled by neocortex, it is reasonable to correlate that expansion with the elaboration of the expansion of the mammalian brain can be accounted for as an expansion of neocortex it is reasonable to correlate that expansion with the elaboration of the knowledge. The evolution of neocortex may then be thought of as the evolution of different ways of knowing the real world, or, more precisely, of constructing a real world.

6. Appendix: Endocast Data

Data are in millimeters for the foramen magnum and square millimeters for other structures. Group codes: A, order Condylarthra (archaic ungulates); C, order Carnivora (progressive carnivores); D, order Creodonta (archaic carnivores); N, orders Notoungulata and Litopterna (neotropical archaic ungulates); P, orders Perissodactyla and Artiodactyla (progressive ungulates). Age is millions of years ago, based on Radinsky (1978) and Savage and Long (1986). Original drawings for the first 55 species are in Radinsky (1973, 1975, 1977, 1981); those for the last 4 are from this chapter and Jerison (1973).

Species	Group	Age	Neocortex	Paleocortex	Hindbrain	Olfactory bulbs	Foramen magnum
Arctocyon	A	60	89	499	714	221	12
Mesonyx	A	48	1105	436	967	124	14
Phenacodus	A	56	179	357	630	142	13
Apterodon	A	32	752	374	925	204	15
Vulpavus	C	50	255	141	209	103	8
Procynodictis	C	40	505	182	478	101	10
Hesperocyon	C	30	397	105	234	81	9
Hoplophoneus	C	30	1268	208	832	117	16
Nimraevus	C	30	2004	106	1152	228	16
Mesocyon	C	22	922	271	625	183	11
Enhydrocyon	C	15	1296	208	888	240	13
Phlaocyon	C	15	1198	247	690	190	13
Cynodesmus	C	12	740	134	528	95	11
Newgenus1	C	8	1448	271	751	215	17

(*continued*)

Species	Group	Age	Neocortex	Paleocortex	Hindbrain	Olfactory bulbs	Foramen magnum
Tomarctus	C	8	1269	277	659	185	12
Pseudaelurus	C	6	1628	299	1096	132	13
Aelurodon A	C	4	2128	351	764	169	14
Aelurodon B	C	4	2819	457	1278	419	16
Osteoborus	C	4	3011	482	1020	232	16
Atilax	C	0	939	87	187	87	9
Bdeogale crassicauda	C	0	517	65	270	78	9
Bdeogale nigripes	C	0	787	83	270	126	11
Canis adjustus	C	0	1050	198	396	178	12
Canis lupus	C	0	2455	578	629	355	15
Canis mesomalis	C	0	1235	163	328	177	9
Cerdocyon	C	0	953	213	325	142	11
Cryptoprocta	C	0	957	122	326	78	10
Cynogale	C	0	918	152	270	78	11
Dusicyon	C	0	786	157	284	117	9
Felis concolor	C	0	2319	330	1091	188	18
Fennecus	C	0	653	71	220	66	7
Fossa fossa	C	0	526	83	196	87	8
Genetta	C	0	531	67	259	62	8
Hemigalus	C	0	539	104	274	100	9
Lycaon	C	0	2847	436	517	213	15
Osbornicus	C	0	692	67	196	98	8
Paradoxurus	C	0	543	135	304	78	7
Prionodon	C	0	375	67	228	45	6
Suricata	C	0	405	52	104	44	7
Viverra civetta	C	0	940	303	616	205	11
Viverra tangalunga	C	0	607	156	393	67	9
Viverricula	C	0	567	134	313	103	9
Thinocyon	D	48	44	108	95	38	6
Cynohyaenodon	D	40	188	264	291	118	8
Humbertia	D	40	499	168	473	79	10
Pterodon	D	40	754	494	932	245	15
Hyaenodon crucians	D	32	450	575	665	270	12
Hyaenodon horridus	D	32	901	880	1078	364	18
Megistotherium	D	17	2599	948	2631	502	27
Rynchippus	N	35	1865	259	1710	461	22
Diadiaphorus	N	20	902	654	1195	170	16
Hegetotherium	N	20	272	239	434	83	11
Proterotherium	N	20	748	620	1068	167	15
Pseudotypotherium	N	5	966	668	1284	469	23
Hyrachyus	P	48	872	588	1124	205	16
Anoplotherium	P	40	1017	1131	1375	244	25
Bathygenys	P	35	256	200	293	33	9
Protoceras	P	25	750	506	656	108	14
Odocoileus	P	0	2103	417	722	222	20

ACKNOWLEDGMENTS. It would not have been possible to write this chapter without the major body of data published by the late L. B. Radinsky, to whom the chapter is dedicated. I thank Professor David Spencer Smith, Dr. Richard Dawkins, and Dr. Paul Harvey for their hospitality when I began this chapter as an Academic Visitor at the Department of Zoology at Oxford, Dr. Alan Charig, Dr. Alan Gentry, and Dr. Jerry Hooker of the British Museum (Natural History) for access to their fossils and their library, and Dr. Herbert Roitblatt of the

University of Hawaii for help with computer analysis of the data. I wish to thank, especially, Professor Zofia Kielan-Jaworowska, now at the Paleontological Museum, University of Oslo, for her kindness and graciousness when I visited her in Warsaw. She allowed me free access to her extensive collection of Mesozoic materials at her laboratory at the Paleobiological Institute of the Polish Academy of Sciences in Warsaw and generously provided photographs and casts of important specimens. On reading a draft of this chapter, she noted that the multituberculates did not become extinct until the Oligocene epoch, about 30 million years ago, which makes them the most long-lived of mammalian orders, with a span of at least 120 million years.

7. References

Baron, G., Frahm, H. D., Bhatnagar, K. P., and Stephan, H., 1983, Comparison of brain structure volumes in Insectivora and primates. III. Main olfactory bulb (MOB), *J. Hirnforsch.* **24:**551–558.

Bauchot, R., and Stephan, H., 1966, Données nouvelles sur l'encéphalisation des Insectivores et des Prosimiens, *Mammalia* **30:**160–196.

Bauchot, R., and Stephan, H., 1967, Encéphales et moulages endocraniens de quelques insectivores et primates actuels, *Colloq. Int. CNRS* **163:**575–587.

Connolly, C. J., 1950, *External Morphology of the Primate Brain,* Thomas, Springfield, Ill.

Cracraft, J., and Eldredge, N. (eds.), 1979, *Phylogenetic Analysis and Paleontology,* Columbia University Press, New York.

Diamond, I. T., 1979, The subdivisions of the neocortex: A proposal to revise the traditional view of sensory, motor, and association areas, *Prog. Psychobiol. Physiol. Psychol.* **8:**1–43.

Edinger, T., 1975, Paleoneurology, 1804–1966: An annotated bibliography, *Adv. Anat. Embryol. Cell Biol.* **49:**12–258.

Falk, D., 1980, A reanalysis of the South African australopithecine natural endocasts, *Am. J. Phys. Anthropol.* **53:**525–539.

Falk, D., 1983a, Cerebral cortices of East African early hominids, *Science* **221:**1072–1074.

Falk, D., 1983b, The Taung endocast: A reply to Holloway, Am. J. Phys. Anthropol. **60:**479–489.

Falzi, G., Perrone, P., and Vignolo, L. A., 1982, Right–left asymmetry in the anterior speech region, *Arch. Neurol.* **39:**239–240.

Geschwind, N., and Galaburda, A. M. (eds.), 1984, *Cerebral Dominance: The Biological Foundations,* Harvard University Press, Cambridge, Mass.

Gurche, J. A., 1982, Early primate brain evolution, in: *Primate Brain Evolution: Methods and Concepts* (E. Armstrong and D. Falk, eds.), Plenum Press, New York, pp. 227–246.

Harman, H. H., 1976, *Modern Factor Analysis,* 3rd ed., University of Chicago Press, Chicago.

Harman, P. J., 1957, Paleoneurologic, neoneurologic, and ontogenetic aspects of brain phylogeny, James Arthur Lecture on the Evolution of the Human Brain, American Museum of Natural History, New York.

Holloway, R. L., 1976, Paleoneurological evidence for language origins, *Ann. N.Y. Academy Sci.* **280:**330–348.

Holloway, R. L., 1981a, Volumetric and asymmetry determinations of recent hominid endocasts: Spy I and II, Djebel IHROUD I, and the Salè *Homo erectus* specimens, with some notes on Neandertal brain size, *Am. J. Phys. Anthropol.* **55:**385–393.

Holloway, R. L., 1981b, Revisiting the South African Taung australopithecine endocast: The position of the lunate sulcus as determined by the stereoplotting technique, *Am. J. Phys. Anthropol.* **56:**43–58.

Holloway, R. L., 1984, The Taung endocast and the lunate sulcus: A rejection of the hypothesis of its anterior position, *Am. J. Phys. Anthropol.* **64:**285–287.

Jerison, H. J., 1973, *Evolution of the Brain and Intelligence,* Academic Press, New York.

Jerison, H. J., 1976, Discussion paper: Paleoneurology and the evolution of language, *Ann. N.Y. Acad. Sci.* **280:**370–382.

Jerison, H. J., 1982, Allometry, brain size, cortical surface, and convolutedness, in: *Primate Brain Evolution: Methods and Concepts* (E. Armstrong and D. Falk, eds.), Plenum Press, New York, pp. 77–84.

Jerison, H. J., 1985a, Animal intelligence as encephalization, *Philos. Trans. R. Soc. London Ser. B* **308:**21–35.
Jerison, H. J., 1985b, Issues in brain evolution, *Oxford Surv. Evol. Biol.* **2:**102–134.
Jerison, H. J., 1986, The perceptual worlds of dolphins, in: *Dolphin Cognition and Behavior: A Comparative Approach* (R. J. Schusterman, J. Thomas, and F. G. Wood, eds.), Erlbaum, Hillsdale, N.J.
Jones, E. G., and Powell, T. P. S., 1970, An anatomical study of converging sensory pathways within the cerebral cortex of the monkey, *Brain* **93:**793–820.
Kielan-Jaworowska, Z., 1983, Multituberculate endocranial casts, *Paleovertebrata (Montpellier)* **13**(1–2)**:**1–12.
Kielan-Jaworowska, Z., 1984, Evolution of the therian mammals in the late Cretaceous of Asia. Part VI. Endocranial casts of eutherian mammals, *Palaeontol. Pol.* **46:**151–171.
Kielan-Jaworowska, Z., 1986, Brain evolution in Mesozoic mammals, in: *G. G. Simpson Memorial Volume: Contributions to Geology* (J. A. Lillegraven, ed.), University of Wyoming, Special Paper **3:**21–34.
Kielan-Jaworowska, Z., Presley, R., and Poplin, C., 1986, The cranial vascular system in taeniolabidoid multituberculate mammals, *Philos. Trans. R. Soc. London Ser. B* **313:**525–602.
Kochetkova, V. I., 1978, *Paleoneurology,* Winston/Wiley, Washington, D.C.
LeMay, M., and Culebras, A., 1972, Human brain: Morphologic differences in the hemispheres demonstrable by carotid arteriography, *N. Eng. J. Med.* **287:**168–170.
Quiroga, J. C., 1980, The brain of the mammal-like reptile *Probainognathus jenseni* (Therapsida, Cynodontia). A correlative paleo-neurological approach to the neocortex at the reptile–mammals transition, *J. Hirnforsch.* **21:**299–336.
Radinsky, L., 1967, Relative brain size: A new measure, *Science* **155:**836–838.
Radinsky, L., 1973, Evolution of the canid brain, *Brain Behav. Evol.* **7:**169–202.
Radinsky, L., 1975, Viverrid neuroanatomy, *J. Mammal.* **56:**130–150.
Radinsky, L., 1976, The brain of *Mesonyx,* a Middle Eocene mesonychid condylarth, *Fieldiana Geol.* **33:**323–337.
Radinsky, L., 1977, Brains of early carnivores, *Paleobiology* **3:**333–349.
Radinsky, L., 1978, Evolution of brain size in carnivores and ungulates, *Am. Nat.* **112:**815–831.
Radinsky, L., 1979, *The Fossil Record of Primate Brain Evolution: The James Arthur Lecture,* American Museum of Natural History, New York.
Radinsky, L., 1981, Brain evolution in extinct South American ungulates, *Brain Behav. Evol.* **18:**169–187.
Rensch. B., 1959, *Evolution Above the Species Level,* Columbia University Press, New York.
Sarnat, H. B., and Netsky, M. G., 1974, *Evolution of the Nervous System,* Oxford University Press, London.
Savage, R. J. G., and Long, M. R., 1986, *Mammal Evolution: An Illustrated Guide,* British Museum (Natural History), London.
Simpson, G. G., 1970, Uniformitarianism: An inquiry into principle, theory, and method in geohistory and biohistory, in: *Essays in Evolution and Genetics in Honor of Theodosius Dobzhansky* (M. K. Hecht and W. C. Steere, eds.), North-Holland, Amsterdam, pp. 43–96.
Stephan, H., Frahm, H., and Baron, G., 1981, New and revised data on volumes of brain structures in insectivores and primates, *Folia Primatol.* **35:**1–29.
Tobias, P. V., 1971, *The Brain in Hominid Evolution,* Columbia University Press, New York.
Trinkaus, E., and Howells, W. W., 1979, The Neandertals, *Sci. Am.* **241**(6)**:**118–133.
Wilson, J. A., 1971, Early Tertiary vertebrate faunas, Vieja Group. Trans-Pecos Texas: Agriochoeridae and Merycoidodontidae, *Tex. Mem. Mus. Bull.* **18:**1–83.

9

Modulatory Events in the Development and Evolution of Primate Neocortex

EDWARD G. JONES

1. Introduction

The cerebral cortex of primates is one of the hallmarks of mammalian brain evolution. The large number and density of neurons in the cortex, their multiple patterns of connectivity, and the highly ordered patterns of cortical lamination and cytoarchitecture present an appearance of great complexity. Such complexity arises, however, from a sequence of orderly and predictable developmental events proceeding from a relatively simple pattern of cellular proliferation and migration. The early phases of neocortical development, during which proliferation and migration are the predominant events, have been the subjects of detailed investigation and a number of reviews (Rakic, 1974). The later events in neocortical ontogenesis, especially those associated with connection formation, the establishment of cytoarchitectonic boundaries, and the genesis of the definitive neuronal phenotypes, have been less thoroughly studied. There are several obvious reasons for this, not the least of which is the lack of ready access to significant numbers of fetal primates for experimental studies. Certain data are emerging, however, which when coupled with those derived from studies of nonprimates, enable what may be the critical events in later neocortical development in the primate to be identified and their implications for the evolution of the primate cortex to be considered. Inevitably, however, the following contains much that is conjectural.

EDWARD G. JONES • Department of Anatomy and Neurobiology, University of California, Irvine, Irvine, California 92717.

2. Intrinsic and Extrinsic Factors in the Establishment of Cytoarchitectonic Order

2.1 Units of Cortical Structure

Any cytoarchitectonic field in the cerebral cortex is distinguished with greater or lesser clarity from its neighbors by differences in overall cell packing density and in cell size, shape, and staining character (Brodmann, 1912; Vogt and Vogt, 1919). Within any one area, individual laminae are distinguished from one another by similar features. In some areas, certain subregions acquire separate cytoarchitectural identities within a common overall pattern. The barrels of the somatic sensory cortex of rodents (Woolsey and Van der Loos, 1970; Welker, 1971, 1976) and the callosally connected part of the human and monkey primary visual area (Shoumura, 1974; Shoumura *et al.*, 1975; Glickstein and Whitteridge, 1976) are well-known examples. In the first, layer IV consists of interrupted patches of densely aggregated granule cells within which subsidiary aggregations, themselves often with a substructure, are found. In the second, the part of area 17 adjoining area 18 possesses a denser concentration of large pyramidal cells in deep layer III than in the remainder of the area.

The fundamental principles that govern the development of cortical cytoarchitecture are probably in large part those that regulate cell aggregation which in turn determines areal and laminar cell packing densities and thus the cytoarchitectonic parcellation of the cortex. To these principles must be added those that regulate cell morphology and connection formation, for these will determine the structure of the neuropil.

Despite the intuitive, cytoarchitectonicist's view that virtually every cortical area differs from its fellows, recent quantitative studies of the neuronal population of a number of cytoarchitectonic areas in macaque monkeys suggest a considerable degree of homogeneity in neuronal organization from area to area (Rockel *et al.*, 1974, 1980; O'Kusky and Colonnier, 1982a; Hendry *et al.*, 1987a; see also Peters, 1987).

Rockel *et al.* (1980) counted the number of Nissl-stained neurons found in 30 μm-wide columns extending through the thickness of the cortex from pia mater to white matter. They discovered that there was a remarkable uniformity in cell number per column (about 110) not only from area to area of a single species but also across species. Even in the mouse the number of cells per column did not differ greatly from that in the somewhat thicker cortex of primates. The only area showing a significant difference from all other areas was area 17 of all primate species examined, including human. In area 17 of primates, the cell numbers per column were more than double those of columns in other areas of primates and approximately double those of columns in the visual areas of nonprimates. Differences in cortical thickness change across species and lead to differences in packing density of neurons from area to area. These differences are not great, however, the average cortical thickness in a mouse being approximately 1–1.5 mm and that of a human 3–4 mm (see Allman, 1990). The figures of Rockel *et al.* thus show that for all cortical areas except the primary visual area of primates, changes in cortical thickness are probably achieved to a large extent by modifying the neuropil and not by changing significantly the number of neurons per unit column. O'Kusky and Colonnier (1982a) in the macaque visual cortex found somewhat fewer cells beneath a 1-mm^2 area of cortical surface

(202,000) than did Rockel *et al.* (352,000) but their figure took tissue shrinkage into account. A figure of 194,000 later reported by Powell and Hendrickson (1981) is close to that of O'Kusky and Colonnier.

The studies of Beaulieu and Colonnier (1983, 1985, 1989), on several areas of the cat cortex, are in general agreement with that of Rockel *et al.*, with some variations in detail, except that in their hands the motor area of the cat shows fewer cells per unit column than do the visual and somatic sensory areas. This discrepancy between the results of the two groups has not been explained.

Hendry *et al.* (1987a) in another study counted the neuronal population in ten areas of monkey cortex and demonstrated the contribution of the subpopulation that showed immunoreactivity for the inhibitory transmitter, GABA, to the total (Tables 1–4). The results confirm the basic similarity of mean cortical neuron number per arbitrary radial column from area to area, area 17 being the

Table 1. Mean Number (±SD) of GABA-Immunoreactive Neurons per 50-μm-Wide Column through the Thickness of Areas 4, 3b, 1–2, 5, 7, 18, and 17 in Macaque Monkeys[a]

	CM 181			CM 187		
Area	A	B	C	A	B	C
4	38.8 ± 3.1	37.6 ± 3.7	39.0 ± 2.9	40.1 ± 2.9	40.7 ± 3.7	39.4 ± 2.7
3b	38.7 ± 2.6	38.0 ± 3.3	39.9 ± 2.8	33.4 ± 4.0	32.0 ± 4.9	32.1 ± 3.9
1–2	41.2 ± 2.9	39.1 ± 3.9	40.1 ± 2.6	39.6 ± 2.7	40.7 ± 3.8	38.4 ± 2.1
5	39.6 ± 2.4	39.1 ± 3.2	40.4 ± 2.2	40.1 ± 3.1	38.1 ± 4.1	39.3 ± 2.4
7	38.3 ± 3.1	41.0 ± 3.1	38.4 ± 2.9	38.1 ± 2.6	41.3 ± 2.2	40.4 ± 3.7
18	39.4 ± 3.0	37.9 ± 2.2	38.1 ± 2.9	38.7 ± 3.9	40.1 ± 2.2	38.6 ± 3.6
17	60.3 ± 5.3	58.2 ± 6.1	59.1 ± 5.7	60.8 ± 5.9	59.1 ± 6.9	61.2 ± 4.6
	CM 183			CM 189		
Area	A	B	C	A	B	C
4	39.1 ± 2.3	38.1 ± 3.6	40.4 ± 3.9	38.1 ± 2.9	39.0 ± 3.5	39.4 ± 2.4
3b	34.2 ± 5.0	35.1 ± 6.3	33.1 ± 5.0	39.4 ± 2.1	39.4 ± 2.5	40.1 ± 3.0
1–2	37.4 ± 3.1	36.2 ± 3.8	38.0 ± 3.3	38.6 ± 3.3	39.1 ± 3.8	38.4 ± 3.3
5	38.7 ± 2.6	39.4 ± 3.1	39.1 ± 2.7	39.7 ± 3.0	38.4 ± 4.3	38.5 ± 3.9
7	39.1 ± 3.0	38.6 ± 4.1	39.9 ± 2.9	40.1 ± 2.7	38.6 ± 3.7	41.4 ± 2.7
18	37.7 ± 3.1	39.1 ± 4.2	38.4 ± 3.9	36.9 ± 3.3	39.7 ± 3.7	41.0 ± 2.1
17	59.7 ± 4.0	60.1 ± 5.3	61.1 ± 4.7	59.4 ± 5.6	60.1 ± 6.7	60.7 ± 7.7
	CM 184					
Area	A	B	C			
4	38.6 ± 3.0	41.4 ± 5.0	40.7 ± 3.4			
3b	33.9 ± 5.1	34.1 ± 5.2	33.0 ± 4.7			
1–2	38.4 ± 2.7	39.7 ± 3.4	40.1 ± 3.0			
5	39.4 ± 2.2	39.0 ± 3.6	40.4 ± 2.0			
7	40.4 ± 2.7	40.1 ± 3.9	40.7 ± 2.8			
18	38.6 ± 3.7	41.4 ± 4.4	40.8 ± 5.0			
17	57.8 ± 4.8	60.0 ± 5.7	61.3 ± 5.0			

[a]Modified from Hendry *et al.* (1987a). Counts were made from photomicrographic montages at a final magnification of either 480 (A) or 960 (B) or directly from the microscope at a magnification of 1250 (C).

Table 2. Mean Number (±SD) of GABA-Immunoreactive Neurons and of All Thionin-Stained Neurons per 50-μm-Wide Column through the Thickness of Areas 4, 3b, 1–2, 5, 7, 18, and 17 in Macaque Monkeys[a]

	CM 181			CM 187		
Area	GABA	Total	%	GABA	Total	%
4	39.0 ± 2.9	158.9 ± 16.1	24.5 ± 2.0	39.4 ± 2.7	159.4 ± 16.3	24.5 ± 1.3
3b	39.9 ± 2.8	152.7 ± 11.9	26.1 ± 1.7	32.1 ± 5.9	154.9 ± 11.7	20.6 ± 1.2
1–2	40.1 ± 2.6	154.8 ± 14.2	25.9 ± 1.8	38.4 ± 4.0	157.9 ± 16.1	24.4 ± 2.1
5	40.4 ± 2.2	163.1 ± 10.6	24.7 ± 2.3	39.2 ± 2.0	160.6 ± 14.1	24.4 ± 1.9
7	38.4 ± 2.9	158.2 ± 9.4	24.1 ± 1.6	40.7 ± 3.7	158.8 ± 14.8	25.1 ± 1.4
18	38.1 ± 2.9	157.3 ± 10.1	24.4 ± 2.1	38.6 ± 3.6	157.9 ± 15.0	24.7 ± 2.1
17	59.1 ± 5.7	315.7 ± 19.4	18.7 ± 2.1	58.2 ± 5.7	309.9 ± 22.7	19.2 ± 2.0

	CM 183			CM 189		
Area	GABA	Total	%	GABA	Total	%
4	40.4 ± 3.9	161.7 ± 15.3	25.0 ± 1.7	39.4 ± 2.4	161.6 ± 17.0	24.3 ± 2.1
3b	33.1 ± 5.0	157.9 ± 12.2	21.0 ± 2.1	40.1 ± 3.0	156.5 ± 12.7	25.6 ± 2.3
1–2	38.0 ± 3.3	154.3 ± 11.6	24.6 ± 2.3	38.4 ± 3.3	158.8 ± 11.9	24.0 ± 1.9
5	39.1 ± 2.7	155.3 ± 9.4	25.2 ± 1.8	38.5 ± 3.4	155.1 ± 12.0	24.9 ± 2.0
7	39.9 ± 2.9	157.6 ± 11.8	25.3 ± 1.3	40.4 ± 2.4	160.1 ± 16.5	25.2 ± 2.4
18	38.4 ± 3.9	152.0 ± 9.1	25.3 ± 1.6	38.1 ± 2.9	157.5 ± 17.5	24.2 ± 1.7
17	61.1 ± 4.7	309.8 ± 17.1	19.6 ± 1.5	59.1 ± 5.7	321.3 ± 23.6	18.5 ± 2.2

	CM 184		
Area	GABA	Total	%
4	40.7 ± 3.4	157.1 ± 15.5	25.2 ± 1.1
3b	32.9 ± 5.7	160.4 ± 11.0	20.2 ± 1.9
1–2	40.1 ± 3.1	154.9 ± 9.7	25.1 ± 1.7
5	40.4 ± 2.0	158.1 ± 13.0	24.9 ± 1.3
7	40.7 ± 2.8	159.9 ± 11.7	24.3 ± 2.0
18	40.8 ± 5.0	158.4 ± 16.2	24.7 ± 1.8
17	60.3 ± 5.3	319.6 ± 21.4	19.1 ± 1.0

[a] Modified from Hendry *et al.* (1987a). Counts of GABA-positive neurons and of all neurons were made at a magnification of 1250 from adjacent sections, and the percentages of GABA cells were calculated from the quotient of the two values.

Table 3. Mean Number (±SD) of GABA-Immunoreactive Neurons and of All Thionin-Stained Neurons in 50-μm-Wide Columns through the Thickness of Area 21 of the Temporal Lobe and Areas of Orbital Frontal and Lateral Frontal Cortex in Macaque Monkeys[a]

	CM 181			CM 189		
	GABA	Total	%	GABA	Total	%
Area 21	38.6 ± 3.2	161.4 ± 10.4	24.1 ± 2.1	38.1 ± 3.6	156.4 ± 11.2	24.4 ± 2.2
Inferior frontal	37.4 ± 2.8	157.6 ± 9.0	23.6 ± 2.0	39.6 ± 4.0	158.9 ± 12.2	24.8 ± 2.9
Lateral frontal	39.0 ± 2.9	163.1 ± 12.7	23.9 ± 1.9	40.6 ± 2.6	160.4 ± 14.1	25.1 ± 2.5

[a] Modified from Hendry *et al.*, (1987a). Counts of GABA-positive neurons and of thionin-stained neurons were made at a magnification of 1250, and the percentages of GABA cells were calculated from the quotient of the two values.

Table 4. Mean Number (±SD) of GABA-Immunoreactive Neurons Calculated to Be Present in a Volume of 1 mm³ or to Underlie Surface 1 mm² or 50 × 50 μm Squares in Macaque Monkeys[a]

	Count/mm³		Count/mm²		Count/50 × 50 μm	
Area	CM 181	CM 187	CM 181	CM 187	CM 181	CM 187
4	10,400 ± 190	11,200 ± 210	22,800 ± 840	24,100 ± 990	57.0 ± 2.1	60.3 ± 2.5
3b	17,200 ± 310	14,900 ± 270	23,600 ± 1100	19,900 ± 970	59.0 ± 3.8	49.8 ± 4.7
1–2	16,100 ± 210	15,800 ± 230	24,200 ± 1200	23,800 ± 1080	60.5 ± 3.3	59.2 ± 3.8
5	15,200 ± 240	14,600 ± 230	23,500 ± 1050	23,100 ± 1160	58.8 ± 1.9	57.8 ± 2.3
17	23,160 ± 390	23,700 ± 400	34,500 ± 1700	36,300 ± 1660	86.3 ± 7.0	90.8 ± 7.7

[a] Modified from Hendry *et al.* (1987a). Calculations were made from measurements in 1-μm-thick plastic sections through the thickness of areas 4, 3b, 1–2, 5, and 17.

exception, with approximately twice the number of cells. The number of GABA neurons per radial column follows a similar pattern, the percentage being about 25% for most areas. Area 17 is the exception. There, despite a large increase in GABA neuron numbers, the GABA cell population falls to about 20% on account of the twofold increase in total neuron number. A similar proportion of GABA neurons was found in the cat visual cortex by Gabbott and Somogyi (1986). There is, in fact, considerable variation in total cell numbers and in the percentage of GABA cells from column to column within the areas of the monkey studied. No repeating patterns of variance could be detected, although this was later reported by Schwartz *et al.* (1988) in the frontal cortex.

These data argue for a fundamental similarity in neuron numbers and in proportions of particular neuronal types from area to area in the primate cerebral cortex. Hence, the development of cytoarchitectonic areas in the cortex, if based upon the assembly of modular groupings of neurons as suggested by Rakic (1988), is likely to consist of the assembly of groupings that contain approximately the same numbers of neurons. The consistency in the proportion of GABA neurons per radial unit in the adult further suggests that relative percentages of cells of different types could be maintained in each hypothetical modular grouping. The other major cortical cell type, that apparently using glutamate as a transmitter and accounting probably for all or most of the cortical cells that possess significant numbers of dendritic spines, has not been quantified in the same manner as the GABA cells. Counts by Conti *et al.* (1987) suggest approximately 30% of cells in the somatosensory cortex of rats and approximately 50% in that of monkeys are glutamate immunoreactive, but these counts have not been done in a manner that permits comparison with the "columnar" counts made in the studies mentioned above. Other cell populations that may vary from area to area are those that express neuroactive peptides. Most of these are contained in GABA cells (Hendry *et al.*, (1984). Despite the apparent consistency in number of GABA cells from area to area, GABA cells coexpressing a particular peptide can vary considerably from area to area in the monkey cortex. Somatostatin-immunoreactive cells, for example, show a fourfold increase in the temporal areas and a sixfold increase in the anterior cingulate area in comparison with the visual area (Campbell *et al.*, 1987). It is not yet clear whether this represents a redistribution of morphological types of GABA neurons or the differential expression of a colocalized peptide (Jones and Hendry, 1986) from area to area.

From the tentative idea of the assembly of cytoarchitectonic fields out of some fundamental, perhaps unitary, grouping of cells, it is not too unreasonable to invoke, as many have, the idea of a unitary pattern of intrinsic connections that intervene between cortical input and cortical output in the mature cerebral cortex. It seems possible that eventually a standard intracortical circuit will be identified that is common to all areas and which represents the basic operating unit of cortical function. If such a circuit exists, *a priori,* it is reasonable to assume that it may be based upon a repeating pattern of radial units extending through the thickness of the cortex, each containing representatives, possibly in fixed proportions, of each cortical cell type. To establish this, further counts of morphologically and immunocytochemically defined cell populations will be essential.

The neocortex develops in a manner that can be interpreted as indicative of a repeating radial pattern of neurons. Young neurons, on being generated in the ventricular or subventricular zones of the neuroepithelium, migrate into the overlying cortical plate, probably following the radial processes of neuroepithelial cells (Rakic, 1972). Because young neurons are often lined up in columns along the radial glial processes, it is possible to conjecture that the migrating column of neurons is a sort of clone, having arisen from a single precursor cell in the neuroepithelium and leading to the formation of a basic columnar unit of cortical architecture (Rakic and Goldman-Rakic, 1982; Mountcastle, 1978; Rakic, 1988). Further conjecture would lead to a view that the size of a cortical area might be governed by the number of radial units allocated to it in the neuroepithelium (Rakic, 1988). Studies involving retrovirus infection of cortical precursor cells and which hold out the hope of determining whether a clonal pattern can account for the columnarity of the cortex, are still in their infancy and results to date can be variously interpreted as either supporting or negating a columnar, developmental pattern (Luskin *et al.*, 1988; Price and Thurlow, 1988; Walsh and Cepko, 1988).

It is an attractive hypothesis that each area of the primate cortex is made up of an aggregation of repeating columnar units. It is a further attractive idea that the extent of a particular cortical area is governed by the number of units assigned to it in development and that the increase in size of the cortex and the appearance in it of new architectonic areas in the course of evolution are determined by the addition of new columnar units. In supposing this, it has to be remembered that the columns in which several investigators have counted cortical neurons are arbitrary conveniences and show a wide range of variation from column to column. They are not anatomically defined entities and definitely not functional units. Similarly, the members of each striking radial array of neurons migrating into the cortex in the course of development are not yet confirmed to remain linked connectionally with one another in the mature cortex. Further, in the absence at the early stages of markers for identifying migrating cells that will become the different types of neurons of the cortex, it is not known if representatives of every cortical cell type are found in each radial array and in the proportions found in adult units of cortical circuitry. These are questions for which answers are sorely needed.

2.2 The Role of Afferent Fibers

There is a wealth of data to indicate a close correlation between the cytoarchitectonic delineation of the adult cerebral cortex and afferent, efferent, and intrinsic connectivity (Rose and Woolsey, 1948, 1949; Powell, 1973). The fiber terminations of many afferent systems are confined within architectonic borders and certain populations of cells projecting to a particular target can be found in one area but not in its neighbors. Corticorubral cells, for example, are not found behind the central sulcus (Jones and Wise, 1977). The role of the periphery and thus of afferent activity in regulating the development of the subareal structure of the rodent somatic sensory cortex is well known. The cytoarchitectonic units called barrels fail to develop if their corresponding vibrissae on the face are removed within a few days of birth (Van der Loos and Woolsey, 1973; Killackey *et al.,* 1978; Jeanmonod *et al.,* 1981) and in mutant mice with supernumerary or absent vibrissae, corresponding barrels form or fail to appear, respectively (Welker and Van der Loos, 1986). To what extent, therefore, is afferent connectivity a determinant of cortical field structure?

Cytoarchitectonic field structure is not discernible from the moment a cortical plate appears in the telencephalic wall, although areas could already be predetermined in the neuroepithelium, since it appears that only a restricted zone of proliferation in the ventricular lining contributes cells to a particular field (Rakic, 1974). In all mammals the oldest (deeper) cortical layers begin to acquire a laminar structure similar to that of the adult before the overlying cortical plate disappears in the formation of the overlying layers, so that the architecture of the layers appears to arise from their individual maturation rather than by interactions between them. The deeper layers, at least, acquire a cytoarchitecture similar to that of the adult before the ingrowth of significant numbers of thalamic or callosal afferent fibers, suggesting that their structural delineation is not the result of an interactive process with these afferents (e.g., Wise and Jones, 1978).

Areal borders also become clearly evident relatively early in development as the cortical plate disappears. In monkeys, the border between areas 17 and 18 and to a large extent the laminar structure of area 17 can be discerned before the invasion of the cortex by geniculocortical afferents around day 110 of gestation (Rakic, 1976). The borders between most other cortical areas in monkeys are also discernible by the 110th day, e.g., that between areas 3a and 4 is readily seen (Huntley *et al.,* 1990). In cats, the cytoarchitectonic fields of the somatosensory and motor areas become discernible at about the time of birth, while in rats they appear in the first 4–5 days of postnatal life (e.g., Wise and Jones, 1976, 1978; Wise *et al.,* 1977); the primary visual cortex of cats is fully delineated within the first week after birth (Shatz and Luskin, 1986). In all of these instances, cytoarchitectonic borders appear before the ingrowth of significant numbers of thalamic or callosal afferents.

Thalamic and callosal axons, nevertheless, arrive relatively early in the vicinity of the developing sensory-motor and visual areas of cats, rats, and monkeys. There they undergo a relatively protracted waiting period in the underlying part of the intermediate zone (called the subplate by many authors) and begin to invade the cortex at about the time the last cells are completing their migration to the cortex, i.e., just before the cortical plate disappears and as cytoarchitectonic borders are becoming established (Wise and Jones, 1976, 1978; Rakic, 1976; Lund and Mustari, 1977; Wise *et al.,* 1977; Shatz and Luskin, 1986; Killackey and Chalupa, 1986). It appears as though axons commence ingrowth after their major cortical target neurons are in place, which may help explain why thalamic fibers commence their ingrowth prior to that of callosal fibers since the latter extend their terminations more superficially in the cortex.

Although on present evidence it seems doubtful that afferent fiber innervation plays a role in setting up cytoarchitectonic field structure and cytoarchitectonic border formations, there is reason to believe that both the internal substructure of an architectonic field and architectonic field size may be determined by afferent connectivity. Intraareal remodeling is demonstrable as the extrinsic innervation becomes established in a variety of sites. Area 4 of the human and monkey fetus possesses a clearly distinct granular layer IV up to birth but becomes agranular as postnatal development proceeds (Ramōn y Cajal, 1900; Huntley *et al.*, 1990). Similarly, the lamination that enables areas 3a, 3b, 1, and 2 to be clearly identifiable only becomes clear-cut after afferent innervation is established. The cytochrome oxidase-stained periodicities visible in layer III of area 17 in the monkey only become detectable late in fetal life (Horton and Hedley-Whyte, 1984), again, after afferent innervation is established. These observations appear to indicate an interactive role of afferent fibers and the definitive cytoarchitectonic pattern in the later stages of cortical development.

Neural activity may be the key to understanding how afferent fibers influence the finer development of architectonic field substructure. For example, barrels of the contralateral somatic sensory cortex fail to develop if related vibrissal follicles on the muzzles of rats and mice are destroyed within an early, critical, postnatal period (Van der Loos and Woolsey, 1973; Killackey *et al.*, 1978). Barrels also fail to appear if the thalamus is destroyed at birth in rats, thus destroying thalamocortical fibers before they have fully penetrated the cortex (Wise and Jones, 1978). However, the broad aggregate of granule cells in layer IV which represents the face as a whole and in which the barrels are normally set, is not lost after these manipulations and appears to have just as clear-cut borders as it does in the normal brain when barrels are present. Similarly, the forelimb representation remains a cytoarchitectonic entity after fetal forelimb removal and despite expansion of the entity representing the hindlimb (Dawson and Killackey, 1987). The supernumerary or absent barrels in the mutant mice of Welker and Van der Loos (1986) seem to be an analogous case. Hence, the thalamic fibers and the influence of the periphery that they normally carry appear to be determinants not of the architectonic borders of the somatic sensory area, but of its subareal structure.

Afferent innervation may also determine the size of an architectonic field. The influence of an afferent innervation on architectonic field size can be seen in experiments in monkeys in which both eyes had been removed during intrauterine life. There, the area 17/18 border was still visible postnatally, although area 17 appeared reduced in size (Rakic, 1988; Dehay *et al.*, 1989). Similarly, localized destruction of parts of the neuroepithelium by cytotoxic drugs in rats is accompanied by a loss of areas and parts of areas, not by adaptation to restore lost areas (Yurkewicz *et al.*, 1984). It is possible, therefore, that a major secondary feature determined by interactions with afferent fibers is cortical field size.

The conclusion to be drawn from the preceding is that cytoarchitectonic borders are likely to be predetermined in the proliferating neuroepithelium and not molded by interactions with afferent fibers in the early cortex. Architectonic borders in the adult cortex may be predetermined by the capacities of adjacent regions of neuroepithelium each to form only cells appropriate to an individual area, since architecture is fundamentally established before the arrival of the major extrinsic afferent connections. Afferent connections appear to regulate subareal architectonic features and may also regulate field size. The potential factors that enter into these more subtle modifications of cortical field structure and organization are only just starting to be revealed.

2.3. Mutability of Cortical Organization during Development

The accumulated weight of evidence up to this point is that the fundamental basis of cytoarchitectonic field structure and especially of cytoarchitectonic border formation, is determined by mechanisms that are "intrinsic" or not dependent on the presence of afferent connections. Connections, mediating the influence of the periphery on sensory areas, appear to determine field size and subareal structure. It would seem doubtful, from the available data, that a portion of the early cortical mantle destined to form visual cortex could ever be converted, en masse, to motor cortex simply by rearranging its connections. A number of recent findings force us, however, to consider the extent to which this is an inviolable rule and whether a region of telencephalic wall after its generation from the neuroepithelium and before its expression of the architectonic structure typical of that region, is indeed mutable.

The first set of observations that force this reevaluation are those showing that afferent fibers, deprived of their normal target nuclei in the thalamus of newborn rodents and carnivores, may densely innervate adjacent, partially deafferented thalamic nuclei, even if these form parts of a different sensory system (Frost, 1981, 1982, 1984, 1986; Sur *et al.*, 1989). This occurs by axonal sprouting and in some cases by maintenance of a preexisting but developmentally transient projection (see Frost, 1986, for references). Functional synapses are formed in the new nucleus (Frost and Metin, 1985; Frost and Campbell, 1987) and these can form the basis for the projection of a new modality to a cortical area that would normally be concerned with a different sensory system. For example, visually evoked unit responses can be obtained in auditory cortex after sprouting of optic tract fibers into the developing medial geniculate nucleus (Sur *et al.*, 1989). These responses have many of the attributes of those normally recorded in the visual cortex.

The second set of experiments indicates that pieces of developing rat visual cortex, taken at a time before the appearance of architectonic field structure, and transplanted to the vicinity of the developing sensory-motor cortex, will be incorporated into the latter area and that certain cells of the transplant may extend efferent axons into regions of the CNS appropriate to the new area, viz. the spinal cord (O'Leary and Stanfield, 1989). These projections appear to represent the maintenance of axons normally transiently extended by layer V cells of the visual cortex to the spinal cord but which later disappear (Stanfield and O'Leary, 1985a,b; O'Leary and Stanfield, 1986).

The third set of experiments involved the transplantation of specific sets of cells in the developing visual cortex of ferrets (McConnell, 1985, 1988). Neurons born on embryonic day E31 and destined to form layers 5 and 6 were dissociated and transplanted to the ventricular zone of a newborn ferret at a time when neurons destined for layers 2 and 3 were being generated. Forty-three percent of the subsequently identified cells had migrated to layers 2 and 3, for them a heterotopic site, while the remainder mainly occupied layers 5 and 6, their normal site, and developed efferent connections typical of that layer. It has been suggested that the transplanted cells that found their way to layers 2 and 3 may be those that underwent a further round of DNA synthesis and mitotic division in the host. Whether they develop connections typical of layers 2 and 3 has not been established.

It is difficult at this time to see these profound modifications as necessarily implying that cortical organization is determined solely by extrinsic afferent influences. Rather, they seem to signify the importance of local cues in determining cell fate and connection formation. They seem to imply that afferents and cells foreign to a particular area are themselves modified under the influence of local conditions: foreign afferents accepted into existing circuitry; heterotopically transplanted cells, committed to the loss of a particular set of axon collaterals being forced to maintain them, possibly even at the expense of others; and heterochronically transplanted cells seemingly committed to one laminar and connectional fate forced into another. These local conditions may be specified in the portion of the neuroepithelium from which the host area arises and can be envisaged as factors that not only govern cell identity, and through it, boundary formation, but which also permit this identity to be transmitted to afferent fibers and to new cells entering the area, either in the course of normal development or in the experimental situations quoted above.

The nature of the molecular factors that govern cell assembly and border formation in the development of a cortical architectonic field are not known. We may envisage some of these factors as being comparable to those proposed by Edelman (1988b) in which cycles of expression of cell and substrate adhesion molecules bring cells together so that they may exchange signals that may in turn regulate further expression. Several studies in the somatic sensory cortex of neonatal rodents have shown that glial-specific antigens, lectin receptors, and glial-related adhesion molecules are expressed transiently in a topographic map that precedes the aggregation of neurons into the cytoarchitectonic elements (barrels) that form the basis of the body map (Cooper and Steindler, 1986, 1989; McCandlish *et al.*, 1989; Steindler *et al.*, 1989). Assembly of the early map appears to be regulated by afferent activity since early whisker removal prevents the formation of the protobarrel units within it (Cooper and Steindler, 1989). It is conceivable that this developmentally regulated expression of glial-related molecules is an indication of a large number of similar changes occurring throughout cortical development. These types of molecular events are likely to be subject to variation and thus to have become targets for natural selection in the course of cortical evolution.

2.4. The Role of the Subplate

The cortex, as we have seen, is colonized by afferent fibers relatively late in its development. Although, from the foregoing, afferent innervation may not be the primary factor in establishing cytoarchitectonic order and border formation, a number of secondary features of cortical structure appear to develop contemporaneously with, and may be profoundly influenced by the arrival of afferent fibers. It is possible, however, that afferent fibers can exert an earlier, but in a sense, remote influence on cortical development by interacting with migrating cells before the arrival of either cells or fibers in the cortex.

Thalamocortical, commissural, and corticocortical axons arriving beneath the cortex accumulate there over a substantial waiting period before making a second growth spurt into the cortex. During this time, migrating cells destined for the middle and upper cortical layers and forming a major proportion of those on which the fibers will ultimately terminate, pass through them, en route to the cortex. The possibility exists for an exchange of signals which may later lead to recognition when the fibers invade the cortex (Wise and Jones, 1976, 1978; Shatz and Luskin, 1986). A population of early generated and early maturing neurons, many of which subsequently die (Kostovic and Rakic, 1980; Luskin and Shatz, 1985b; Chun *et al.*,1987; Valverde and Facal-Valverde, 1988), resides in the subplate region immediately deep to the cortex throughout cortical development, from before the cortical plate is formed to the establishment of cortical innervation. This population of cells has been conjectured as playing a key role in allowing later-arriving, permanent cells and afferent axons to recognize one another and may play a role in laying down pathways for early growing cortical efferent fibers (Shatz *et al.,* 1988; McConnell, 1988; McConnell *et al.,* 1989).

The extent to which accumulating fibers, subplate cells, and migrating young cortical cells recognize one another and whether information is exchanged between them are currently unknown. Synapselike formations and other forms of intercellular contacts have been described in the subplate region (Kristt and Molliver, 1976) but it is not clear whether these are between accumulating axons and cells migrating to the cortex, between axons and the population of subplate neurons (Shatz *et al.,* 1988), or between the cells themselves. Apart from the morphological identification of synapselike structures, the appearance of $GABA_A$ receptors and synaptic vesicle antigens in the subplate region (and in layer I) before their appearance in the cortical plate also suggests the presence of a synaptic zone in the waiting compartment beneath the cortical plate (Chun and Shatz, 1988a; Huntley *et al.,* 1990). Immunoreactivity for the $GABA_A$ receptor and the synaptic vesicle antigens disappears from the subplate as afferent fibers invade the cortex and it appears in the cortex proper as afferent innervation is being established. An exchange of signals in the subplate relevant to subsequent re-recognition and synapse formation in the cortex is thus a tantalizing prospect but not one that has been completely verified.

3. Establishment of Afferent Innervation

3.1. Thalamic Innervation

In a previous section (2.2), the waiting period of afferent fibers beneath the cortex was described. In monkeys, thalamic fibers to the visual cortex arrive beneath the immature cortex in the middle of gestation (approximately on the 80th day) and invade the cortex on or subsequent to the 110th day, achieving their final laminar disposition by the 124th to 144th day (Rakic, 1976, 1981). Studies based on the development of acetylcholinesterase staining patterns suggest a comparable timing of ingrowth and laminar distribution in the prestriate and prefrontal areas (Kostovic and Rakic, 1984; Kostovic and Goldman-Rakic, 1983) and a similar time scale is probable for the sensory-motor cortex (Killackey and Chalupa, 1986). Fiber ingrowth appears initially to be somewhat diffuse, with eye dominance columns appearing in the visual cortex 2 or 3 weeks before birth, just after the definitive laminar distribution of the fibers is established and definitely before the onset of extrauterine visual experience. The final columnar distribution, however, is achieved only under the influence of normal postnatal visual experience (Section 4.1).

In rats and cats, arriving thalamocortical axons start to accumulate in the intermediate zone beneath the cortex before birth and commence invading the cortex at or just after birth (Wise and Jones, 1976; Lund and Mustari, 1977; Wise *et al.*, 1977; Shatz and Luskin, 1986). As they commence innervating the visual cortex of the cat and the somatic sensory cortex of the rat, the fibers enter the cortex in a diffuse manner and it is only with time that ocular dominance domains and barrel terminations crystallize out. In the cat visual cortex, the transformation from a diffuse to an ocular dominance pattern occurs over the third to sixth weeks of life (LeVay *et al.*, 1978; Shatz and Stryker, 1978). In the rat somatic sensory cortex, segregated barrel terminations are detectable only after the fifth postnatal day (Wise and Jones, 1978). Normal innervation of the facial vibrissae is required to ensure the development of the definitive segregated distribution of thalamocortical fibers among the barrels of the rat somatic sensory cortex (Killackey and Belford, 1979). In the cat visual cortex, segregation of thalamic fiber terminations bearing inputs from the two eyes is dependent on normal visual experience in the first 3 months of life (Section 4).

The fine resolution of the patterns of thalamocortical innervation in monkeys, cats, and rats is therefore definitely dependent upon normal sensory experience. However, the appearance of rudimentary ocular dominance columns in monkeys before the onset of visual experience (Rakic, 1976, 1981) suggests that there is already a degree of topographic order built into the system of thalamocortical fibers and their terminations from the earliest times. Such ordering may also be operational in the innervation of other areas of the cortex by developing thalamocortical fibers. Irrespective of this inbuilt predisposition, however, the onset of normal neural activity is a powerful epigenetic influence on the subsequent establishment of fine topographic order in the higher mammalian cerebral cortex, as much as it is in the developing and regenerating retinotectal systems of nonmammalian vertebrates (e.g., Meyer, 1982, 1983; Reh and Constantine-Paton, 1985). Conceivably, order is imposed from below, since evidence suggests that the sensory pathways mature from the periphery centrally (Killackey and Belford, 1979; Killackey and Shinder, 1981). We shall return to this topic in Section 4.

3.2. Commissural and Corticocortical Innervation

Accumulating beneath the cortical plate with the thalamocortical fibers are the corticocortical and callosal fibers. Many of these, especially those that are inappropriate for the overlying area, may never enter the cortex (Clarke and Innocenti, 1986). Those that do, appear to be under a temporal control different from that of the thalamic fibers, for they make their second growth spurt into the overlying cortex a short time after the thalamic fibers. In rats and cats, their ingrowth follows that of the thalamic fibers by 2 or 3 days (Wise and Jones, 1976, 1978; Wise *et al.*, 1977); in monkeys, possibly by as much as 20 days (Killackey and Chalupa, 1986). The initial ingrowth of these other two fiber systems has also been reported as diffuse although this has recently been challenged (Price and Zumbroich, 1989). There is, in any case, significant redistribution of the fibers to lead to the patchy, disjunctive distribution patterns of corticocortical and callosal fiber terminations seen in many cortical areas in the adult. This presumably involves interactions with the thalamocortical fibers as they are setting up their definitive patterns of distribution.

The nature of any interactiveness between the different sets of afferent fibers in the newly maturing cortex are not known but can be assumed from available data. Interference with one set of afferents, e.g., the visual thalamocortical system by removal of an eye, is reported to lead in cats to changes in the distribution of callosally projecting cells and their terminal fiber ramifications so that the diffuse pattern of the immature cortex is retained in many areas (Innocenti and Frost, 1979, 1980; Cusick and Lund, 1982; Olavarria and Van Sluyters, 1984). This may not occur, however, in monkeys (Dehay *et al.,* 1989). In the disruptions of the visual pathway in the Siamese cat (Hubel and Wiesel, 1971; Kaas and Guillery, 1973; Shatz, 1977a,b) and associated with the induction of a strabismus in normal infant cats (Lund *et al.,* 1978; Innocenti and Frost, 1979), the patterns of retinotopic organization and callosal fiber distribution imply developmental interactions between thalamic and callosal axons. These observations suggest some form of interaction at the cortical level, possibly a competition for synaptic space in which corticocortical and callosal fibers under perturbed conditions survive in inappropriate regions by gaining an advantage over nonfunctional thalamocortical fibers. Again, activity may be the key to the understanding of this phenomenon and we take up this important issue in the next section.

4. Activity-Dependent Regulation of Cortical Organization

4.1. Activity-Dependent Regulation of Cortical Structure

It is now well established that a normal pattern of afferent activity is essential for the maturation of neuronal receptive fields in the visual cortex of cats and monkeys (reviewed in Hubel and Wiesel, 1970; Buisseret and Imbert, 1976; Blakemore *et al.,* 1978; Vital-Durand *et al.,* 1978; Cynader *et al.,* 1980; Movshon and Van Sluyters, 1981; Sherman and Spear, 1982; Fregnac and Imbert, 1984; Mitchell and Timney, 1984). There is a critical period of early postnatal development (approximately 4 months in the cat and 6 months in the macaque monkey) during which perturbations of visual experience can seriously compromise the establishment of the normal complement of cortical receptive field types (Hubel and Wiesel, 1970; Wiesel and Hubel, 1963; Olson and Freeman, 1978; Blakemore and Van Sluyters, 1975). The effect is much greater in the first few weeks of this period. In kittens, as little as 1 day of monocular deprivation can have physiological consequences for visual cortical neurons (Olson and Freeman, 1975; Movshon and Dursteler, 1977).

Experience-dependent modifications of visual cortex physiology during the critical period of development have been more thoroughly investigated in cats, but there are sufficient data from monkeys to suggest that comparable plasticity occurs in the primate visual cortex as well (Wiesel and Hubel, 1974; Crawford *et al.,* 1975; Hubel and Wiesel, 1977; Hubel *et al.,* 1977; Des Rosiers *et al.,* 1978; LeVay *et al.,* 1980). It is clear that a cat, monocularly deprived for the first 3 months of life never acquires normal vision with the deprived eye and that this is consequent upon a radical shift in the responsivity of cells in the visual cortex in the direction of the nondeprived eye (Dews and Wiesel, 1970; Wiesel and Hubel, 1963, 1965a,b; reviewed in references cited above). Where, in a normal cat, approximately 85% of the cells recorded in layer IV of the visual cortex are binocularly driven, in a cat monocularly deprived in the first few weeks of life, 90% of the cells become responsive primarily to the nondeprived eye (Wiesel and Hubel, 1963) although they still receive some synaptic input from the deprived eye (Olson and Freeman, 1978). If vision is perturbed in other ways, such as by manipulating the eyes to produce images that are not in register (Hubel and Wiesel, 1965; Baker *et al.*, 1974; Blakemore *et al.*, 1978; Blakemore and Van Sluyters, 1975; Blasdel and Pettigrew, 1979; Freeman *et al.*, 1983; Kratz and Spear, 1976) or by rearing in a visual environment consisting of single orientations (Pettigrew, 1974; Cynader and Mitchell, 1977; Buisseret *et al.*, 1978), shifts in the normal distribution or responsivity of visual cortical cells can be similarly engendered. During the critical period of the first 3 months in cats, the majority of these effects can be partially reversed by the restoration of normal, binocular visual experience and reverse suturing of the eyelids is especially effective (Movshon, 1976; Olson and Freeman, 1978). In monkeys in which ocular dominance shifts have also been reported but few of the other phenomena, the critical period for the thalamocortical plasticity appears to end in the fifth month of life (LeVay *et al.*, 1980) but effect on layer III cells have been reported as late as 16 months (Blakemore *et al.*, 1978).

Many perturbations of sensory experience during the early critical period can have morphologically detectable effects. Removal of one eye, deactivation of one retina by injections of tetrodotoxin (TTX), or the reduction by eyelid suture of patterned stimuli entering one eye can lead in either species to expansion of the cortical ocular dominance columns related to the normal eye and the constriction of those related to the perturbed eye. The expansion of the non deprived columns, first detected by single unit mapping (see Hubel *et al.*, 1977), has an anatomical correlate in the more extensive distribution of the terminal ramifications of the thalamocortical fibers related to the normal eye and a reciprocal, less extensive distribution of those related to the perturbed eye (Hubel *et al.*, 1977; LeVay *et al.*, 1980). A comparable situation may occur in the rat somatic sensory cortex after removal of whiskers or section of the infraorbital nerve (Jensen and Killackey, 1987). It is customary to interpret these results as indicative of the fibers with normal patterns of afferent activity gaining an advantage over those with perturbed activity in the competition for synaptic space in the maturing cortex. The expanded terminal ramifications related to the normal eye are customarily thought to have failed to withdraw from a more expanded earlier territory rather than to have "sprouted." That is to say, in the course of normal visual cortex development, it is assumed that right-eye and left-eye thalamocortical fibers withdraw from substantially overlapping territories into their own ocular dominance domains as the result of competitive interactions in which each has an equal advantage. The effect appears to be activity-dependent since bilateral blockade of retinal ganglion cell activity by injections of TTX during the critical period in kittens results in a failure of segregation of geniculocortical cortical fiber ramifications into ocular dominance columns and cells recorded in layer IV remain largely binocular (Stryker and Harris, 1986). Differences in the timing of inputs from the two eyes seem to be the key to the normal formation of ocular dominance columns (Wilson *et al.*, 1977; Stryker and Strickland, 1984) since rearing in the dark or bilateral eyelid suture in cats and monkeys still results in a considerable degree of anatomical and physiological ocular dominance segregation (Wiesel and Hubel, 1974; Stryker and Harris, 1986). The spontaneous activity of the retinae *in utero* (Mastronade, 1983) may be the initiating effect, for as pointed out in Section 3, the segregation of geniculocortical afferents into ocular dominance columns begins in the last few weeks of fetal life in monkeys (Rakic, 1976, 1981). Confirmatory evidence that the development of ocular dominance segregation depends upon the timing of influences from each eye comes from experiments in which a surgically imposed strabismus in kittens, although leading to a loss of cortical responses to stimulation of one eye, leads to ocular dominance columns with unusually sharp borders (Shatz *et al.*, 1977; Kalil, 1982).

In order to effect the changes occurring in normal development and those occurring under perturbed conditions in the critical period, it seems necessary for afferent activity to enter the cortex and to be synchronized with the discharges of postsynaptic cortical cells. Diffuse, temporally modulated illumination of one eye (which does not excite cortical neurons) in the critical period of kittens does not result in an ocular dominance shift (Wilson *et al.*, 1977; Singer *et al.*, 1977). Disruption of visual cortical activity by chronic infusion of glutamate in monocularly deprived kittens during the critical period also prevents ocular dominance shifts or changes in other physiological parameters (Shaw and Cynader, 1984). It is thought that this effect results from the enhancement of cortical neuronal discharges by glutamate in the absence of a concomitant increase in geniculocortical afferent discharge. A further set of experiments shows that the postsynaptic cell is also involved in the phenomenon of visual cortical plasticity. After recovering from the inhibition caused by chronic intracortical infusion of the GABA agonist, muscimol, during monocular deprivation in the critical period, cells in the visual cortex of kittens prefer the formerly closed eye (Reiter and Stryker, 1988). That is, less active inputs have become dominant, presumably because conductance or polarization changes are effective even in the absence of action potentials.

Two significant points emerge from the large number of anatomical and physiological studies that have been conducted on the plasticity of the maturing visual cortex. First, the effect is dependent upon the weakening and strengthening of synaptic inputs to cortical neurons. The most overt demonstration of this is the failure of geniculocortical afferent fibers to segregate completely into ocular dominance domains in the cortex of deprived animals, with the concomitant changes in cellular responsivity referred to above. But at the single cell level in these animals, the presence of reduced inward currents associated with EPSPs evoked in the deprived cells (Mitzdorf and Singer, 1980) and the reduction in AMP-dependent phosphorylation of a structural protein, MAP2, in dark-reared animals (Aoki and Siekevitz, 1985), suggest that deprived synapses have become less effective and that structural modifications have been induced. According to Tieman (1988), the terminals of deprived geniculocortical terminals are smaller than normal and end on dendritic spines of smaller than usual size.

The second important fact to be derived from the studies of visual cortex plasticity in development is that it depends upon the presence of a behavioral context. Ocular dominance and orientation changes do not occur, for example, if a monocularly deprived kitten is paralyzed and anesthetized (Buisseret *et al.*,

1978; Freeman and Bonds, 1979; Rauschecker and Singer, 1982). This probably reveals the role not only of the principal afferent systems in the activity-dependent regulation of normal cortical development but also of those afferent systems that are thought to govern state-dependent cortical activity (Section 5.1).

4.2. Activity-Dependent Changes in Cortical Maps

Just as the development of normal cortical function requires sensory experience (Section 4.1), so the maintenance of cortical topographic organization throughout life appears to depend on sensory activity. This has been amply revealed by multiunit mapping of the somatic sensory and auditory areas and by microstimulation studies of the motor area in a number of species, particularly primates (Rasmusson, 1982; Kaas *et al.*, 1983; Merzenich *et al.*, 1983a,b, 1984, 1988; Kelahan and Doetsch, 1984; McKinley and Kruger, 1988; Wall and Cusick, 1986; Calford and Tweedle, 1988; Sanes *et al.*, 1988; Robertson and Irvine, 1989). After removal of a portion of the receptive periphery, such as a finger or part of the cochlea, previously silent representations of adjacent peripheral regions, such as parts of adjacent fingers, are revealed within the expected silent part of the map. These effects occur sufficiently rapidly, often in a matter of hours, that it is unlikely that they are the result of axon sprouting and new synapse formation. Rather, they suggest the uncovering of or strengthening of connections that had been previously suppressed, at least under the conditions of the experiments.

Mutability of somatic sensory cortical maps in the absence of peripheral sensory perturbations may be a normal feature of cortical organization. Studies in which a peripheral representation in the cortex has been mapped over time in single animals (Merzenich *et al.*, 1988) have shown that the representation of a single body region can change dramatically in size and position. That such plastic changes depend upon the patterns of neural activity emanating from the periphery has been revealed in experiments by Clark *et al.* (1988) in which correlated activity of sensory receptors in the skin of two adjacent fingers was obtained by a surgically imposed syndactyly. In these circumstances, the multiunit-derived map of the first somatic sensory area showed a reorganization indicative of a continuous rather than the normal disjunctive mapping of receptive fields across the two fingers. Clearly, as a young primate commences using its fingers as sensory appendages and in a goal-directed manner, similar influences could play a powerful role in modulating cortical maps of the sensory and motor periphery.

The basis for these rapid, activity-dependent changes in cortical maps in the adult primate is likely to be found in the presence of far-flung (Jones, 1990) terminations of single thalamocortical axons that are normally unrevealed by conventional extracellular mapping (Snow *et al.*, 1988) and in the sensitive activity-dependent regulation of transmitter levels, particularly of GABA, in cortical neurons (Hendry and Jones, 1986, 1988). Each of these points is taken up in the following paragraphs.

In adult monkeys, single thalamocortical axons have recently been shown to terminate in area 3b of the somatic sensory cortex in three or more focal domains each approximately 500μm in extent but widely separated from one another and collectively spreading over a cortical domain of 5–10 mm (Jones, 1990). Indirect evidence from retrograde labeling studies indicates that any 500 μm focal zone in area 3b will receive axon terminations from several score thalamic neurons with the same place and modality properties (Jones and Friedman, 1982; Jones *et al.,* 1982a). The input to a cortical focus is thus from a bundle of thalamic axons, not a single axon, and a single thalamic cell can have axon branches in more than one bundle. Given these facts and the far-flung focal terminations of individual axons, the opportunities for multiple, reduplicated, and overlapping representations of the same body part are considerable.

That certain branches of these thalamocortical axons may not reveal themselves in the somatic sensory cortex in conventional multiunit mapping studies is shown by the experiments of Snow *et al.* (1988) on the somatic sensory thalamocortical system of the cat. Here, thalamic neurons with identified receptive fields were shown by antidromic activation from the previously mapped cortex, to project to a wider area of cortex than that containing neurons with receptive fields on the same part of the body as the thalamic neurons.

The powerful influence of the inhibitory neurotransmitter GABA on adult cortical neurons has been demonstrated in both the somatic sensory and visual areas (Sillito, 1975, 1979; Duffy *et al.,* 1976; Tsumoto *et al.,* 1979; Sillito *et al.,* 1980; Hicks and Dykes, 1983; Dykes *et al.,* 1984; Alloway and Burton, 1986). In both areas, when GABA receptors are blocked by the antagonist bicuculline, elements of receptive field structure indicative of normally suppressed inputs from other sources, including other regions of the receptive periphery, are revealed. In the cerebral cortex of normal adult monkeys (Hendry and Jones, 1988) and in the barrel cortex of rodents (Welker *et al.,* 1989), the production of the enzyme involved in GABA synthesis, glutamic acid decarboxylase (GAD), is clearly activity-dependent. In monkeys, blockade of action potentials in the optic nerve by injection of TTX into an eye leads within 4 days to a 50% reduction in cells immunoreactive for GAD and GABA in the relevant eye dominance columns of the visual cortex. This effect is reversible upon restoration of binocular vision (Hendry and Jones, 1986, 1988). It seems inevitable that similar activity-dependent regulation of GABA will occur in other areas of the primate cortex and during development, and that this should have a profound influence not only upon the receptive fields of single neurons but also upon multiunit-derived sensory and motor maps overall.

GABA-mediated inhibition is present early in the neonatal kitten cortex (Sato and Tsumoto, 1984; Tsumoto and Sato, 1985; Shaw *et al.,* 1984; Wolf *et al.,* 1986; Albus and Wolf, 1984). In infant kittens, chronic infusion or iontophoretic application of bicuculline to the visual cortex during the critical period, results in reductions in orientation selectivity, unusually large receptive fields, a loss of direction selectivity, and certain other receptive field changes on the part of visual cortex neurons (Ramoa *et al.,* 1988; Wolff *et al.,* 1984). This seems to be unaccompanied by significant changes in the number of cells stainable for GAD (Bear *et al.,* 1985). Hence, a rearrangement of GABA synapses may have occurred. In rats, chronic whisker trimming from birth leads to enhanced responsiveness of somatosensory cortical neurons, accompanied by reduced direction or angular selectivity (Simons and Land, 1987) but also apparently without changes in the number of GABA cells (unpublished observations). This too seems to indicate a rearrangement or reduced efficacy of inhibitory synapses under the influence of reduced afferent activity.

The implications of these various sets of data indicating the activity-dependent mutability of cortical maps and its possible basis in thalamocortical anatomy and neurotransmitter plasticity, are that similar effects are likely to play a profound role in the shaping of the cortex during its development and maturation. As development proceeds and as neuronal function becomes established and is brought under the influence of peripheral inputs, the effects of the phenomena revealed in the adult would seem to be major determinants of connectional stabilization and the development of normal cortical representations. A potential for representational plasticity is also likely to have been a major factor in the evolution of the cortex, especially in the formation of multiple maps of the receptive periphery.

5. Neurotransmitter-Dependent Effects in Neocortical Plasticity

5.1. Noradrenaline and Acetylcholine

The changes in synaptic efficiency and the permissive role of behavioral context in the phenomenon of visual cortex plasticity (Section 4.2) during postnatal maturation suggest that afferent systems that exert a modulatory effect over cortical synapses, and afferent systems that control behavioral state, should be critically involved. This has led a number of workers to focus on the developmental role of the brain stem and basal forebrain systems that appear to be involved in state-dependent behavior, that innervate all cortical areas, and that release transmitters such as noradrenaline and acetylcholine which seem to have normally a modulatory effect upon cortical synapses (Videen *et al.*, 1984; Singer, 1988).

The permissive role of the brain stem systems in visual cortical plasticity seems confirmed by the studies of Rauschecker and Singer (1982) who found that monocular stimulation of paralyzed kittens, formerly reared in darkness, led to an ocular dominance shift in favor of the stimulated eye only if combined with high-frequency electrical stimulation of the mesencephalic reticular formation and/or medial thalamus. Visual stimulation alone or stimulation of the central core structures alone was ineffectual in promoting the ocular dominance shift in the paralyzed, anesthetized state. This suggests that normally the coincidence of increased inputs from a nondeprived eye and facilitatory effects from the so-called nonspecific afferent systems may be critical in increasing the efficacy of synapses driven by that eye at the expense of synapses driven by the other, deprived eye.

The influence of the neurotransmitter agents released by the diffusely distributed brain stem and basal forebrain pathways on visual cortical plasticity in cats has become a somewhat controversial subject. Initially, Kasamatsu and Pettigrew (1976, 1979) reported that the depletion of noradrenaline in the visual cortex by the intraventricular or intracortical infusion of 6-hydroxydopamine (6-OHDA) during the critical period prevented the expected ocular dominance shift of monocular deprivation. To a considerable extent, however, plasticity could be restored by the infusion of noradrenaline after that of 6-OHDA (Kasamatsu *et al.*, 1979). This effect, implying a critical involvement of the noradrenergic system in cortical plasticity, cannot, however, be duplicated by other forms of noradrenergic depletion such as section of the dorsal noradrenergic bundle or destruction of the locus coeruleus (Daw *et al.*, 1984, 1985a,b; Bear *et al.*, 1983) and only local infusion of 6-OHDA or other agents that deplete noradrenaline seems confirmed as being effective in preventing plasticity (Daw *et al.*, 1983, 1985b,c; Adrien *et al.*, 1985; Bear *et al.*, 1983; Paradiso *et al.*, 1983). If, however, both the locus coeruleus and the cholinergic pathways arising in the basal forebrain are destroyed so as to deplete both noradrenaline and acetylcholine in the cortex, the ocular dominance shift caused by monocular deprivation is indeed prevented (Bear and Singer, 1986). This implies that both the cholinergic and noradrenergic afferent systems may interact in the selective synaptic stabilization that seems to be at the root of experience-dependent cortical plasticity.

5.2. Other Transmitters

Both the inhibitory transmitter GABA and an excitatory, acidic amino acid transmitter acting at the NMDA receptor have also been implicated in the phenomenon of experience-dependent plasticity of the maturing visual cortex in cats. Gene transcription for the GABA-synthesizing enzyme, GAD, is increased during the critical period in kittens (Neve and Bear, 1989) and chronic blockade of GABA receptors by bicuculline during this period appears to affect the normal development of orientation selectivity, center–surround antagonism, and receptive field size in visual cortical neurons (Ramoa *et al.*, 1988). A certain percentage of visual cortical cells that become driven only by the deprived eye after monocular deprivation in the critical period in cats, can be shown to retain a suppressed, deprived-eye input under the influence of bicuculline (Sillito *et al.*, 1981). This may imply that GABA-mediated inhibition is enhanced in deprived eye columns. However, continuous application of the GABA agonist muscimol, when combined with monocular deprivation in kittens, leads to a shift in ocular dominance toward the deprived eye (Reiter and Stryker, 1988).

The blockade of the NMDA receptor by infusion of the antagonist APV into the visual cortex of kittens has also been shown to prevent the ocular dominance shift normally induced by monocular deprivation in the critical period (Kleinschmidt *et al.*, 1987) and APV-sensitive glutamate binding sites increase markedly in the kitten visual cortex during the critical period, falling as it ends (Bode-Gruel and Singer, 1989). APV infusion in kittens also prevents (in part) the reversal of the shift that occurs when, during the critical period, a deprived eye is opened and the formerly nondeprived eye closed (Gu *et al.*, 1989). These observations suggest that many different neurotransmitter systems in the maturing cortex may exert a permissive effect on the plasticity of synaptic stabilization that is present during the critical period. Given the capacity for afferent activity to regulate several neurotransmitter and neuropeptide systems in the adult cortex (see Section 4.2), it is perhaps not surprising to learn that neurotransmitter-dependent modulation of developing cortical synapses is multidimensional, interactive, and probably not based on a single central core afferent system as formerly supposed.

5.3. Potential Inductive Influences of Neurotransmitters and Neuropeptides

Changes occur over time in the appearance and distributions of cortical neurons exhibiting immunoreactivity for particular neurotransmitters, their receptors and neuropeptides during the normal fetal and postnatal development of the monkey cortex (Huntley and Jones, 1990; Huntley *et al.*, 1987, 1988a,b, 1990; Naus *et al.*, 1988). Changes have been described in the distributions of comparable cells and receptors in the cortex of other species (Jonsson and Kasamatsu, 1983; Shaw *et al.*, 1984; Chun *et al.*, 1987; Wilkinson *et al.*, 1983; Shaw and Cynader, 1985, 1986; Cavanaugh and Parnavelas, 1988; Foster and Schultzberg, 1984; Wahle *et al.*, 1987; Shaw *et al.*, 1989). In the visual cortex of monkeys, although cytoarchitecture, as revealed in Nissl stains, is virtually adultlike by the end of the second third of gestation, the distributions of GABA and peptide cells change over time both in laminar pattern and in numbers (Figs. 1–3). Substantial populations of tachykinin cells, for example, are found early in layer V but decline there and show increased numbers in layer IVc. Moreover, certain populations, notably those showing immunoreactivity for choline acetyltransferase or proenkephalin, and not found in adults, have a transient appearance in the developing cortex (Hendry *et al.*, 1987b; Huntley *et al.*, 1988a,b). Also evident at the early stages are substantial populations of neurons immunoreactive for neuropeptide Y, somatostatin, proenkephalin, and GABA in the subplate or white matter beneath the developing cortex. Although never completely disappearing, their numbers decline as development proceeds.

It seems significant that these defined populations of cortical neurons become detectable by immunocytochemistry some time before thalamocortical fibers commence entering the cortex and that the major changes in their numbers and distribution occur as cortical innervation is being established. Results of diverse studies in other systems make it possible that the onset of transmitter release may be associated with inductive influences on cortical cells. The presence of certain monoamine neurotransmitters or acetylcholine has been shown to prevent neurite outgrowth from mammalian neurons *in vitro* while others such as glutamate may promote neurite outgrowth (Lipton, 1988; Mattson, 1988; Patterson, 1988). In other situations, *in vivo,* however, glutamate or a similar acidic amino acid acting via the NMDA receptor may prevent sprouting (Cline *et al.*, 1987). Glutamate in high doses can also induce cell death. These effects appear to be mediated by changes in intracellular calcium levels (Kater *et al.*, 1988; Choi, 1987).

Neuropeptides have often been assumed to play an important role in modulating neural development because of the *in vitro* effects of some (e.g., substance P) on promoting the growth and mitosis of nonneuronal cells (Nilsson *et al.*, 1985) and in eliciting neurite outgrowth from neurons (Bulloch, 1987; Grimm-Jorgensen, 1987). A tantalizing speculation is that the appearance of a partially transient population of neuropeptide cells in the subplate beneath the cortex as afferent axons undergo their waiting period, could be correlated with an inhibitory effect of the peptides on axonal growth (Shatz *et al.*, 1988) which is later released allowing the fibers to enter the cortex as described in Section 3.

There may be a close correlation between the onset of activity in a particular set of cortical afferents or cells and the release of developmentally relevant molecules in the cerebral cortex. In *in vitro* studies, the presence of both neural activity and certain peptides seems necessary to ensure neuronal survival. For example, in cultures of dissociated mouse spinal cord neurons, TTX causes a loss of neurons if applied at a critical developmental period. This effect can be blocked by the addition of vasoactive intestinal peptide (VIP) (Brenneman and Eiden, 1986). The VIP effect is prevented by the addition of a peptide that prevents VIP stimulation of adenylate cyclase. It is assumed that normally the

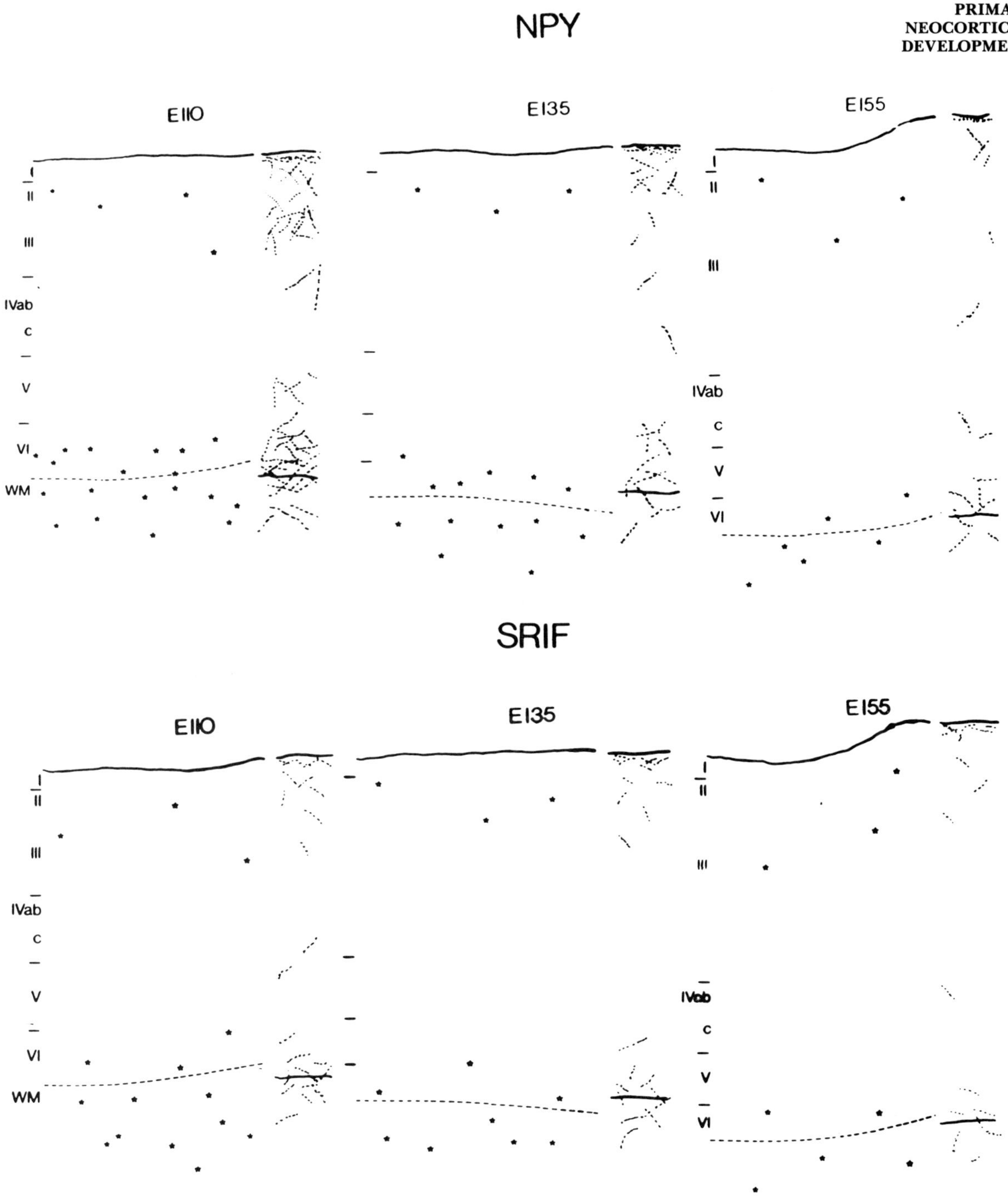

Figure 1. Camera lucida drawings through area 17 of fetal monkeys at E110, E135, and E155, showing the temporal changes in the distribution of somata and fibers immunoreactive for neuropeptide Y (NPY) (top) or somatostatin (SRIF) (bottom), plotted from single 15-μm-thick sections. Bar = 100 μm. From Huntley *et al.* (1988a).

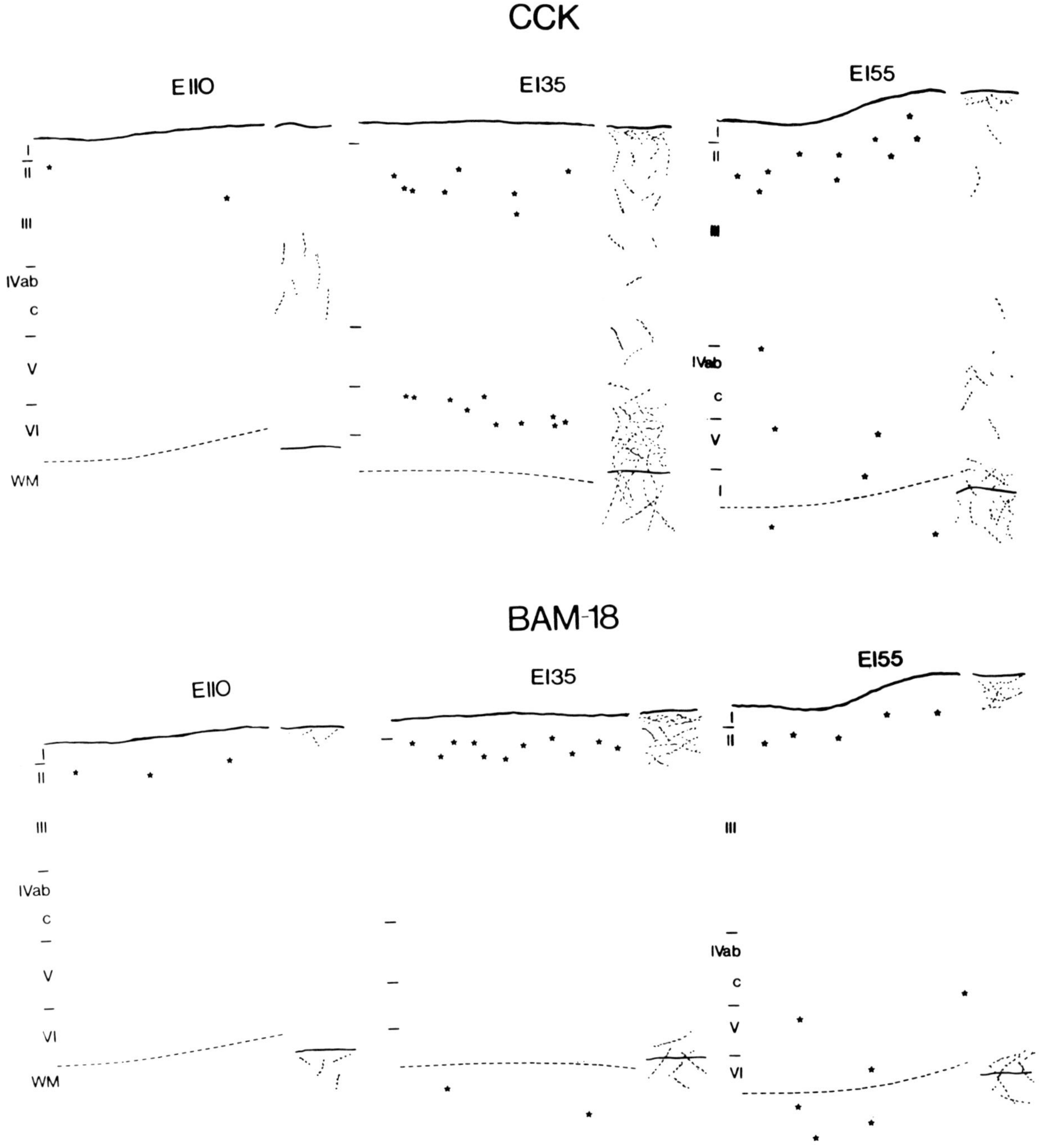

Figure 2. Camera lucida drawings through area 17 of fetal monkeys at E110, E135, and E155, showing the temporal changes in the distribution of somata and fibers immunoreactive for cholecystokinin (CCK) (top) or proenkephalin (BAM-18) (bottom), plotted from single 15-μm-thick sections. Bar = 100 μm. From Huntley *et al.* (1988a).

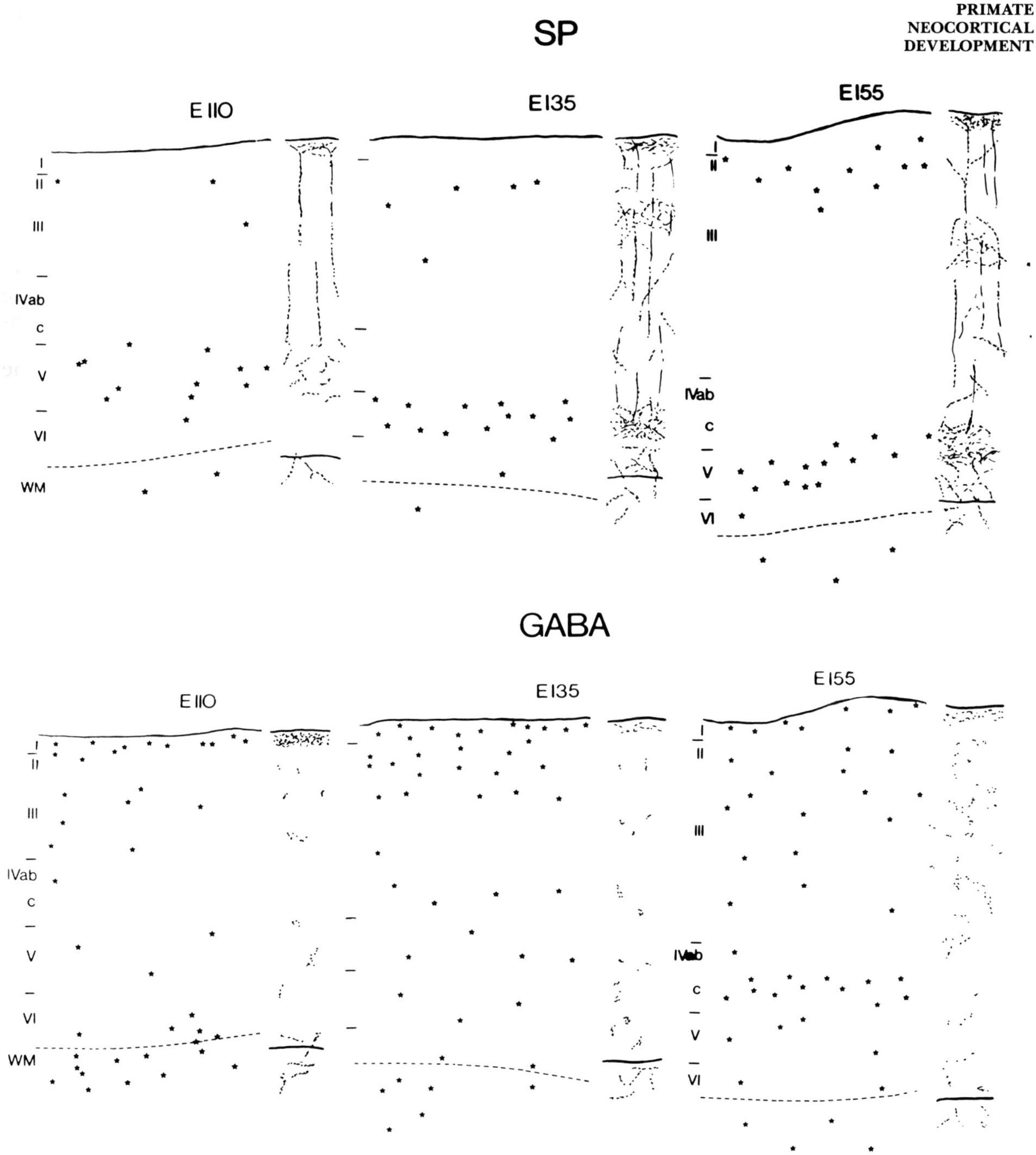

Figure 3. Camera lucida drawings through area 17 of fetal monkeys at E110, E135, and E155, showing the temporal changes in the distribution of somata and fibers immunoreactive for substance P (SP) (top) or GABA (bottom), plotted from single 15-μm-thick sections. Bar = 100 μm. From Huntley *et al.* (1988a).

survival of the cells would be ensured by the release of VIP or some similar molecule from physiologically active nerve terminals.

It is probably too early to identify a precise inductive role for any particular cortical transmitter or neuropeptide in the early development of the cerebral cortex. It is clear, however, that neurotransmitters play a role in the plasticity and stabilization of the cortex during the later, critical periods. Transmitter expression in the mature cortex is activity dependent (Section 4) and in the developing cortex the maturation of neurotransmitter- and neuropeptide-expressing cells is coincident with the arrival of afferent fibers. There is also growing evidence from nonmammalian and *in vitro* systems that not only neural activity but also associated influences mediated by neurotransmitters and possibly neuropeptides can have a profound effect on neural morphogenesis and connection formation. Thus, the interplay of neuronal activity, gene expression for transmitters, peptides, and their receptors, and interactions between these are likely to play major roles in the early establishment of patterns of cortical connectivity and to be potential targets for natural selection in the evolution of the cortex.

6. The Development of Neuronal Phenotype

6.1. The Nature of Neuronal Phenotype

There are multiple expressions of neuronal phenotype. These include somadendritic and axonal morphology, the connections made, and the transmitter synthesized. Certain neurons can also be identified as specific types by the expression of unique antigens (Hendry *et al.*, 1987a; Arimatsu *et al.*, 1987; Watanabe *et al.*, 1989), different calcium binding proteins (Hendry *et al.*, 1989), and specific types of neuronal kinases (Hendry and Kennedy, 1986).

6.2. Cell Morphology

There are many classical descriptions of the development of neuronal morphology in the neocortex. Most of these are focused, however, on the pyramidal neurons. The nonpyramidal, intrinsic, neurons have received relatively less attention.

The development of the cortical pyramidal cells has been repeatedly described in species such as rodents and other laboratory animals. The original description of Ramón y Cajal (1891, 1911) has been expanded somewhat but the basic facts remain unchanged. Ramón y Cajal believed that all pyramidal neurons first appeared in the intermediate zone as piriform-shaped cells, having lost their primitive bipolar form by resorption of the ventricular and aventricular processes. The axon was said to be the first new process to grow out but was quickly followed by the apical dendrite and the migrating cell became bipolar again. This phase, Ramón y Cajal stated, is established for virtually all pyramidal cells in the human cortex by the third month of gestation. The apical dendrite begins to form a tuft of branches on reaching layer I. Basal dendrites develop next, followed by oblique branches of the apical dendrite. Dendritic spines are the last to appear, most being formed after birth in rodents. Axon collaterals arise late in development, commencing on the axons of the larger deep pyramids of layer V, the earliest being those that arise closest to the soma. Collateral formation has not begun on many superficial pyramids at birth in rodents and those of the largest pyramids are still rudimentary. Growth, lengthening, and branching of collaterals occur over a protracted postnatal period.

Ramón y Cajal's succinct description of the ontogenesis of the pyramidal cell has not been bettered and subsequent studies have generally added little more than quantification of points that he made qualitatively. Ramón y Cajal (1894, 1911) was ahead of his time in suggesting that the major influence upon dendritic extension and spine and synapse formation was interaction with afferent axons. He, in fact, saw this as an active process continuing throughout the life of an individual under the influence of functional activity. Pyramidal cells are clearly establishing their fundamental form long before thalamic, associational, or callosal afferents enter the cortex. Thus, the afferents cannot be regarded as the initiators of pyramidal cell process outgrowth. It is logical, however, to see the afferents as potential modulators of the definitive form of tertiary branching and spine formation (Valverde, 1967; Valverde and Esteban, 1968; Ruiz-Marcos and Valverde, 1969; Harris and Woolsey, 1979; Steffen and Van der Loos, 1980), by analogy with the effects demonstrable in deafferentation studies of cerebellar Purkinje cells (Sotelo and Arsenio-Nunez, 1976; Sotelo *et al.*, 1975).

Before thalamic, associational, and callosal fibers enter, the rodent cortex at least, possesses very few synapses. These are located exclusively on each side of the cortical plate in the intermediate and marginal zones (Kristt and Molliver, 1976). There are none in the cortical plate itself in which the young pyramidal neurons are growing their apical dendrites. The origins of the earliest cortical synapses are not clear. Marin-Padilla (1971, 1988) in the cortex of fetal cats and humans shows extrinsic afferent fibers in the marginal and intermediate zones from the earliest times when the cortical plate is first forming. It is possible that these are catecholamine-containing fibers arising in the brain stem. Certainly the latter appear in the cortex earlier than other extrinsic afferents (Schlumpf *et al.*, 1980; Caviness and Korde, 1981; Lidov and Molliver, 1982; Verney *et al.*, 1982) and they or the transmitters they release could exert an inductive influence over dendrite formation in pyramidal cells. Other sets of fibers that have been thought to influence the apical dendritic development of the pyramidal cells are those forming the considerable plexus that occupies layer I (the marginal zone) from the earliest times. These fibers arise in part from the Cajal–Retzius cells in layer I itself and partly from the ascending axons of Martinotti cells situated in the intermediate zone beneath the cortical plate (Ramón y Cajal, 1891, 1911; Marin-Padilla, 1971, 1984). The apical dendrite of every pyramidal cell grows rapidly to layer I and forms a considerable tuft of branches there, before the development of significant basal and oblique branches. Hence, it is easy to conjecture that the layer I axons and early generated layer I neurons (Section 6) have an inductive influence over the apical dendrites.

The maturation of secondary and tertiary dendritic branches and dendritic spines appears to proceed contemporaneously with the arrival of thalamic, associational, and callosal fibers and the establishment of afferent innervation. In all species studied, including primates, layer V pyramidal neurons appear to be in advance of layer III pyramidal neurons (Marin-Padilla, 1969, 1971; Morest, 1969; Boothe *et al.*, 1979; Wise *et al.*, 1979). Dendrite outgrowth appears to be laminar specific in the sense that it reflects that to be expected for cells of a particular layer and does not proceed from an initial overproduction that is subsequently pruned. In the monkey visual cortex it seems to be complete by postnatal week 8 (Boothe *et al.,* 1979), but in the human motor cortex it may not be complete until postnatal month 8 (Marin-Padilla, 1970b); in rats and cats it is complete by the end of postnatal week 2 or 3 (Wise *et al.,* 1979; Miller, 1986).

The other indication of pyramidal cell dendritic maturation is the production of dendritic spines, the presence of which may be an index of synaptic development since they represent the major postsynaptic site on adult pyramidal cells. Their appearance seems to follow the induction of secondary and tertiary dendrites; the number is initially small but this is followed by a rapid overproduction that is reduced to the normal adult numbers over a protracted period, probably under the influence of afferent activity (Valverde, 1967; Valverde and Esteban, 1968; Ruiz-Marcos and Valverde, 1969).

In the visual cortex of monkeys, spine production by pyramidal cells commences prenatally but increases rapidly only after birth, reaching a maximum with numbers some 40% greater than adult levels at 8 weeks of age. Thereafter, there is a sudden rapid decline followed by a more gradual loss continuing up to postnatal month 9 (Boothe *et al.,* 1979). A comparable sequence, occurring over the first month of postnatal life, occurs in the rat somatic sensory cortex (Wise *et al.,* 1979). In the rat motor cortex, the differences in spine distribution that mark slow and fast pyramidal tract neurons are established by the end of postnatal week 4 (Oka *et al.,* 1985).

Apart from the "special" or Cajal–Retzius cells of layer I, Ramón y Cajal and his successors paid little attention to the morphological differentiation of the nonpyramidal intrinsic cortical neurons. Ramón y Cajal mentioned that nonpyramidal cells in layers II–VI probably developed from an early apolar stage and that their development proceeded in parallel with that of the pyramidal neurons with which they begin establishing connections. A similar conclusion has been drawn by Marin-Padilla (1969, 1970a,b) in his study of the development of the human motor cortex.

Recent studies also tend to support this conclusion. The small spiny nonpyramidal cell of layer IV in monkeys and rats undergoes a sequence of dendritic and dendritic spine development, including an initial overproduction, over a time frame similar to that of the pyramidal cells and its axonal development appears to proceed over the same time period (Boothe *et al.,* 1979; Meyer and Ferres-Torres, 1984). The several varieties of nonpyramidal cell also appear to acquire their definitive dendritic and axonal forma *pari passu* with the pyramidal cells of the layer in which they lie, hence the development of layer V cells in general precedes that of layer III cells. However, the large basket cells of both layers III and V may be somewhat in advance of other morphological forms (Marin-Padilla, 1970a,b; Meyer and Ferres-Torres, 1984).

The setting up of definitive synaptic connectivity by the nonpyramidal neurons may lag slightly behind that of the afferent innervation. In neonatal cat motor cortex, for example, EPSPs can be generated in pyramidal tract neurons by stimulation of the deep cerebellar nuclei from birth but longer-latency IPSPs, which can be taken to be due to cortical interneurons, can only be demonstrated after 3 days (Oka *et al.,* 1985).

One feature recognized by all modern authors and evident in Ramón y Cajal's drawings is the presence of significant numbers of dendritic spines on nonpyramidal cells that will eventually be spine-free in the adult (Marin-Padilla,

1970a, b; Jones, 1975; Lund *et al.*, 1977; Boothe *et al.*, 1979). These appear to be lost over the same time period as the normal, developmentally regulated loss of spines on spiny neurons.

In his studies on the development of the cat cortex, Marin-Padilla (1969, 1970a,b, 1971, 1972, 1988) reported that nonpyramidal neurons appearing early in development in layer I and in what would now be called the subplate zone are the derivatives of a primitive set of young neurons that colonize the marginal zone before being separated into two bands by the arrival of the successive waves of subsequent neurons to form the cortical plate. [^{3}H]-Thymidine studies in rats and cats (Rickman *et al.*, 1977; Konig *et al.*, 1975; Raedler and Raedler, 1978; Raedler *et al.*, 1980; Luskin and Shatz, 1985a,b; Peduzzi, 1988) generally support the view that the earliest neurons to arrive in the primordial cortex are split into superficial and deep strata by the later-arriving cortical plate cells which follow the now classical inside-out pattern of migration from the ventricular and subventricular zones (Rakic, 1974). In the visual cortex of fetal cats the early arriving neurons are generated at E24–E30 whereas cortical plate cells are generated at and subsequent to E30. In monkeys, the earliest neurons probably reach the cortex prior to E70.

The dendritic and axonal maturation of the layer I and subplate zone cells was reported by Marin-Padilla to be far in advance of that of cortical plate neurons and the deep group, which Marin-Padilla considers remain in the white matter into adulthood have a very close similarity to the neuropeptide Y-immunoreactive cells that are found there in large numbers in the fetus and adult (Hendry *et al.*, 1984; Chun *et al.*, 1987; Huntley *et al.*, 1988,a,b) and which probably represent cells of the subplate (Chun *et al.*, 1987). Luskin and Shatz (1985a,b) have reported that many of the latter die in the course of development in the cat because [^{3}H]thymidine labeling of these cells, achieved by injections at E24, does not persist to any extent beyond 1 month postnatally. It is these cells that we have previously considered (Section 2.4) as potential inductive influences on cortical cell and connectional maturation and as guides to fiber outgrowth.

The cells of layer I are also reported to undergo a reduction in numbers with age. It is customary to regard this as reflecting the putative loss of Cajal–Retzius cells with age. The cells described in layer I by Marin-Padilla, however, fall into two groups: Cajal–Retzius cells and more typical nonpyramidal cells. These each have different developmental histories (Ramón y Cajal, 1911; Marin-Padilla, 1984) and the latter clearly persist into adulthood as GABAergic neurons (Hendry *et al.*, 1987a). The fate of the Cajal–Retzius cells is still, however, something of a mystery. After birth in monkeys they cannot be stained for the calcium binding proteins that distinguish them throughout fetal life (Huntley *et al.*, 1990a) and they cannot be detected by Golgi stains. They must either disappear, undergo substantial changes in morphological phenotype, or become so widely dispersed as to be undetectable (Marin-Padilla, 1984, 1988).

The early appearance and early maturation of a set of nonpyramidal neurons on either side of the cortical plate, the projection of axons of cells in the deep stratum to the superficial stratum, the association of the two strata with the early arriving, presumed brain stem, afferents (Marin-Padilla, 1984, 1988; and see Section 2.4), the appearance of the earliest cortical synapses in the two strata (Molliver *et al.*, 1973; Kostovic and Molliver, 1974; Kristt and Molliver, 1976), plus the earlier maturation of the pyramidal cell dendrites on reaching layer I, are all facts that raise the issue of whether the early arriving and maturing subplate and layer I cells exert an inductive influence over the morphogenesis of neurons that arrive later and develop from the cortical plate proper.

One indication of the potential inductive influence of layer I and the subplate cells may come from the transient, early expression of the extracellular matrix molecule, fibronectin, in these sites, for such a molecule could influence both neuronal assembly and process elongation (Chun and Shatz, 1988b). The possibility that many of the early formed groups of cells in layer I and the subplate die makes conjectures of their inductive role doubly attractive. It is incorrect, however, to assume that all marginal and subplate cells die, for many neurons persist in layer I and the subcortical white matter into adulthood, especially in primates, and both contribute to the circuitry of the cortex (Hendry *et al.,* 1984, 1987; Jones *et al.,* 1988; Chun and Shatz, 1989b). The remaining white matter cells represent an integral part of the cerebral cortex and may with justification be referred to as its deepest layer. To see either these or the layer I neurons as a completely transitory phenomenon concerned only with regulating the development of other cells does not seem justified, although this could undoubtedly be one of their functions during development.

6.3. Connection Formation

6.3.1. Laminar Specificity of Efferent Connections

Cortical cell phenotype, as expressed in the connections of efferent cells, rather than in their morphology, is laminar-specific. In adults, although there are some species and areal differences, a pyramidal neuron of a particular layer can be predicted to have a limited number of targets (e.g., Jones, 1984). Subcortical projections, for examples, arise only from infragranular cells: layer VI neurons project to thalamus or claustrum, layer V neurons project to other subcortical sites. This laminar specificity appears to be fixed early in fetal life and may be specified even before postmitotic cells leave the neuroepithelium. In rats and cats, the precursors of many layer VI neurons send their axons to the thalamus while the majority of layer V neurons, although migrating more or less contemporaneously, send theirs to other subcortical sites. In monkeys, neurons destined for the supragranular layers of the prefrontal cortex can have an axon in the contralateral hemisphere even before the cell has completed its migration (Schwartz and Goldman-Rakic, 1986). In rats, cells destined for layer VI can have an axon in the thalamus before the neurons arrive in that layer (Schreyer and Jones, 1982). No mistakes are made. Aberrant, subcortically projecting cells are never found in the supragranular layers of normal fetal or adult animals. Even when cortical development is perturbed, cells appear to make efferent connections appropriate for their predestined lamina. Malpositioning and distorted dendritic morphology can be brought about in cortical neurons by destroying glial guides to their migration by cytotoxic drugs or irradiation, or by transplanting fetal neurons to a host, but the neurons form extrinsic connections appropriate for the layer in which they would have normally found themselves (Jones *et al.,* 1982a; Yurkewicz *et al.,* 1985; Floeter and Jones, 1985; Jensen and Killackey, 1987). Similarly, the majority of heterochronically transplanted neurons in the ferret cortex (Section 2.3) develop connections appropriate for the layer for which they were originally destined (McConnell, 1988) and surviving layer V and VI cells never reconstitute layers II–IV when the precursors of the latter have been destroyed by cytotoxic drugs (Yurkewicz *et al.*, 1984).

The early growth of axons toward their targets, even before migration is complete, makes it unlikely that afferent influences conveyed by thalamic, callosal, or corticocortical fibers play a role in inducing the formation of these projections. This is reinforced by the fact that corticofugal fibers from the monkey visual cortex innervate the lateral geniculate nucleus, pulvinar, and tectum some 10–20 days prior to the growth of geniculocortical axons into the visual cortex (Shatz and Rakic, 1981). It cannot be ruled out, however, that the so-called nonspecific afferents from the brain stem and basal forebrain and which appear to reach the developing cortex very early (Schlumpf *et al.*, 1980; Caviness and Korde, 1981; Lidov and Molliver, 1982; Verney *et al.*, 1982; Crandall and Caviness, 1984) influence the outgrowth of cortical efferent axons, especially since they appear to be concentrated in the subplate.

6.3.2. Collateral Elimination

The connectional phenotype of extrinsically projecting cortical neurons can be changed to a certain extent in development. Layer III neurons projecting through the corpus callosum in the sensory-motor, auditory, and visual areas of developing cats, rats, and monkeys, for example, are more numerous and more widely distributed than in the adult (Innocenti *et al.*, 1977; Innocenti and Caminiti, 1980; Innocenti, 1981; Ivy and Killackey, 1981, 1982; Feng and Brugge, 1983; Cornwell *et al.*, 1984; Innocenti and Clarke, 1984b; Killackey and Chalupa, 1986; Dehay *et al.*, 1986; Chalupa and Killackey, 1989; Chalupa *et al.*, 1989). Corticocortical cells in the developing auditory cortex of the cat or monkey may send axons atypically to the vicinity of the ipsi- and contralateral visual cortex (Innocenti and Clarke, 1984a; Dehay *et al.*, 1984). Layer V cells in areas not definitively containing corticospinal neurons can contribute axons to the corticospinal tract in developing rodents (Stanfield and O'Leary, 1985a,b; Crandall *et al.*, 1985; Schreyer and Jones, 1988a,b). As discussed earlier (Section 2.3), the cells appear to break all such aberrant projections as development proceeds, but the cells survive and appear to be preserved from atrophy by the preservation of ipsilateral corticocortical axon branches (O'Leary *et al.*, 1986; Innocenti *et al.*, 1986; Chalupa and Killackey, 1989).

It has become clear that there is a substantially greater number of cortical efferent axons growing toward a particular target in the developing brain that is found in the comparable tracts of the mature brain and spinal cord. In the corpus callosum of monkeys and cats, for example, 70% of the axons are reported to disappear between birth and adulthood (LaMantia and Rakic, 1990). In the corticospinal tract of rats (Schreyer and Jones, 1988; Gorgels *et al.*, 1989) and the pyramidal tract of hamsters (Reh and Kalil, 1981, 1982), corresponding losses of axons occur in the later phases of development. These appear to be mainly reflections of the breaking of inappropriate collaterals since they occur after the major phases of cortical cell death are over.

Although from the foregoing it seems that afferent activity can be ruled out as a determining factor in directing cortical cell axon outgrowth and target finding, it may play a substantial role in the stabilization of these connections and in influencing the number of neurons that express a particular extrinsic connectional phenotype in the adult. In the visual cortex of cats, visual experience in the critical period appears to influence the number of callosally projecting cells. Removal of one eye, monocular eyelid suture, or induction of a strabismus appears to stabilize transient callosal projections that would otherwise be lost from parts of the visual field representation outside that of the central field (Shatz, 1977a,b; Innocenti and Frost, 1979, 1980; Lund and Mitchell, 1979a; Cusick and Lund, 1982; Olavarria and Van Sluyters, 1985). This is accompanied by a more widespread distribution of callosal fiber terminals (Lund *et al.*, 1978). Bilateral visual deprivation by dark rearing or bilateral eyelid suture, results in a reduction of callosally projecting cells to as much as 50% less than normal and a corresponding reduction in callosal axon terminations (Lund and Mitchell, 1979b; Innocenti *et al.*, 1985). Prior visual experience can diminish or prevent the experimentally induced effect. These are very powerful indications of the maintenance of an inappropriate collateral, possibly by the formation or stabilization of synapses that would otherwise not have formed or been transient. In monkeys, comparable experiments have not been performed. It is reported, however, that bilateral eye removal in fetal monkeys, although resulting in an area 17 that is smaller than normal, does not affect the development of normal distribution of callosally projecting neurons (Dehay *et al.*, 1989).

Within the cortex itself, some redistribution of neurons that form the intrinsic long horizontal connections of a cortical area (mostly formed by collaterals of pyramidal cell axons) may occur. This is indicated by the development of the typical patchy distribution of the cells forming such connections over the first 3 weeks of postnatal development in the cat visual cortex (Price and Blakemore, 1985, 1986; Luhmann *et al.*, 1986).

The extent to which a cell was originally committed to, say, the corticocortical instead of the commissural of corticospinal fate and the extent to which epigenetic factors led it to adopt the former, are still matters for conjecture. There is recent evidence to suggest that in the case of corticocortical neurons in the cat auditory cortex, the axon branches to "appropriate" sites may be the primary ones and the axon branches to inappropriate sites, such as the vicinity of the visual cortex, never enter the gray matter (Clarke and Innocenti, 1986). Hence, such an axon branch may never be in a position to make synaptic connections.

It is possible that these early projections toward atypical targets are simply aberrant reflections of an early growing axon's propensities to branch and to follow its neighbors or some pathfinder axon, in which case the branches might be intrinsically incapable of making connections in the atypical site. On the other hand, they might be actively prevented from doing so by the early formation of synapses in an appropriate area by the appropriately directed axon branch or branches. Aberrant branches could also be kept out of an inappropriate target by being at a competitive disadvantage with the normal afferents to that target. The evidence that inappropriate subcortical collaterals may be preserved by transplantation of pyramidal cells to a new site (Section 2.3). is an indication that one constraint on the development of connectional phenotype may be intrinsic to the cortical area in which a pyramidal cell lies. However, the evidence from Section 2.3 that thalamic afferent fibers may, when given a competitive advantage, colonize an inappropriate thalamic nucleus and establish functional connections, suggests that under appropriate circumstances a preserved collateral of a subcortically projecting axon may be able to form functional connections.

6.3.3. The Nature of Axon Guidance from the Developing Cortex

This innate programming of cortical axon growth toward the appropriate target is undoubtedly influenced by axon–substrate interactions which can currently only be guessed at. Morphologically definable guidance channels for axons, formed by specifically oriented neuroglial cells or their processes (Silver *et al.*, 1987), do not seem relevant at the stages of early outgrowth and their relevance even at later stages has been questioned (Valentino and Jones, 1981; Valentino *et al.*, 1983; Schreyer and Jones, 1982). What seems more likely is some kind of interaction between the axonal membrane and molecular constituents of the growth substrate which, if temporally regulated, could serve to direct bundles of growing axons from each newly generated wave of young cortical neurons toward a particular target.

Undoubtedly, the early environment in which migratory and newly arrived cortical neurons find themselves represents a favorable substrate for axon and the dendrite extension. This may be provided by the cells themselves and/or local neuroglial cells, including the population of subplate cells, reflecting the permissive role of epithelial cells in early neurite extension in many nonmammalian nervous systems and in other parts of the mammalian nervous system. The relevant molecules probably include various kinds of cell adhesion molecules such as NCAM and N-cadhedrin (Edelman, 1988a; Takeichi, 1988) and possibly molecules such as laminin and fibronectin (Akers *et al.*, 1981; Hatten *et al.*, 1982; Rogers *et al.*, 1983; Smalheiser *et al.*, 1984; Cohen *et al.*, 1986; Hall *et al.*, 1987; Riggott and Moody, 1987; Letourneau *et al.*, 1988). Of these, fibronectin or a similar molecule has been found to be expressed in the subplate of the cat and mouse visual cortex (Stewart and Pearlman, 1987; Chun and Shatz, 1988b), although apparently at a time after the earliest wave of cortical axons has grown out.

Growth cone protrusion and route finding by growing axons is also likely to be determined by molecules associated with the substrate and/or released by adjacent cells. These molecules may include transiently expressed glycoproteins or changes in their molecular composition, e.g., in the degree of sialation of NCAM (Edelman *et al.*, 1983; Hoffman and Edelman, 1983; Chuong and Edelman, 1984; Sunshine *et al.*, 1987; Rutishauser *et al.*, 1988). Some of them may represent permissive or nonpermissive substrates that require growth cone recognition and selection of a route, or contact inhibition and direction into a new route. Indirectly, they may be expected to affect the production and assembly of cytoskeletal and membrane proteins in the growing axon and growth cone (Spiegelman *et al.*, 1979; Bovolenta and Mason, 1987; Gordon-Weeks, 1989). None of these factors have yet been explored in the outgrowing cortical efferent systems.

An interplay of mechanical and molecular factors may be expected as axon growth proceeds (see Dodd and Jessell, 1988). Mechanical factors could include guidepost cells which, when recognized, direct growing axons into a particular path, and the presence of pathfinder axons or other processes upon which later growing axons can selectively fasciculate (Jessell, 1988; Rubin, 1988). There is now good evidence that in the growth of cortical efferent pathways, a few axons precede the main wave of growing axons and that the later axons fasciculate onto them (Schreyer and Jones, 1982; McConnell *et al.*, 1989; Hogan and Berman, 1989). The earliest path of this kind may be formed by the processes of the early maturing subplate neurons which in cats extend down into the internal capsule, thalamus, and superior colliculus, providing routes that seem to be followed by the axons emanating from cells of layers V and VI (McConnell *et al.*, 1989).

Growth to a target may require the presence of trophic factors comparable, for example, to nerve growth factor, and recognition of the target which may also be effected by chemotrophic guidance (e.g., Lumsden and Davies, 1983, 1986). Similarly, the development of topographic order and map assembly at the target will necessitate the presence of positional cues of some kind. The final events in cortical efferent axon growth are obviously axon–cell recognition, synaptogenesis, and synaptic maintenance and stabilization (Changeaux and Danchin, 1976; Schmidt, 1985).

Although temporal changes in the expression of a very small number of molecules could be instrumental in controlling all of these events, it is likely that the processes outlined are multifactorial and later expression of a critical factor may depend on effective recognition and interchange of signals at an earlier stage. Very little work has been done in this field to date and none has been conducted on the cortical efferent systems.

6.4. Transmitter Status

Cortical cell phenotype as manifested by transmitter expression, like connectional phenotype, also appears to be specified early in development. GABAergic and peptidergic neurons can be identified by immunocytochemistry in the cortex some time before all cellular migration has ceased (Chun *et al.*, 1987; Fairén *et al.*, 1985; Huntley *et al.*, 1988a,b) and even migratory cells may be immunoreactive for peptides (Schwartz and Goldman-Rakic, 1986). GABA cells of a particular layer in the mouse and ferret are generated contemporaneously with non-GABA cells of the same layer (Fairén *et al.*, 1986; Peduzzi, 1988) and presumably migrate together. This would further imply that at least the large GABAergic component of the cortical interneuronal population arises sequentially with projection neurons of the same layer and not, as sometimes suggested, from a separate later wave of proliferation and migration.

Neuropeptide cells display some redistribution even in the later stages of cortical development in fetal monkeys and cats (Chun *et al.*, 1987; Huntley *et al.*, 1988a,b, 1989; Yamashita *et al.*, 1989). The concentration of neuropeptide Y cells is higher in the fetal white matter or subplate than in that of the adult. From their morphology and concentration in the white matter, these neuropeptide Y cells appear very similar to the early appearing cells of the intermediate zone which can be identified before the formation of a cortical plate (Marin-Padilla, 1971). As mentioned earlier (Section 2.4), a significant proportion of these cells has been shown to die during development of the cat visual cortex (Chun *et al.*, 1987). Tachykinin-immunoreactive neurons in the monkey cortex first appear in layer V and remain there in most areas but in the visual cortex they are later much more predominant in other layers (Huntley *et al.*, 1988a). It seems doubtful that the cells of the visual area undergo a late secondary migration, so they may either die or cease producing the peptide in detectable amounts. Given the fact that postnatal cell death appears to affect mainly the supragranular layers (Section 6.5), the latter seems the more likely possibility. It is possible that in these and other cells there may be a switching of transmitter production: choline acetyltransferase, tyrosine hydroxylase, and proenkephalin-immunoreactive neurons, for example, are present in fetal monkey cortex between E110 and E150 but they have not been demonstrated in adults (Hendry *et al.*, 1987b; Huntley *et al.*, 1988a,b). The extent to which the onset of neuronal activity could modulate the expression of neuroactive substances in developing cortical neurons has not been explored. Since relatively modest forms of visual deprivation in adult monkeys can lead to significant alterations in tachykinin expression in the visual cortex (Hendry *et al.*, 1988b; Section 4.2), similar effects could be responsible for a fine tuning of cortical cell transmitter expression as function is becoming established during development.

A number of neurotransmitter receptors and related potential markers of synaptic circuitry undergo transient or changing expression in the developing and maturing cerebral cortex of several species. The transient expression of acetylcholinesterase activity in the somatic sensory and visual cortex of rats (Kristt and Waldman, 1982; Kristt and Kasper, 1983; Robertson *et al.*, 1985; Prusky *et al.*, 1988), the visual cortex of cats (Bear, 1985), and the prestriate cortex of monkeys (Kostovic and Rakic, 1984) is well known. In the developing cortex of rats and cats, its distribution coincides with a similar transient expression of nicotinic acetylcholine receptors (Prusky *et al.*, 1988a,b). In the kitten visual cortex, a band of acetylcholinesterase activity and [^{3}H]nicotine binding coincides with layer IV, the major layer of developing thalamic fiber terminations. Both markers disappear when the cortex is deafferented but survive destruction of cortical neurons by quinolinic acid lesions, suggesting that the nicotinic sites may be on afferent fiber terminals (Prusky *et al.*, 1988a,b).

Nicotinic binding increases in layer IV as binding for markers of muscarinic cholinergic sites declines. At the peak of the critical period, when activity ensures the normal consolidation of cortical circuitry, the two are about equal in layer IV (Shaw *et al.*, 1984). Thereafter, muscarinic binding declines steeply in the supra- and infragranular layers. Nicotinic binding declines in layer IV in the later weeks of early postnatal development, coinciding with a similar decline in acetylcholinesterase activity. These results strongly suggest a correlation between the presence of cholinergic innervation and the stabilization of geniculocortical innervation under activity-dependent conditions during the critical period and may be a basis for the permissive role exerted by cholinergic innervation under deprivation-induced cortical plasticity (Section 4).

Developmentally regulated alterations occur in the distributions of many other neurotransmitter receptors in the visual cortex of kittens during the critical period (Shaw *et al.*, 1986). Similar changes have not yet been reported in primates. At present, the significance of these changes in normal neocortical development are uncertain but, given the powerful effect mediated by monoamine and cholinergic afferent systems in experience-dependent plasticity, the evidence that GABA and acidic amino acids may also effect normal maturation, and the putative roles of peptides, it is to be assumed that these developmentally regulated changes are important elements in the establishment of normal cortical circuitry.

6.5. Cell Death in Cortical Development

The degree of cell death that occurs during development of the primate cortex is not fully known. Even in other common experimental animals, it has been variously estimated (Finlay and Slattery, 1983). In a study of the monkey visual cortex, O'Kusky and Colonnier (1982b) reported a 16% loss of neurons, primarily from layers I, II, IVc, V, and VI, between birth and maturity. Experimental data from other systems would suggest that there is likely to be an even greater loss at about the time connections are being formed (see Cowan, 1973). In monkeys, this would be well into the prenatal period. In hamsters between postnatal day 5, when the cortex is just acquiring laminae and cell migration is ending, and maturity, there is a 20–25% cell loss in the upper layers and approximately 5% in the deeper layers. The deeper layers mature first so a major wave of anticipated cell death could occur in them prior to day 5.

In rodents, a considerable degree of cell death has been reported in the first few days of postnatal life. The loss of cells may be as high as 50% in mice and is said to affect predominantly the superficial two or three layers of the cortex (Leuba *et al.*, 1977; Heumann *et al.*, 1973, 1978; Finlay and Slattery, 1983). This may simply reflect the later maturation of the superficial layers. Similar reductions in cell numbers may have occurred prenatally in the deeper layers. Cell loss of the kind reported in the cortex of rodents may reflect preprogrammed cell death since the few manipulations of connectivity that have been carried out have not demonstrated enhanced cell death due to destruction of the afferent or efferent pathways of the cortex (Ramirez and Kalil, 1985; Windrem *et al.*, 1988; Finlay and Pallas, 1989).

We are obliged to conclude, therefore, that although histogenetic cell death is as prevalent in the developing cortex as elsewhere in the developing brain, its role in the development of cortical architectonic structure and the establishment of connectional order is essentially unknown. Variance in the degree of cell death and its dependence on the establishment of connectivity would, however, be a major target for evolutionary processes that determined the extent of the cortex.

7. The Development of the Neuropil

In examining a section of the cerebral cortex of a fetal mammal, it is impossible to avoid noticing that the cell density is particularly high. The numerical density of cells in the primate cerebral cortex has not been quantified in the critical prenatal stages of development. However, between 6 months of age and maturity, neuronal density in the developing visual cortex of the macaque monkey decreases by approximately 30% without a corresponding loss in neuronal numbers (O'Kusky and Colonnier, 1982a,b). Zilles *et al.*, (1986) have recently shown how the volume of the neuropil increases markedly during the ontogeny of areas 17 and 18 in the human visual cortex.

The neuropil obviously expands as dendrites and axons grow and branch and as synapses are formed. The regulation of the growth of dendrites and axons has been considered above. As this proceeds, the density of synapses undergoes an enormous increase which in the monkey continues well into postnatal life (O'Kusky and Colonnier, 1982b; Rakic *et al.*, 1986). The density of synapses in the macaque monkey visual cortex increases by 34% from birth to 6 months of age as synaptogenesis presumably occurs but falls off with age, and the density is 95% less at maturity than at 6 months of age (O'Kusky and Colonnier, 1982b).

In the study of Rakic *et al.* (1986) which included visual, sensory-motor, and prefrontal areas, the density of synapses increased rapidly in the last 2 months of gestation, reaching 15–20 per 100 μm^2 at birth. This continued to rise for several months after birth, between 2 and 4 months reaching 30 per 100 μm^2 in most areas and 40 per 100 μm^2 in area 17. By maturity, the density in all areas had fallen again to that seen at birth. The initiation and rate of progression of this process are not affected by visual experience (Bourgeois *et al.*, 1989). An interesting feature of this study is the evidence that the growth in synaptic numbers proceeds concurrently and at the same rate in all areas and in all layers of any area. It therefore does not appear to follow the sequence of maturation of the cortical layers or that of ingrowth of the fibers, although the initiation of synaptogenesis may be expected to follow such a sequence. Similarly, these data show that synaptogenesis in the association areas does not lag behind that in the primary sensory areas, despite the well-known delayed onset of myelination in the former areas (Flechsig, 1905; Vogt and Vogt, 1919), which has often been considered to betoken a sequence of developmental maturation proceeding from the primary to the association areas.

An increase followed by a decrease in synapse density has also been reported in the visual cortex of nonprimates (Winfield, 1981). These changes in synaptic density appear to be susceptible to activity-dependent modulations. In the visual cortex of adult cats raised with binocular deprivation from birth, there are significant reductions in synaptic density manifested as a prolongation of the initial increase and a reduced rate of decrease so that the density in adults is greater than normal (Winfield, 1983). Monocular deprivation results in no changes in synaptic density, presumably because of compensatory effects of the nondeprived eye.

The apparent overproduction of synapses is accompanied by a progressive increase in the number of dendritic spines on pyramidal neurons (Section 6.3). This has been quantified in a number of species and in the rat the onset of increased spine production has been shown to coincide with the arrival of extrinsic afferent fibers at a particular layer. Thus, layer V pyramidal cells are in advance of layer III pyramidal cells (Wise *et al.*, 1979). However, spine density continues to rise long after extrinsic afferent connectivity is established. This may be indicative of the continuing synaptogenesis that underlies the vast increase in cortical synapses as development proceeds. Spine density declines with time and is 40% less in adult monkeys than at 8 weeks of age (Boothe *et al.*, 1979). This may be further indicative of the reduction of synapse density with age.

Spine synapses show the greatest reduction in the developmentally regulated loss of synapses that occurs in infant monkeys (Rakic *et al.*, 1986). In the prefrontal cortex at birth the ratio of asymmetrical to symmetrical synapses is 4 to 1. It becomes 7 to 1 at 4 months of age but is reduced to 4 to 1 again in young adulthood. In the motor cortex the ratio increases from 7 to 1 to 24 to 1 and then declines to 7 to 1 again. Afferent activity may be one of the major constraints on definitive spine numbers since eye removal in infant animals can significantly reduce spine numbers in the visual cortex (Valverde, 1967; Valverde and Esteban, 1968; Ruiz-Marcos and Valverde, 1969; Rothblatt and Schwartz, 1979).

The influences operating to reduce synaptic density as the cerebral cortex matures must be preeminently those associated with the shaping of topographic patterns and occurring under the influence of afferent activity. Much has been written about the nature of competition for synaptic space during development and the stabilization of synapses with the onset of function (reviewed in Changeaux and Danchin, 1976). It is doubtful that the cerebral cortex will prove to be different from other brain regions in regard to these fundamental mechanisms. The magnitude of the changes effected in synaptic density during cortical development and maturation is further evidence of how manifest is the influence of afferent activity over cortical structure.

8. Developmental Constraints on the Evolution of Neocortical Structure

Undoubtedly, the evolution of primate neocortical structure has been brought about under the influence of the three fundamental evolutionary forces: a tendency to morphological variation; the genetic regulation of basic developmental processes; and Darwinian natural selection. The interplay of the first two of these in the evolution of biological structure has been addressed by Edelman (1988b). The extent to which the workings of any of these factors can be deduced from current knowledge of cortical organization and development is extremely limited. It is clear that developing systems have considerable intrinsic self-organizing and pattern-generating properties and these are clearly evident during the development of the neocortex. Such intrinsic properties should undoubtedly restrict the number of phenotypic variants that may result from evolutionary influences operating at the developmental level. The range of cortical cell types, for example, seems fixed across all mammals and what distinguishes cells in the cortex of, say, humans and hedgehogs appears to be not so much a difference in fundamental morphological- or transmitter-defined types as differences in size, dendritic branching complexity, and perhaps density and extent of axon terminations. The evidence presented in Section 2, so for as it goes, suggests that the proportions of cortical cell types do not vary greatly among species of different orders, that the thickness of the cortex does not change greatly, and that the number of cells per columnar "unit" is relatively fixed except for the primary visual area of primates. Contrary to earlier predictions, therefore, the evolutionary advance in the functional complexity of the cortex may not be attended by increasing diversity of cell types.

Neuronal phenotype, however, is expressed in a variety of ways and if somal and dendritic morphology and transmitter or neuropeptide status are not major variables in the course of cortical evolution, that other expression, the nature of the axonal connections effected by a cell, is still a potential source of variance. The major long tract connections of any mammalian brain are fundamentally similar in terms of sites of origin and termination. Differences at the gross connectional level are primarily those of a quantitative nature. However, at finer levels of resolution, greater differences may be revealed. Ramón y Cajal (1891,

1911), for example, was of the opinion that the human brain had acquired a new population of cortical efferent cells whose axons were targeted on a single extracortical target without the extensive collateralization to multiple targets that he detected for cells in the same layers in small mammals and in a second population of cortical efferent cells in humans. The extensive but transient collateralization of the axons of developing cortical efferent cells (Section 6.4) is an example of morphological variance constrained by place- and time-dependent developmental processes which when perturbed could lead to evolutionary changes in patterns of cortical efferent connections that could strengthen, weaken, or make more or less amenable to precise control the connections formed as part of a typical mammalian long CNS tract.

Perturbations in place- and time-constrained developmental processes are perhaps the most likely sources of variance in cortical surface structure, leading ultimately, for example, to the different sulcal and gyral patterns of the ungulates and carnivores on the one hand and of the primates on the other. But here, too, there is a considerable intrinsic tendency to the formation of relatively few types of sulcal and gyral folding. The ungulates and carnivores which derive from totally different roots have the same basic ectosylvian pattern and New World and Old World monkeys which diverged perhaps 40 million years ago share a common fundamental pattern.

The most overt expression of neocortical evolution in primates is the massive increase in surface area of the cortex. To this, many would add an increasing degree of areal parcellation, a view that I basically share although one must remember that it is the large size of the human, ape, and Old World monkey cortex that affords the increased resolution whereby cortical areas can be identified. The potential alternative view, that every mammal, even that with the smallest lissencephalic cortex, contains a few cortical columns with connections appropriate for each of the cortical areas of higher primates, has never been disproved.

Probably the prevailing view of neocortical expansion over evolutionary time is that it has occurred as the result of reduplication of units that some would call columns but which are probably best described in more neutral terms as a constellation of interconnected cortical cells, perhaps linked by some common pattern of inputs and outputs. Such reduplication would account for the expansion of a cortex made up of the same fundamental neuronal elements linked in identical columnar arrays. If the germ of every cortical area of the advanced primate brain is present in every cortex, then reduplication of elements would easily explain the pari- passu expansion of individual cortical areas. If not, then the germ of a new area would first have to be induced as a new formation presumably in the neuroepithelium. In either case, however, and in the case of the expansion of the cortex as a whole, this must be dependent on a complex interplay of developmental events commencing with cell proliferation in the neuroepithelium, and passing through cell migration, cell aggregation, and connection formation and probably ending with histogenetic cell death. At all points it can be assumed that time- and place-dependent developmental constraints will apply and that perturbations can lead to morphological evolution.

The unexplored areas in neocortical development that may yield clues to the points at which morphological variance, expressed as increased numbers of cortical units, can be induced, are those dealing with cell aggregation and pattern formation. We may anticipate, from evidence in other developing systems

(Chuong and Edelman, 1985a,b; Chuong *et al.*, 1987; Raphael *et al.*, 1988), that cell aggregation in the cortex is dependent on cyclical expression of cell and substrate adhesion molecules. It is these that will presumably get arrays of cells in order to add to the extent of an area and thus to the cortex as a whole. By bringing cells together that will ultimately become functionally interrelated in a cortical column, the exchange of signals important for the regulation of gene expression for numerous structural and transmitter-related molecules and for the establishment and maintenance of appropriate synaptic contacts will be effected.

One of the most powerful epigenetic effects on cortical development appears to be neuronal activity. It is evident (Section 2) that the developing cortex has an intrinsic predisposition to form maps and that maps have been part of the evolutionary history of the mammalian neocortex from earliest times. Monotremes, for example, have well-defined somatosensory and visual cortical maps (Bohringer and Rowe, 1977). It is also evident, however, that this genetically determined capacity for self-organization is subject to profound activity-dependent influences that continue throughout the life of the animal. Current opinion would seem to favor the idea that activity-dependent modifications in cortical maps result from the uncovering and strengthening of previously unrevealed synapses. A genetically determined morphological variant occurring in development could thus be reinforced by the strong epigenetic influence of arriving afferent and developing neural activity. Perhaps an overt manifestation of this is the variation in barrel field structure of rodents with supernumerary or deficient whiskers (Welker and Van der Loos, 1986). But similar influences could underlie the evolutionary reduplication of cortical maps and the expansion and proliferation of cortical fields that accompanies this.

ACKNOWLEDGMENTS. Personal work reported herein was supported by Grants NS 21377 and EY 07193 from the National Institutes of Health, United States Public Health Service. I am grateful for the suggestions of Dr. S. H. C. Hendry and Mr. G. W. Huntley.

9. References

Adrien, J., Blanc, G., Buisseret, P., Fregnac, Y., Gary-Bobo, E., Imbert, M., Tassin, J. P., and Trotter, Y., 1985, Noradrenaline and functional plasticity in kitten visual cortex: A reexamination, *J. Physiol. (London)* **367:**73–98.

Akers, R. M., Moshier, D. F., and Lilien, J. E., 1981, Promotion of retinal neurite outgrowth by substratum-bound fibronectin, *Dev. Biol.* **86:**179–188.

Albus, K., and Wolf, W., 1984, Early post-natal development of neuronal function in the kitten's visual cortex: A laminar analysis, *J. Physiol. (London)* **348:**153–185.

Allman, J., 1990, Evolution in the neocortex, in: *Cerebral Cortex,* Volume 8A (E. G. Jones and A. Peters, eds.), Plenum Press, New York, in press.

Alloway, K. D., and Burton, H., 1986, Bicuculline-induced alterations in neuronal responses to controlled tactile stimuli in the second somatosensory cortex of the cat: A microiontophoretic study, *Somatosens. Res.* **3:**197–211.

Aoki, C., and Siekevitz, P., 1985, Ontogenetic changes in the cyclic adenosine 3′, 5′ monophosphate-stimulatable phosphorylation of cat visual cortex proteins, particularly of microtubule-associated protein 2 (MAP-2): effects of normal and dark rearing and of exposure to light, *J. Neurosci.* **5:**2465–2483.

Arimatsu, Y., Naegele, J. R., and Barnstable, C. J., 1987, Molecular markers of neuronal subpopulations in layers 4, 5, and 6 of cat primary visual cortex, *J. Neurosci.* **7:**1250–1263.

Baker, F. H., Grigg, P., and von Noorden, G. K., 1974, Effects of visual deprivation and strabismus on the response of neurons in the visual cortex of the monkey, including studies in the striate and prestriate cortex in the normal animal, *Brain Res.* **66:**185–208.

Bear, M. F., 1985, Postnatal changes in the distribution of acetylcholinesterase in kitten striate cortex, *J. Comp. Neurol.* **237:**519–532.

Bear, M. F., and Singer, W., 1986, Modulation of visual cortical plasticity by acetylcholine and noradrenaline, *Nature* **320:**172–176.

Bear, M. F., Paradiso, M. A., Schwartz, M., Nelson, S. B., Carnes, K. M., and Daniels, J. D., 1983, Two methods of catecholamine depletion in kitten visual cortex yield different effects on plasticity, *Nature* **302:**245–257.

Bear, M. F., Schmechel, D. E., and Ebner, F. F., 1985, Glutamic acid decarboxylase in the striate cortex of normal and monocularly deprived kittens, *J. Neurosci.* **5:**1262–1275.

Beaulieu, C., and Colonnier, M., 1983, The number of neurons in the different laminae of the binocular and monocular regions of area 17 in the cat, *J. Comp. Neurol.* **217:**337–344.

Beaulieu, C., and Colonnier, M., 1985, A comparison of the number of neurons in individual laminae of cortical areas 17, 18 and posteromedial suprasylvian (PMLS) area in the cat, *Brain Res.* **339:**166–179.

Beaulieu, C., and Colonnier, M., 1989, Number of neurons in individual laminae of areas 3B, 4gamma, and 6aalpha of the cat cerebral cortex: A comparison with major visual areas, *J. Comp. Neurol.* **279:**228–234.

Berbel, P., and Innocenti, G. M., 1988, The development of the corpus callosum in cats: A light- and electron-microscopic study, *J. Comp. Neurol.* **276:**132–156.

Blakemore, C., and Van Sluyters, R. C., 1975, Innate and environmental factors in the development of the kitten's visual cortex, *J. Physiol. (London)* **248:**663–716.

Blakemore, C., Garey, L. J., and Vital-Durand, F., 1978, The physiological effects of monocular deprivation and their reversal in monkey's visual cortex, *J. Physiol. (London)* **283:**223–262.

Blasdel, G. G., and Pettigrew, J. D., 1979, Degree of interocular synchrony required for maintenance of binocularity in kitten's visual cortex, *J. Neurophysiol.* **42:**1692–1710.

Bode-Gruel, K. M., and Singer, W., 1989, The development of N-methyl-D-aspartate receptors in cat visual cortex, *Dev. Brain Res.* **466:**197–204.

Bohringer, R. C., and Rowe, M. J., 1977, The organization of the sensory and motor areas of cerebral cortex in the platypus (*Ornithorhynchus anatinus*), *J. Comp. Neurol.* **174:**1–14.

Boothe, R. G., Greenough, W. T., Lund, J. S., and Wrege, K. 1979, A quantitative investigation of spine and dendrite development of neurons in visual cortex (area 17) of *Macaca nemestrina* monkeys, *J. Comp. Neurol.* **186:**473–490.

Bourgeois, J. -P., Jastreboff, P. J., and Rakic, P., 1989, Synaptogenesis in visual cortex of normal and preterm monkeys: Evidence for intrinsic regulation of synaptic overproduction, *Proc. Natl. Acad. Sci. USA* **86:**4297–4301.

Bovolenta, P., and Mason, C., 1987, Growth cone morphology varies with position in the developing mouse visual pathway from retina to first targets, *J. Neurosci.* **7:**1447–1460.

Brenneman, D. E., and Eiden, L. E., 1986, Vasoactive intestinal peptide and electrical activity influence neuronal survival, *Proc. Natl. Acad. Sci. USA* **83:**1159–1162.

Brodmann, K., 1912, Neue Ergebnisse über die vergleichende histologische Lokalisation der Grosshirnrinde mit Berücksichtigung des Stirnhirns, Anat.Anz. **41** (Suppl.)**:**157–216.

Buisseret, P., and Imbert, P., 1976, Visual cortical cells: Their developmental properties in normal and dark-reared kittens, *J. Physiol. (London)* **255:**511–525.

Buisseret, P., Gary-Bobo, E., and Imbert, M., 1978, Ocular motility and recovery of orientational properties of visual cortical neurones in dark-reared kittens, *Nature* **272:**816–817.

Bulloch, A. G. M., 1987, Somatostatin enhances neurite outgrowth and electrical coupling of regenerating neurons in Helisoma, *Brain Res.* **412:**6–17.

Calford, M. B., and Tweedle, R., 1988, Immediate and chronic changes in responses of somatosensory cortex in adult flying-fox after induced sensory deficits, *Nature* **332:**446–448.

Campbell, M. J., Lewis, D. A., Benoit, R., and Morrison, J. H., 1987, Regional heterogeneity in the distribution of somatostatin-28- and somatostatin-28 1–12 -immunoreactive profiles in monkey neocortex, *J. Neurosci.* **7:**1133–1144.

Cavanaugh, M. E., and Parnavelas, J. G., 1988, Development of somatostatin immunoreactive neurons in the rat occipital cortex: A combined immunocytochemical–autoradiographic study, *J. Comp. Neurol.* **268:**1–12.

Caviness, V. S., Jr., and Korde, M. G., 1981, Monoaminergic afferents to the neocortex: A developmental histofluorescence study in normal and reeler mouse embryos, *Brain Res.* **209:**1–9.

Chalupa, L. M., and Killackey, H. P., 1989, Process elimination underlies ontogenetic changes in the distribution of callosal projection neurons in the postcentral guyrus of the fetal rhesus monkey, *Proc. Natl. Acad. Sci. USA* **86:**1076–1079.

Chalupa, L. M., Killackey, H. P., Snider, C. J., and Lia, B., 1989, Callosal projection neurons in area 17 of the fetal rhesus monkey, *Dev. Brain Res.* **46:**303–308.

Changeux, J. P., and Danchin, 1976, Selective stabilisation of developing synapses as a mechanism for the specification of neuronal networks, *Nature* **264:**705–712.

Choi, D. W., 1987, Ionic dependence of glutamate neurotoxicity, *J. Neurosci.* **7:**369–379.

Chun, J. J. M., and Shatz, C. J., 1988a, Redistribution of synaptic vesicle antigens is correlated with the disappearance of a transient synaptic zone in the developing cerebral cortex, *Neuron* **1:**297–310.

Chun, J. J. M., and Shatz, C. J., 1988b, A fibronectin-like molecule is present in the developing cat cerebral cortex and is correlated with subplate neurons, *J. Cell Biol.* **106:**857–872.

Chun, J. J. M., and Shatz, C. J., 1989a, The earliest-generated neurons of the cat cerebral: Characterization by MAP2 and neurotransmitter immunohistochemistry during fetal life, *J. Neurosci.* **9**(5)**:**1648–1667.

Chun, J. J. M., and Shatz, C. J., 1989b, Interstitial cells of the adult neocortical white matter are the remnant of the early-generated subplate neuron population, *J. Comp. Neurol.* **282:**555–569.

Chun, J. J. M., Nakamura, M. J., and Shatz, C. J., 1987, Transient cells of the developing mammalian telencephalon are peptide immunoreactive neurons, *Nature* **325:**617–620.

Chuong, C. -M., and Edelman, G. M., 1984, Alterations in neural cell adhesion molecules during development of different regions of the nervous system, *J. Neurosci.* **4:**2354–2368.

Chuong, C. -M., and Edelman, G. M., 1985a, Expression of cell adhesion molecules in embryonic induction. I. Morphogenesis of nesting feathers, *J. Cell Biol.* **101:**1009–1026.

Chuong, C. -M., and Edelman, G. M., 1985b, Expression of cell adhesion molecules in embryonic induction. II. Morphogenesis of adult feathers, *J. Cell Biol.* **101:**1027–1043.

Chuong, C. -M., Crossin, K. L., and Edelman, G. M., 1987, Sequential expression and differential function of multiple-adhesion molecules during the formation of cerebellar cortical layers, *J. Cell Biol.* **104:**331–342.

Clark, S. A., Allard, T., Jenkins, W. M., and Merzenich, M. M., 1988, Receptive fields in the body-surface map in adult cortex defined by temporally correlated inputs, *Nature* **332:**444–445.

Clarke, S., and Innocenti, G. M., 1986, Organization of immature intrahemispheric connections, *J. Comp. Neurol.* **251:**1–22.

Cline, H. T., Debski, E. A., and Constantine-Paton, M., 1987, N-Methyl-D-aspartate receptor antagonist desegregates eye-specific stripes, *Proc. Natl. Acad. Sci. USA* **84:**4342–4345.

Cohen, J., Burne, J. F., Winter, J., and Bartlett, P., 1986, Retinal ganglion cells lose response to laminin with maturation, *Nature* **322:**465–467.

Conti, F. A., Rustioni, A., Petrusz, P., and Towle, A. C., 1987, Glutamate-positive neurons in the somatic sensory cortex of rats and monkeys, *J. Neurosci.* **7:**1887–1901.

Cooper, N. G. F., and Steindler, D. A., 1986, Lectins demarcate the barrel subfield in the somatosensory cortex of the early postnatal mouse, *J. Comp. Neurol.* **249:**157–169.

Cooper, N. G. F., and Steindler, D. A., 1989, Critical period-dependent alterations of the transient body image in the rodent cerebral cortex, *Brain Res.* **489:**167–176.

Cornwell, P., Ravizza, R., and Payne, B., 1984, Extrinsic visual and auditory connections in the 4-day old kitten, *J. Comp. Neurol.* **229:**97–120.

Cowan, W. M., 1973, Neuronal death as a regulative mechanism in the control of cell number in the nervous system, in: *Development and Aging in the Nervous System* (M. Rockstein and M. L. Sussman, eds.), Academic Press, New York, pp. 19–41.

Crandall, J. E., and Caviness, V. S., Jr., 1984, Axon strata of the cerebral wall in embryonic mice, *Dev. Brain Res.* **14:**185–196.

Crandall, J. E., Whitcomb, J. M., and Caviness, V. S., Jr., 1985, Development of the spinal–medullary projection from the mouse barrel field, *J. Comp. Neurol.* **239:**205–215.

Crawford, M. L. J., Blake, R., Cool, S. J., and von Noorden, G. K., 1975, Physiological consequences of unilateral and bilateral eye closure in macaque monkeys: Some further observations, *Brain Res.* **84:**150–154.

Cusick, C. G., and Lund, R. D., 1982, Modification of visual callosal projections in rats, *J. Comp. Neurol.* **212:**385–398.

Cynader, M., and Mitchell, D. E., 1977, Monocular astigmatism effects on kitten visual cortex development, *Nature* **270:**177–178.
Cynader, M., Timney, B. N., and Mitchell, D. E., 1980, Period of susceptibility of kitten visual cortex to the effects of monocular deprivation extends beyond six months of age, *Brain Res.* **191:**545–550.
Daw, N. W., Rader, R. K., Robertson, T. W., and Ariel, M., 1983, Effects of 6-hydroxydopamine on visual deprivation in the kitten striate cortex, *J. Neurosci.* **3:**907–914.
Daw, N. W., Robertson, T. W., Rader, R. K., Videen, T. O., and Coscia, C. J., 1984, Substantial reduction of cortical noradrenaline by lesions of adrenergic pathway does not prevent effects of monocular deprivation, *J. Neurosci.* **4:**1354–1360.
Daw, N. W., Robertson, T. W., and Rader, R. K., 1985a, DSP-4 (N-(2-chloroethyl)-N-ethyl-2-bromobenzylamine) depletes noradrenaline in kitten visual cortex without altering the effects of monocular deprivation, *J. Neurosci.* **5:**1925–1933.
Daw, N. W., Videen, T. O., Robertson, T., and Rader, R. K., 1985b, An evaluation of the hypothesis that noradrenaline affects plasticity in the developing visual cortex, in: *The Visual System,* Liss, New York, pp. 133–144.
Daw, N. W., Videen, T. O., Rader, R. K., Robertson, T. W., and Coscia, C. J., 1985c, Substantial reduction of noradrenaline in kitten visual cortex by intraventricular injections of 6-hydroxydopamine does not always prevent ocular dominance shifts after monocular deprivation, *Exp. Brain Res.* **59:**30–35.
Dawson, D. R., and Killackey, H. P., 1987, The organization and mutability of the forepaw and hindpaw representations in the somatosensory cortex of the neonatal rat, *J. Comp. Neurol.* **256:**246–256.
Dehay, C., Bullier, J., and Kennedy, H., 1984, Transient projections from the fronto parietal and temporal cortex to areas 17, 18, and 19 in kitten, *Exp. Brain Res.* **57:**208–212.
Dehay, C., Kennedy, H., and Bullier, J., 1986, Callosal connectivity of areas V1 and V2 in the newborn monkey, *J. Comp. Neurol.* **254:**20–33.
Dehay, C., Kennedy, H., Bullier, J., and Berland, M., 1988, Absence of interhemispheric connections of area 17 during development in the monkey, *Nature* **331:**348–350.
Dehay, C., Horsburgh, G., Berland, M., Killackey, H. P., and Kennedy, H., 1989, Maturation and connectivity of the visual cortex in monkey is altered by prenatal removal of retinal input, *Nature* **337:**265–267.
Des Rosiers, M., Sakurada, O., Jehle, J., Shinohara, M., Kennedy, C., and Sokoloff, L., 1978, Demonstration of functional plasticity in the immature striate cortex of the monkey by means of the [^{14}C] deoxyglucose method, *Science* **200:**447–449.
Dews, P. B., and Wiesel, T. N., 1970, Consequences of monocular deprivation on visual behavior in kittens, *J. Physiol. (London)* **206:**437–455.
Dodd, J., and Jessell, T. M., 1988, Axon guidance and the patterning of neuronal projections in vertebrates, *Science* **242:**692–700.
Duffy, F. H., Snodgrass, S. R., Burchfiel, J. L., and Conway, J. L., 1976, Bicuculline reversal of deprivation amblyopia in the cat, *Nature* **260:**256–257.
Dykes, R. W., Landry, P., Metherate, R., and Hicks, T. P., 1984, Functional role of GABA in cat primary somatosensory cortex: Shaping receptive fields and cortical neurons, *J. Neurophysiol.* **52:**1066–1093.
Edelman, G. M., 1988a, Morphoregulatory molecules, *Biochemistry* **17:**3533–3543.
Edelman, G. M., 1988b, *Topobiology: An Introduction to Molecular Embryology,* Basic Books, New York.
Edelman, G. M., Gallin, W. J., Delouvee, A., Cunningham, B. A., and Thiery, J. -P., 1983, Early epochal maps of two different cell adhesion molecules, *Proc. Natl. Acad. Sci. USA* **80:**4384–4388.
Fairén, A., Cobas, A., and Fonseca, M., 1986, Times of generation of glutamic acid decarboxylase immunoreactive neurons in mouse somatosensory cortex, *J. Comp. Neurol.* **251:**67–83.
Feng, J. Z., and Brugge, J. F., 1983, Postnatal development of auditory callosal connections in the kitten, *J. Comp. Neurol.* **214:**416–426.
Finlay, B. L., and Pallas, S. L., 1989, Control of cell number in the developing mammalian visual system *Prog. Neurobiol.* **32:**207–234.
Finlay, B. L., and Slattery, M., 1983, Local differences in the amount of early cell death in neocortex predict adult local specializations, *Science* **219:**1349–1351.
Flechsig, P., 1905, Einige Bemerkungen über die Untersuchungens methoden der Grosshirnrinde, insbesondere des Menschen, *Arch.Anat. Entwicklungs.* **1905:**337–444.
Floeter, M. K., and Jones, E. G., 1985, Transplantation of fetal postmitotic neurons to rat cortex:

Survival, early pathway choices and long-term projections of outgrowing axons, *Dev. Brain Res.* **22:**19–38.

Foster, G. A., and Schultzberg, M., 1984, Immunohistochemical analysis of the ontogeny of neuropeptide Y immunoreactive neurons in the foetal rat brain, *Int. J. Dev. Neurosci.* **2:** 387–407.

Freeman, R. D., and Bonds, A. B., 1979, Cortical plasticity in monocularly deprived immobilized kittens depends on eye movement, *Science* **206:**1093–1095.

Freeman, R. D., Sclar, G., and Ohzawa, I., 1983, An electrophysiological comparison of convergent and divergent strabismus in the cat: Visual evoked potentials, *J. Neurophysiol.* **49:**227–237.

Fregnac, Y., and Imbert, M., 1984, Development of neuronal selectivity in primary visual cortex of cat, *Physiol. Rev.* **64:**325–434.

Frost, D. O., 1981, Orderly anomalous retinal projections to the medial geniculate, ventrobasal, and lateral posterior nuclei of the hamster, *J. Comp. Neurol.* **203:**227–256.

Frost, D. O., 1982, Anomalous visual connections to somatosensory and auditory systems following brain lesions in early life, *Dev. Brain Res.* **3:**627–636.

Frost, D. O., 1984, Axonal growth and target selection during development: Retinal projections to the ventrobasal complex and other "nonvisual" structures in neonatal Syrian hamsters, *J. Comp. Neurol.* **230:**576–592.

Frost, D. O., 1986, Development of anomalous retinal projections to nonvisual thalamic nuclei in Syrian hamsters: A quantitative study, *J. Comp. Neurol.* **252:**95–105.

Frost, D. O., and Campbell, G., 1987, Target-controlled differentiation of axon terminals and synaptic organization, *Proc. Natl. Acad. Sci. USA* **84:**6929–6933.

Frost, D. O., and Metin, C., 1985, Induction of functional retinal projections to the somatosensory system, *Nature* **317:**162–164.

Gabbott, P. L. A., and Somogyi, P., 1986, Quantitative distribution of GABA-immunoreactive neurons in the visual cortex (area 17) of the cat, *Exp. Brain Res.* **61:**323–331.

Glickstein, M., and Whitteridge, D., 1976, Degeneration of layer III pyramidal cells in area 18 following destruction of callosal input, *Brain Res.* **104:**148–151.

Gordon-Weeks, P. R., 1989, Growth at the growth cone, *Trends Neurosci.* **12:**238–240.

Gorgels, T. G. M. F., De Kort, E. J. M., Van Aanholt, H. T. H., and Nieuwenhuys, R., 1989, A quantitative analysis of the development of the pyramidal tract in the cervical spinal cord in the rat, *Anat. Embryol.* **179:**377–385.

Grimm-Jorgensen, Y., 1987, Somatostatin and calcitonin stimulate neurite regeneration of molluscan neurons in vitro, *Brain Res.* **403:**121–126.

Gu, Q., Bear, M. F., and Singer, W., 1989, Blockade of NMDA receptors prevents ocularity change in kitten visual cortex after reversed monocular deprivation, *Dev. Brain Res.* **47:**281–288.

Hall, D. E., Neugebauer, K. M., and Reichardt, L. F., 1987, Embryonic neural retinal cell response to extracellular matrix proteins: Developmental changes and effects of the cell substratum attachment antibody (CSAT), *J. Cell Biol.* **104:**623–634.

Harris, R. M., and Woolsey, T. A., 1979, Morphology of Golgi-impregnated neurons in mouse cortical barrels following vibrissae damage at different postnatal stages, *Brain Res.* **161:**143–149.

Hatten, M. E., Furie, M. B., and Rifkin, D. B., 1982, Binding of developing mouse cerebellar cells to fibronectin: possible mechanism for the formation of the external granular layer, *J. Neurosci.* **2:**1195–1206.

Hendry, S. H. C., and Jones, E. G., 1986, Reduction in number of immunostained GABAergic neurones in deprived-eye dominance columns of monkey area 17, *Nature* **320:**750–753.

Hendry, S. H. C., and Jones, E. G., 1988, Activity-dependent regulation of GABA expression in the visual cortex of adult monkeys, *Neuron* **1:**701–712.

Hendry, S. H. C., and Kennedy, M. B., 1986, Immunoreactivity for a calmodulin-dependent protein kinase is selectively increased in macaque striate cortex after monocular deprivation, *Proc. Natl. Acad. Sci. USA,* **83:**1536–1540.

Hendry, S. H. C., Jones, E. G., and Emson, P. C., 1984, Morphology distribution and synaptic relations of somatostatin and neuropeptide Y-immunoreactive neurons in rat and monkey neocortex, *J. Neurosci.* **4:**2497–2517.

Hendry, S. H. C., Schwark, H. D., Jones, E. G., and Yan, J., 1987a, Numbers and proportions of GABA-immunoreactive neurons in different areas of monkey cerebral cortex, *J. Neurosci.* **7:**1503–1519.

Hendry, S. H. C., Jones, E. G., Killackey, H. P., and Chalupa, L. M., 1987b, Choline acetyl transferase immunoreactive neurons in fetal monkey cerebral cortex, *Dev. Brain Res. 37:*313–317.

Hendry, S. H. C., Jones, E. G., Hockfield, S., and McKay, R. D. G., 1988a, Neuronal populations stained with the monoclonal antibody Cat-301 in the mammalian cerebral cortex and thalamus, *J. Neurosci.* **8:**518–542.

Hendry, S. H. C., Jones, E. G., and Burstein, N., 1988b, Activity-dependent regulation of tachykinin-like immunoreactivity in neurons of monkey visual cortex, *J. Neurosci.* **8:**1225–1238.

Hendry, S. H. C., Jones, E. G., Emson, P. C., Lawson, D. E. M., Heizmann, C. W., and Streit, P., 1989, Two classes of cortical GABA neurons defined by differential calcium binding protein immunoreactivities, *Exp. Brain Res.* **76:**467–472.

Heumann, D., Leuba, G., and Rabinowicz, T., 1977, Postnatal development of the mouse cerebral cortex. II. Quantitative cytoarchitectonics of visual and auditory areas, *J. Hirnforsch.* **18:**483–500.

Heumann, D., Leuba, G., and Rabinowicz, T., 1978, Postnatal development of the mouse cerebral cortex. IV. Evolution of the total cortical volume, of the population of neurons and glial cells, *J. Hirnforsch.* **19:**385–393.

Hicks, T. P., and Dykes, R. W., 1983, Receptive field size for certain neurons in primary somatosensory cortex is determined by GABA-mediated intracortical inhibition, *Brain Res.* **274:**160–164.

Hoffman, S., and Edelman, G. M., 1983, Kinetics of homophilic binding by embryonic and adult forms of the neural cell adhesion molecule, *Proc. Natl. Acad. Sci. USA* **80:**5762–5766.

Hogan, D., and Berman, N. E. J., 1989, Axon trajectory and growth cone morphology in the developing corpus callosum, *Soc. Neurosci. Abstr.* **15:**874.

Horton, J. C., and Hedley-Whyte, E. T., 1984, Mapping of cytochrome oxidase patches and ocular dominance columns in human visual cortex, *Philos. Trans. R. Soc. London Ser. B* **304:**255–272.

Hubel, D. H., and Wiesel, T. N., 1965, Binocular interaction in striate cortex of kittens reared with artificial squint, *J. Neurophysiol.* **28:**1041–1059.

Hubel, D. H., and Wiesel, T. N., 1970, The period of susceptibility to the physiological effects of unilateral eye closure in kittens, *J. Physiol. (London)* **206:**419–436.

Hubel, D. H., and Wiesel, T. N., 1971, Aberrant visual projections in the Siamese cat, *J. Physiol. (London)* **218:**33–62.

Hubel, D. H., and Wiesel, T. N., 1977, Functional architecture of macaque monkey visual cortex, *Proc. R. Soc. London Ser. B.* **198:**1–59.

Hubel, D. H., Wiesel, T. N., and LeVay, S., 1977, Plasticity of ocular dominance columns in monkey striate cortex, *Philos. Trans. R. Soc. London Ser. B* **278:**377–409.

Huntley, G. W., and Jones, E. G., 1990, Cajal-Retzius neurons in developing monkey neocortex show immunoreactivity for calcium binding proteins, *J. Neurocytol.*, in press.

Huntley, G. W., Hendry, S. H. C., Jones, E. G., Chalupa, L. M., and Killackey, H. P., 1987, GABA neuropeptide and ChaT expression in neurons of the fetal monkey sensory-motor cortex, *Soc. Neurosci. Abstr.* **13:**76.

Huntley, G. W., Hendry, S. H. C., Killackey, H. P., Chalupa, L. M., and Jones, E. G., 1988a, Temporal sequence of neurotransmitter expression by developing neurons of fetal monkey visual cortex, *Dev. Brain Res.* **43:**69–96.

Huntley, G. W., Hendry, S. H. C., Killackey, H. P., Chalupa, L. M., and Jones, E. G., 1988b, GABA neuropeptide, and tyrosine hydroxylase immunoreactivity in the frontal cortex of fetal monkeys, *Soc. Neurosci. Abstr.* **14:**1021.

Huntley, G. W., Jones, E. G., and DeBlas, A. I., 1990, $GABA_A$ receptor immunoreactivity in adult and developing monkey sensory-motor cortex, *Exp. Brain Res.*, in press.

Innocenti, G. M., 1981, Growth and reshaping of axons in the establishment of visual callosal connections, *Science* **212:**824–827.

Innocenti, G. M., and Caminiti, R., 1980, Postnatal shaping of callosal connections from sensory areas, *Exp. Brain Res.* **38:**381–394.

Innocenti, G. M., and Clarke, S., 1984a, Bilateral transitory projection to visual areas from auditory cortex in kittens, *Dev. Brain Res.* **14:**143–148.

Innocenti, G. M., and Clarke, S., 1984b, The organization of immature callosal connections, *J. Comp. Neurol.* **230:**287–309.

Innocenti, G. M., and Frost, D. O., 1979, Effects of visual experience on the maturation of the efferent system to the corpus callosum, *Nature* **280:**231–233.

Innocenti, G. M., and Frost, D. O., 1980, The postnatal development of visual callosal connections in the absence of visual experience of the eyes, *Exp. Brain Res.* **39:**365–375.

Innocenti, G. M., Fiore, L., and Caminiti, R., 1977, Exuberant projection into the corpus callosum from the visual cortex of newborn cats, *Neurosci. Lett.* **4:**237–242.

Innocenti, G. M., Frost, D. O., and Illers, J., 1985, Maturation of visual callosal connections in visually deprived kittens: A challenging critical period, *J. Neurosci.* **5:**255–267.

Innocenti, G. M., Clarke, S., and Kraftsik, R., 1986, Interchange of callosal and association projections in the developing visual cortex, *J. Neurosci.* **6:**1384–1409.

Ivy, G. O., and Killackey, H. P., 1981, The ontogeny of the distribution of callosal projection neurons in the rat parietal cortex, *J. Comp. Neurol.* **195:**367–389.

Ivy, G. O., and Killackey, H. P., 1982, Ontogenetic changes in the projections of neocortical neurons, *J. Neurosci.* **2:**735–743.

Jeanmonod, D., Rice, F. L., and Van der Loos, H., 1981, Mouse somatosensory cortex: Alterations in the barrelfield following receptor injury at different early postnatal ages, *Neuroscience* **6:**1503–1535.

Jensen, K. F., and Killackey, H. P., 1987, Terminal arbors of axons projecting to the somatosensory cortex of the adult rat. II. The altered morphology of thalamocortical afferents following neonatal infraorbital nerve cut, *J. Neurosci.* **7:**3544:3553.

Jessell, T. M., 1988, Adhesion molecules and the hierarchy of neural development, *Neuron* **1:**3–13.

Jones, E. G., 1975, Varieties and distribution of non-pyramidal cells in the somatic sensory cortex of the squirrel monkey, *J. Comp. Neurol.* **160:**205–268.

Jones, E. G., 1984, Laminar Distribution of cortical efferent cells, in: *Cerebral Cortex,* Volume 1 (E. G. Jones and A. Peters, eds.), Plenum Press, New York, pp. 521–555.

Jones, E. G., 1990, Cellular organization in the primate postcentral gyrus, *Prog. Brain Res.* in press.

Jones, E. G., and Hendry, S. H. C., 1986, Co-localization of GABA and neuropeptides in neocortical neurons, *Trends Neurosci.* **9:**71–76.

Jones, E. G., and Wise, S. P., 1977, Size laminar and columnar distribution of efferent cells in the sensory-motor cortex of monkeys, *J. Comp. Neurol.* **175:**391–438.

Jones, E. G., Valentino, K. L., and Fleshman, J. W., 1982a, Adjustment of connectivity in rat neocortex after prenatal destruction of precursor cells of layers II–IV, *Dev. Brain Res.* **2:**425–431.

Jones, E. G., DeFelipe, J., Hendry, S. H. C., and Maggio, J. E., 1988, A study of tachykinin-immunoreactive neurons in monkey cerebral cortex, *J. Neurosci.* **8:**1206–1224.

Jonsson, G., and Kasamatsu, T., 1983, Maturation of monoamine neurotransmitters, and receptors in cat occipital cortex during postnatal critical period, *Exp. Brain Res.* **50:**449:458.

Kaas, J. H., and Guillery, R. W., 1973, The transfer of abnormal visual field representations from the dorsal lateral geniculate nucleus to the visual cortex in Siamese cats, *Brain Res.* **59:**61–96.

Kaas, J. H., Merzenich, M. M., and Killackey, H. P., 1983, The reorganization of the somatosensory cortex following peripheral nerve damage in adult and developing mammals, *Annu. Rev. Neurosci.* **6:**325–356.

Kalil, R. E., 1982, Development of ocular dominance columns in cats reared with binocular deprivation or strabismus, *Soc. Neurosci. Abstr.* **8:**4.

Kasamatsu, T., and Pettigrew, J. D., 1976, Depletion of brain catecholamines: Failure of ocular dominance shift after monocular occlusion in kittens, *Science* **194:**206–209.

Kasamatsu, T., and Pettigrew, J. D., 1979, Preservation of binocularity after monocular deprivation in the striate cortex of kittens treated with 6-hydroxydopamine, *J. Comp. Neurol.* **185:**139–162.

Kasamatsu, T., Pettigrew, J. D., and Ary, M., 1979, Restoration of visual cortical plasticity by local microperfusion of norepinephrine, *J. Comp. Neurol.* **185:**163–182.

Kater, S. B., Mattson, M. P., Cohan, C., and Connor, J., 1988, Calcium regulation of the neuronal growth cone, *Trends Neurosci.* **11:**315–321.

Kelahan, A. M., and Doetsch, G. S., 1984, Time-dependent changes in the functional organization of somatosensory cerebral cortex following digit amputation in adult raccoons, *Somatosens. Res.* **2:**49–81.

Killackey, H. P., and Belford, G. R., 1979, The formation of afferent patterns in the somatosensory cortex of the neonatal rat, *J. Comp. Neurol.* **183:**285–304.

Killackey, H. P., and Chalupa, 1986, Ontogenetic change in the distribution of callosal projection neurons in the postcentral gyrus of the fetal rhesus monkey, *J. Comp. Neurol.* **244:**331–348.

Killackey, H. P., and Shinder, A., 1981, Central correlates of peripheral pattern alterations in the trigeminal system of the rat. II. The effect of nerve section, *Dev. Brain Res.* **1:**121–126.

Killackey, H., Ivy, G. O., and Cunningham, T. J., 1978, Anomalous organization of SMI somatotopic map consequent to vibrissae removal in the newborn rat, *Brain Res.* **155:**136–140.

Kleinschmidt, A., Bear, M. F., and Singer, W., 1987, Blockade of "NMDA" receptors disrupts experience-dependent plasticity of kitten striate cortex, *Science* **238:**355–358.

Konig, N., Roch, G., and Marty, R., 1975, The onset of synaptogenesis in rat temporal cortex, *Anat. Embryol.* **148:**73–87.

Kostovic, I., and Goldman-Rakic, P. S., 1983, Transient cholinesterase staining in the mediodorsal nucleus of the thalamus and its connections in the developing human and monkey brain, *J. Comp. Neurol.* **219:**431–447.

Kostovic, I., and Molliver, M. E., 1974, A new interpretation of the laminar development of cerebral cortex: Synaptogenesis in different layers of neopallium in the human fetus, *Anat. Rec.* **178:**395.

Kostovic, I., and Rakic, P., 1980, Cytology and time of origin of interstitial neurons in the white matter in infant and adult human and monkey telencephalon, *J. Neurocytol.* **9:**219–242.

Kostovic, I., and Rakic, P., 1984, Development of prestriate visual projections in the monkey and human fetal cerebrum revealed by transient cholinesterase staining, *J. Comp. Neurol.* **4:**25–42.

Kratz, K. E., and Spear, P. D., 1976, Effects of visual deprivation and alterations in binocular competition on responses of striate cortex neurons in the cat, *J. Comp. Neurol.* **170:**141–152.

Kristt, D. A., and Kasper, E. K., 1983, High density of cholinergic muscarinic receptors accompanies high intensity acetylcholinesterase-staining in layer IV of infant rat somatosensory cortex, *Dev. Brain Res.* **8:**373–376.

Kristt, D. A., and Molliver, M. E., 1976, Synapses, in newborn rat cerebral cortex: A quantitative ultrastructural study, *Brain Res.* **108:**180–186.

Kristt, D. A., and Waldman, J. V., 1982, Developmental reorganization of acetylcholinesterase-rich inputs to the somatosensory cortex of the mouse, *Anat. Embryol.* **164:**331–342.

LaMantia, A. S., and Rakic, 1990, Cytological and quantitative characteristics of four cerebral commissures in the rhesus monkey, *J. Comp. Neurol.* **291:**520–537.

Letourneau, P. C., Madsen, A. M., Palm, S. L., and Furcht, L. T., 1988, Immunoreactivity for laminin in the developing ventral longitudinal pathway of the brain, *Dev. Biol.* **125:**135–144.

Leuba, G., Heumann, D., and Rabinowicz, T., 1977, Postnatal development of the mouse cerebral neocortex. I. Quantitative cytoarchitectonics of some motor and sensory areas, *J. Hirnforsch.* **18:**461–481.

LeVay, S., Stryker, M. P., and Shatz, C. J., 1978, Ocular dominance columns and their development in layer IV of cat's visual cortex. A quantitative study, *J. Comp. Neurol.* **179:**223–244.

LeVay, S., Wiesel, T. N., and Hubel, D. H., 1980, The development of ocular dominance columns in normal and visually deprived monkeys, *J. Comp. Neurol.* **191:**1–51.

Lidov, H. G. W., and Molliver, M. E., 1982, An immunohistochemical study of serotonin neuron development in the rat: Ascending pathways and terminal fields, *Brain Res. Bull.* **8:**389–430.

Lipton, S. A., 1988, Spontaneous release of acetylcholine affects the physiological nicotinic responses of rat retinal ganglion cells in culture, *J. Neurosci.* **8:**3857–3868.

Lipton, S. A., and Kater, S. B., 1989, Neurotransmitter regulation of neuronal outgrowth, plasticity and survival, *Trends Neurosci.* **12:**265–269.

Lubke, J., and Albus, K., 1989, The postnatal development of layer VI pyramidal neurons in the cat's striate cortex, as visualized by intracellular Lucifer yellow injections of aldehyde-fixed tissue, *Dev. Brain Res.* **45:**29–38.

Luhmann, H. J., Martinez Millan, L., and Singer, W., 1986, Development of horizontal intrinsic connections in cat striate cortex, *Exp. Brain Res.* **144:**169–172.

Lumsden, A. G. S., and Davies, A. M., 1983, Earliest sensory nerve fibers are guided to peripheral targets by attractants other than nerve growth factor, *Nature* **306:**786–788.

Lumsden, A. G. S., and Davies, A. M., 1986, Chemotropic effect of specific target epithelium in the developing mammalian nervous system, *Nature* **323:**538–539.

Lund, J. S., Boothe, R. G., and Lund, R. D., 1977, Development of neurons in the visual cortex (area 17) of the monkey (Macaca nemestrina): A Golgi study from fetal day 127 to postnatal maturity, *J. Comp. Neurol.* **176:**149.

Lund, R. D., and Mitchell, D. E., 1979a, Asymmetry in the visual callosal connections of strabismic cats, *Brain Res.* **167:**176–179.

Lund, R. D., and Mitchell, D. E., 1979b, The effects of dark-rearing on visual callosal connections of cats, *Brain Res.* **167:**172–175.

Lund, R. D., and Mustari, M. J., 1977, Development of the geniculocortical pathway in rats, *J. Comp. Neurol.* **173:**289–306.

Lund, R. D., Mitchell, D. E., and Henry, G. H., 1978, Squint-induced modification of callosal connections in cats, *Brain Res.* **144:**169–172.

Luskin, M. B., and Shatz, C. J., 1985a, Neurogenesis of the cat's primary visual cortex, *J. Comp. Neurol.* **242:**611–631.

Luskin, M. B., and Shatz, C. J., 1985b, Studies of the earliest generated cells of the cat's visual cortex: Cogeneration of subplate and marginal zones, *J. Neurosci.* **5:**1062–1075.

Luskin, M. B., Pearlman, A. L., and Sanes, J. R., 1988, Cell lineage in the cerebral cortex of the mouse studied in vivo and in vitro with a recombinant retrovirus, *Neuron* **1:**635–647.

McCandlish, C., Waters, R. S., and Cooper, N. G. F., 1989, Early development of the representation of the body surface in SI cortex barrel field in neonatal rats as demonstrated with peanut agglutinin binding: Evidence for differential development within the rattunculus, *Exp. Brain Res.* **77:**425–431.

McConnell, S. K., 1985, Migration and differentiation of cerebral cortical neurons after transplantation into the brains of ferrets, *Science* **229:**1268–1271.

McConnell, S. K., 1988, Fates of visual cortical neurons in the ferret after isochronic and heterochronic transplantation, *J. Neurosci.* **8:**945–974.

McConnell, S. K., Ghosh, A., and Shatz, C. J., 1989, Subplate neurons pioneer the first axon pathway from the cerebral cortex, *Science* **245:**978–981.

McKinley, P. A., and Kruger, L., 1988, Nonoverlapping thalamocortical connections to normal and deprived primary somatosensory cortex for similar forelimb receptive fields in chronic spinal cats, *Somatosens. Res.* **5:**311–323.

Marin-Padilla, M., 1969, Origin of the percellular baskets of the pyramidal cells of the human motor cortex, *Brain Res.* **14:**633–646.

Marin-Padilla, M., 1970a, Prenatal and early postnatal ontogenesis of the human motor cortex: A Golgi study. I. The sequential development of the cortical layers, *Brain Res.* **23:**167–183.

Marin-Padilla, M., 1970b, Prenatal and early postnatal ontogenesis of the human motor cortex: A Golgi study. II. The basket pyramidal system, *Brain Res.* **23:**185–191.

Marin-Padilla, M., 1971, Early prenatal ontogenesis of the cerebral cortex (neocortex) of the cat (Felis domestica). A Golgi study, *Z. Anat. Entwicklungsgesch.* **134:**117–145.

Marin-Padilla, M., 1984, Neurons of layer I: A developmental analysis, in: *Cerebral Cortex,* Volume 1 (A. Peters and E. G. Jones, eds.), Plenum Press, New York, pp. 447–478.

Marin-Padilla, M., 1988, Early ontogenesis of the human cerebral cortex, in: *Cerebral Cortex,* Volume 7 (A. Peters and E. G. Jones, eds.), Plenum Press, New York, pp. 1–34.

Mastronade, D. N., 1983, Correlated firing of ct retinal ganglion cells. I. Spontaneously active inputs to X- and Y-cells, *J. Neurophysiol.* **49:**303–324.

Mattson, M. P., 1988, Neurotransmitters in the regulation of neuronal cytoarchitecture, *Brain Res. Rev.* **13:**179–212.

Merzenich, M. M., Kaas, J. H., Wall, J. T., Sur, M., Nelson, R. J., and Felleman, D. J., 1983a, Progression of change following median nerve section in the cortical representation of the hand in areas 3b and 1 in adult owl and squirrel monkeys, *Neuroscience* **10:**639–666.

Merzenich, M. M., Kaas, J. H., Wall, J., Nelson, R. J., Sur, M., and Felleman, D., 1983b, Topographic reorganization of somatosensory cortical areas 3B and 1 in adult monkeys following restricted deafferentation, *Neuroscience* **8:**33–56.

Merzenich, M. M., Nelson, R. J., Stryker, M. P., Cynader, M. S., Schoppmann, A., and Zook, J. M., 1984, Somatosensory cortical map changes following digit amputation in adult monkeys, *J. Comp. Neurol.* **224:**591–605.

Merzenich, M. M., Recanzone, G., Jenkins, W. M., Allard, T., and Nudo, R. J., 1988, Cortical representational plasticity, in: *The Neural and Molecular Basis for Learning* (J. P. Changeux and M. Konishi, eds.), Wiley, New York.

Meyer, G., and Ferres-Torres, R., 1984, Postnatal maturation of nonpyramidal neurons in the visual cortex of the cat, *J. Comp. Neurol.* **228:**226–244.

Meyer, R. L., 1982, Tetrodotoxin blocks the formation of ocular dominance columns in goldfish, *Science* **218:**589–591.

Meyer, R. L., 1983, Tetrodotoxin inhibits the formation of refined retinotopography in goldfish, *Dev. Brain Res.* **6:**293–298.

Miller, M. W., 1985, Cogeneration of retrogradely labeled corticocortical projection and GABA-immunoreactive local circuit neurons in cerebral cortex, *Dev. Brain Res.* **23:**187–192.

Miller, M. W., 1986, Maturation of rat visual cortex. III. Postnatal morphogenesis and synaptogenesis of local circuit neurons, *Dev. Brain Res.* *25:*271–285.

Mitchell, D. E., and Timney, B., 1984, Postnatal development of function in the mammalian visual system, in: *Nervous System III, Handbook of Physiology,* Volume III, Section I, American Physiological Society, Washington, D.C., pp. 507–555.

Mitzdorf, U., and Singer,W. 1980, Monocular activation of visual cortex in normal and monocularly deprived cats: An analysis of evoked potentials, *J. Physiol. (London)* **304:**

Molliver, M. E., Kostovic, I., and Van der Loos, H., 1973, The development of synapses in cerebral cortex of the human fetus, *Brain Res.* **50:**403–407.

Morest, D. K., 1969, The growth of dendrites in the mammalian brain, *Z. Anat. Entwicklungsgesch.* **128:**290–317.

Mountcastle, V. B., 1978, An organizing principle for cerebral function: The unit module and the distributed system, in: *The Mindful Brain: Cortical Organization and the Group-Selective Theory of higher Brain Function* by G. M. Edelman and V. B. Mountcastle, MIT Press, Cambridge, Mass. pp. 7–50.

Movshon, J. A., 1976, Reversal of the physiological effects of monocular deprivation in the kitten's visual cortex, *J. Physiol (London)* **261:**125–174.

Movshon, J. A., and Dursteler, M. R., 1977, Effects of brief periods of unilateral eye closure on the kitten's visual system, *J. Neurophysiol.* **40:**1255–1265.

Movshon, J. A. and Van Sluyters, R. C., 1981, Visual neuronal development, *Annu. Rev. Psychol.* **32:**477–522.

Mower, G. D., Caplan, C. J., Christen, W. G., and Duffy, F. H., 1985, Dark-rearing prolongs physiological but not anatomical plasticity of the cat visual cortex, *J. Comp. Neurol.* **235:**448–466.

Mower, G. D., Rustad, R., and White, W. F., 1988, Quantitative comparisons of gamma-aminobutyric acid neurons and receptors in the visual cortex of normal and dark-reared cats, *J. Comp. Neurol.* **272:**293–302.

Nakanishi, S., 1983, Extracellular matrix during laminar pattern formation of neocortex in normal and reeler mutant mice, *Dev. Biol.* **95:**305–316.

Naus, C. C. G., Miller, F. D., Morrison, J. H., and Bloom, F. E., 1988, Immunohistochemical and in situ hybridization analysis of the development of the rat somatostatin-containing neocortical neuronal system, *J. Comp. Neurol.* **269:**448–463.

Neve, R. L., and Bear, M. F., 1989, Visual experience regulates gene expression in the developing striate cortex, *Proc. Natl. Acad. Sci. USA* **86:**4781–4784.

Nilsson, J., von Euler, A. M., and Dalsgaard, C. -J., 1985, Stimulation of connective tissue cell growth by substance P and substance K, *Nature* **315:**61–63.

Oka, H., Samejima, A., and Yamamoto, T., 1985, Post-natal development of pyramidal tract neurones in kittens, *J. Physiol. (London)* **363:**481–499.

O'Kusky, J., and Colonnier, M., 1982a, A laminar analysis of the number of neurons, glia, and synapses in the visual cortex (area 17) of adult macaque monkeys, *J. Comp. Neurol.* **210:** 278–290.

O'Kusky, J., and Colonnier, M., 1982b, Postnatal changes in the number of neurons and synapses in the visual cortex (area 17) of the macaque monkey: A stereological analysis in normal and monocularly deprived animals, *J. Comp. Neurol.* **210:**291–505.

Olavarria, J., and Van Sluyters, R., 1985, Organization and postnatal development of callosal connections in the visual cortex of the rat, *J. Comp. Neurol.* **239:**1–26.

O'Leary, D. D. M., and Stanfield, B. B., 1986, A transient pyramidal tract projection from the visual cortex in the hamster and its removal by selective collateral elimination, *Dev. Brain Res.* **27:**87–100.

O'Leary, D. D. M., and Stanfield, B. B., 1989, Selective elimination of axons extended by developing cortical neurons is dependent on regional locale: Experiments utilizing fetal cortical transplants, *J. Neurosci.* **9:**2230–2246.

Olson, C. R., and Freeman, R. D., 1975, Progressive changes in kitten striate cortex during monocular vision, *J. Neurophysiol.* **38:**26–32.

Olson, C. R., and Freeman, R. D., 1978, Monocular deprivation and recovery during sensitive period in kittens, *J. Neurophysiol.* **41:**65–74.

Paradiso, M. A., Bear, M. F., and Daniels, J. D., 1983, Effects of intracortical infusion of 6-hydroxydopamine on the response of kitten visual cortex to monocular deprivation, *Exp. Brain Res.* **51:**413–422.

Patterson, P. H., 1988, On the importance of being inhibited, or saying no to growth cones, *Neuron* **1:**263–267.

Peduzzi, J. D., 1988, Genesis of GABA-immunoreactive neurons in the ferret visual cortex, *J. Neurosci.* **8:**920–931.

Peters, A., 1987, Number of neurons and synapses in primary visual cortex, in: *Cerebral Cortex,* Volume 6 (E. G. Jones and A. Peters, eds.), Plenum Press, New York, pp. 267–294.

Peters, A., Kara, D. A., and Harriman, K. M., 1985, The neuronal composition of area 17 of rat visual cortex. III. Numerical considerations, *J. Comp. Neurol.* **238:**263–274.

Pettigrew, J. D., 1974, The effect of visual experience on the development of stimulus specificity by kitten cortical neurones, *J. Physiol. (London)* **237:**249–274.

Powell, T. P. S., 1973, The organization of the major functional areas of the cerebral cortex, *Symp. Zool. Soc. London* **33:**235–252.

Powell, T. P. S., and Hendrickson, A. E., 1981, Similarity in number of neurons through the depth of the cortex in the binocular and monocular part of area 17 of the monkey, *Brain Res.* **216:**409–413.

Price, D. J., and Blakemore, C., 1985, The postnatal development of the association projection from visual cortical area 17 to area 18 in the cat, *J. Neurosci.* **5:**2443–2452.

Price, D. J., and Blakemore, C., 1986, The postnatal development of clustered intrinsic connections in area 18 of the visual cortex in kittens, *Dev. Brain Res.* **24:**31–38.

Price, J., and Thurlow, L., 1988, Cell lineage in the rat cerebral cortex: A study using retroviral-mediated gene transfer, *Development* **104:**473–482.

Price, D. J., and Zumbroich, T. J., 1989, Postnatal development of corticocortical efferents from area 17 in the cat's visual cortex, *J. Neurosci.* **9:**600–613.

Prusky, G. T., Arbuckle, J. M., and Cynader, M. S., 1988a, Transient concordant distributions of nicotinic receptors and acetylcholinesterase activity in infant rat visual cortex, *Dev. Brain Res.* **39:**154–159.

Prusky, G. T., Shaw, C., and Cynader, M. S., 1988b, The distribution and ontogenesis of [^{3}H]nicotine binding sites in cat visual cortex, *Dev. Brain Res.* **39:**161–176.

Raedler, E., and Raedler, A., 1978, Autoradiographic study of early neurogenesis in rat neocortex, *Anat. Embryol.* **158:**267–284.

Raedler, E., Raedler, A., and Feldhaus, S., 1980, Dynamical aspects of neocortical histogenesis in the rat, *Anat. Embryol.* **158:**253–269.

Rakic, P., 1972, Mode of cell migration to the superficial layers of fetal monkey neocortex, *J. Comp. Neurol.* **145:**61–84.

Rakic, P., 1974, Neurons in rhesus monkey visual cortex: Systematic relation between time of origin and eventual disposition, *Science* **183:**425–427.

Rakic, P., 1976, Prenatal genesis of connections subserving ocular dominance in the rhesus monkey, *Nature* **261:**467–471.

Rakic, P., 1981, Development of visual centers in the primate brain depends on binocular competition before birth, *Science* **214:**928–931.

Rakic, P., 1988, Specification of cerebral cortical areas, *Science* **241:**170–176.

Rakic, P., and Goldman-Rakic, P. S., 1982, Development and modifiability of the cerebral cortex, *Neurosci. Res. Progr. Bull.* **20:**433–606.

Rakic, P., Bourgeois, J. -P. Eckenhoff, M. F., Zecevic, N., and Goldman-Rakic, P. S., 1986, Concurrent overproduction of synapses in diverse regions of the primate cerebral cortex, *Science* **231:**232–235.

Ramirez, L. F., and Kalil, K., 1985, Critical stages for growth in the development of cortical neurons, *J. Comp. Neurol.* **237:**506–518.

Ramoa, A. S., Paradiso, M. A., and Freeman, R. D., 1988, Blockade of intracortical inhibition in kitten striate cortex: Effects on receptive field properties and associated loss of ocular dominance plasticity, *Exp. Brain Res.* **73:**285–298.

Ramón y Cajal, S., 1891, Sur la structure de l'écorce cérébrale de quelques mammiféres, *La Cellule* **7:**125–176.

Ramón y Cajal, S., 1894, The Croonian Lecture: La fine structure des centres nerveaux, *Proc. R. Soc. London* **55:**444–467.

Ramón y Cajal. S., 1900, Estudios sobre la corteza cerebral humana II: corteza motríz (conclusion), *Rev. Trim. Micrograf.* **5:**1–11.

Ramón y Cajal, S., 1911 Histologie du système nerveux de l'homme et des vertébrés, Consejo Superior de Investigaciones Cientificas, Volume 2, p. 993.

Raphael, Y., Volk, T., Crossin, K. L. Edelman, G. M., and Geiger, G., 1988, The modulation of cell adhesion molecule expression and intercellular junction formation in the developing avian inner ear, *Dev. Biol.* **128:**222–235.

Rasmusson, D. D., 1982, Reorganization of raccoon somatosensory cortex following removal of the fifth digit, *J. Comp. Neurol.* **205:**313–326.

Rauschecker, J. P., and Singer, W., 1982, The effects of early visual experience on the cat's visual cortex and their possible explanation by Hebb synapses, *J. Physiol. (London)* **310:**215–239.

Reh, T. A., and Constantine-Paton, M., 1985, Eye-specific segregation requires neural activity in three-eyed Rana pipiens, *J. Neurosci.* **5:**1132–1143.

Reh, T., and Kalil, K., 1981, Development of the pyramidal tract in the hamster: I. A light microscopic study, *J. Comp. Neurol.* **200:**55–67.

Reh, T., and Kalil, K., 1982, Development of the pyramidal tract in the hamster: II. An electron microscopic study, *J. Comp. Neurol.* **205:**77–88.

Reiter, H. O., and Stryker, M. P., 1988, Neural plasticity without postsynaptic action potentials: Less-active inputs become dominant when kitten visual cortical cells are pharmacologically inhibited, *Proc. Natl. Acad. Sci. USA* **85:**3623–3627.

Rickman, M., Chronwall, B. M., and Wolff, J. R., 1977, On the development of nonpyramidal neurons and axons outside the cortical plate: The early marginal zone as a pallial anlage, *Anat. Embryol.* **151**:285–307.
Riggett, M. J., and Moody, S. A., 1987, Distribution of laminin and fibronectin along peripheral trigeminal axon pathways in the developing chick, *J. Comp. Neurol.* **258**:580–596.
Robertson, D., and Irvine, D. R. F., 1989, Plasticity of frequency organization in auditory cortex of guinea pigs with partial unilateral deafness, *J. Comp. Neurol.* **282**:456–471.
Robertson, R. T., Tijerina, A., and Gallivan, M. E., 1985, Transient patterns of acetylcholinesterase activity in visual cortex of the rat: Normal development and the effects of neonatal monocular enucleation, *Dev. Brain Res.* **21**:203–214.
Rockel, A. J., Hiorns, R. W., and Powell, T. P. S., 1974, Numbers of neurons through full depth of neocortex, *J. Anat.* **118**:371.
Rockel, A. J., Hiorns, R. W., and Powell, T. P. S. Powell, 1980, The basic uniformity in structure of the neocortex, *Brain* **103**:221–244.
Rogers, S. L., Letourneau, P. C., Palm, S. L., McCarthy, J., and Furcht, L. T., 1983, Neurite extension by peripheral and central nervous system neurons in response to substratumbound fibronectin and laminin, *Dev. Biol.* **98**:212–220.
Rose, J. E., and Woolsey, C. N., 1948, Structure and relations of limbic cortex and anterior thalamic nuclei in rabbit and cat, *J. Comp. Neurol.* **89**:279–347.
Rose, J. E., and Woolsey, C. N., 1949, The relations of thalamic connections, cellular structure and evocable electrical activity in the auditory region of the cat, *J. Comp. Neurol.* **91**:441–466.
Rothblatt, L. A., and Schwartz, M. L., 1979, The effect of monocular deprivation on dendritic spines in visual cortex of young and adult albino rats: Evidence for a sensitive period, *Brain Res.* **161**:156–161.
Rubin, G. M., 1988, Drosophila melanogaster as an experimental organism, *Science* **240**:1453–1459.
Ruiz-Marcos, A., and Valverde, F., 1969, The temporal evolution of the distribution of dendritic spines in the visual cortex of normal and dark reared mice, *Exp. Brain Res.* **8**:284–294.
Rutishauser, U., Acheson, A., Hall, A. K., Mann, D. M., and Sunshine, J., 1988, The neural cell adhesion molecule (NCAM) as a regulator of cell–cell interactions, *Science* **240**:53–57.
Sanes, J. N., Suner, S., Lando, J. F., and Donoghue, J. P., 1988, Rapid reorganization of adult rat motor cortex somatic representation patterns after motor nerve injury, *Proc. Natl. Acad. Sci. USA* **85**:2003–2007.
Sato, H., and Tsumoto, T., 1984, GABAergic inhibition already operates on a group of neurons in the kitten visual cortex at the time of eye opening, *Dev. Brain Res.* **12**:311–316.
Schlumpf, M., Shoemaker, W. J., and Bloom, F. E., 1980 Innervation of embryonic rat cerebral cortex by catecholamine-containing fibers, *J. Comp. Neurol.* **192**:361–376.
Schmidt, J. T., 1985, Formation of retinotopic connections: Selective stabilization by an activity dependent mechanism, *Cell. Mol. Neurobiol.* **5**:65–84.
Schreyer, D. J., and Jones, E. G., 1982, Growth and target finding by axons of the corticospinal tract in the prenatal and postnatal rat, *Neuroscience* **7**:1837–1853.
Schreyer, D. J., and Jones, E. G., 1988a, Axon elimination in the developing corticospinal tract of the rat, *Dev. Brain Res.* **38**:103–119.
Schreyer, D. J., and Jones, E. G., 1988b, Topographic sequence of outgrowth of corticospinal axons in the rat: A study using retrograde axonal labeling with fast blue, *Dev. Brain Res.* **38**:89–101.
Schwartz, M. L., and Goldman-Rakic, P. S., 1986, Some callosal neurons of the fetal monkey frontal cortex have axons in the contralateral hemisphere prior to the completion of migration, *Soc. Neurosci. Abstr.* **12**:1211.
Schwartz, M. L., Zheng, D. -S., and Goldman-Rakic, P. S., 1988, Periodicity of GABA-containing cells in primate prefrontal cortex, *J. Neurosci.* **8**:1962–1970.
Shatz, C., 1977a, A comparison of visual pathways in Boston and midwestern Siamese cats, *J. Comp. Neurol.* **171**:205–228.
Shatz, C., 1977b, Abnormal interhemispheric connections in the visual system of Boston Siamese cats: A physiological study, *J. Comp. Neurol.* **171**:229–246.
Shatz, C. J., and Luskin, M. B., 1986, The relationship between the geniculocortical afferents and their cortical target cells during development of the cat's primary visual cortex, *J. Neurosci.* **6**:3655–3668.
Shatz, C. J., and Rakic, P., 1981, The genesis of efferent connections from the visual cortex of the fetal rhesus monkey, *J. Comp. Neurol.* **196**:287–308.
Shatz, C. J., and Stryker, M. P., 1978, Ocular dominance in layer IV of the cat's visual cortex and the effects of monocular deprivation, *J. Physiol. (London)* **281**:267–284.

Shatz, C. J., Lindstrom, S., and Wiesel, T. N., 1977, The distribution of afferents representing the right and left eyes in the cat's visual cortex, *Brain Res.* **131**:103–106.
Shatz, C. J., Chun, J. J. M., and Luskin, M. B., 1988, The role of the subplate in the development of the mammalian telencephalon, in: *Cerebral Cortex,* Volume 7 (A. Peters and E. G. Jones, eds.), Plenum Press, New York, pp. 35–58.
Shaw, C., and Cynader, M., 1984, Disruption of cortical activity prevents ocular dominance changes in monocularly deprived kittens, *Nature* **308**:731–734.
Shaw, C., and Cynader, M., 1985, Alterations in the laminar distribution of pentagastrin binding sites in cat visual cortex during postnatal development, *Dev. Brain Res.* **20**:132–136.
Shaw, C., and Cynader, M., 1986, Laminar distribution of receptors in monkey (Macaca fascicularis) geniculostriate system, *J. Comp. Neurol.* **248**:301–312.
Shaw, C., Needler, M. C., and Cynader, M., 1984, Ontogenesis of muscimol binding sites in cat visual cortex, *Brain Res. Bull.* **13**:331–334.
Shaw, C., Wilkinson, M., Cynader, M., Needler, M. C., Aoki, C., and Hall, S. E., 1986, The laminar distributions and postnatal development of neurotransmitter and neuromodulator receptors in cat visual cortex, *Brain Res. Bull.* **16**:661–671.
Shaw, C., Prusky, G., van Huizen, F., and Cynader, M., 1989, Differential effects of quinolinic acid lesions on muscarinic acetylcholine receptors in cat visual cortex during postnatal development, *Brain Res. Bull.* **22**:771–776.
Sherman, S. M., and Spear, P. D., 1982, Organization of visual pathways in normal and visual deprived cats, *Physiol. Rev.* **62**:738–855.
Shoumura, K., 1974, An attempt to relate the origin and distribution of commissural fibers to the presence of large and medium pyramids in layer III in the cat's visual cortex, *Brain Res.* **67**:13–25.
Shoumura, K., Ando, T., and Kato, K., 1975, Structural organization of 'callosal' OBg in human corpus callosum agenesis, *Brain Res.* **93**:241–252.
Sillito, A. M., 1975, The contribution of inhibitory mechanisms to the receptive field properties of neurones in the striate cortex of the cat, *J. Physiol. (London)* **250**:305–329.
Sillito, A. M., 1979, Inhibitory mechanisms influencing complex cell orientation selectivity and their modification at high resting discharge levels, *J. Physiol. (London)* **289**:33–53.
Sillito, A. M., Kemp, J. A., and Milson, J. A., and Berardi, N., 1980, A re-evaluation of the mechanisms underlying simple cell orientation selectivity, *Brain Res.* **194**:517–520.
Sillito, A. M., Kemp, J. A., and Blakemore, C., 1981, The role of GABAergic inhibition in the cortical effects of monocular deprivation, *Nature* **291**:318–320.
Silver, J., Poston, M., and Rutishauser, U., 1987, Axon pathway boundaries in the developing brain. I. Cellular and molecular determinants that separate the optic and olfactory projections, *J. Neurosci.* **7**:2264–2272.
Simons, D. J., and Land, P. W., 1987, Early experience of tactile stimulation influences organization of somatic sensory cortex, *Nature,* **326**:694–697.
Singer, W., 1979, Central-core control of visual-cortex functions, in: Neurosciences 4th Study Program, MIT Press, Cambridge, Mass., pp. 1093–1110.
Singer, W., 1988, Pattern recognition and self-organization in biological systems, in: *Processing Structures for Perception and Action* (H. Marko, G. Hauske, and A. Struppler, eds.), VCH Verlagsgesellschaft, Weinheim, pp. 1–18.
Singer, W., Rauschecker, J., and Werth, R., 1977, The effect of monocular exposure to temporal contrasts on ocular dominance in kittens, *Brain Res.* **134**:568–572.
Smalheiser, N. R., Crain, S. M., and Reid, L. M., 1984, Laminin as a substrate for retinal axons in vitro, *Dev. Brain Res.* **12**:136–140.
Snow, P. J., Nudo, R. J., Rivers, W., Jenkins, W. M., and Merzenich, M. M., 1988, Somatotopically inappropriate projections from thalamocortical neurons to SI cortex of the cat demonstrated by the use of intracortical microstimulation, *Somatosens. Res.* **5**:349–372.
Sotelo, C., and Arsenio-Nunez, M. L., 1976, Development of Purkinje cells in absence of climbing fibers, *Brain Res.* **111**:389–395.
Sotelo, C., Hillman, D. E., Zamora, A. J., and Llinás, R., 1975, Climbing fiber deafferentation: Its action on Purkinje cell dendritic spines, *Brain Res.* **98**:574–581.
Spiegelman, B. M., Lopata, M. A., and Kirschner, M. W., 1979, Aggregation of microtubule initiation sites preceding neurite outgrowth in mouse neuroblastoma cells, *Cell* **16**:253.
Stanfield, B. B., and O'Leary, D. D. M., 1985a, The transient corticospinal projection from the occipital cortex during the postnatal development of the rat, *J. Comp. Neurol.* **238**:236–248.

Stanfield, B. B., and O'Leary, D. D. M., 1985b, Fetal occipital cortical neurones transplanted to the rostral cortex can extend and maintain a pyramidal tract axon, *Nature* **313:**135–137.
Stanfield, B. B., O'Leary, D. D. M., and Fricks, C., 1982, Selective collateral elimination in early postnatal development restricts cortical distribution of rat pyramidal tract neurones, *Nature* **298:**371–373.
Steffen, H., and Van der Loos, H., 1980, Early lesions on mouse vibrissal follicles: Their influence on dendrite orientation in the cortical barrelfield, *Exp. Brain Res.* **40:**419–431.
Steindler, D. A., Cooper, N. G. F., Faissner, A., and Schachner, M., 1989, Boundaries defined by adhesion molecules during development of the cerebral cortex: The J1/tenascin glycoprotein in the mouse somatosensory cortical barrelfield, *Dev. Biol.* **131:**243–260.
Stewart, G. R., and Pearlman, A. L., 1987, Fibronectin-like immunoreactivity in the developing cerebral cortex, *J. Neurosci.* **7:**3325–3333.
Stryker, M. P., and Harris, W. A., 1986, Binocular impulse blockade prevents the formation of ocular dominance columns in cat visual cortex, *J. Neurosci.* **6:**2117–2132.
Stryker, M. P., and Strickland, S. L., 1984, Physiological segregation of ocular dominance columns depends on the pattern of afferent electrical activity, *Invest. Ophthalmol. Suppl.* **25:**278.
Sunshine, J., Balak, K., Rutishauser, U., and Jacobson, M., 1987, Changes in neural cell adhesion molecule (NCAM) structure during vertebrate neural development, *Proc. Natl. Acad. Sci. USA* **84:**5986–5990.
Sur, M., Garraghty, P., and Roe, A. W., 1989, Experimentally induced visual cortical projections into auditory thalamus and cortex, *Science* **242:**1437–1441.
Takeichi, M., 1988, The cadhedrins: Cell-cell adhesion molecules controlling morphogenesis, *Development* **102:**639–655.
Tieman, S. B., 1988, Effects of monocular deprivation on geniculocortical synapses in the cat, *J. Comp. Neurol.* **222:**166–176.
Tsumoto, T., and Sato, H., 1985, GABAergic inhibition and orientation selectivity of neurons in the kitten visual cortex at the time of eye opening, *Vision Res.* **25:**383–388.
Tsumoto, T., Eckart, W., and Creutzfeldt, 1979, Modification of orientation sensitivity of cat visual cortex neurons by removal of GABA mediation inhibition, *Exp. Brain Res.* **34:**351–365.
Valentino, K. L., and Jones, E. G., 1981, The early formation of the corpus callosum: A light and electron microscopic study in fetal and neonatal rats, *J. Neurocytol.* **11:**583–609.
Valentino, K. L., Jones, E. G., and Kane, S. A., 1983, Expression of GFAP immunoreactivity during development of long fiber tracts in the rat CNS, *Dev. Brain Res.*
Valverde, F., 1967, Apical dendritic spines of the visual cortex and light deprivation in the mouse, *Exp. Brain Res.* **3:**337–352.
Valverde, F., and Esteban, M. E., 1968, Peristriate cortex of mouse: Location and the effects of enucleation on the number of dendritic spines, *Brain Res.* **9:**145–148.
Valverde, F., and Facal-Valverde, M. V., 1988, Postnatal development of interstitial (subplate) cells in the white matter of the temporal cortex of kittens: A correlated Golgi and electron microscopic study, *J. Comp. Neurol.* **269:**168–192.
Van der Loos, H., and Woolsey, T. A., 1973, Somatosensory cortex: Structural alterations following early injury to sense organs, *Science* **179:**395–398.
Verney, C., Berger, B., Adrien, J., Vigny, A., and Gay, M., 1982, Development of the dopaminergic innervation of the rat cerebral cortex. A light microscopic immunocytochemical study using anti-tyrosine hydroxylast antibodies, *Dev. Brain Res.* **5:**41–52.
Videen, T. O., N. W., Daw, and R. K., Rader, 1984, The effect of norepinephrine on visual cortical neurons in kittens and adult cats, *J. Neurosci.* **4:**1607–1617.
Vital-Durand, F., Garey, L. J., and Blakemore, C., 1978, Monocular and binocular deprivation in the monkey: Morphological effects and reversibility, *Brain Res.* **158:**45–64.
Vogt, C., and Vogt, O., 1919, Allgemeinere Ergebnisse unserer Hirnforschung, *J. Psychol. Neurol.* **25:**277–462.
Wahle, P., Meyer, G., Wu, J. -Y, and Albus, K., 1987, Morphology and axon terminal pattern of glutamate decarboxylase-immunoreactive cell types in the white matter of the cat occipital cortex during early postnatal development, *Dev. Brain Res.* **36:**53–62.
Wall, J. T., and Cusick, C. G., 1986, The representation of peripheral nerve inputs in the S-I hindpaw cortex of rats raised with incompletely innervated hindpaws, *J. Neurosci.* **6:**1129–1147.
Walsh, C., and Cepko, C. L., 1988, Clonally related cortical cells show several migration patterns, *Science* **241:**1342–1345.
Watanabe, E., Fujita, S. C., Murakami, F., Hayashi, M., and Matsumura, M., 1989, A monoclonal

antibody identifies a novel epitope surrounding a subpopulation of the mammalian central neurons, *Neuroscience* **29**(3):645–658.
Welker, C., 1971, Microelectrode delineation of five grain somatotopic organization of SmI cerebral neocortex in albino rat, *Brain Res.* **26**:259–276.
Welker, C., 1976, Receptive fields of barrels in the somatosensory neocortex of the rat, *J. Comp. Neurol.* **166**:173–190.
Welker, E., and Van der Loos, H., 1986, Quantitative correlation between barrel field size and the sensory innervation of the whiskerpad: A comparative study in six strains of mice bred for different patterns of mystacial vibrissae, *J. Neurosci.* **6**:3355–3373.
Welker, E., Soriano, E., and Van der Loos, H., 1989, Plasticity in the barrel cortex of the adult mouse: Effects of peripheral deprivation of GAD-immunoreactivity, *Exp. Brain Res.* **74**:441–452.
Wiesel, T. N., and Hubel, D. H., 1963, Single cell responses in striate cortex of kittens deprived of vision in one eye, *J. Neurophysiol.* **26**:1003–1017.
Wiesel, T. N., and Hubel, D. H., 1965a, Extent of recovery from the effects of visual deprivation in kittens, *J. Neurophysiol.* **28**:1059–1072.
Wiesel, T. N., and Hubel, D. H., 1965b, Comparison of the effects of unilateral and bilateral closure on cortical unit responses in kittens, *J. Neurophysiol.* **28**:1029–1040.
Wiesel, T. N., and Hubel, D. H., 1974, Ordered arrangement of orientation columns in monkeys lacking visual experience, *J. Comp. Neurol.* **158**:307–318.
Wilkinson, M., Shaw, C., Khan, I., and Cynader, M., 1983, Ontogenesis of beta-adrenergic binding sites in kitten visual cortex and the effects of visual deprivation, *Dev. Brain Res.* **7**:349–352.
Wilson, J. R., Webb, S. V., and Sherman, S. M., 1977, Conditions for dominance of one eye during competitive development of central connections in visually deprived cats, *Brain Res.* **136**:277–287.
Windrem, M. S., Jan de Beur, S., and Finlay, B. L., 1988, Control of cell number in the developing neocortex. II. Effects of corpus callosum section, *Dev. Brain Res.* **43**:13–22.
Winfield, D. A., 1981, The postnatal development of synapses in the visual cortex of the cat and the effects of eyelid closure, *Brain Res.* **206**:166–171.
Winfield, D. A., 1983, The postnatal development of synapses in different laminae of the visual cortex in the normal kitten and in kittens with eyelid suture, *Dev. Brain Res.* **9**:155–169.
Wise, S. P., and Jones, E. G., 1976, The organization and postnatal development of the commissural projection of the rat somatic sensory cortex, *J. Comp. Neurol.* **163**:313–344.
Wise, S. P., and Jones, E. G., 1978, Developmental studies of thalamocortical and commissural connections in the rat somatic sensory cortex, *J. Comp. Neurol.* **178**:187–208.
Wise, S. P., Hendry, S. H. C., and Jones, E. G., 1977, Prenatal development of sensorimotor cortical projections in cats, *Brain Res.* **138**:538–544.
Wise, S. P., Fleshman, J. W., and Jones, E. G., 1979, Maturation of pyramidal cell form in relation to developing afferent and efferent connections of rat somatic sensory cortex, *Neuroscience* **4**:1275–1297.
Wolf, W., Hicks, T. P., and Albus, K., 1986, The contributions of GABA-mediated inhibitory mechanisms to visual response properties of neurons in the kitten's striate cortex, *J. Neurosci.* **6**:2779–2795.
Wolff, J. R., Bottcher, H., Zetzshe, T., Oertel, W. H., and Chronwall, B. M., 1984, Development of GABAergic neurons in rat visual cortex as identified by glutamate decarboxylase-like immunoreactivity, *Neurosci. Lett.* **47**:207–212.
Woolsey, T. A., and Van der Loos, H., 1970, The structural organization of layer IV in the somatosensory region (SI) of mouse cerebral cortex, *Brain Res.* **17**:205–242.
Woolsey, T. A., and Wann, J. R., 1976, Areal changes in mouse cortical barrels following vibrissal damage at different postnatal ages, *J. Comp. Neurol.* **170**:53–66.
Yamashita, A., Hayashi, M., Shimizu, K., and Oshima, K., 1989, Ontogeny of somatostatin in cerebral cortex of macaque monkey: An immunohistochemical study, *Dev. Brain Res.* **45**:103–112.
Yurkewicz, L., Valentino, K. L., Floeter, M. K., Fleshman, J. W., Jr., and Jones, E. G., 1984, Effects of cytotoxic deletions of somatic sensory cortex in fetal rats, *Somatosens. Res.* **1**:303–327.
Zilles, K., Werners, R., Busching, U., and Schleicher, A., 1986, Ontogenesis of the laminar structure in areas 17 and 18 of the human visual cortex. A quantitative study, *Anat. Embryol.* **174**:339–353.

Index